Land-Atmosphere Interactions and Effects on the Climate of the Tibetan Plateau and Surrounding Regions III

Guest Editors

Yaoming Ma
Institute of Tibetan
Plateau Research
Chinese Academy of Sciences
Beijing
China

Li Jia
Aerospace Information
Research Institute
Chinese Academy of Sciences
Beijing
China

Massimo Menenti
Department of Geoscience
and Remote Sensing
Technische Universiteit Delft
Delft
The Netherlands

Lei Zhong
School of Earth and
Space Sciences
University of Science and
Technology of China
Hefei
China

Editorial Office
MDPI AG
Grosspeteranlage 5
4052 Basel, Switzerland

This is a reprint of the Special Issue, published open access by the journal *Remote Sensing* (ISSN 2072-4292), freely accessible at: https://www.mdpi.com/journal/remotesensing/special_issues/WLN1UV1KJ4.

For citation purposes, cite each article independently as indicated on the article page online and as indicated below:

Lastname, A.A.; Lastname, B.B. Article Title. *Journal Name* **Year**, *Volume Number*, Page Range.

ISBN 978-3-7258-3371-9 (Hbk)
ISBN 978-3-7258-3372-6 (PDF)
https://doi.org/10.3390/books978-3-7258-3372-6

Cover image courtesy of Yaoming Ma

Contents

About the Editors vii

Preface ix

Zhuqoun Li, Siqiong Luo, Xiaoqing Tan and Jingyuan Wang
Trend Analysis of High-Resolution Soil Moisture Data Based on GAN in theThree-River-Source Region During the 21st Century
Reprinted from: *Remote Sens.* **2024**, *16*, 4367, https://doi.org/10.3390/rs16234367 1

Xinyu Chen, Minhong Song, Yaqi Wang and Tongwen Wu
Evaluation of the Predictive Capability of CMA Climate Prediction System Model for Summer Surface Heat Source on the Tibetan Plateau
Reprinted from: *Remote Sens.* **2024**, *16*, 4118, https://doi.org/10.3390/rs16214118 28

Jiantao Dang, Xinrui Xie and Xiaohang Wen
Evaluation of Boundary Layer Characteristics at Mount Si'e Based on UAV and Lidar Data
Reprinted from: *Remote Sens.* **2024**, *16*, 3816, https://doi.org/10.3390/rs16203816 45

Lu Li, Xiaohua Dong, Yaoming Ma, Hanyu Jin, Chong Wei and Bob Su
Relationship between Tibetan Plateau Surface Heat Fluxes and Daily Heavy Precipitation in the Middle and Lower Yangtze River Basins (1980–2022)
Reprinted from: *Remote Sens.* **2024**, *16*, 3779, https://doi.org/10.3390/rs16203779 65

Qiuyu Wu, Jinghua Chen and Yan Yin
Characterizing the Supercooled Cloud over the TP Eastern Slope in 2016 via Himawari-8 Products
Reprinted from: *Remote Sens.* **2024**, *16*, 3643, https://doi.org/10.3390/rs16193643 85

Artem Y. Shikhovtsev, Chun Qing, Evgeniy A. Kopylov, Sergey A. Potanin and Pavel G. Kovadlo
Vertical Distribution of Optical Turbulence at the Peak Terskol Observatory and Mount Kurapdag
Reprinted from: *Remote Sens.* **2024**, *16*, 2102, https://doi.org/10.3390/rs16122102 96

Junru Jia, Massimo Menenti, Li Jia, Qiting Chen and Anlun Xu
Retrieval of At-Surface Upwelling Radiance and Albedo by Parameterizing Cloud Scattering and Transmittance over Rugged Terrain
Reprinted from: *Remote Sens.* **2024**, *16*, 1723, https://doi.org/10.3390/rs16101723 114

Xiaohang Wen, Jie Ma and Mei Chen
Analysis of Atmospheric Boundary Layer Characteristics on Different Underlying Surfaces of the Eastern Tibetan Plateau in Summer
Reprinted from: *Remote Sens.* **2024**, *16*, 1645, https://doi.org/10.3390/rs16091645 141

Yangxingyi Du, Dong Zheng, Yijun Zhang, Wen Yao, Liangtao Xu and Xianggui Fang
Spatiotemporal Distributions of the Thunderstorm and Lightning Structures over the Qinghai–Tibet Plateau
Reprinted from: *Remote Sens.* **2024**, *16*, 468, https://doi.org/10.3390/rs16030468 159

Lin Wang, Hong-Li Ren, Xiangde Xu, Li Gao, Bin Chen, Jian Li, et al.
Improving Predictions of Tibetan Plateau Summer Precipitation Using a Sea Surface Temperature Analog-Based Correction Method
Reprinted from: *Remote Sens.* **2023**, *15*, 5669, https://doi.org/10.3390/rs15245669 180

Chenxia Zhu, Shijie Li, Daniel Fiifi Tawia Hagan, Xikun Wei, Donghan Feng, Jiao Lu, et al.
Long-Term Characteristics of Surface Soil Moisture over the Tibetan Plateau and Its Response to Climate Change
Reprinted from: *Remote Sens.* **2023**, *15*, 4414, https://doi.org/10.3390/rs15184414 . . . . **198**

Luhan Li, Xuelong Chen, Yaoming Ma, Wenqing Zhao, Hongchao Zuo, Yajing Liu, et al.
Implications for Validation of IMERG Satellite Precipitation in a Complex Mountainous Region
Reprinted from: *Remote Sens.* **2023**, *15*, 4380, https://doi.org/10.3390/rs15184380 . . . . **219**

Minghong Liu, Hong-Li Ren, Run Wang, Jieru Ma and Xin Mao
Distinct Impacts of Two Types of Developing El Niño–Southern Oscillations on Tibetan Plateau Summer Precipitation
Reprinted from: *Remote Sens.* **2023**, *15*, 4030, https://doi.org/10.3390/rs15164030 . . . . **246**

Liyang Wu and Hui Zheng
Regional Climate Effects of Irrigation under Central Asia Warming by 2.0 °C
Reprinted from: *Remote Sens.* **2023**, *15*, 3672, https://doi.org/10.3390/rs15143672 . . . . **262**

Xingxu Zhou, Chao Zhang, Yunying Li and Zhiwei Zhang
Comparison of Spring Wind Gusts in the Eastern Part of the Tibetan Plateau and along the Coast: The Role of Turbulence
Reprinted from: *Remote Sens.* **2023**, *15*, 3655, https://doi.org/10.3390/rs15143655 . . . . **279**

Yuanhong You, Chunlin Huang and Yuhao Zhang
Assessing the Sensitivity of Snow Depth Simulations to Land Surface Parameterizations within Noah-MP in Northern Xinjiang, China
Reprinted from: *Remote Sens.* **2024**, *16*, 594, https://doi.org/10.3390/rs16030594 . . . . **295**

About the Editors

Yaoming Ma

Prof. Dr. Yaoming Ma is a distinguished Professor and Research Fellow at the Institute of Tibetan Plateau Research in the Chinese Academy of Sciences . He obtained a Doctorate degree in Atmospheric Physics from Okayama University in Japan and a Doctorate degree in Environmental Science from Wageningen University in the Netherlands. He is also the Academic Director of the National Observation and Research Station for Qomolongma Special Atmospheric Processes and Environmental Changes. His research interests include atmospheric boundary layer observation and satellite remote sensing application. He has been responsible for nearly 20 national major scientific research programs, including key programs of the National Natural Science Foundation of China, the National Science Fund for Distinguished Young Scholars, major international cooperation programs, etc. Also, he is one of the two chief coordinators of the major international cooperation research programs 'GAME/Tibet' and 'CAMP/Tibet'.

Li Jia

Prof. Dr. Li Jia is a Full Professor and Director of the Laboratory for Earth Observation for Terrestrial Water Cycle and Climate Change at the Aerospace Information Research Institute (AIR) in the Chinese Academy of Sciences (CAS). She received her PhD degree in Environmental Science from Wageningen University in the Netherlands in 2004. Her main research disciplines are quantitative remote sensing and remote sensing applications to hydrometeorology, eco-hydrology, water resources, carbon assimilation, drought process and monitoring, and climate change. She has more than 300 publications, with more than 90 as peer-reviewed ISI journal papers. She is a member of the WCRP/GEWEX Hydroclimatology Panel (GHP) (since 2019) and an IEEE member. She is the co-chair of the working group on Drought Monitoring and Evaluation of the AOGEO (Asia–Oceania Group on Earth Observations) (since 2017). She acts as an Editor/Associate Editor of several professional journals.

Massimo Menenti

Prof. Dr. Massimo Menenti is an internationally renowned scientist in the fields of earth observation and global terrestrial water cycle. He has held senior research positions in the Netherlands, France, USA, China, and Italy, received research awards in the Netherlands, France, USA, and China, and led large projects with participants from Europe, Asia, America, and Africa. His best-known achievements have been attained in the aspects of surface parameter retrievals from remote sensing, remote sensing-based evapotranspiration (ET) estimation, time series analysis of remote sensing data, and the application of remote sensing technology in hydrology and climate models. Prof. Menenti initiated the use of RS to assess and monitor crop water requirements and irrigation performance in the late 1980s. He is one of the earliest researchers in using laser technology to measure surface aerodynamic roughness. He initiated the use of the time series analysis of satellite data. He presented the surface energy balance index (SEBI) theory for ET estimation, which is the prototype of the subsequent S-SEBI, SEBS, and SEBAL models.

Lei Zhong

Prof. Dr. Lei Zhong is a Professor at the School of Earth and Space Sciences in the University of Science and Technology of China. He received his PhD degree in Natural Geography in 2008 from the Institute of Tibetan Plateau Research in the Chinese Academy of Sciences. His main research interests focus on land–atmosphere interactions, energy and water cycles, the application of remote sensing, and monsoon climate. He won the National Natural Science Foundation for Distinguished Young Scholars Award in 2015. He received first prize for the Tibet Autonomous Region Science and Technology Award in 2022. He has been actively involved in major international cooperation programs, such as 'CAMP/Tibet' and ESA-MOST Dragon program. He is a Director of the China Association for Scientific Exploration. He is also a member of the IAMAS Youth Working Group.

Preface

This reprint focuses on recent advances in land–atmosphere interactions and their effects on climate change in the Tibetan Plateau and surrounding regions using multisource remote sensing data and in situ measurements.

The retrieval of land surface variables and surface heat fluxes, as well as change monitoring in snow, glaciers, lakes, and other land surface covers, is of particular interest. Special attention is given to quantitative assessments of land surface temperature, sensible heat flux, soil moisture, vegetation and drought indices, groundwater storage, runoff, condensation, and desublimation, as well as the distinct surface processes of lakes and glaciers coupled with climate warming.

This Special Issue was jointly funded by the Second Tibetan Plateau Scientific Expedition and Research (STEP) Program (Grant No. 2019QZKK0103), the National Natural Science Foundation of China (Grant Nos. 42230610, 42425502, U2442213 and 42375071) and DTE-CLIMATE in the framework of the ESA-MOST Dragon 6 Programme (Grant ID 95357).

Yaoming Ma, Li Jia, Massimo Menenti, and Lei Zhong
Guest Editors

remote sensing

MDPI

Article

Trend Analysis of High-Resolution Soil Moisture Data Based on GAN in the Three-River-Source Region During the 21st Century

Zhuoqun Li [1,2], Siqiong Luo [1,*], Xiaoqing Tan [1,2] and Jingyuan Wang [1]

1 Key Laboratory of Cryospheric Science and Frozen Soil Engineering, Northwest Institute of Eco-Environment and Resources, Chinese Academy of Sciences, Lanzhou 730000, China; lizhuoqun221@mails.ucas.ac.cn (Z.L.); tanxiaoqing@nieer.ac.cn (X.T.); wangjingyuan@lzb.ac.cn (J.W.)

2 University of Chinese Academy of Sciences, Beijing 100049, China

* Correspondence: lsq@lzb.ac.cn

Abstract: Soil moisture (SM) is a crucial factor in land-atmosphere interactions and climate systems, affecting surface energy, water budgets, and weather extremes. In the Three-River-Source Region (TRSR) of China, rapid climate change necessitates precise SM monitoring. This study employs a novel UNet-Gan model to integrate and downscale SM data from 17 CMIP6 models, producing a high-resolution (0.1°) dataset called $\text{CMIP6}_{\text{UNet-Gan}}$. This dataset includes SM data for five depth layers (0–10 cm, 10–30 cm, 30–50 cm, 50–80 cm, 80–110 cm), four Shared Socioeconomic Pathways (SSP1-2.6, SSP2-4.5, SSP3-7.0, SSP5-8.5). The UNet-Gan model demonstrates strong performance in data fusion and downscaling, especially in shallow soil layers. Analysis of the $\text{CMIP6}_{\text{UNet-Gan}}$ dataset reveals an overall increasing trend in SM across all layers, with higher rates under more intense emission scenarios. Spatially, moisture increases vary, with significant trends in the western Yangtze and northeastern Yellow River regions. Deeper soils show a slower response to climate change, and seasonal variations indicate that moisture increases are most pronounced in spring and winter, followed by autumn, with the least increase observed in summer. Future projections suggest higher moisture increase rates in the early and late 21st century compared to the mid-century. By the end of this century (2071–2100), compared to the Historical period (1995–2014), the increase in SM across the five depth layers ranges from: 5.5% to 11.5%, 4.6% to 9.2%, 4.3% to 7.5%, 4.5% to 7.5%, and 3.3% to 6.5%, respectively.

Keywords: soil moisture; statistical downscaling; CMIP6; Generative Adversarial Network; deep learning; Three-River-Source Region

Citation: Li, Z.; Luo, S.; Tan, X.; Wang, J. Trend Analysis of High-Resolution Soil Moisture Data Based on GAN in the Three-River-Source Region During the 21st Century. *Remote Sens.* **2024**, *16*, 4367. https://doi.org/10.3390/rs16234367

Academic Editor: Gabriel Senay

Received: 30 August 2024
Revised: 14 November 2024
Accepted: 19 November 2024
Published: 22 November 2024

1. Introduction

Soil moisture (SM) is a critical variable in the land-atmosphere interactions and plays a vital role in various aspects of the Earth's climate system. It influences surface energy and water budgets, including sensible and latent heat fluxes, radiation exchange, and vegetation growth [1]. SM affects local and regional climate patterns by modulating evaporation and precipitation processes, contributing significantly to weather extremes and long-term climate behavior [2,3]. Given its central role in climate dynamics, understanding and predicting SM variations is essential for managing water resources, agricultural productivity, and responding to climate-related challenges [4,5].

The Three-River-Source Region (TRSR) in China, located at the headwaters of the Yangtze, Yellow, and Lancang rivers, is particularly sensitive to climate change [6,7]. The region experiences warming at rates more than twice the global average and has witnessed significant increases in precipitation since the early 21st century [8,9]. This accelerated warming and altered precipitation patterns have profound effects on soil temperatures, permafrost stability, and freeze-thaw processes in the TRSR [10,11]. Accurate monitoring and modeling of SM in this region are crucial for understanding its climate impacts and for developing effective adaptation strategies to mitigate adverse effects.

Recent advancements in climate modeling, particularly the Coupled Model Intercomparison Project Phase 6 (CMIP6), provide valuable tools for predicting future SM conditions under various emission scenarios [12,13]. CMIP6 offers a suite of climate models and scenarios, including the Shared Socioeconomic Pathways (SSP), which represent diverse pathways of future socio-economic and environmental development [14]. Evaluating future SM projections from CMIP6 models is essential for anticipating changes in drought characteristics, heatwave patterns, and overall water availability, which are critical for planning and implementing climate adaptation strategies [15,16].

High-resolution CMIP6 data are crucial for understanding and predicting localized climate impacts, such as extreme events and changes in agricultural productivity. However, CMIP6 models typically produce data at relatively coarse spatial resolutions, limiting their ability to capture fine-scale variations in climate variables. To address this limitation, various statistical downscaling methods have been developed to enhance the spatial resolution and accuracy of CMIP6 simulations [17,18]. Traditional methods, such as bilinear interpolation, are simple but often fail to capture high-frequency details, leading to potential inaccuracies in fine-scale predictions [19]. More advanced methods, like CDF-t (probabilistic downscaling) [20] and Support Vector Machines [21], offer improved performance but still have limitations in preserving spatial details and handling complex climate variables [22]. Among these, Convolutional Neural Networks (CNN) have shown promise in generating high-resolution climate data by leveraging deep learning techniques to enhance spatial resolution and model non-linear relationships in climate variables [22,23].

However, CNN models may still have limitations in capturing complex patterns and reducing artifacts, leading to potential inaccuracies in high-resolution climate projections. To overcome these challenges, this study introduces a novel approach using a Generative Adversarial Network (GAN) [24] for downscaling CMIP6 SM data. GAN, compared to a single CNN, offers several advantages, including improved generation of high-resolution data by learning from both real and synthetic samples through adversarial training [25]. This method can better capture complex patterns and reduce artifacts, leading to more accurate and reliable high-resolution SM projections [26]. For instance, GANs have been successfully applied in various remote sensing applications, demonstrating their ability to enhance image quality and detail [27,28].

In this paper, we use the UNet-Gan model to merge and downscale data from 17 CMIP6 models, resulting in the $\text{CMIP6}_{\text{UNet-Gan}}$ dataset [29]. This dataset includes SM data for five depth layers (0–10 cm, 10–30 cm, 30–50 cm, 50–80 cm, 80–110 cm), four SSP (SSP1-2.6, SSP2-4.5, SSP3-7.0, SSP5-8.5), and a spatial resolution of 0.1 degrees. The dataset will be used to conduct a trend analysis for the TRSR based on the obtained fine-resolution SM data. In Section 2, we first introduce the study area and provide detailed information about the data used, including ERA5-Land, AMSMQTP data, data from 17 CMIP6 models, and in situ data. We then describe the structure and equations of the UNet-Gan model used in this study, as well as the novel in situ loss function based on in situ soil observations that we introduce for the first time. This section also covers the Mann–Kendall trend analysis method used for trend analysis and the specific implementation details of UNet-Gan. In Section 3, we begin by assessing the accuracy of ERA5-Land and AMSMQTP in the TRSR and validate the accuracy of the $\text{CMIP6}_{\text{UNet-Gan}}$ data. We then analyze the spatial distribution of SM change trends and the time series of SM. Finally, we provide the mean values and rates of change of SM for different periods in the TRSR: early 21st century (2021–2040), mid-21st century (2041–2070), and late 21st century (2071–2100).

2. Materials and Methods

2.1. Study Area

The study area is the TRSR, located in the interior of the Tibetan Plateau (TP), China, with coordinates ranging from 89°24′ to 103°30′E and 31° to 37°12′N [6] (seen Figure 1 below). TRSR, also known as the "Chinese Water Tower", contains the headwaters of the Yellow, Yangtze, and Lancang rivers [30]. The TRSR lies at the transition zone between permafrost and seasonally frozen ground, making it a critical area for studying the impacts of permafrost degradation [7]. Covering an area of approximately 363,000 km^2, TRSR includes multiple counties across four Tibetan Autonomous Prefectures in Qinghai Province [31]. The region supports a diverse ecosystem and serves as a vital source of water for millions of people downstream [32]. However, it is also an area where the effects of climate change are most pronounced, leading to shifts in the ecosystem and water availability [8].

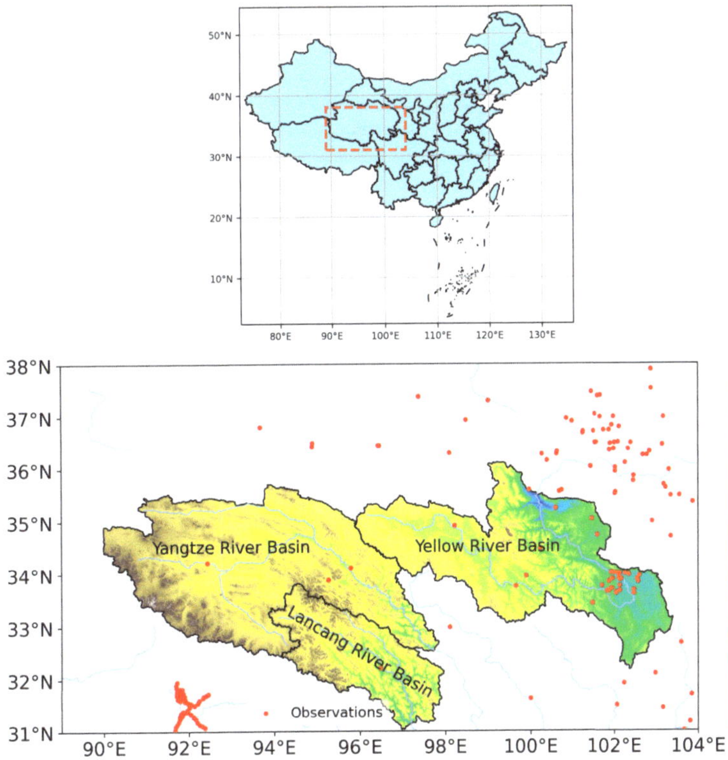

Figure 1. Above is the location of the study area in China, **below** is the elevation of the study area and the location of the in-site.

2.2. Data

2.2.1. ERA5-Land Data

ERA5-Land is a high-resolution reanalysis dataset developed by the European Centre for Medium-Range Weather Forecasts (ECMWF) [33] (https://cds.climate.copernicus.eu/datasets/reanalysis-era5-land-monthly-means?tab=download, accessed on 1 August 2024). It provides a comprehensive and consistent record of land surface variables with a spatial resolution of 9 km and a temporal resolution of one hour, covering the period from 1950 to the present [33]. The use of ERA5-Land data is advantageous in scientific research that requires high-resolution and long-term consistent land surface observations [34]. Its fine spatial and temporal resolutions enable detailed analysis of climate variability, hydrological cycles, and land-atmosphere interactions [35]. Therefore, this study selected monthly SM data from three layers: 0–7 cm, 7–28 cm, and 28–100 cm, as the target variables for the model. The unit of measurement is m^3/m^3, covering the time period from 1950 to 2014. The spatial extent is defined by a rectangular area between 89°E to 104°E and 31°N to 38°N, with 71 grid cells in the latitude and 151 grid cells in the longitude.

2.2.2. AMSMQTP Data

The AMSMQTP dataset is constructed using an automated machine learning algorithm that integrates multi-source meteorological and land surface data, alongside in situ SM observations from the International Soil Moisture Network (ISMN) and the China Meteorological Administration (CMA) [36] (http://www.ncdc.ac.cn/portal/metadata/4d94e0b3-a10c-43f6-b971-23e2df62c8ff, accessed on 1 August 2024). Covering the period from 2000 to 2021, AMSMQTP provides daily SM data for five depth layers ranging from 0 to 110 cm with a spatial resolution of 0.1 degrees. Compared to other datasets, AMSMQTP exhibits the lowest root mean square error (RMSE) and unbiased RMSE (ubRMSE) and the highest correlation coefficient (R) for the TP. The dataset is available in three versions, with AMSMQTP_ensemble being used in this study due to its outstanding robustness and minimal spatial artifacts. For the model fine-tuning phase, we selected SM data from the five layers as target variables. The raw data were averaged monthly and then resampled to match the ERA5-Land grid, covering the time range from 2000 to 2021.

2.2.3. CMIP6 Data

The Coupled Model Intercomparison Project Phase 6 (CMIP6) represents the latest iteration of a collaborative effort to improve climate model projections and enhance our understanding of climate change [12] (https://cds.climate.copernicus.eu/datasets/projections-cmip6?tab=download, accessed on 1 August 2024). CMIP6 includes a comprehensive suite of climate models from various international research institutions, offering a broad range of scenarios to explore future climate conditions [37]. The project aims to provide a robust framework for evaluating model performance, understanding climate variability, and projecting future climate changes. The CMIP6 dataset includes multiple representative concentration pathways (RCPs) and SSP, providing a comprehensive view of possible future climate scenarios [38,39]. This extensive range of scenarios and model outputs facilitates a more nuanced understanding of climate dynamics and helps inform adaptation and mitigation strategies.

This study selects monthly data for the variable "The mass of water in all phases in the upper 10 cm of the soil layer" from 17 CMIP6 model simulations, with the variable unit in kg/m^2. The model details are provided in Table 1. We use the historical data from 1950 to 2014 to train the model on the mapping relationship between CMIP6 data and either ERA5-Land or AMSMQTP multi-layer SM data. Subsequently, data from four SSPs (SSP1-2.6, SSP2-4.5, SSP3-7.0, SSP5-8.5) were selected for model inference. These SSPs were chosen to represent different socio-economic development paths, ranging from sustainable development to high-emission scenarios, and can reflect the impact of different policy and market responses on climate change. Initially, we convert the unit of CMIP6 SM data to m^3/m^3. Due to significant differences in spatial resolution among the 17 CMIP6 models, we perform linear interpolation of the data to match the ERA5-Land grid for ease of subsequent modeling.

Table 1. Details of 17 CMIP6 models.

Model Name	Modelling Centre	Resolution (Lon × Lat)	Used Member
ACCESS-CM2	CSIRO-ARCCSS	1.875° × 1.25°	r1i1p1f1
BCC-CSM2-MR	BCC	1.125° × 1.1215°	r1i1p1f1
CanESM5-CanOE	CCCMA	2.8125° × 2.7906°	r1i1p2f1
CESM2	NCAR	1.25° × 0.9424°	r1i1p1f1
CMCC-CM2-SR5	CMCC	1.25° × 0.9424°	r1i1p1f1
CNRM-CM6-1-HR	CNRM-CERFACS	0.5° × 0.5°	r1i1p1f2
CNRM-ESM2-1	CNRM-CERFACS	1.4° × 1.4°	r1i1p1f2
EC-Earth3-Veg-LR	EC-Earth-Consortium	1.125° × 1.1215°	r1i1p1f1
KACE-1-0-G	NIMS-KMA	1.875° × 1.25°	r1i1p1f1
MIROC6	MIROC	1.4° × 1.4°	r1i1p1f1
MIROC-ES2L	MIROC	2.8125° × 2.7906°	r1i1p1f2

Table 1. *Cont.*

Model Name	Modelling Centre	Resolution (Lon × Lat)	Used Member
MPI-ESM1-2-LR	MPI-M AWI	1.875° × 1.865°	r1i1p1f1
MRI-ESM2-0	MRI	1.125° × 1.1215°	r1i1p1f1
NorESM2-LM	NCC	2.5° × 1.895°	r1i1p1f1
NorESM2-MM	NCC	1.25° × 0.9424°	r1i1p1f1
TaiESM1	AS-RCEC	1.25° × 0.9424°	r1i1p1f1
UKESM1-0-LL	MOHC, NERC, NIMS-KMA, NIWA	1.875° × 1.25°	r1i1p1f2

2.2.4. In Situ Data

The ISMN is a global database that collects, compiles, and provides access to in situ SM measurements from various ground-based observation networks around the world [40,41] (https://ismn.earth/en/, accessed on 1 August 2024). The network comprises over 2600 stations across different climates and land cover types, ensuring a diverse and comprehensive representation of SM conditions. The data provided by ISMN include measurements at various soil depths, these in situ observations are essential for cross-validating satellite-based SM products, enhancing model simulations, and improving our understanding of SM dynamics on a global scale.

In this study, we selected ISMN in situ data within the rectangular area defined by the ERA5-Land dataset, covering the period from May 2008 to September 2019 (with 94 ISMN stations distributed across three observation networks: the CTP_SMTMN [42] network with 56 stations, the MAQU Dente et al. [43] network with 27 stations, and the NAQU [44] network with 11 stations). Additionally, we utilized in situ data from the CMA within the same area from 2015 to 2018 (with 93 CMA stations). The in situ data cover soil depths from 0 to 110 cm, and we averaged the data on a monthly scale. In addition to being used to validate the monthly scale ERA5-Land and AMSMQTP data errors, the in situ data also contribute to the calculation of a portion of the model training loss (in situ loss) for backpropagation.

2.3. Method

In this study, we employ a GAN model, where the historical surface SM data from 17 CMIP6 models are used as the model input (referred to as I^{CMIP6}, with 17 channels), and the five-layer SM data from AMSMQTP (with 5 channels) are used as the model's target output. By learning the mapping relationship between CMIP6 and AMSMQTP during the historical period, the trained model is then applied to infer CMIP6 data under four emission scenarios (SSP1-2.6, SSP2-4.5, SSP3-7.0, SSP5-8.5). This results in five-layer SM data (0–10 cm, 10–30 cm, 30–50 cm, 50–80 cm, 80–110 cm) for the TRSR from 2015 to 2100, with a temporal resolution of one month and a spatial resolution of 0.1 degrees, referred to as $CMIP6_{UNet\text{-}Gan}$.

To enhance the model's performance and robustness, our training process is divided into two stages. The first stage is the pretraining phase, where ERA5-Land data are used as the target output. The second stage is the fine-tuning phase, with AMSMQTP data as the target output. During the fine-tuning phase, the initialization values of the model parameters, except for the last convolutional layer, are derived from the parameters obtained in the pre-training phase, while the initial values of the last convolutional layer are randomly initialized. It is noteworthy that during the fine-tuning phase, we also calculate the in situ loss, as illustrated in Figure 2. Finally, we employ the Mann–Kendall trend analysis [45] to examine the trends in $CMIP6_{UNet\text{-}Gan}$ under the four emission scenarios.

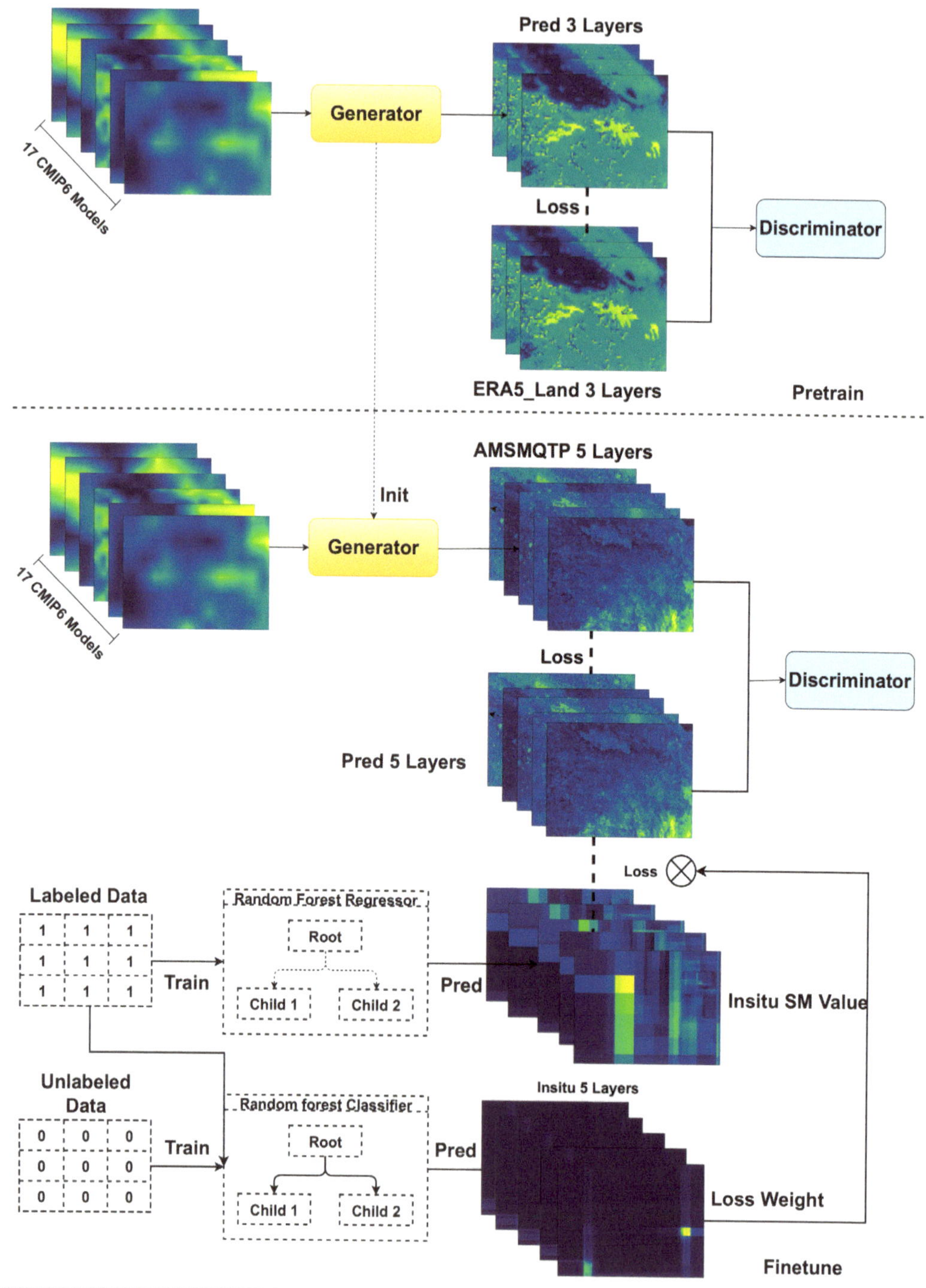

Figure 2. Training details of the UNet-GAN network.

2.3.1. UNet-Gan

GAN is a class of deep learning models consisting of two neural networks: a generator and a discriminator, which are trained simultaneously through an adversarial process. The generator aims to produce data that mimic the real data distribution, while the discriminator's role is to distinguish between real data and the data generated by the generator. During training, the generator learns to create increasingly realistic data by trying to fool the discriminator, and the discriminator improves its ability to differentiate between real and synthetic data. This adversarial setup leads to the generator producing high-quality, realistic outputs.

In the context of this study, we utilize a GAN architecture where the generator is based on a UNet structure, which is well-suited for tasks requiring detailed pixel-level predictions, such as SM mapping [46]. The generator takes input from CMIP6 historical SM data and attempts to generate corresponding high-resolution SM data, while the discriminator evaluates the generated outputs against the target data. The goal is to train the generator to produce SM outputs that are indistinguishable from the real data, ensuring that the model captures the complex relationships between the input and output datasets. This method allows for the generation of accurate future projections of SM under various climate scenarios. Our optimization objective is an adversarial min-max problem, where (D_{θ_D}) is the discriminator, (G_{θ_G}) is the generator, (I_{CMIP6}) represents the CMIP6 data, and (I_{Target}) represents ERA5-Land during the pretraining phase and AMSMQTP during the fine-tuning phase. The formula is as follows:

$$\min_{\theta_G} \max_{\theta_D} \mathbb{E}_{I^{Target} \sim p_{\text{train}}(I^{Target})}[\log D_{\theta_D}(I^{Target})] + \mathbb{E}_{I^{CMIP6} \sim p_G(I^{CMIP6})}[\log(1 - D_{\theta_D}(G_{\theta_G}(I^{CMIP6})))] \tag{1}$$

- **Generator**
  The U-Net architecture is a convolutional neural network (CNN) designed for image segmentation tasks, particularly in biomedical imaging [47]. It has a symmetric architecture comprising two main parts: a contracting path (encoder) and an expansive path (decoder). The contracting path captures context by downsampling the input image through successive convolutional and pooling layers, effectively extracting high-level features. The expansive path then upsamples these features, reconstructing the spatial resolution while combining them with corresponding high-resolution features from the contracting path via skip connections. These skip connections help preserve spatial information that might otherwise be lost during the downsampling process, allowing for precise localization of features.
  In this study, we employ a U-Net architecture as the generator within a GAN framework; the model structure is shown in Figure 3. The U-Net's ability to accurately capture and reconstruct spatial features makes it particularly suitable for generating high-resolution SM data from lower-resolution inputs. By leveraging the U-Net structure, the model can effectively learn the mapping between coarse CMIP6 inputs and the corresponding high-resolution AMSMQTP or ERA5-Land outputs, ensuring that fine-scale details are preserved in the generated SM predictions. The optimization objective for the generator is expressed by the following equation:

$$\hat{\theta}_G = \arg\min_{\theta_G} \frac{1}{N} \sum_{n=1}^{N} l^G(G_{\theta_G}(I_n^{CMIP6}), I_n^{Target}, I_n^{In-situ}) \tag{2}$$

  The l^G loss function is a weighted average of several losses, including content loss, total variation Loss [48], adversarial loss [49], and a novel auxiliary loss method specifically designed for the Earth sciences, referred to as in situ loss. The calculation methods for each type of loss will be described in detail in the following sections.

- **Discriminator**

  The model structure of the discriminator is shown in Figure 4. It first processes the input through a convolutional layer and a leaky ReLU activation, followed by seven convolutional blocks. The output is then mapped to a one-dimensional value through a dense layer and a sigmoid layer, representing the probability that an image is real or fake, with values ranging from 0 to 1. The optimization objective for the discriminator is as follows:

$$\begin{aligned}\hat{\theta}_D = \arg\min_{\theta_D} - log(D_{\theta_D}(I^{Target}) - \mathbb{E}[D_{\theta_D}(G_{\theta_G}(I^{CMIP6}))]) \\ - log(1 - (D_{\theta_D}(G_{\theta_G}(I^{CMIP6}))) - \mathbb{E}[D_{\theta_D}(I^{Target})])\end{aligned} \quad (3)$$

  The above formula calculates the relative loss between I^{CMIP6} and the I^{Target}, where the expectation operation $\mathbb{E}$ is computed over each minibatch [49].

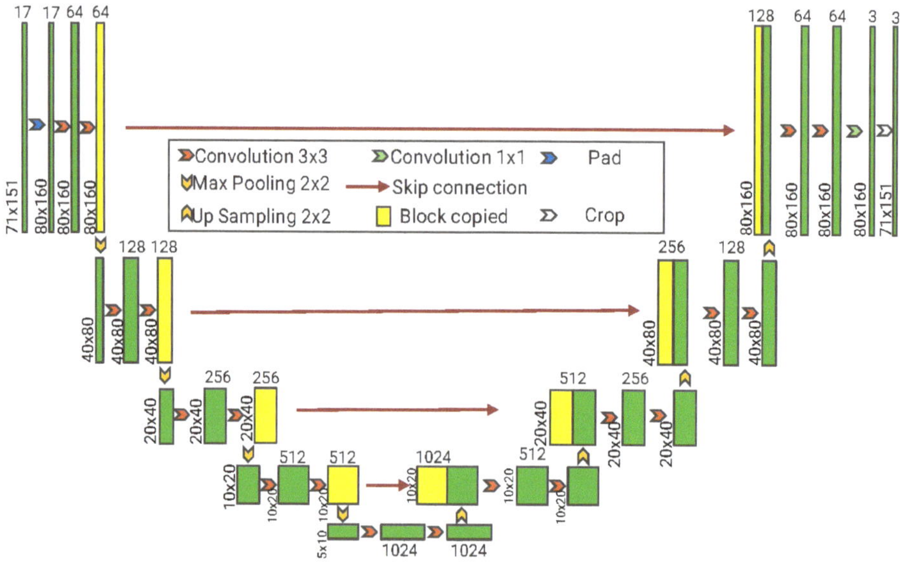

Figure 3. The model structure of U-Net (with three output channels for the pretraining phase and five output channels for the fine-tuning phase).

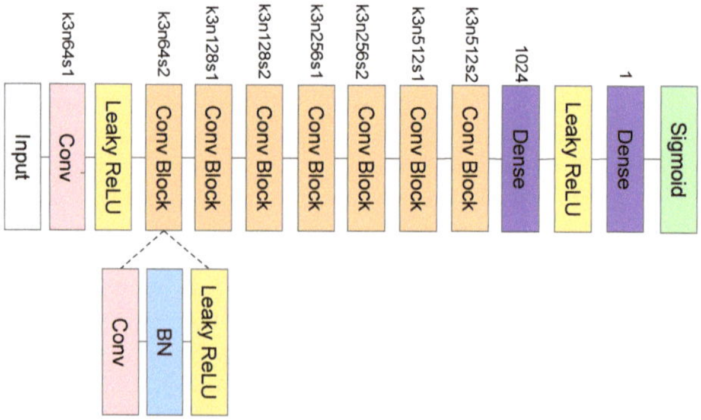

Figure 4. The model structure of the discriminator. The number following k indicates the convolution kernel size, the number following n indicates the number of output channels, and the number following s indicates the stride.

2.3.2. Generator's Loss Function

The loss function for the generator is given by the following formula (all losses are computed only within the vector boundaries of the TRSR):

$$l^G = \alpha_1 l^G_{Content} + \alpha_2 l^G_{Adv} + \alpha_3 l^G_{In-situ} + \alpha_4 l^G_{TV} \quad (4)$$

- **Content loss**
  Mean squared error (MSE) loss measures the average of the squared differences between predicted and target values, providing a direct assessment of prediction accuracy. It is commonly used in regression tasks within generator models, as it penalizes larger errors more heavily, encouraging the model to produce outputs closer to the true values. In the following formula, H represents the number of latitude pixels, W represents the number of longitude pixels, and C represents the number of soil layers.

$$l^G_{MSE} = \frac{1}{CWH} \sum_{x=1}^{W} \sum_{y=1}^{H} \sum_{z=1}^{C} (I^{Target}_{x,y,z} - G_{\theta_G}(I^{CMIP6})_{x,y,z})^2 \quad (5)$$

  VGG Loss, based on the VGG network, captures perceptual differences by comparing high-level features between generated and target images, helping to produce visually more realistic results [48]. It is particularly effective in preserving texture and fine details, as it measures the similarity in feature space rather than directly comparing pixel values [48]. In this study, the pre-trained VGG16 model is used, where i and j represent the starting and ending layers of the pre-trained model, respectively. $\phi_{i,j}$ denotes a portion of the pre-trained model, and $H_{i,j}$, $W_{i,j}$, and $C_{i,j}$ represent the height, width, and number of channels of the feature map produced by $\phi_{i,j}$. In this paper, i and j are set to 0 and 30 (both inclusive), respectively. The formula is as follows:

$$\begin{aligned} l^G_{VGG/i.j} = \frac{1}{W_{i,j}H_{i,j}C_{i,j}} \sum_{x=1}^{W_{i,j}} \sum_{y=1}^{H_{i,j}} \sum_{z=1}^{C_{i,j}} (\phi_{i,j}(I^{Target})_{x,y,z} \\ - \phi_{i,j}(G_{\theta_G}(I^{CMIP6}))_{x,y,z})^2 \end{aligned} \quad (6)$$

  The content loss is obtained by a weighted combination of the two loss functions mentioned above, as expressed in the following formula:

$$l^G_{Content} = \beta_1 l^G_{MSE} + \beta_2 l^G_{VGG/i,j} \quad (7)$$

- **Adversarial loss**
  The formula for the adversarial loss is similar to that of the discriminator loss, but the optimization objective is completely opposite. We aim for $G_{\theta_G}(I^{CMIP6})$ to be closer to real data than I^{Target}. The formula is as follows:

$$l^G_{Adv} = -log(D_{\theta_D}(G_{\theta_G}(I^{CMIP6})) - \mathbb{E}[D_{\theta_D}(I^{Target})]) \quad (8)$$

- **In-situ loss**
  In situ observations are generally considered the most accurate means of measuring SM, which is why this paper introduces a novel in situ loss as an auxiliary loss. The main process is illustrated in Figure 2. Our key idea is to convert point-based in situ observations into spatial grids. We first use Random Forest [50] Regression, training it with five simple features (soil layer, latitude, longitude, year, and month) to infer $I^{In-situ}$. Additionally, we employ a Random Forest Classifier using the same five features, where labeled data are assigned a class of 1, the pseudocode is shown in Algorithm 1. We then randomly sample unlabeled data from the spatial domain (matching the number of labeled data points for class balance) and assign them a class of 0. The classifier is trained and used to infer I^{Prob}, which indicates the reliability of the data, with values ranging from 0 to 1. The formula for calculating in situ loss is as follows:

$$l_{In-situ}^{G} = \frac{1}{CWH}\sum_{x=1}^{W}\sum_{y=1}^{H}\sum_{z=1}^{C}[I_{x,y,z}^{Prob}(I_{x,y,z}^{In-situ} - G_{\theta_G}(I^{CMIP6})_{x,y,z})^2] \tag{9}$$

Algorithm 1 Random Forest Classifier for In Situ Observations

Input: Labeled dataset $D_l = \{(x_i, 1)_{i=1...n}\}$, where x_i are the features. Unlabeled dataset $D_u = \{(x_j, 0)\}$

Output: Inferred class probabilities I^{Prob} for unlabeled data

1: Randomly sample n unlabeled data points
2: Assign class 0 to sampled points
3: $RF \leftarrow$ RandomForestClassifier($n\_estimators = k$)
4: RF.fit($D_{combined}$)
5: $I^{Prob} \leftarrow RF$.predict_proba(D_u)
6: **return** I^{Prob}

- **Total Variation Loss**
  Total Variation (TV) Loss is incorporated into the generator's loss function to promote spatial smoothness in the generated outputs while preserving essential image structures. By minimizing the differences between neighboring pixels, TV Loss reduces noise and prevents the introduction of artifacts, ensuring that the generated SM maps maintain coherent and natural-looking spatial patterns. This loss is particularly effective in enhancing the visual quality of generated images, making them more realistic and consistent with physical processes.

$$l_{TV}^{G} = \sum_{x=1}^{W}(I_{x,y} - I_{x-1,y})^2 + \sum_{y=1}^{H}(I_{x,y} - I_{x,y-1})^2 \tag{10}$$

2.3.3. Mann–Kendall Trend Analysis

The Mann–Kendall test is a non-parametric statistical method used to identify trends in time series data, particularly for environmental and climate data [6]. It assesses whether a monotonic trend exists in a dataset without assuming any specific distribution for the data. The test is widely used due to its robustness and minimal assumptions, making it suitable for various types of time series data, including those with missing values or outliers.

The Mann–Kendall test evaluates the null hypothesis H_0 that there is no trend in the data against the alternative hypothesis H_1 that a trend exists. The test statistic S is calculated as follows:

$$S = \sum_{i=1}^{n-1}\sum_{j=i+1}^{n} \text{sgn}(X_j - X_i) \tag{11}$$

where $sgn(x)$ is the sign function, which returns 1 if $x > 0$, -1 if $x < 0$, and 0 if $x = 0$. X_i and X_j are the data values at time i and j, respectively. n is the number of observations.

The variance of S under the null hypothesis is given by the following:

$$\text{Var}(S) = \frac{n(n-1)(2n+5) - \sum_{t=1}^{k}(n_t(n_t-1)(2n_t+5))}{18} \tag{12}$$

where k is the number of tied groups, and n_t is the number of observations in the t-th tied group.

To determine the significance of the trend, the standardized test statistic Z is computed:

$$Z = \begin{cases} \frac{S-1}{\sqrt{\text{Var}(S)}} & \text{if } S > 0 \\ 0 & \text{if } S = 0 \\ \frac{S+1}{\sqrt{\text{Var}(S)}} & \text{if } S < 0 \end{cases} \tag{13}$$

The value of Z is compared against critical values from the standard normal distribution to assess statistical significance. Typically, a significance level of 0.05 is used, with a positive Z indicating an upward trend and a negative Z indicating a downward trend.

The significance of the trend is evaluated based on the magnitude of Z. If the absolute value of Z exceeds the critical value from the normal distribution (e.g., ± 1.96 for a 95% confidence level), the null hypothesis H_0 is rejected, indicating a statistically significant trend in the data. If $|Z|$ is less than the critical value, the trend is not considered statistically significant.

To quantify the magnitude of the trend, Sen's slope estimator Q is used. It is computed as:

$$Q = \text{median}\left(\frac{X_j - X_i}{j - i}\right) \tag{14}$$

where the median is taken over all pairs (i, j) with $i < j$. This estimator provides a robust measure of the trend's rate.

In this study, the Mann–Kendall test is applied to assess the significance of trends in the generated SM data under various emission scenarios. This analysis helps identify significant changes in SM over time, contributing to a better understanding of the impacts of climate change on SM dynamics.

2.3.4. Evaluation Metrics

MAE measures the average magnitude of errors between predicted and observed values, providing an intuitive indication of the model's overall prediction accuracy.

$$\text{MAE}(y, \hat{y}) = \frac{1}{n_{\text{samples}}} \sum_{i=0}^{n_{\text{samples}}-1} |y_i - \hat{y}_i|. \tag{15}$$

RMSE quantifies the square root of the average squared differences between predicted and observed values, emphasizing larger errors and giving a sense of how well the model captures extreme deviations.

$$\text{RMSE}(y, \hat{y}) = \frac{1}{n_{\text{samples}}} \sum_{i=0}^{n_{\text{samples}}-1} \sqrt{(y_i - \hat{y}_i)^2}. \tag{16}$$

MAPE expresses the error as a percentage of the observed values, allowing for a normalized measure of prediction accuracy, particularly useful when comparing errors across different scales.

$$\text{MAPE}(y, \hat{y}) = \frac{1}{n_{\text{samples}}} \sum_{i=0}^{n_{\text{samples}}-1} \frac{|y_i - \hat{y}_i|}{\max(\epsilon, |y_i|)} \tag{17}$$

In the above equations, ϵ is a small but strictly positive number introduced to avoid undefined results when y is zero.

Pearson R assesses the linear correlation between predicted and observed values, indicating how well the model captures the strength and direction of the relationship between the two datasets.

$$\text{R}(y, \hat{y}) = \frac{\sum_{i=0}^{n_{\text{samples}}-1}(y_i - \overline{y_i})(\hat{y}_i - \overline{\hat{y}_i})}{\sqrt{\sum_{i=0}^{n_{\text{samples}}-1}(y_i - \overline{y_i})^2}\sqrt{\sum_{i=0}^{n_{\text{samples}}-1}(\hat{y}_i - \overline{\hat{y}_i})^2}} \tag{18}$$

2.3.5. Implementation Details

In the pretrain stage, data from 1950 to 1999 were used as the training set, data from 2000–2007 were used as the validation set, and data from 2008 to 2014 were used as the test set. In the fine-tuning stage, the data spanned from 2000 to 2014, and we employed five-fold cross-validation. Our model was implemented using the PyTorch 2.2 [51] framework, and all experiments were conducted on an NVIDIA Tesla V100-SXM2 32 GB GPU. We manually searched for hyperparameter settings and ultimately chose a batch size of 100, 200 epochs, a learning rate of 5×10^{-4} for the generator, and 1×10^{-6} for the discriminator. The size of all convolution kernels in the UNet was set to 3. We applied early stopping, with the patience set to 10 epochs, and also used a cosine learning rate decay strategy. The loss weights were set as follows:

$$\beta_1 = 1, \beta_2 = 0.006, \alpha_1 = 1, \alpha_2 = 0.001, \alpha_3 = 0.01, \alpha_4 = 4 \times 10^{-8} \quad (19)$$

3. Results and Analysis

3.1. Accuracy Comparison of ERA5-Land and AMSMQTP Data in the TRSR

As shown in Figure 5, by comparing the errors and accuracy between the in situ SM and the corresponding grid SM from AMSMQTP and ERA5-Land, AMSMQTP exhibits higher accuracy across all layers compared to ERA5-Land (which is also why we use AMSMQTP in the fine-tuning stage). Additionally, based on the R metric, we observe that for both datasets, the median R decreases as depth increases. As shown in Table S1, the median R for Layer 1 of AMSMQTP is 0.918, and for Layer 5, it is 0.658; the median R for Layer 1 of ERA5-Land is 0.659, and for Layer 5, it is 0.250. We also find that although AMSMQTP's data accuracy is significantly better than that of ERA5-Land, the distribution of its evaluation metrics exhibits a long-tail distribution (i.e., there are some outliers), indicating that AMSMQTP still has lower accuracy at certain points. This is one of the reasons we introduce in situ loss in this paper.

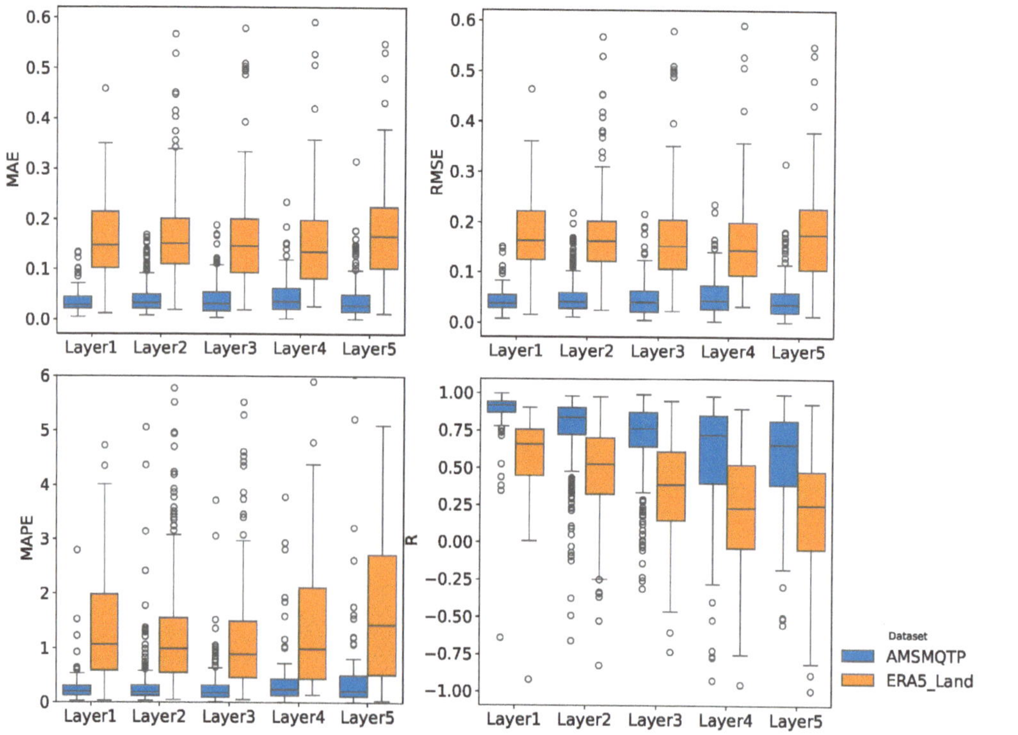

Figure 5. Box plots of evaluation metrics for SM from different layers of ERA5-Land and AMSMQTP compared to in situ observations (for MAE, RMSE, and MAPE, smaller values are better, and for R, larger values are better).

3.2. Evaluation of CMIP6$_{UNet\text{-}Gan}$

Figure 6 demonstrates the performance of UNet-Gan in data fusion and downscaling for the 17 CMIP6 datasets (with the actual values being the AMSMQTP data). We observe that UNet-Gan effectively integrates and downscales CMIP6 data for the five SM layers. However, data accuracy shows a decreasing trend with increasing depth, except for the fourth layer. The R^2 values are 0.819, 0.802, 0.694, 0.778, and 0.541, respectively, and the R values are 0.905, 0.895, 0.833, 0.882, and 0.736, respectively. Figure 7 shows a comparison of the data predicted by the UNet-Gan model for the first layer in the test set with the data from 17 CMIP6 models and their Ensemble, against the AMSMQTP data. We found that UNet-Gan achieved the highest correlation coefficient, and its standard deviation is very close to that of AMSMQTP, indicating similar variability. Additionally, UNet-Gan also achieved the lowest RMSE. We also observed that five CMIP6 models had correlation coefficients less than 0, and the Ensemble did not achieve the highest correlation coefficient, except for UNet-Gan. This may be due to some CMIP6 models having poor accuracy in the TSRS.

As shown in Figures 6 and 7, UNet-Gan achieved the best performance. We applied the bagging method to the five models obtained from five-fold cross-validation and performed data fusion and downscaling for the 17 CMIP6 datasets across historical periods and four emission scenarios. This resulted in the TSRS five-layer SM data for historical periods (1950–2014) and the four future emission scenarios (2015–2100), with a spatial resolution of 0.1 degrees and a temporal resolution of one month. We refer to this dataset as CMIP6$_{UNet\text{-}Gan}$.

3.3. Spatial Distribution of SM Trends Under Four Emission Scenarios

As shown in Figure 8, SM in Layer 1 exhibits a significant increasing trend within the TRSR under all four emission scenarios. The areas with the highest rates of moisture increase are concentrated in the western part of the Yangtze River and the southeastern part of the Yellow River. Layer 2 shows a trend similar to that of Layer 1. In Layer 3, in addition to the two regions with high moisture increase rates seen in Layer 1, the northeastern part of the Yellow River also emerges as a high-value area. For Layer 4, regions with high moisture increase rates are only found in small areas of the northeastern and southeastern parts of the Yellow River. In Layer 5, under the SSP1-2.6 scenario, the high moisture increase rate is concentrated in much of the Yangtze River area, while the other three emission scenarios do not show significant concentrated high-value areas. Across all five layers, there is a consistent pattern where moisture increase rates rise with higher SSP scenarios, and the rate of increase slows with increasing depth. Figures corresponding to Layer 2 through Layer 5, which depict similar content to Figure 8, are provided in the Supplementary Materials.

As shown in Figure 9, Layer 1 exhibits a significant increasing trend in SM during winter across the entire TRSR under all four emission scenarios, with the highest moisture increase located in the southeastern part of the Yellow River. In spring, the entire TRSR also shows a significant moisture increase, with the highest rates mainly concentrated in much of the Yangtze River region and a small part of the southern Yellow River area. During summer, the regions with significant moisture increases are much smaller; under the SSP1-2.6 scenario, only a small portion of the northern Yangtze River passes the significance test. As the SSP increases, the area passing the significance test gradually expands, but the highest moisture increase is still confined to the northwestern part of the Yangtze River. In autumn, the entire TRSR exhibits a significant increasing trend, with the highest moisture increase concentrated in most of the Yangtze River region.

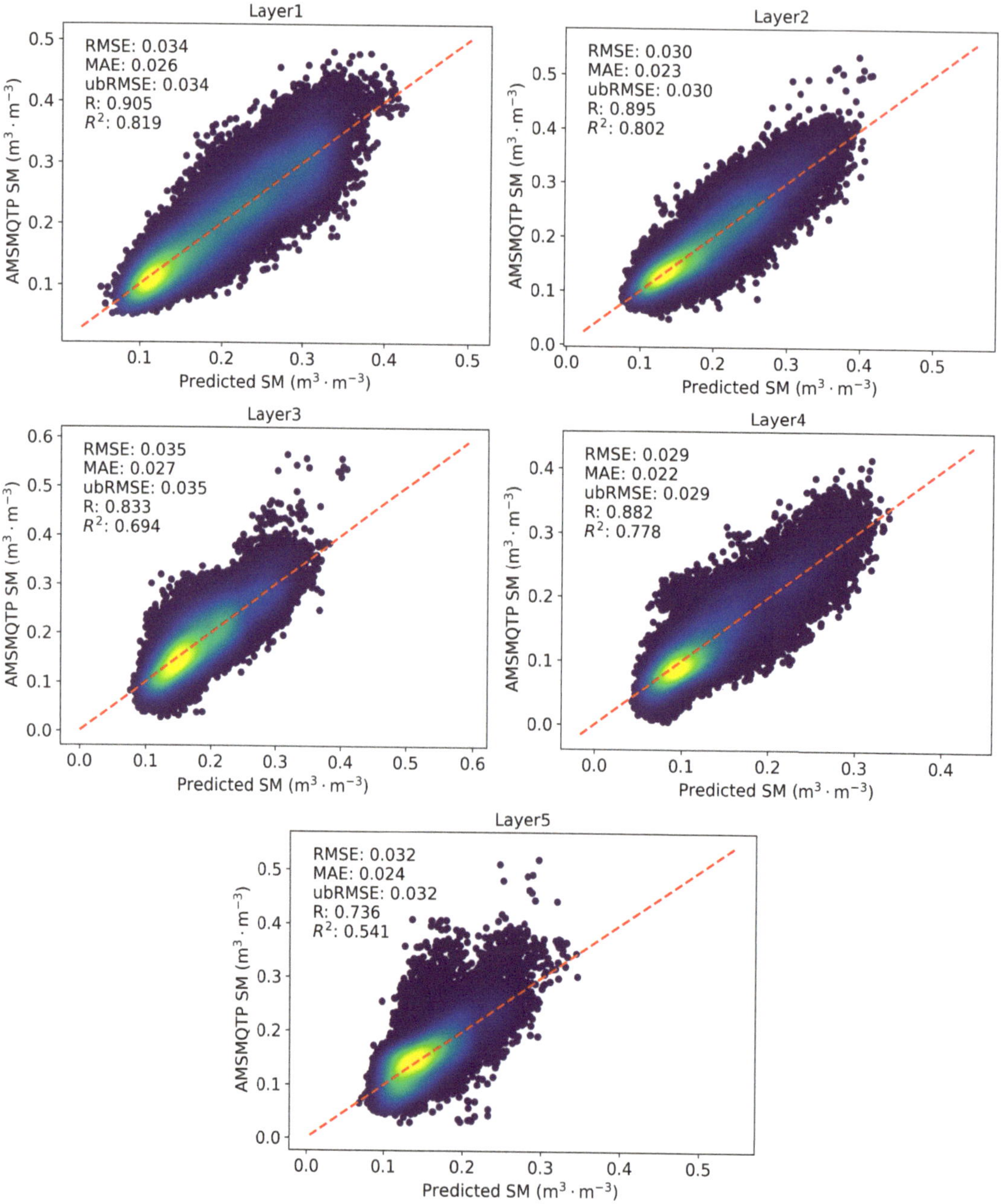

Figure 6. Scatter plot of predicted values versus true values on the test set for five-fold cross-validation (The red dotted line represents a 45-degree diagonal, and the color intensity indicates the density of the points).

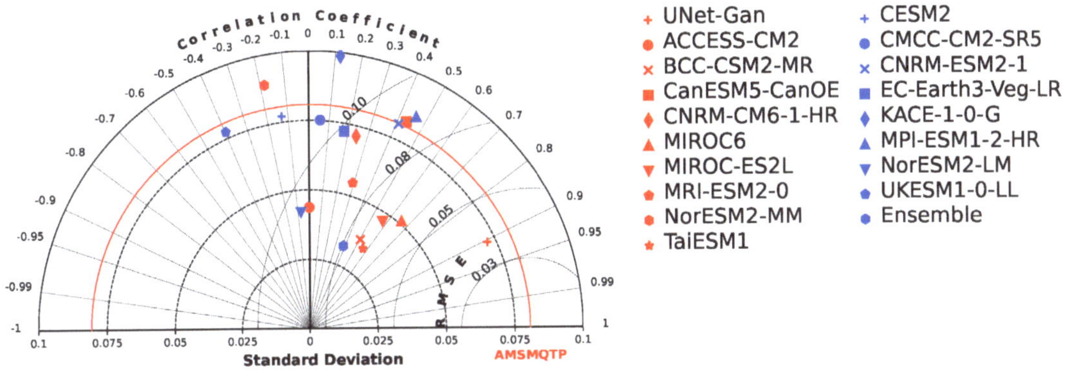

Figure 7. A comparison of the data predicted by the UNet-Gan model for the first layer in the test set with the data from 17 CMIP6 models and their Ensemble, against the AMSMQTP data. The diagram includes three metrics: RMSE, standard deviation, and correlation coefficient.

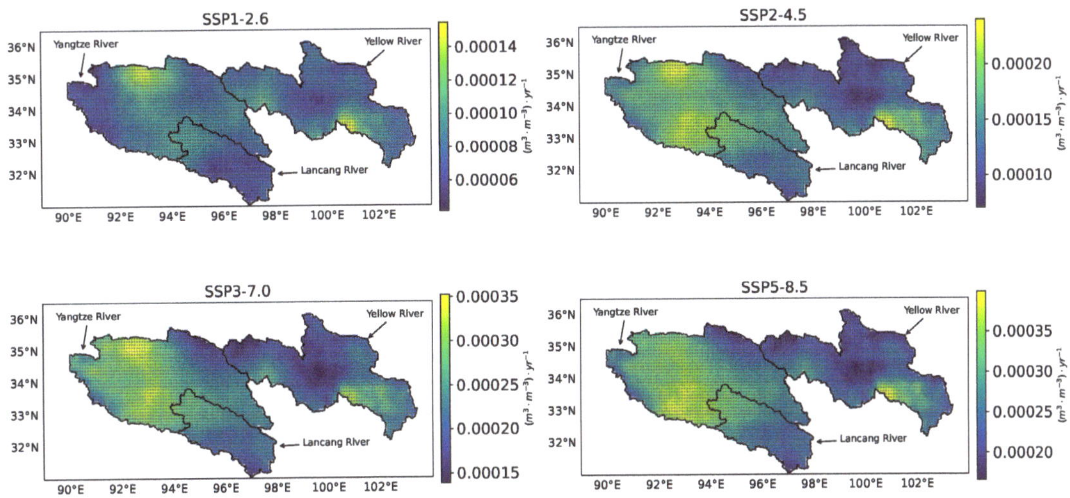

Figure 8. Spatial distribution of the annual mean SM trends for Layer 1 under four emission scenarios, with dotted areas indicating regions significant at the 95% confidence level.

From Figure 9, we observe that across all four seasons, SM increase rates rise with higher SSP scenarios. The summer season shows the smallest area and rate of significant moisture increase, while the other three seasons display a significant moisture increase across the entire TRSR, with winter showing the highest increase rate. Layers 2 and 3 show similar seasonal trends to Layer 1, while Layers 4 and 5 have smaller areas that pass the significance test, and the moisture increase rate in deeper layers slows compared to the upper layers across all seasons. Figures corresponding to Layers 2 through 5, which depict similar content to Figure 9, are provided in the Supplementary Materials. We focus only on regions that pass the significance test, disregarding those that do not, and find that nearly all areas that pass the significance test exhibit a trend of increasing moisture.

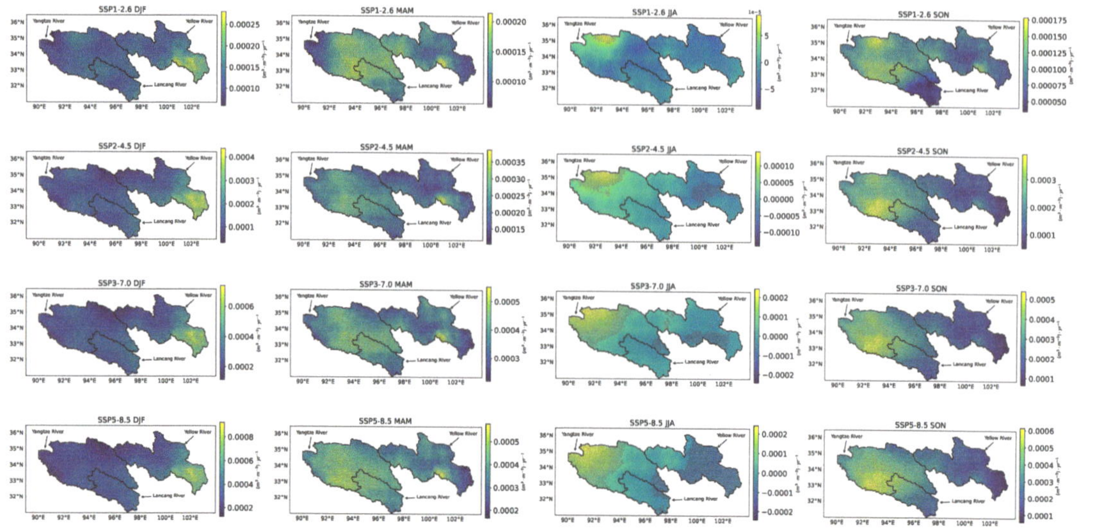

Figure 9. Spatial distribution of annual average SM trend changes in Layer 1 under four emission scenarios for different seasons, with highlighted areas indicating significance at the 95% confidence level (DJF represents winter, MAM represents spring, JJA represents summer, and SON represents autumn).

3.4. Time Series of SM in the TRSR

As shown in Figure 10, the annual average SM for Layer 1 in the TRSR region and three sub-regions exhibits a significant increasing trend under both historical and four emission scenarios. The moisture increase rates for the Historical scenario are similar to those for SSP1-2.6. Specifically, the increase rate for Historical is slightly lower than SSP1-2.6 in the TRSR and Yangtze River regions, while it is slightly higher in the Yellow River and Lancang River regions. Additionally, the increase rate grows with higher SSP scenarios, with Layer 1 showing increase rates of 8.49×10^{-5}, 1.54×10^{-4}, 2.43×10^{-4}, and 2.73×10^{-4} $(m^3 \cdot m^{-3}) \cdot yr^{-1}$, respectively. Similar results are observed for the other layers. Figures for Layers 2 through 5, which are similar to Figure 10, are provided in the Supplementary Materials. We also find that as depth increases, the moisture increase rates in the TRSR and three sub-regions show a decreasing trend under both the Historical and the four emission scenarios.

As shown in Figures 11 and 12, under the four future emission scenarios, Layer 1 SM exhibits a significant increasing trend in winter, spring, and autumn, with the highest increase rate observed in spring. As SSP increases, the moisture increase rates in the TRSR region are as follows: 1.53×10^{-4}, 2.33×10^{-4}, 3.89×10^{-4}, and 4.02×10^{-4} $(m^3 \cdot m^{-3}) \cdot yr^{-1}$ for spring; 1.39×10^{-4}, 1.58×10^{-4}, 3.01×10^{-4}, and 3.41×10^{-4} $(m^3 \cdot m^{-3}) \cdot yr^{-1}$ for winter; and 1.02×10^{-4}, 2.07×10^{-4}, 2.86×10^{-4}, and 3.36×10^{-4} $(m^3 \cdot m^{-3}) \cdot yr^{-1}$ for autumn. In summer, no significant increase trend is observed in the TRSR region. As shown in the time series in Figure 12a, the summer series under the four emission scenarios remains stable, fluctuating around a certain mean. Layer 2 shows a similar time series to Layer 1. However, from Layer 3 onward, significant moisture increase rates are observed in summer under SSP3-7.0 and SSP5-8.5 scenarios. The increase rates are as follows: 3.96×10^{-5} and 2.52×10^{-5} $(m^3 \cdot m^{-3}) \cdot yr^{-1}$ for Layer 3; 4.01×10^{-5} and 2.15×10^{-5} $(m^3 \cdot m^{-3}) \cdot yr^{-1}$ for Layer 4; and 4.42×10^{-5} and 2.74×10^{-5} $(m^3 \cdot m^{-3}) \cdot yr^{-1}$ for Layer 5. For these three layers, the rate under SSP5-8.5 is lower than that under SSP3-7.0. For winter, autumn, and summer, the time series for Layers 3 to 5 is similar to that of Layer 1, with a decreasing trend in moisture increase rate as depth increases. Figures for Layers 2 through 5, similar to Figures 11 and 12, are provided in the Supplementary Materials.

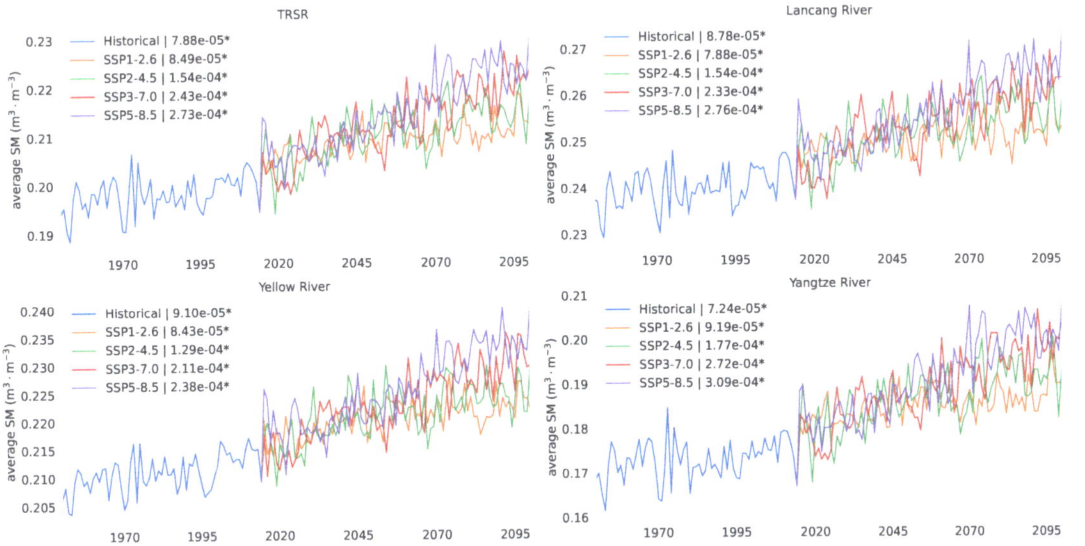

Figure 10. Time series of annual average SM for Layer 1 under different historical and future emission scenarios. The legend provides the rate of change, with units in m^3/m^3 per year, and asterisks indicate significance at the 95% confidence level.

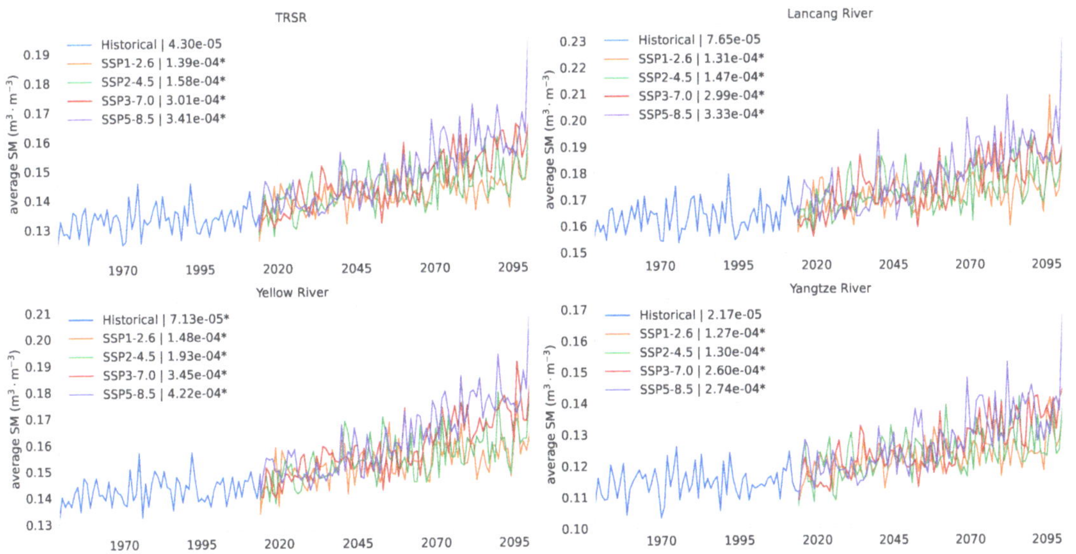

(**a**) DJF

Figure 11. *Cont.*

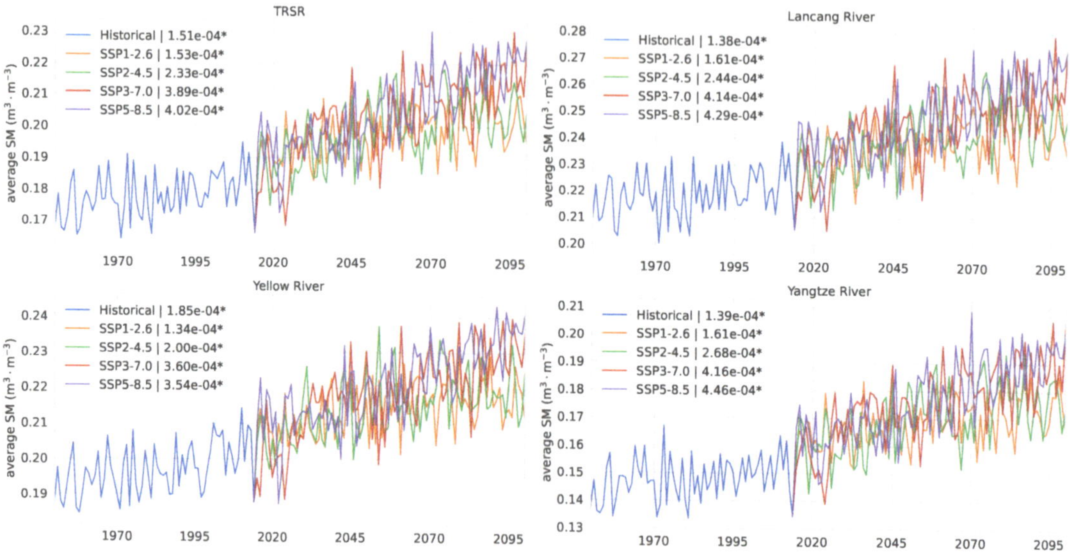

(**b**) MAM

Figure 11. Time series of annual average SM for Layer 1 under historical and future emission scenarios in different seasons, with the legend showing the rate of change in units of m^3/m^3 per year. Asterisks indicate significance at the 95% confidence level (DJF represents winter, MAM represents spring, JJA represents summer, and SON represents autumn).

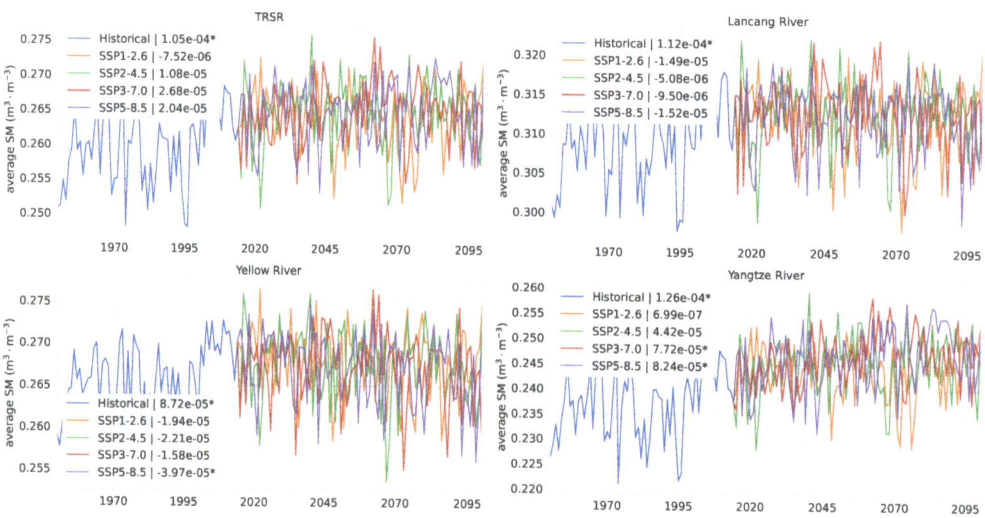

(**a**) JJA

Figure 12. *Cont.*

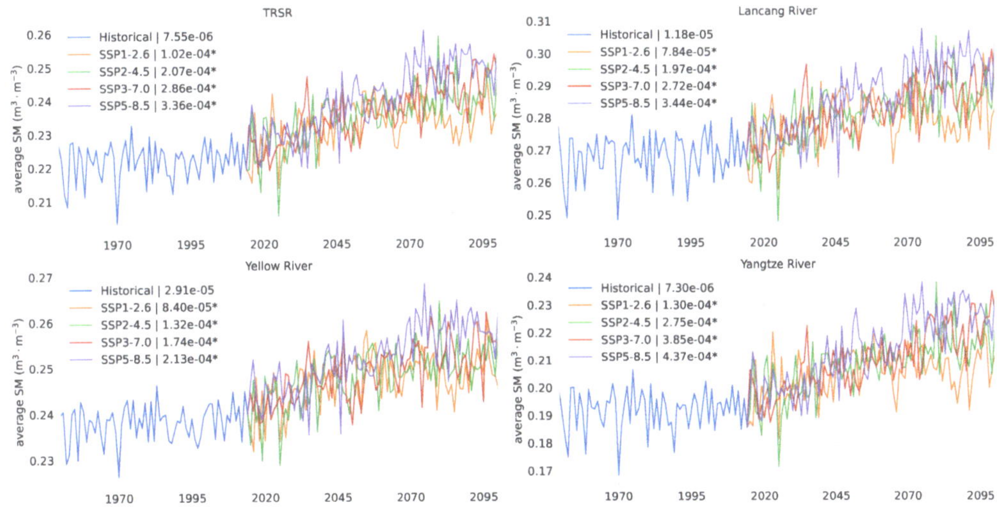

(**b**) SON

Figure 12. Same as Figure 11, but for JJA and SON.

3.5. Temporal TRSR SM and Its Rate of Change

For the Historical period, we selected the years 1995–2014, and for the four future scenarios, we chose the periods 2021–2040, 2041–2070, and 2071–2100. The reasons for studying these three sub-periods are as follows: First, the impacts of climate change may vary across different time periods. By dividing the 21st century into three sub-periods, we can more clearly analyze the effects of climate change on SM, observe how these effects evolve over time, and identify potential turning points. Second, different time periods may correspond to the implementation of different policies and adaptation measures. For instance, measures taken in the early 21st century may differ from those in the mid and late periods. Analyzing SM change trends across different time periods helps assess the effectiveness of policies and adjust future response strategies.

As shown in Figure 13, the average annual SM shows a clear upward trend by the end of this century. For the surface layer (Layer 1), the increase between 1995 and 2014 and 2071–2100 ranges from 5.5% to 11.5%, with significant rises in spring (10.5–19.9%) and winter (9.8–20.3%), while autumn (5.4–12.1%) and summer (0.8–1.1%) exhibit more modest increases. A similar pattern is seen in Layer 2, where the overall increase varies from 4.6% to 9.2%. Spring and winter again show the most notable changes (8.2–14.8% for MAM and 8.1–16.2% for DJF), while summer remains relatively stable (0.8–1.3%) and autumn experiences a moderate increase (3.7–8.4%).

The trend continues in deeper layers. In Layer 3, SM rises by 4.3–7.5%, with the largest increases observed in spring (6.1–11.2%) and winter (6.3–12.0%). Autumn and summer increases are smaller, particularly in summer (0.9–1.9% for JJA). In Layer 4, the rise ranges from 4.5% to 7.5%, with spring and winter increases again leading (7.1–12.7% for MAM and 7.1–14.3% for DJF), and summer contributing the least (1.3–2.0% for JJA). Finally, in Layer 5, the increase is between 3.3% and 6.5%, with the greatest growth observed in spring (5.4–9.5%) and winter (5.8–10.8%), while autumn and summer experience relatively smaller increases (1.9–4.4% for SON and 1.8–2.4% for JJA).

For all layers, spring and winter exhibit the strongest increases in SM, with autumn showing moderate changes and summer exhibiting the smallest gains. Furthermore, as the depth of the soil increases, the growth rate of SM tends to diminish. By the end of the century, the mean SM under the SSP5-8.5 scenario is projected to be the highest across all layers, whereas SSP1-2.6 results in the lowest values.

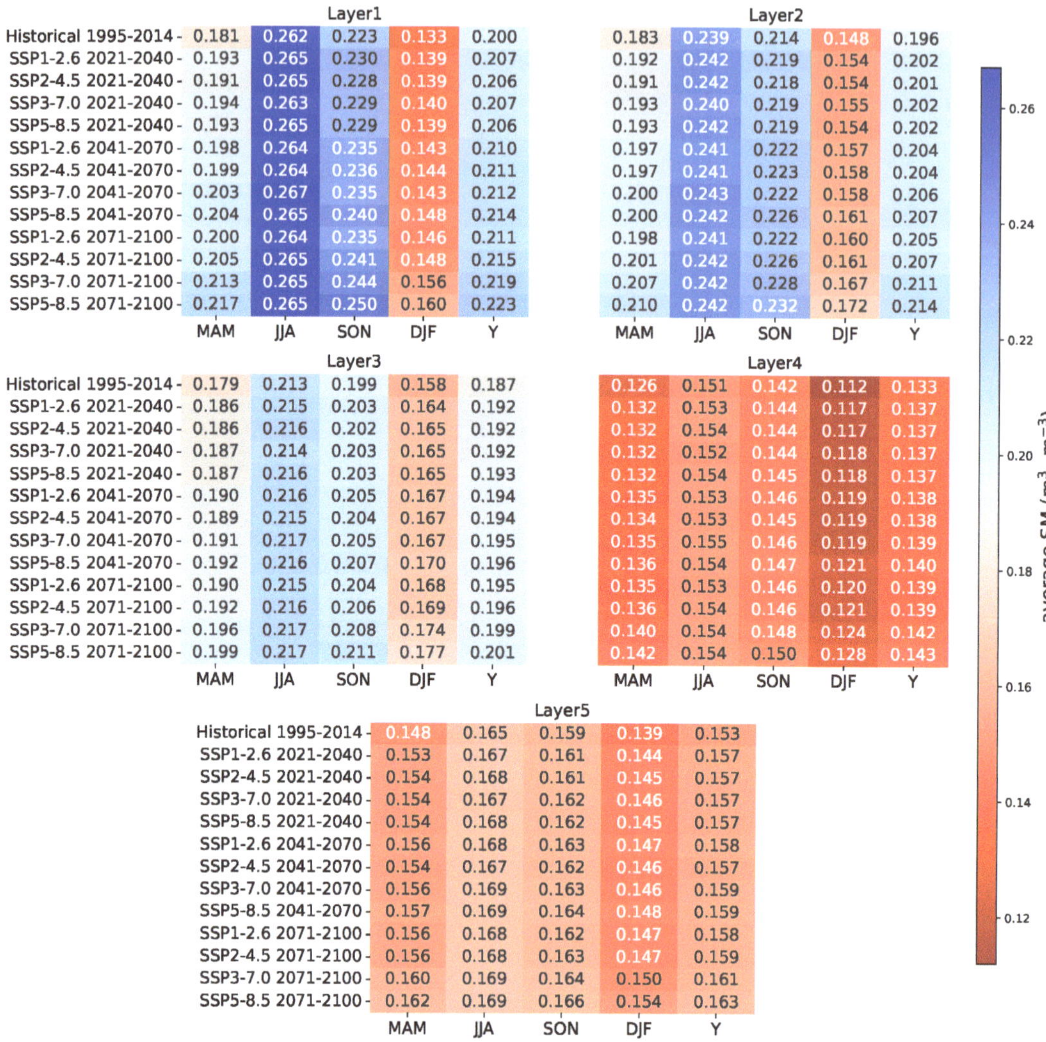

Figure 13. The mean SM of the five layers in TRSR over different periods, with the Historical period covering 1995–2014, and the four emission scenarios covering 2021–2040, 2041–2070, and 2071–2100, respectively. MAM, JJA, SON, and DJF represent mean SM values for spring, summer, autumn, and winter, respectively, while Y represents the mean annual SM.

Figure 14 shows the percentage increase in SM for Layer 1 under four emission scenarios at the end of the 21st century (2071–2100) compared to the beginning of the century (1995–2014). The northwest corner of the Yangtze River shows a significant moisture increase, ranging from approximately 12% to 20% under higher SSP scenarios. This can be attributed to several factors. Regional climate change is driving shifts in atmospheric circulation, leading to more moisture being transported to the area and an increase in precipitation frequency and intensity. Warmer temperatures also raise the atmosphere is water vapor holding capacity, promoting more rainfall and higher SM. Precipitation patterns in the region may change, with more rainfall during the growing season, particularly influenced by the East Asian summer monsoon. Additionally, the region's topography, including mountains, may enhance moisture accumulation through the orographic effect, trapping moisture in the area. The spatial distribution maps for Layers 2–5 (provided in the Supplementary Materials) show that moisture increases extend to deeper soil layers but diminish with depth. By Layers 4 and 5, moisture increases become more uniform across regions, with a small moisture increase observed in the southeast corner of the Yellow River, particularly in Layers 4 (2% to 4%) and 5 (1% to 2%).

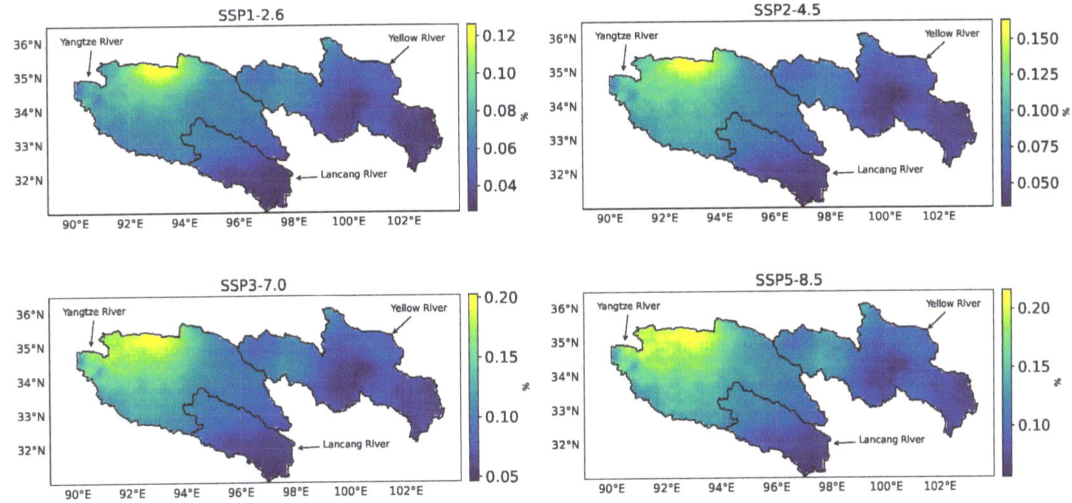

Figure 14. The spatial distribution of the percentage increase in Layer 1 SM under four emission scenarios for the end of the 21st century (2071–2100) compared to the beginning of the century (1995–2014).

Figure 15 shows the SM change rates that passed the 95% significance test. From 1995 to 2014, only the top three layers exhibited a wetting trend, which decreased with depth (annual average rates are 3.303×10^{-4}, 2.437×10^{-4}, and 1.205×10^{-4} $(m^3 \cdot m^{-3}) \cdot yr^{-1}$, respectively). Layers 4 and 5 did not pass the significance test. In the future four emission scenarios, the spring wetting rate for the five layers in the SSP3-7.0 scenario from 2021 to 2040 was the highest, at 1.522×10^{-3}, 1.052×10^{-3}, 6.513×10^{-4}, 4.832×10^{-4}, and 3.377×10^{-4} $(m^3 \cdot m^{-3}) \cdot yr^{-1}$, respectively, and decreased with depth. From Figure 15, it can be inferred that for all five soil layers, the wetting rate is generally highest in spring. Regarding the annual SM change rate, for all five layers, significant wetting mainly occurs in the early 21st century (2021–2040) and the late 21st century (2071–2100), with almost no significant trends in the mid-21st century (2041–2070). Additionally, the wetting rate in the early 21st century is generally higher than in the late 21st century. Seasonally, significant wetting trends are mainly concentrated in spring and winter, while significant wetting trends are less common in summer and autumn, and significant wetting trends decrease with depth.

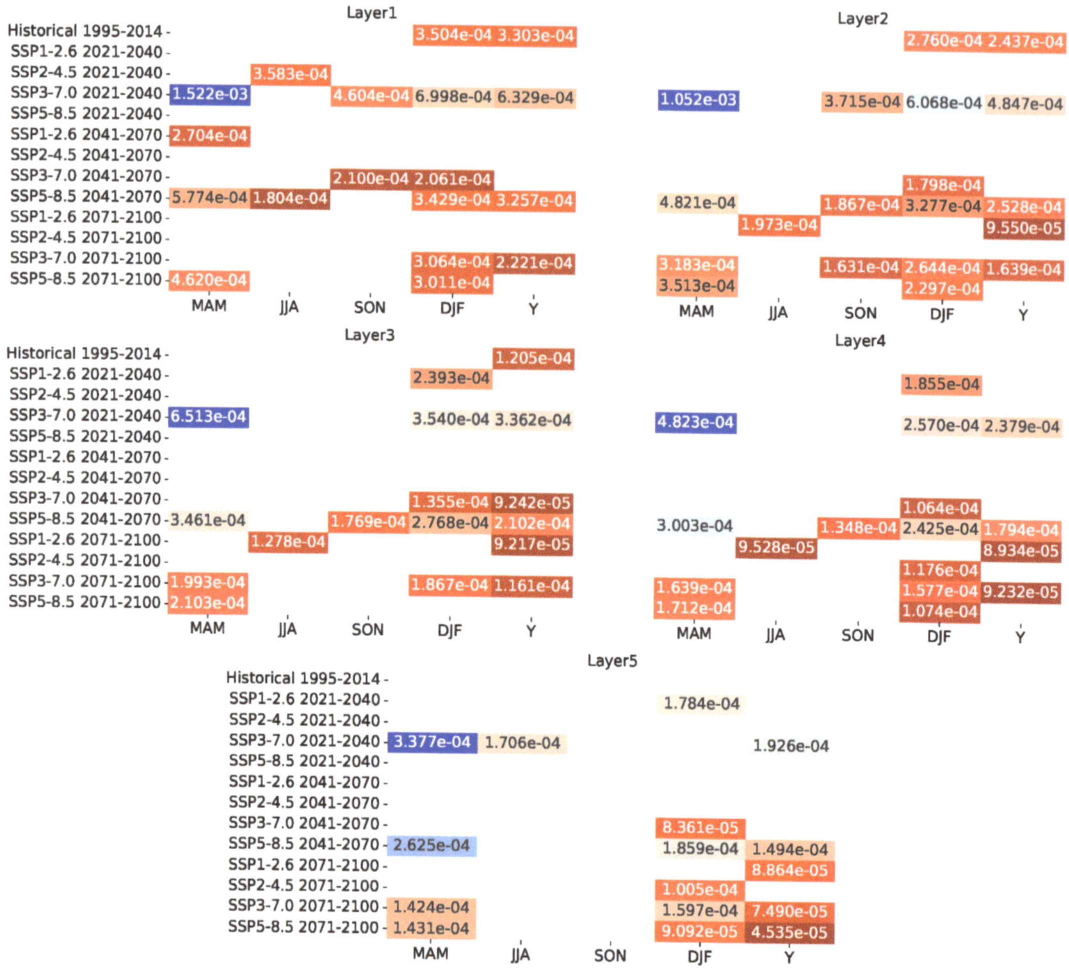

Figure 15. The rate of change in SM for the five layers in TRSR over different periods, with the Historical period covering 1995–2014, and the four emission scenarios covering 2021–2040, 2041–2070, and 2071–2100, respectively. Rates that did not pass the 95% significance test are not shown in the figure. MAM, JJA, SON, and DJF represent the rate of SM change for spring, summer, autumn, and winter, respectively, while Y represents the annual rate of SM change, with units in m^3/m^3 per year.

4. Discussion

This study investigated the spatial and temporal trends of SM change in the TRSR region under four emission scenarios (SSP1-2.6, SSP2-4.5, SSP3-7.0, SSP5-8.5) using UNet-Gan downscaled climate data. The results revealed several key findings, which we now discuss in light of existing knowledge and potential future directions.

The observed slower response of deeper soil layers to climate change, compared to shallower layers, can be attributed to several factors. One key reason is the dampening effect of the upper layers. Changes in surface temperature and precipitation primarily impact the near-surface SM. This moisture then percolates downward through the soil profile, but the process is often slow and subject to several constraints. The infiltration rate is influenced by factors such as soil texture, structure, and porosity. For example, finer soils, like clay, retain water more tightly, which can slow the downward movement of moisture. Coarser soils, such as sandy ones, may allow quicker infiltration but can lose moisture more rapidly at depth due to lower retention capacity [52].

Moreover, rainfall infiltration is also heavily influenced by external factors such as terrain, precipitation intensity, and duration. In hilly or mountainous regions, runoff can be significant, leading to less moisture infiltration into the soil, especially on slopes where gravity accelerates water movement over the surface. The intensity and duration of rainfall also play crucial roles: heavy, short-duration rains can cause surface runoff before the moisture has a chance to penetrate deeply, whereas light, steady rainfall may allow more effective infiltration, although it may not always reach deeper layers if the soil surface becomes too dry [53]. The interaction of these factors complicates the percolation process and contributes to the slower response of deeper soil layers to changes in surface conditions.

Additionally, the higher thermal inertia of deeper soils further dampens their response to climate change. Deeper soils experience slower temperature fluctuations compared to shallower layers, which results in a delayed response to changes in surface conditions. This slower thermal response leads to a dampened effect on moisture exchange processes, as heat transfer is more gradual at greater depths [54].

In terms of seasonal variations, SM increases significantly in spring and winter, followed by autumn, with the least increase observed in summer. This can be explained by several factors. In spring and winter, lower evaporation rates combined with increased precipitation—especially snowmelt in winter, as future projections indicate earlier melting and delayed freezing [55] in the TRSR region—help retain more moisture in the soil. In contrast, during summer, rising future temperatures lead to higher evapotranspiration [56], reducing the availability of significant SM increases, even if precipitation levels might be higher during the rainy season. This interaction between precipitation and temperature drives the observed seasonal variations.

This study presents a comprehensive analysis of SM trends under four SSP emission scenarios, but several limitations remain. First, the study does not provide detailed insights into the potential biases introduced by land surface conditions, such as surface lithology or geomorphology, although this may not be as important in deep learning, and these factors may play a crucial role in shaping SM dynamics. Moreover, the interaction between local precipitation and temperature, which directly influences SM, lacks in-depth discussion. Although this study obtained SM data for the future TRSR region at a 0.1-degree spatial resolution, which is a significant improvement over the spatial resolution of individual CMIP6 models, future research should focus on addressing these limitations by using higher-resolution satellite products and more reliable validation methods to improve the spatial accuracy of SM estimates [57,58]. Additionally, future studies should explore the influence of surface characteristics (such as vegetation cover, lithology, and geomorphic types) on SM distribution to gain a more detailed understanding of the observed spatial trends. Moreover, future work could delve into the impacts of SM changes under different SSP scenarios on the regional water cycle, vegetation dynamics, and carbon cycle, providing a more comprehensive perspective on the broader environmental effects of SM trends.

5. Conclusions

First, by comparing the SM data accuracy between ERA5-Land and AMSMQTP in the TRSR, we found that AMSMQTP data exhibit superior accuracy across all depth layers compared to ERA5-Land. However, despite AMSMQTP's overall better performance, its evaluation metrics still show a long-tail distribution, indicating lower accuracy at certain locations. This suggests that, although AMSMQTP offers notable advantages, additional correction measures, such as the in situ loss method proposed in this study, are necessary to further improve data accuracy in practical applications.

Second, based on the evaluation of the UNet-Gan model, we observed its outstanding performance in data fusion and downscaling. UNet-Gan successfully integrated and downscaled 17 CMIP6 datasets, demonstrating high-accuracy predictive performance across five SM layers. Particularly in Layer 1, UNet-Gan achieved the highest correlation coefficient and showed a standard deviation very close to the actual data, further demonstrating its potential in SM data processing. However, as the depth increases, the data accuracy tends to decline, which requires further attention and optimization in future research.

Third, by utilizing the UNet-Gan model to generate more refined and accurate future data at a 0.1-degree spatial resolution for four emission scenarios (SSP1-2.6, SSP2-4.5, SSP3-7.0, SSP5-8.5) ($CMIP6_{UNet\text{-}Gan}$), the following observations were made:

- Overall Increasing Trend in SM: SM in the TRSR and its sub-regions shows a significant increasing trend across all five layers, with higher emission scenarios (SSP5-8.5) leading to faster increases.
- Spatial Distribution Characteristics: There is spatial heterogeneity in SM increase rates among different soil layers in various regions of the TRSR. For instance, the high-value areas for moisture increase rates in Layer 1 and Layer 2 are mainly concentrated in the western part of the Yangtze River and the southeastern part of the Yellow River. In deeper layers (Layer 3 to Layer 5), the distribution of high-value areas changes, particularly with high-value areas emerging in the northeastern part of the Yellow River.
- Decrease in SM Increase Rate with Depth: The rate of moisture increase decreases with increasing depth from Layer 1 to Layer 5. This phenomenon is significant throughout the year and across all seasons, indicating that deeper soils respond more slowly to climate change compared to shallower soils.
- Seasonal Variations in Moisture Increase Trends: Among the four seasons, spring and winter exhibit the most significant moisture increase trends, with the highest increase rates, while summer and autumn show fewer significant moisture increase trends. Particularly in summer, the significant moisture increase areas and rates are the smallest, with some scenarios even showing no significant increase trend.
- Temporal Distribution of Future Moisture Increase Rates: In the early 21st century (2021–2040) and late 21st century (2071–2100), the rate of significant SM increase is higher, while there is almost no significant increase trend in the mid-21st century (2041–2070). Additionally, the moisture increase rate in the early 21st century is generally higher than that in the late 21st century.

Supplementary Materials: The following supporting information can be downloaded at https://www.mdpi.com/article/10.3390/rs16234367/s1. Table S1; Figures S1–S20.

Author Contributions: Conceptualization, Z.L., S.L., J.W. and X.T.; methodology, Z.L.; software, Z.L.; validation, Z.L.; formal analysis, Z.L.; investigation, Z.L., S.L., J.W. and X.T.; resources, S.L.; data curation, Z.L.; writing—original draft preparation, Z.L. and S.L.; writing—review and editing, Z.L. and S.L.; visualization, Z.L.; supervision, S.L.; project administration, S.L.; funding acquisition, S.L. All authors have read and agreed to the published version of the manuscript.

Funding: This work was supported jointly by the National Natural Science Foundation of China (U20A2081), West Light Foundation of the Chinese Academy of Sciences (xbzg-zdsys-202102), and the program of the Key Laboratory of Cryospheric Science and Frozen Soil Engineering,CAS (CSFSE-ZZ-2410).

Data Availability Statement: The $CMIP6_{UNet\text{-}Gan}$ data can be downloaded from the following link: https://figshare.com/articles/dataset/UNet-Gan_based_soil_moisture_data_of_Three_River_Source_Region_in_history_and_four_future_emission_scenarios/26827576 (accessed on 1 August 2024).Our code is open-sourced at https://github.com/lizhuoq/UNet-Gan (accessed on 1 August 2024). The ERA5-Land data used in this study can be downloaded from https://cds.climate.copernicus.eu/datasets/reanalysis-era5-land-monthly-means?tab=download (accessed on 1 August 2024). The AMSMQTP data are available at http://www.ncdc.ac.cn/portal/metadata/4d94e0b3-a10c-43f6-b971-23e2df62c8ff (accessed on 1 August 2024). The CMIP6 data can be accessed via https://cds.climate.copernicus.eu/datasets/projections-cmip6?tab=download (accessed on 1 August 2024), and the ISMN data are available for download at https://ismn.earth/en/ (accessed on 1 August 2024).

Acknowledgments: This study utilized the code from the pyMannKendall repository (https://github.com/mmhs013/pyMannKendall, accessed on 1 August 2024) for the Mann-Kendall trend analysis. We thank the contributors to this repository for their efforts.

Conflicts of Interest: The authors declare no conflict of interest.

References

1. Rodriguez-Iturbe, I.; D'odorico, P.; Porporato, A.; Ridolfi, L. On the spatial and temporal links between vegetation, climate, and soil moisture. *Water Resour. Res.* **1999**, *35*, 3709–3722. [CrossRef]
2. Wasko, C.; Nathan, R.; Peel, M.C. Changes in antecedent soil moisture modulate flood seasonality in a changing climate. *Water Resour. Res.* **2020**, *56*, e2019WR026300. [CrossRef]
3. Humphrey, V.; Berg, A.; Ciais, P.; Gentine, P.; Jung, M.; Reichstein, M.; Seneviratne, S.I.; Frankenberg, C. Soil moisture–atmosphere feedback dominates land carbon uptake variability. *Nature* **2021**, *592*, 65–69. [CrossRef] [PubMed]
4. Wu, D.; Wang, T.; Di, C.; Wang, L.; Chen, X. Investigation of controls on the regional soil moisture spatiotemporal patterns across different climate zones. *Sci. Total Environ.* **2020**, *726*, 138214. [CrossRef]
5. Li, W.; Migliavacca, M.; Forkel, M.; Denissen, J.M.; Reichstein, M.; Yang, H.; Duveiller, G.; Weber, U.; Orth, R. Widespread increasing vegetation sensitivity to soil moisture. *Nat. Commun.* **2022**, *13*, 3959. [CrossRef]
6. Tan, X.; Luo, S.; Li, H.; Hao, X.; Wang, J.; Dong, Q.; Chen, Z. Investigating the effects of snow cover and vegetation on soil temperature using remote sensing indicators in the three river source region, China. *Remote Sens.* **2022**, *14*, 4114. [CrossRef]
7. Luo, S.; Wang, J.; Tan, X.; Meng, X.; Shang, L.; Li, H.; Zhou, B.; Chen, Q. Characteristics of ground surface heat flux for alpine vegetation in freeze-thaw cycles in the three river source region. *Renew. Energy* **2024**, *221*, 119793. [CrossRef]
8. Liang, L.; Li, L.; Liu, C.; Cuo, L. Climate change in the Tibetan Plateau Three Rivers Source Region: 1960–2009. *Int. J. Climatol.* **2013**, *33*, 2900–2916. [CrossRef]
9. Li, H.; Yan, L.; Wen, T.; Feng, X. Characteristics of Climate Change and Its Impact Assessment in the Three-River Regions. *Plateau Meteorol.* **2022**, *41*, 306–316.
10. Luo, S.; Fang, X.; Lyu, S.; Jiang, Q.; Wang, J. Interdecadal changes in the freeze depth and period of frozen soil on the Three Rivers Source Region in China from 1960 to 2014. *Adv. Meteorol.* **2017**, *2017*, 5931467. [CrossRef]
11. Fang, X.; Luo, S.; Lyu, S.; Cheng, C.; Li, Z.; Zhang, S. Numerical modeling of the responses of soil temperature and soil moisture to climate change over the Tibetan Plateau, 1961–2010. *Int. J. Climatol.* **2021**, *41*, 4134–4150. [CrossRef]
12. O'Neill, B.C.; Tebaldi, C.; Van Vuuren, D.P.; Eyring, V.; Friedlingstein, P.; Hurtt, G.; Knutti, R.; Kriegler, E.; Lamarque, J.F.; Lowe, J.; et al. The scenario model intercomparison project (ScenarioMIP) for CMIP6. *Geosci. Model Dev.* **2016**, *9*, 3461–3482. [CrossRef]
13. Zelinka, M.D.; Myers, T.A.; McCoy, D.T.; Po-Chedley, S.; Caldwell, P.M.; Ceppi, P.; Klein, S.A.; Taylor, K.E. Causes of higher climate sensitivity in CMIP6 models. *Geophys. Res. Lett.* **2020**, *47*, e2019GL085782. [CrossRef]
14. Tebaldi, C.; Debeire, K.; Eyring, V.; Fischer, E.; Fyfe, J.; Friedlingstein, P.; Knutti, R.; Lowe, J.; O'Neill, B.; Sanderson, B.; et al. Climate model projections from the scenario model intercomparison project (ScenarioMIP) of CMIP6. *Earth Syst. Dyn.* **2021**, *12*, 253–293. [CrossRef]
15. Cook, B.I.; Mankin, J.S.; Marvel, K.; Williams, A.P.; Smerdon, J.E.; Anchukaitis, K.J. Twenty-first century drought projections in the CMIP6 forcing scenarios. *Earth's Future* **2020**, *8*, e2019EF001461. [CrossRef]
16. Hirsch, A.L.; Ridder, N.N.; Perkins-Kirkpatrick, S.E.; Ukkola, A. CMIP6 MultiModel evaluation of present-day heatwave attributes. *Geophys. Res. Lett.* **2021**, *48*, e2021GL095161. [CrossRef]
17. Gebrechorkos, S.; Leyland, J.; Slater, L.; Wortmann, M.; Ashworth, P.J.; Bennett, G.L.; Boothroyd, R.; Cloke, H.; Delorme, P.; Griffith, H.; et al. A high-resolution daily global dataset of statistically downscaled CMIP6 models for climate impact analyses. *Sci. Data* **2023**, *10*, 611. [CrossRef]
18. Kreienkamp, F.; Lorenz, P.; Geiger, T. Statistically downscaled CMIP6 projections show stronger warming for Germany. *Atmosphere* **2020**, *11*, 1245. [CrossRef]
19. Tian, J.; Zhang, Z.; Ahmed, Z.; Zhang, L.; Su, B.; Tao, H.; Jiang, T. Projections of precipitation over China based on CMIP6 models. *Stoch. Environ. Res. Risk Assess.* **2021**, *35*, 831–848. [CrossRef]
20. Noël, T.; Loukos, H.; Defrance, D.; Vrac, M.; Levavasseur, G. Extending the global high-resolution downscaled projections dataset to include CMIP6 projections at increased resolution coherent with the ERA5-Land reanalysis. *Data Brief* **2022**, *45*, 108669. [CrossRef]
21. Dey, A.; Sahoo, D.P.; Kumar, R.; Remesan, R. A multimodel ensemble machine learning approach for CMIP6 climate model projections in an Indian River basin. *Int. J. Climatol.* **2022**, *42*, 9215–9236. [CrossRef]
22. Feng, D.; Wang, G.; Wei, X.; Amankwah, S.O.Y.; Hu, Y.; Luo, Z.; Hagan, D.F.T.; Ullah, W. Merging and downscaling soil moisture data from CMIP6 projections using deep learning method. *Front. Environ. Sci.* **2022**, *10*, 847475. [CrossRef]
23. Soltani, K.; Amiri, A.; Ebtehaj, I.; Cheshmehghasabani, H.; Fazeli, S.; Gumiere, S.J.; Bonakdari, H. Advanced Forecasting of Drought Zones in Canada Using Deep Learning and CMIP6 Projections. *Climate* **2024**, *12*, 119. [CrossRef]
24. Goodfellow, I.; Pouget-Abadie, J.; Mirza, M.; Xu, B.; Warde-Farley, D.; Ozair, S.; Courville, A.; Bengio, Y. Generative adversarial networks. *Commun. ACM* **2020**, *63*, 139–144. [CrossRef]

25. Creswell, A.; White, T.; Dumoulin, V.; Arulkumaran, K.; Sengupta, B.; Bharath, A.A. Generative adversarial networks: An overview. *IEEE Signal Process. Mag.* **2018**, *35*, 53–65. [CrossRef]
26. Wang, K.; Gou, C.; Duan, Y.; Lin, Y.; Zheng, X.; Wang, F.Y. Generative adversarial networks: Introduction and outlook. *IEEE/CAA J. Autom. Sin.* **2017**, *4*, 588–598. [CrossRef]
27. Jozdani, S.; Chen, D.; Pouliot, D.; Johnson, B.A. A review and meta-analysis of generative adversarial networks and their applications in remote sensing. *Int. J. Appl. Earth Obs. Geoinf.* **2022**, *108*, 102734. [CrossRef]
28. Zhang, H.; Song, Y.; Han, C.; Zhang, L. Remote sensing image spatiotemporal fusion using a generative adversarial network. *IEEE Trans. Geosci. Remote Sens.* **2020**, *59*, 4273–4286. [CrossRef]
29. Ronneberger, O.; Fischer, P.; Brox, T. U-Net: Convolutional networks for biomedical image segmentation. In *Medical Image Computing and Computer-Assisted Intervention—MICCAI 2015, Proceedings of the 18th International Conference, Munich, Germany, 5–9 October 2015*; Proceedings, Part III 18; Springer: Cham, Switzerland, 2015; pp. 234–241.
30. Luo, S.; Fang, X.; Lyu, S.; Ma, D.; Chang, Y.; Song, M.; Chen, H. Frozen ground temperature trends associated with climate change in the Tibetan Plateau Three River Source Region from 1980 to 2014. *Clim. Res.* **2016**, *67*, 241–255. [CrossRef]
31. Goodman, D.S. Qinghai and the emergence of the West: Nationalities, communal interaction and national integration. *China Q.* **2004**, *178*, 379–399. [CrossRef]
32. Yeh, E.T. Green governmentality and pastoralism in western China: 'Converting pastures to grasslands'. *Nomadic Peoples* **2005**, *9*, 9–30. [CrossRef]
33. Muñoz-Sabater, J.; Dutra, E.; Agustí-Panareda, A.; Albergel, C.; Arduini, G.; Balsamo, G.; Boussetta, S.; Choulga, M.; Harrigan, S.; Hersbach, H.; et al. ERA5-Land: A state-of-the-art global reanalysis dataset for land applications. *Earth Syst. Sci. Data* **2021**, *13*, 4349–4383. [CrossRef]
34. Wang, Y.R.; Hessen, D.O.; Samset, B.H.; Stordal, F. Evaluating global and regional land warming trends in the past decades with both MODIS and ERA5-Land land surface temperature data. *Remote Sens. Environ.* **2022**, *280*, 113181. [CrossRef]
35. Gomis-Cebolla, J.; Rattayova, V.; Salazar-Galán, S.; Francés, F. Evaluation of ERA5 and ERA5-Land reanalysis precipitation datasets over Spain (1951–2020). *Atmos. Res.* **2023**, *284*, 106606. [CrossRef]
36. Luo, S.; Li, Z.; Tan, X.; Hao, X. *The Dataset of Daily Soil Moisture at 0.1-Degree Resolution over Five Layers on the Qinghai–Tibet Plateau from 2000 to 2021*; National Cryosphere Desert Data Center: Lanzhou, China, 2024. [CrossRef]
37. Turnock, S.T.; Allen, R.J.; Andrews, M.; Bauer, S.E.; Deushi, M.; Emmons, L.; Good, P.; Horowitz, L.; John, J.G.; Michou, M.; et al. Historical and future changes in air pollutants from CMIP6 models. *Atmos. Chem. Phys.* **2020**, *20*, 14547–14579. [CrossRef]
38. Su, B.; Huang, J.; Mondal, S.K.; Zhai, J.; Wang, Y.; Wen, S.; Gao, M.; Lv, Y.; Jiang, S.; Jiang, T.; et al. Insight from CMIP6 SSP-RCP scenarios for future drought characteristics in China. *Atmos. Res.* **2021**, *250*, 105375. [CrossRef]
39. Olén, N.B.; Lehsten, V. High-resolution global population projections dataset developed with CMIP6 RCP and SSP scenarios for year 2010–2100. *Data Brief* **2022**, *40*, 107804. [CrossRef]
40. Dorigo, W.; Himmelbauer, I.; Aberer, D.; Schremmer, L.; Petrakovic, I.; Zappa, L.; Preimesberger, W.; Xaver, A.; Annor, F.; Ardö, J.; et al. The International Soil Moisture Network: Serving Earth system science for over a decade. *Hydrol. Earth Syst. Sci.* **2021**, *25*, 5749–5804. [CrossRef]
41. Dorigo, W.; Xaver, A.; Vreugdenhil, M.; Gruber, A.; Dostálová, A.; Sanchis-Dufau, A.D.; Zamojski, D.; Cordes, C.; Wagner, W.; Drusch, M. Global Automated Quality Control of In Situ Soil Moisture Data from the International Soil Moisture Network. *Vadose Zone J.* **2013**, *12*, vzj2012.0097. [CrossRef]
42. Yang, K.; Qin, J.; Zhao, L.; Chen, Y.; Tang, W.; Han, M.; Zhu, l.; Chen, Z.; Lv, N.; Ding, B.; et al. A Multi-Scale Soil Moisture and Freeze-Thaw Monitoring Network on the Third Pole. *Bull. Am. Meteorol. Soc.* **2013**, *94*, 1907–1916. [CrossRef]
43. Dente, L.; Su, Z.; Wen, J. Validation of SMOS soil moisture products over the Maqu and Twente regions. *Sensors* **2012**, *12*, 9965–9986. [CrossRef] [PubMed]
44. Su, Z.; Wen, J.; Dente, L.; Velde, R.; Wang, L.; Ma, Y.; Yang, K.; Hu, Z. The Tibetan Plateau observatory of plateau scale soil moisture and soil temperature (Tibet-Obs) for quantifying uncertainties in coarse resolution satellite and model products. *Hydrol. Earth Syst. Sci.* **2011**, *15*, 2303–2316. [CrossRef]
45. Hussain, M.; Mahmud, I. pyMannKendall: A python package for non parametric Mann Kendall family of trend tests. *J. Open Source Softw.* **2019**, *4*, 1556. [CrossRef]
46. Huang, H.; Lin, L.; Tong, R.; Hu, H.; Zhang, Q.; Iwamoto, Y.; Han, X.; Chen, Y.W.; Wu, J. Unet 3+: A full-scale connected unet for medical image segmentation. In Proceedings of the ICASSP 2020—2020 IEEE International Conference on Acoustics, Speech and Signal Processing (ICASSP), Barcelona, Spain, 4–8 May 2020; pp. 1055–1059.
47. Li, Z.; Liu, F.; Yang, W.; Peng, S.; Zhou, J. A survey of convolutional neural networks: Analysis, applications, and prospects. *IEEE Trans. Neural Netw. Learn. Syst.* **2021**, *33*, 6999–7019. [CrossRef] [PubMed]
48. Ledig, C.; Theis, L.; Huszár, F.; Caballero, J.; Cunningham, A.; Acosta, A.; Aitken, A.; Tejani, A.; Totz, J.; Wang, Z.; et al. Photo-realistic single image super-resolution using a generative adversarial network. In Proceedings of the IEEE Conference on Computer Vision and Pattern Recognition, Honolulu, HI, USA, 21–26 July 2017; pp. 4681–4690.
49. Wang, X.; Yu, K.; Wu, S.; Gu, J.; Liu, Y.; Dong, C.; Qiao, Y.; Change Loy, C. Esrgan: Enhanced super-resolution generative adversarial networks. In Proceedings of the European Conference on Computer Vision (ECCV) Workshops, Munich, Germany, 8–14 September 2018.
50. Breiman, L. Random forests. *Mach. Learn.* **2001**, *45*, 5–32. [CrossRef]

51. Paszke, A.; Gross, S.; Massa, F.; Lerer, A.; Bradbury, J.; Chanan, G.; Killeen, T.; Lin, Z.; Gimelshein, N.; Antiga, L.; et al. Pytorch: An imperative style, high-performance deep learning library. *Adv. Neural Inf. Process. Syst.* **2019**, *32*.
52. Dong, J.; Ochsner, T.E. Soil texture often exerts a stronger influence than precipitation on mesoscale soil moisture patterns. *Water Resour. Res.* **2018**, *54*, 2199–2211. [CrossRef]
53. Ran, Q.; Wang, F.; Gao, J. The effect of storm movement on infiltration, runoff and soil erosion in a semi-arid catchment. *Hydrol. Process.* **2020**, *34*, 4526–4540. [CrossRef]
54. Zhou, S.; Williams, A.P.; Lintner, B.R.; Berg, A.M.; Zhang, Y.; Keenan, T.F.; Cook, B.I.; Hagemann, S.; Seneviratne, S.I.; Gentine, P. Soil moisture–atmosphere feedbacks mitigate declining water availability in drylands. *Nat. Clim. Chang.* **2021**, *11*, 38–44. [CrossRef]
55. Xu, M.; Hou, Z.; Kang, S.; Wu, X.; Han, H.; Wang, P. Projection of snowfall and precipitation phase changes over the Northwest China based on CMIP6 multimodels. *J. Hydrol.* **2024**, *641*, 131743. [CrossRef]
56. Zhang, J.; Wu, T.; Xin, X.; Lu, Y.; Yan, J.; Zhang, F.; Liu, L.; Zhao, H. Surface air temperature trend over the Tibetan Plateau in CMIP6 and its constraint in future projection. *J. Geophys. Res. Atmos.* **2023**, *128*, e2023JD039527. [CrossRef]
57. Cui, Y.; Zeng, C.; Zhou, J.; Xie, H.; Wan, W.; Hu, L.; Xiong, W.; Chen, X.; Fan, W.; Hong, Y. A spatio-temporal continuous soil moisture dataset over the Tibet Plateau from 2002 to 2015. *Sci. Data* **2019**, *6*, 247. [CrossRef] [PubMed]
58. Brocca, L.; Zhao, W.; Lu, H. High-resolution observations from space to address new applications in hydrology. *Innovation* **2023**, *4*, 100437. [CrossRef] [PubMed]

remote sensing

Article

Evaluation of the Predictive Capability of CMA Climate Prediction System Model for Summer Surface Heat Source on the Tibetan Plateau

Xinyu Chen [1], Minhong Song [1,*], Yaqi Wang [1] and Tongwen Wu [2]

1 Plateau Atmosphere and Environment Key Laboratory of Sichuan Province, School of Atmospheric Sciences, Chengdu University of Information Technology, Chengdu 610225, China; cxy18116544991@outlook.com (X.C.); 3230101080@stu.cuit.edu.cn (Y.W.)
2 Earth System Numerical Prediction Center, China Meteorological Administration, Beijing 100081, China
* Correspondence: songmh@cuit.edu.cn

Abstract: Surface heat source (SHS) is a crucial factor affecting local weather systems. Particularly SHS on the Tibetan Plateau (TP) significantly influences East Asian atmospheric circulation and global climate. Accurate prediction of summer SHS on the TP is of urgent demand for economic development and local climate change. To evaluate the performance of SHS on the TP, the observed SHS data from the eleven sites on the TP verified against CRA40-land (CRA) is evidenced significantly better than ERA5-land (ERA5), another widely used reanalysis. The predictive capability of the CMA Climate Prediction System Model (CMA-CPS) for SHS on the TP was assessed using multiple scoring methods, including the anomaly correlation coefficient and temporal correlation coefficient, among others. Furthermore, relative variability and trend analysis were conducted. Finally, based on these assessments, the causes of the biases were preliminarily discussed. The CMA-CPS demonstrates a reasonable ability to predict the spatial distribution patterns of SHS, sensible heat (SH), and latent heat (LH) on the TP in summer. Specifically, the prediction results of SHS and LH exhibit an "east-high and west-low" distribution, while the distribution of the predicted SH is opposite. Nevertheless, the predicted values are generally lower than CRA, particularly in interannual variations and trends. Among the predictions, LH exhibits the highest temporal correlation coefficients, consistently above 0.6, followed by SHS, while SH predictions are less accurate. The spatial distribution and skill scores indicate that LH on the TP contributes more significantly to SHS than SH in summer. Furthermore, discrepancies in the predictions of surface temperature gradients, ground wind speed, and humidity on the TP may partly explain the biases in SHS and their components.

Keywords: Tibetan Plateau; CMA Climate Prediction System Model; surface heat source; predictive capability

Citation: Chen, X.; Song, M.; Wang, Y.; Wu, T. Evaluation of the Predictive Capability of CMA Climate Prediction System Model for Summer Surface Heat Source on the Tibetan Plateau. *Remote Sens.* **2024**, *16*, 4118. https://doi.org/10.3390/rs16214118

Academic Editor: Carmine Serio

Received: 26 August 2024
Revised: 16 October 2024
Accepted: 1 November 2024
Published: 4 November 2024

1. Introduction

The Tibetan Plateau (TP) is the highest plateau in the world, with an average elevation exceeding 4000 m, earning it the titles "Roof of the World" and "Third Pole". The surface of the TP is situated in the middle troposphere, and its strong thermal effects in summer significantly influence the atmospheric circulation, monsoons, precipitation patterns in East Asia, and global climate [1–4]. Consequently, the thermal effects and SHS over the TP in summer have been extensively studied [5–9].

SH and LH are critical components of SHS, particularly in complex regions such as TP. Numerous studies have sought to estimate SH using observational station data combined with bulk transfer formulas. These studies have highlighted interannual variation trends and seasonal transition characteristics of SH in TP environments [10–12]. In particular, Yang et al. [13] proposed a physical scheme for estimating the daily average SH based on data from China Meteorological Administration stations. However, the uneven distribution of these stations across the TP often results in an overestimation of SH. Conversely, LH is predominantly analyzed using reanalysis data, with observational station precipitation data utilized as validation tools. This approach has highlighted a significant correlation between LH in the TP and precipitation patterns across China [14–17]. Recent advancements in observational technology have significantly enhanced research on SHS over the TP, utilizing reanalysis data, satellite data [18–23], and model simulation results [24–30]. Multiple reanalysis datasets were assessed alongside observational station data by Chen et al. [19] and Wu et al. [20], concluding that the ERA-Interim reanalysis data most accurately represented the climatic distribution and seasonal variations of SH on the TP.

The CMA Climate Prediction System Model (CMA-CPS), developed by the National Climate Center of the China Meteorological Administration, has been widely implemented and has generated significant social benefits and international impact [31]. Current evaluations of this prediction model primarily focus on atmospheric circulation, precipitation, temperature, extreme events, marine heatwaves, and greening trends on the TP, largely relying on the consistent dynamic framework of the CMA-CSM model [32–39]. Recent research has made significant strides in enhancing climate prediction capabilities for East Asia, particularly regarding summer precipitation and circulation patterns. Notably, one study focused on seasonal predictions of summer precipitation in the middle-lower reaches of the Yangtze River using the NUIST-CFS1 framework, aiming to improve forecasting accuracy and facilitate effective water resource management [40]. In parallel, another investigation explored the abilities of seasonal prediction models to capture East Asian summer circulation and their responses to tropical oceanic changes, elucidating critical mechanisms influencing the regional climate [41]. Furthermore, the ECCC monthly forecasting system demonstrated robust capabilities for sub-seasonal predictions, effectively capturing boreal summer climate variations and examining performance across various climate patterns [42]. Complementarily, an assessment of the Community Earth System Model—Large Ensemble CESM−LE highlighted notable uncertainties in simulating temperature and precipitation trends, particularly extreme values, underscoring the significance of understanding these uncertainties for reliable climate predictions and strategic planning [43]. Lastly, research within the Sub-seasonal to Seasonal Prediction Project database evaluated multiple climate models for their capacity to simulate the Asian summer monsoon's intensity, duration, and spatial distribution, revealing significant model discrepancies that warrant further investigation to enhance sub-seasonal forecasting accuracy [44]. Together, these findings highlight the critical role of improving climate prediction models in understanding the East Asian climate system. The diverse frameworks and methodologies examined, including the NUIST-CFS1 and ECCC monthly forecasting system, reveal both strengths and uncertainties. However, relatively limited attention has been given to the SHS and its components. Based on SHS over the TP playing a critical role in influencing the weather and climate of East Asia and worldwide, this study aims to bridge this gap in accurate prediction. By utilizing measured SHS data over the TP, a comprehensive comparison is made between ERA5 and CRA reanalysis data. The prediction capability of CMA-CPS specifically for SHS over the TP in summer is assessed using various evaluation methods. Furthermore, this study seeks to identify potential reasons for any observed prediction biases, which will provide essential insights for enhancing and refining the CMA-CPS. Ultimately, the results aim to contribute to a better understanding of surface heat dynamics over the TP, which are crucial for improving climate prediction models and their applications.

2. Data and Methods

2.1. Data Description

The site data are provided by the National Tibetan Plateau Data Center, which offers a high-resolution (hourly) comprehensive observation dataset of land-atmosphere interaction processes on the TP. This dataset includes 12 field observation stations distributed across different regions of the TP, covering typical TP surface types such as alpine meadows, alpine steppes, desert gobi, desert steppes, mountain forests, and TP lakes, effectively representing the complex surface types over the TP [45]. Detailed information about each observation station can be found in the study [45]. Based on the completeness and geographical location of the observation station data, eleven stations, excluding the MAWORS station, were selected. The selected period is July and August of each year from the establishment of the stations up to 2020.

The prediction data come from the third-generation sub-seasonal to seasonal to interannual integrated climate model prediction system developed by the National Climate Center of the China Meteorological Administration. In recent years, the National Climate Center has made significant progress in model development, with new versions including the Earth System Model BCC-ESM1.0, the medium-resolution climate system model BCC-CSM2-MR, and the high-resolution climate system model BCC-CSM2-HR [46]. All three models include four components: atmosphere, land surface, ocean, and sea ice. BCC-CSM2-HR has successfully addressed scientific issues related to vertical resolution and the stratosphere by improving multiple physical process parameterization schemes, developing ensemble prediction methods, and establishing the CMA-CPS. The resolution of each subsystem module has been enhanced and optimized, with the atmospheric model reaching $0.45^\circ \times 0.45^\circ$, vertical height extending from 40 km to 65 km, and the ocean model's horizontal resolution improving to $0.25^\circ \times 0.25^\circ$ with a vertical depth of 5.5 km. This study utilizes SHS data with a horizontal resolution of $0.45^\circ \times 0.45^\circ$, predicted by the CMA-CPS starting in May for the summer months from 2001 to 2021, analyzed on both a monthly and daily basis. At the same time, reanalysis data with a horizontal resolution of $0.1^\circ \times 0.1^\circ$ from ERA5 provided by the European Centre for Medium-Range Weather Forecasts (ECMWF) and reanalysis data with a horizontal resolution of $0.25^\circ \times 0.25^\circ$ from CRA provided by the National Meteorological Information Center of the China Meteorological Administration were selected for the same period.

Using a direct algorithm [47], the SH and LH over the TP were summed to obtain the SHS for the corresponding period. For ease of evaluation, all grid point data were uniformly processed to a horizontal resolution of $0.5^\circ \times 0.5^\circ$.

2.2. Methods

The skill score methods used include [48–50] anomaly correlation coefficient (*ACC*) and temporal correlation coefficient (*TCC*), which, respectively, reflect the spatial similarity between predicted and observed values and better characterize the model's ability to forecast anomalies at each grid point. In this paper, *ACC* scores with positive values and *TCC* scores greater than 0.4 are defined as high-skill scores. Mean Absolute Error (*MAE*) describes the average absolute deviation between predicted and observed values; correlation coefficient (r) describes the degree of correlation between predicted and observed values; Root Mean Square Error (*RMSE*) describes the deviation between predicted and observed values; and Relative Root Mean Square Error (*RRMSE*) describes the degree of deviation between predicted and observed values. Additionally, relative variability (the ratio of standard deviation to the absolute value of the mean) [50] and trend rates are used to analyze their trends [11].

The formulas for each skill score method are as follows:

$$ACC = \frac{\sum\limits_{i=1}^{m} (\Delta f_i - \Delta \overline{f_i})(\Delta o_i - \Delta \overline{o_i})}{\sqrt{\sum\limits_{i=1}^{m} (\Delta f_i - \Delta \overline{f_i})^2 \sum\limits_{i=1}^{m} (\Delta o_i - \Delta \overline{o_i})^2}} \tag{1}$$

$$TCC = \frac{\sum\limits_{i=1}^{m} (o_i - \bar{o}_i)(f_i - \bar{f}_i)}{\sqrt{\sum (o_i - \bar{o}_i)^2 \sum (f_i - \bar{f}_i)^2}} \tag{2}$$

$$MAE = \frac{1}{M}\sum_{i=1}^{m}|f_i - o_i| \tag{3}$$

$$RRMSE = \frac{\sqrt{\sum_{i=1}^{m}(o_i - f_i)^2}}{\overline{o_i}} \tag{4}$$

$$r = \frac{\sum_{i=1}^{m}(o_i - \overline{o_i})(f_i - \overline{f_i})}{\sum_{i=1}^{m}(o_i - \overline{o_i})^2(f_i - \overline{f_i})^2} \tag{5}$$

$$RMSE = \sqrt{\frac{1}{m}\sum_{i=1}^{m}(f_i - o_i)^2} \tag{6}$$

In the formula, f and o represent the predicted value and the observed value, respectively. i denotes the sample grid point, and N is the total number of grid points participating in the evaluation.

3. Results

3.1. Model Simulation Evaluation

Figure 1 shows the correlation coefficients (r), root mean square errors ($RMSE$), and standard deviation ratios (SD) between the daily heat flux observation data from 11 observation stations on the TP during July–August and the data from ERA5, CRA, and CMA-CPS. For SH, the site number of positive correlations for CRA is shown at 7 sites, while that for ERA5 is at 9 sites, with overall r values ranging between 0.1 and 0.3. The best correlation for CRA is at NAMORS, approaching 0.4. The RMSE for CRA ranges between 0.8 and 1.6 W/m^2, with standard deviation ratios close to 1. The RMSE for ERA5 is greater than 1.6 W/m^2, with SD generally between 1.5 and 3. For LH, both CRA and ERA5 show positive correlations at 7 sites, with overall r values between 0.1 and 0.3. The best correlation for CRA is at QOMS, approaching 0.4. The RMSE for CRA ranges between 0.8 and 2 W/m^2, with SD generally between 1 and 2; ERA5 shows RMSE greater than 2 W/m^2 at 4 sits, with SD also greater than 2 at 4 stations. For SHS, the correlation indices for CRA and ERA5 are relatively poor compared to SH and LH, with overall r values between 0.06 and 0.28 but with considerable dispersion. For the CMA-CPS, although the dispersion is comparable to that of CRA, the number of positive correlations ranges from 5 to 7, and overall r values lie between 0 and 0.17. The correlation is generally weak in Figure 1, possibly due to the complex environment and harsh climate conditions. However, in some years, the r values exceed 0.5, reaching up to 0.61 (SH in 2008 of CRA at QOMS). Overall, the r values for CRA are slightly better than those for ERA5, with significantly less dispersion. Therefore, CRA is selected as the comparative data to evaluate the prediction ability of the CMA-CPS on the TP.

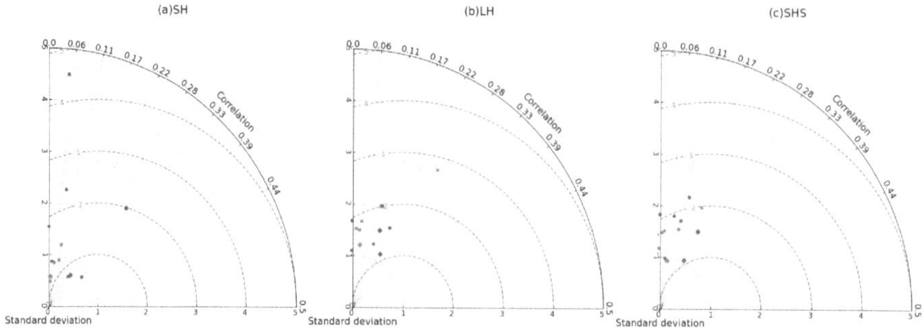

Figure 1. *Cont.*

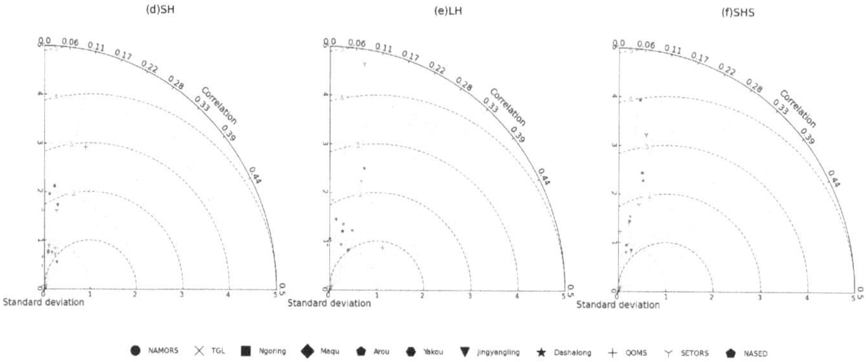

Figure 1. Taylor diagrams of SH (**a**,**d**), LH (**b**,**e**), and SHS (**c**,**f**) from different data sources at 11 observation stations on the TP during July–August. Red represents ERA5 data, blue represents CRA data, green represents CMA−CPS data, and the dashed line indicates the root mean square error.

3.2. Evaluation of Predictive Capability for SH

According to the CMA-CPS predictions, the spatial distribution of summer SH on the TP exhibits a pattern of higher values in the west and lower values in the east. Specifically, the high-value center of SH is situated in the southern part of the Tianshan Mountains, located in the northwest of the TP, with a peak value of approximately 110 W/m^2. Conversely, the SH near the Kunlun Mountains in the western part of the TP is relatively low, falling below 10 W/m^2. In the central part of the TP, the SH generally averages around 40 W/m^2, with higher values observed in the northern TP and the Qaidam Basin, both exceeding 90 W/m^2. Notably, the Qaidam Basin shows a distinct spatial distribution characterized by higher values in the center and lower values around the edges (Figure 2a). The CMA-CPS predictions for June and July (Figure 2b,c) are generally consistent with the summer distribution, with SH hovering around 30 W/m^2 in most parts of the TP, except for the Tianshan Mountains, northern TP, and Qaidam Basin. The predicted SH for August shows a slight increase, with high-value areas concentrated in the Tianshan Mountains and Qaidam Basin (Figure 2d). Similarly, the CRA reanalysis data also show a spatial distribution of summer SH with higher values in the west and lower values in the east. On the TP, there are three high-value centers of summer SH: the Tianshan Mountains, the northern TP, and the vicinity of the Qaidam Basin. The highest values are recorded in the central and western parts of the northern TP, reaching up to 120 W/m^2, while the lowest values are around near the Kunlun Mountains, dipping to about 20 W/m^2 (Figure 2e). The SH distribution in June (Figure 2f) shows an overall increase compared to the summer, with particularly significant values in the northern TP, reaching up to 150 W/m^2. The SH distribution in July and August (Figure 2g,h) generally mirrors the summer average, with CRA values being overall higher than those of CMA-CPS. Specifically, the values in July are higher than the summer average, while the values in August are lower. In terms of relative variability, only a few areas of CMA-CPS's SH exceed 10%, whereas the areas with less than 10% variability in CRA are smaller and primarily located in the northern TP, indicating that the interannual variability of SH predicted by CMA-CPS is relatively modest.

The summer SH trends predicted by the CMA-CPS on the TP range between -10 and 10 $W/m^2 \cdot (10a)^{-1}$. Notably, there is a significant increase in the northern part of the TP, exceeding 10 $W/m^2 \cdot (10a)^{-1}$, albeit confined to a small area. Conversely, the surrounding regions show a significant decrease. The primary areas of increase are located in the northwest, northern, and southern TP, with trends around 5–10 $W/m^2 \cdot (10a)^{-1}$. Certain northern TP regions surpass the significance test (Figure 3a). The spatial distribution of SH trends predicted by the CMA-CPS for June and July generally aligns with the summer results, with differences primarily in the eastern TP. In June, the enhanced area in the southeastern TP is more extensive than during the summer, while in July, the southeastern TP experiences a significant decrease. In August, there were no significant areas of increase in the northern TP, and the trend in the southern TP is decreasing, with the enhanced area in the eastern TP being larger than in summer (Figure 3b–d). The CRA reanalysis data also indicate an area of increase in the northern TP during summer, albeit with lesser intensity. The SH in the northeastern TP and west of 85°E exhibits an overall significant decrease, with the trends ranging from -5 to -15 $W/m^2 \cdot (10a)^{-1}$ in the former and exceeding -15 $W/m^2 \cdot (10a)^{-1}$ in the latter. The southern TP east of 85°E shows an overall increase (Figure 3e). The SH trend distribution in June is generally consistent with summer, but the weakening intensity in the Karakoram Mountains and northern TP exceeds 40 $W/m^2 \cdot (10a)^{-1}$, significantly stronger than in summer. In July, the area of significant decrease in the western TP is larger than in summer, whereas the decrease in the northeastern TP is smaller. In August, the area of increase in the central TP is more extensive than in other periods. The comparison reveals that the overall SH trend predicted by the CMA-CPS is smaller, aligning with the relative variability results in Figure 1, further indicating that the interannual variability of SH predicted by the CMA-CPS is relatively modest.

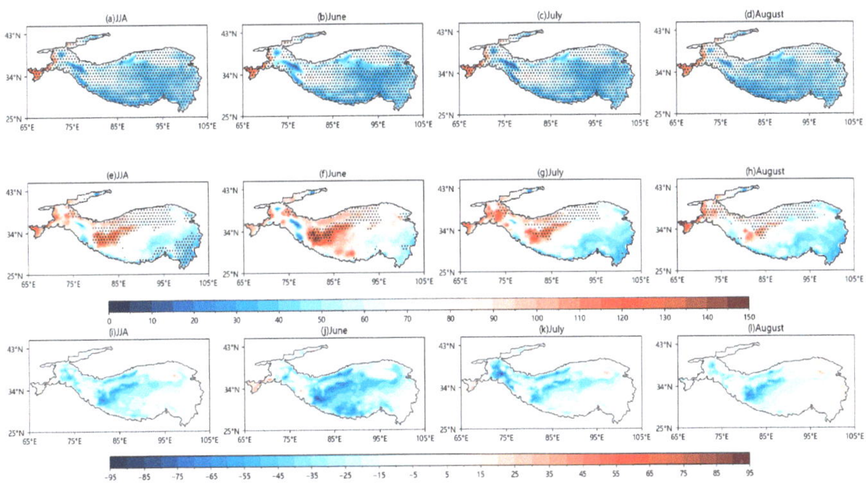

Figure 2. Spatial distribution of SH on the TP from CMA−CPS (**top**) and CRA (**middle**) from 2001 to 2021, and their differences (**bottom**) (unit: W/m^2). (**a**,**e**,**i**) Summer, (**b**,**f**,**j**) June, (**c**,**g**,**k**) July, (**d**,**h**,**l**) August. The dotted areas represent regions with a relative variability of 0–10%.

The *MAE* of SH predicted by CMA-CPS for the TP region in June is significantly higher than in other periods, with July and summer exhibiting comparable values and August recording the lowest. All periods display noticeable fluctuations. The *RRMSE* results are consistent with *MAE* (Figure 4a,b), indicating that the prediction bias of SH by CMA-CPS is the greatest in June and least in August. The proportion of years with high *ACC* scores (Figure 4c) is comparable for summer and June, at 15/21 and 16/21, respectively, followed by July (13/21), and August with the lowest proportion (10/21). This suggests that the spatial similarity between SH predicted by CMA-CPS and CRA is better in summer and June, especially in June 2008 and 2012, where the *ACC* values are exceptionally high, both exceeding 0.4, indicating a high level of agreement between CMA-CPS predictions and observed results for June SH on the TP in these years, with relatively superior prediction performance. The prediction performance in August shows significant fluctuations, with *ACC* values below 0 for more than half of the years, particularly in 2003 and 2014, where the *ACC* values in August are the lowest, below −0.2, further highlighting the need for improvement in prediction accuracy. As shown in Figure 4d, the proportion of years with high *TCC* scores from highest to lowest is August (19/21), summer (17/21), July (16/21), and June (12/21). CMA-CPS demonstrates better predictive capability for grid point anomalies of SH in August and the worst in June. The *TCC* values for SH on the TP in August display relatively minor fluctuations, while the *TCC* values in June show substantial fluctuations, reflecting the sensitivity of SH predictions across different months. Overall, the *TCC* values for summer from 2009 to 2021 exhibit significant fluctuations, which may be attributed to the complexity and variability of the summer TP climate, making the prediction of SH on the TP more challenging.

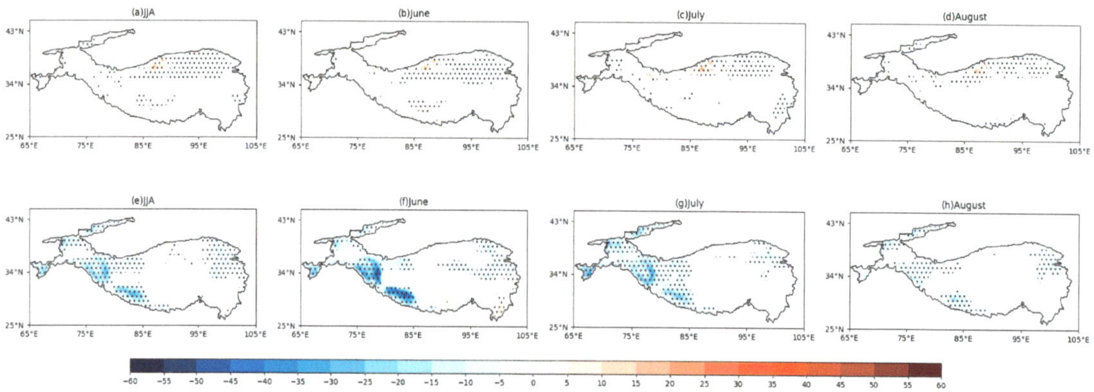

Figure 3. Spatial distribution of the climatic trend rate of SH on the TP from CMA−CPS (**top**) and CRA (**bottom**) from 2001 to 2021 (unit: $W/m^2 \cdot (10a)^{-1}$). (**a**,**e**) Summer, (**b**,**f**) June, (**c**,**g**) July, (**d**,**h**) August. The dotted areas indicate regions that passed the 90% significance test.

Figure 4. Time series of skill scores for SH on the TP predicted by CMA−CPS from 2001 to 2021. (**a**) *MAE*, (**b**) *RRMSE*, (**c**) *ACC*, (**d**) *TCC*.

The spatial distribution of *MAE* for SH predicted by CMA-CPS during different periods in summer shows conclusions consistent with Figure 4a. In summer, high-value areas are predominantly located in the Tianshan Mountains and the northern TP, with *MAE* values exceeding 60 W/m^2. Conversely, low-value areas are mainly concentrated in the eastern TP, with specific regions in the Qaidam Basin and the southeastern TP exhibiting *MAE* values ranging from 0 to 5 W/m^2 (Figure 5a). In June, the high-value area of SH *MAE* extends from the northern TP to the southern side of the Qaidam Basin and the Gangdise Mountains, covering a significantly larger area compared to other periods. The spatial distribution of SH *MAE* in July mirrors that of summer, while in August, the high-value area in the northern TP diminishes compared to other periods, and the *MAE* in the central and eastern TP falls below 45 W/m^2 (Figure 5b–d). The spatial distribution of *RRMSE* for SH during different periods in summer is similar to that of *MAE*, with the primary distinction observed in the Karakoram Mountains, where the scope and values of the high *RRMSE* area are larger. Analyzing the spatial distribution of *ACC* scores for SH in summer (Figure 5i) reveals that the high-score areas are mainly in the northwestern TP, the Tanggula Mountains, and the northeastern TP. Conversely, the negative value areas are predominantly found in the northern TP and the Hengduan Mountains. The spatial distribution of *ACC* scores in June is similar to that of summer, but in June, the Hengduan Mountains exhibit high scores. In July, the high-score area is slightly narrow compared to summer and June, primarily located in the eastern TP, with the central TP largely featuring negative values. In August, the areas with positive and negative values are roughly equal, with positive values significantly smaller than in other periods (Figure 5j–l). The *MAE*, *RRMSE*, and *ACC* scores for SH during various summer periods show good temporal and spatial consistency, whereas the *TCC* scores are relatively weaker. The spatial distribution of *TCC* scores for SH across different periods remains similar, with high-score areas mainly in the northwestern and central-eastern TP. However, in June, the high-score area in the northern TP is significantly larger than in other periods, with minimal differences observed in other regions. Overall, the scoring effect is most favorable in June, followed by July and summer, with August demonstrating the least favorable outcomes (Figure 5m–p).

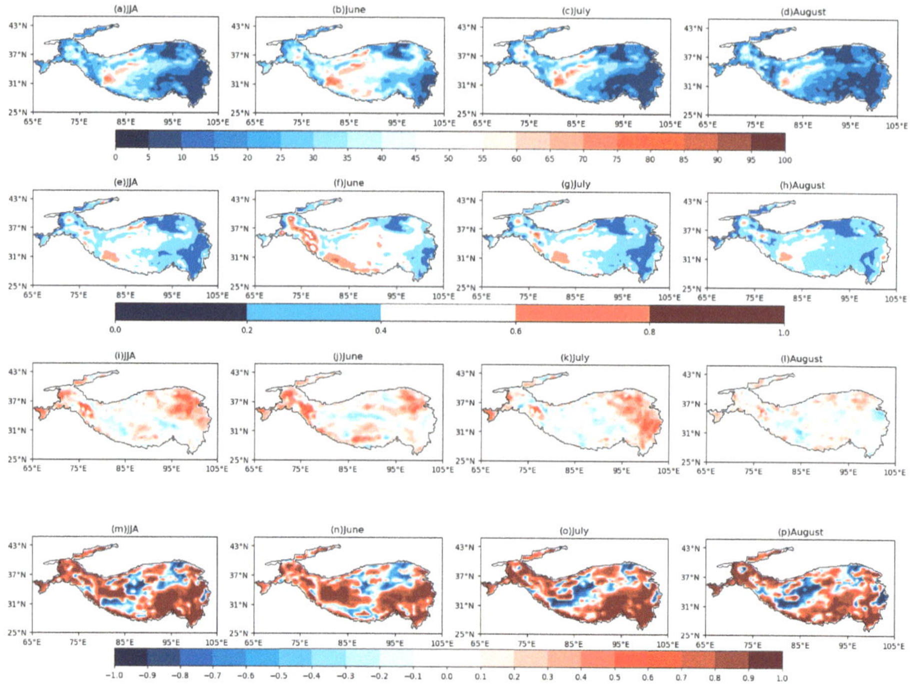

Figure 5. Spatial distribution of skill scores for SH on the TP predicted by CMA−CPS during different periods from 2001 to 2021. (**a–d**) *MAE*, (**e–h**) *RRMSE*, (**i–l**) *ACC*, (**m–p**) *TCC*.

3.3. Evaluation of Predictive Capability for LH

According to the CMA-CPS predictions, the spatial distribution of summer LH on the TP shows a pattern of higher values in the east and lower values in the west (Figure 6a). Specifically, the LH high-value center is located in the southeastern part of the TP, with a peak value approximating 100 W/m^2. Conversely, the LH in the northern TP is relatively low, less than 30 W/m^2, with the Karakoram Mountains and the Qaidam Basin exhibiting values of merely around 5 W/m^2. The distribution of LH predicted by CMA-CPS for June, July, and August (Figure 6b–d) is consistent with the summer distribution. Similarly, the CRA reanalysis data reveal a spatial distribution of summer LH that mirrors the east-high, west-low pattern, with the high-value area in the southeastern TP reaching up to 120 W/m^2 and the Qaidam Basin's low-value center dropping to approximately 5 W/m^2 (Figure 6e). In June, a decline in LH was observed in both the northern and southeastern TP compared to summer levels (Figure 6f). Conversely, in July and August, the LH in the southeastern TP increases compared to summer, with July exhibiting a more significant range and magnitude of increase. The spatial distribution of LH in August generally mirrors the summer pattern, with comparable values (Figure 6g,h). Overall, the LH values derived from CRA are higher than those from CMA-CPS. The areas where CMA-CPS predictions exceed actual values are in the northern TP, with June experiencing a significantly larger range of overestimation compared to other periods. In terms of relative variability, areas with less than 10% variability are mainly located in the high-value areas for both CMA-CPS and CRA. However, the range is broader for CMA-CPS, indicating that the interannual variability of LH predicted by CMA-CPS is relatively smaller.

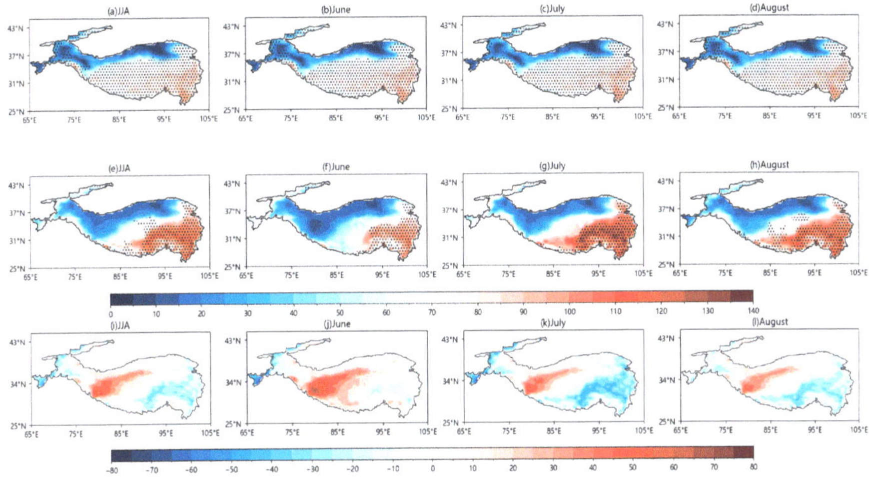

Figure 6. Spatial distribution of LH on the TP from CMA−CPS (**top**) and CRA (**middle**) from 2001 to 2021, and their differences (**bottom**) (unit: W/m^2). (**a**,**e**,**i**) Summer, (**b**,**f**,**j**) June, (**c**,**g**,**k**) July, (**d**,**h**,**l**) August. The dotted areas represent regions with a relative variability of 0–10%.

The trend of summer LH on the TP predicted by CMA-CPS ranges from −5 to 15 $W/m^2{\cdot}(10a)^{-1}$, indicating an overall increasing trend. Notably, the regions experiencing significant increases are concentrated primarily in the northern TP, whereas decreases are also observed, albeit predominantly in the northern TP (Figure 7a). The spatial distribution of LH trends in June, July, and August mirrors that of summer, with the notable exception of discrepancies in the distribution of significantly increased areas in the southeastern TP (Figure 7b–d). According to the CRA reanalysis data, the overall trend of summer LH on the TP also shows an increasing trend, with significant increments in the western and northeastern TP, peaking at 40 $W/m^2{\cdot}(10a)^{-1}$. Conversely, decreases are observed mainly in the southeastern TP, with values hovering around 0–5 $W/m^2{\cdot}(10a)^{-1}$ (Figure 7e). In June and August, the trend of LH in the northern TP is weaker compared to summer, while in July and August, the trend in the northeastern TP is similarly subdued relative to the summer trend (Figure 7f–h). Comparatively, the trend of LH predicted by CMA-CPS is generally smaller than that of CRA, with significant differences in the central locations.

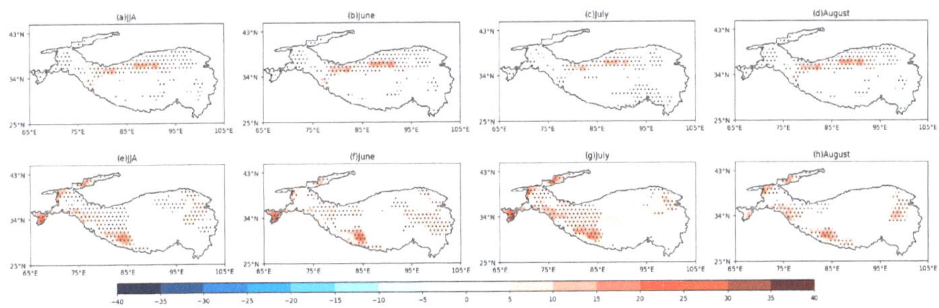

Figure 7. Spatial distribution of the climatic trend rate of LH on the TP from CMA−CPS (**top**) and CRA (**bottom**) from 2001 to 2021 (unit: $W/m^2{\cdot}(10a)^{-1}$). (**a**,**e**) Summer, (**b**,**f**) June, (**c**,**g**) July, (**d**,**h**) August. The dotted areas indicate regions that passed the 90% significance test.

Figure 8 presents the predictive capabilities of CMA-CPS for LH over the TP during various summer periods, evaluated through diverse scoring methodologies. The *MAE* for LH predictions in July surpasses that of other periods, generally ranging between

25 and 30 W/m^2, with only 2001 exhibiting an *MAE* below 25 W/m^2. June and summer predictions are comparable, while August boasts the lowest *MAE* values, all showing noticeable fluctuations. Similarly, the *RRMSE* mirrors the *MAE* trends, with June's *RRMSE* surpassing summer's in certain years from 2005 to 2016, showing significant fluctuations (Figure 8a,b). This indicates that the prediction bias for LH by CMA-CPS is the most pronounced in July and the smallest in August. The proportion of years achieving high *ACC* scores is 14/21 for summer, 15/21 for June, 13/21 for July, and 15/21 for August, indicating that the spatial similarity of LH predicted by CMA-CPS is comparable across these periods (Figure 8c). Notably, *ACC* values for June and summer in 2001 and 2012, as well as for August in 2012, 2018, and 2019, are exceptionally close to 0.4, indicating relatively superior prediction performance during these years. Conversely, all *ACC* values for 2021 across periods are negative, with June and summer dipping below −0.2, highlighting a need for performance enhancement. While the proportion of years achieving high *TCC* scores reaches 100% across periods, June's score is lower, indicating that CMA-CPS's predictive prowess for grid point anomalies of latent heat in June is poor and requires improvement, with fluctuations in other periods remaining relatively small.

The high-*MAE* zones for summer LH predicted by CMA-CPS are concentrated in the western Tianshan Mountains, northern TP, and southeastern TP (figure omitted). In June and July, LH *MAE* values in the northern and eastern TP soar up to 80 W/m^2, with the high-value area on the western side of the Tianshan Mountains in July being significantly larger than in June. The overall *MAE* in August is lower than in summer, with only the northern TP nearing 50 W/m^2. The spatial distribution of LH *RRMSE* aligns with *MAE*. Summer LH *ACC* scores feature negative values only in the northern TP and southern Tianshan Mountains, with the rest of the TP scoring high, indicating a good spatial similarity between CMA-CPS predictions and observed LH values. The *ACC* spatial distribution in June, July, and August resembles summer's, with fewer negative value areas in June compared to summer, while July and August exhibit similar but more extensive negative areas. The overall high scores in August are lower, indicating poorer spatial similarity of LH compared to other periods. Negative *TCC* score areas for LH are primarily located in the northern TP, with the size of these areas for each period consistent with the scoring effect in Figure 8d. The remaining regions score high, indicating good predictive ability of CMA-CPS for grid point anomalies of LH on the TP overall, albeit with weaker performance in the northern TP.

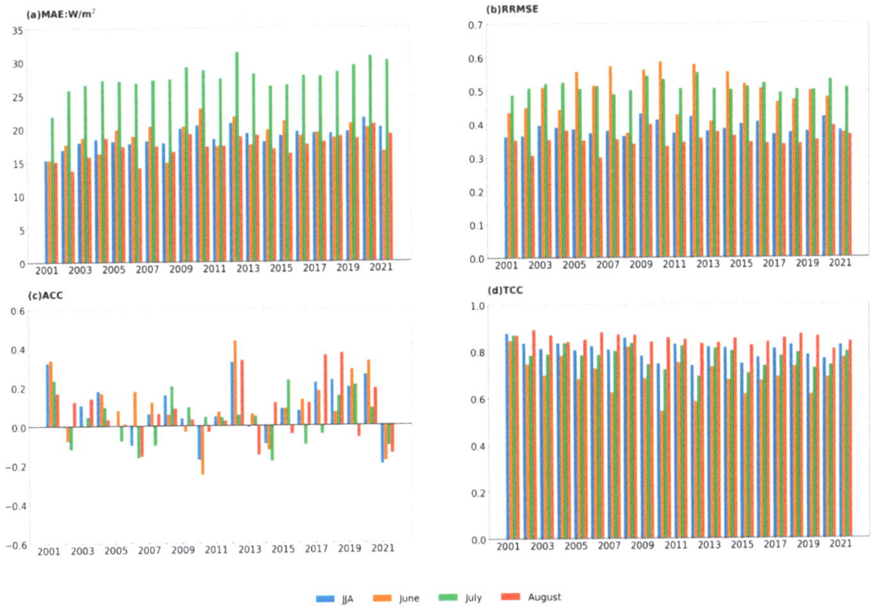

Figure 8. Time series of skill scores for LH on the TP predicted by CMA−CPS from 2001 to 2021. (**a**) *MAE*, (**b**) *RRMSE*, (**c**) *ACC*, (**d**) *TCC*.

3.4. Evaluation of Predictive Capability for SHS

In Figure 9, the spatial distribution pattern of the predicted SHS on the TP is consistent with that of LH in summer. The low-value areas of the predicted SHS are located in the northern TP and the Qaidam Basin, with the lowest values in the northern TP reaching 5 W/m^2. The predicted SHS in the northern TP is around 80 W/m^2, whereas, in other regions, the values are approximately 120 W/m^2. The predicted SHS in the southern Qaidam Basin is lower than that in summer, June, and July and lower in the western northern TP in August. The spatial distribution of SHS from CRA is consistent with that from CMA-CPS, but the overall SHS values are higher, with the center of the high-value area reaching up to 180 W/m^2 and the low-value area down to 40 W/m^2. The relative variability of SHS on the TP for both CMA-CPS and CRA is generally below 10%, possibly because the SHS is derived through direct algorithms and the opposing trends of SH and LH offset each other, weakening the trend.

The scoring results for the predicted SHS are as follows: *MAE* and *RRMSE* are similar to the results for LH in time and space. The proportion of years with high *ACC* scores is best in June (15/21), similar in July (12/21) and August (12/21), and worst in summer (10/21). The high-score areas for SHS in summer, July, and August are significantly larger than the negative value areas, mainly located in the Tianshan Mountains, Karakoram Mountains, and northern TP, while in June, the negative value areas are more numerous, mainly in the central TP. The proportion of years with high *TCC* scores for SHS is consistent with LH, with overall values above 0.6.

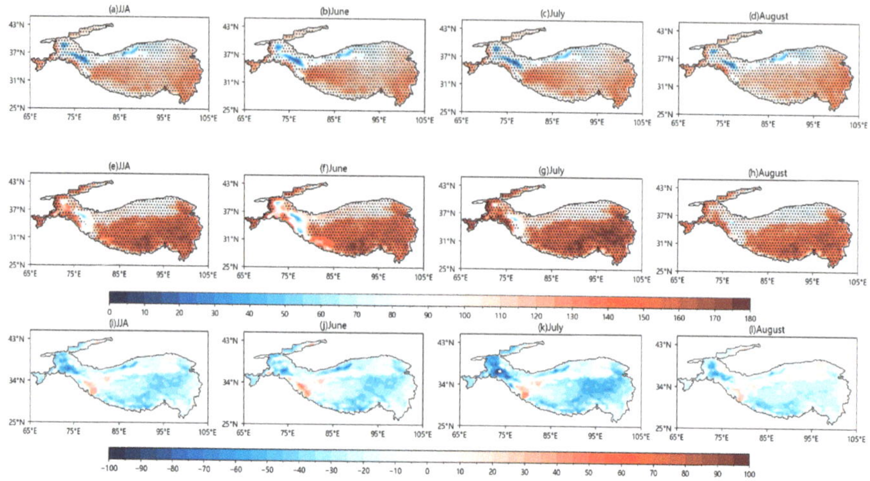

Figure 9. Spatial distribution of SHS on the TP (unit: W/m^2) from CMA−CPS (**top**) and CRA (**middle**) from 2001 to 2021, and their differences (**bottom**). (**a**,**e**,**i**) Summer, (**b**,**f**,**j**) June, (**c**,**g**,**k**) July, (**d**,**h**,**l**) August. The dotted areas represent regions with a relative variability of 0–10%.

4. Discussion

Many studies have shown that the temperature difference between the ground and the atmosphere, surface wind speed, and humidity on the TP have a significant impact on SH and LH [10–16,51,52]. To preliminarily explore the reasons for CMA-CPS's prediction bias, an initial assessment of the temperature difference between the ground and the atmosphere, surface wind speed, and specific humidity on the TP was conducted for both CMA-CPS and CRA.

This study highlights significant variations in SH across the Tibetan Plateau TP, with distinct factors influencing each region. Notably, the elevated SH levels recorded in the eastern TP may stem from an overestimation of surface wind speed, which, when combined with a pronounced temperature differential between the ground and the atmosphere, likely contributes to these high values. In contrast, the lower SH observed in the central TP can be attributed to an underestimation of surface wind speed, suggesting a need for refined measurements in this area. Furthermore, the diminished SH in the western TP appears to be closely associated with an underestimation of the temperature differential between the ground and the atmosphere, as illustrated in Figure 10a–h. In addition to these regional disparities, the spatial distribution map of specific humidity differences, predicted by the CMA-CPS, reveals a complementary pattern: elevated values in the west and reduced values in the east. This spatial distribution closely aligns with the LH differences presented in Figure 10i–l, indicating a correlation that warrants further investigation to elucidate the underlying mechanisms responsible for these observations. Moreover, several characteristics of the TP's underlying surface significantly influence heat dynamics and contribute to the complexity of SH variations. For instance, changes in snow cover, alongside feedback mechanisms related to ice and snow albedo, can markedly alter surface temperature fluctuations across the region. Additionally, shifting permafrost conditions further complicate the thermal dynamics at the surface. Furthermore, surface vegetation exerts a dual influence on SH. It not only mitigates heat penetration into the shallow soil, resulting in lower surface temperatures, but also contributes to a reduction in surface wind speed. Therefore, a comprehensive examination of these underlying surface factors is essential for gaining a nuanced understanding of the observed variations in SH and their broader climatic implications for the TP.

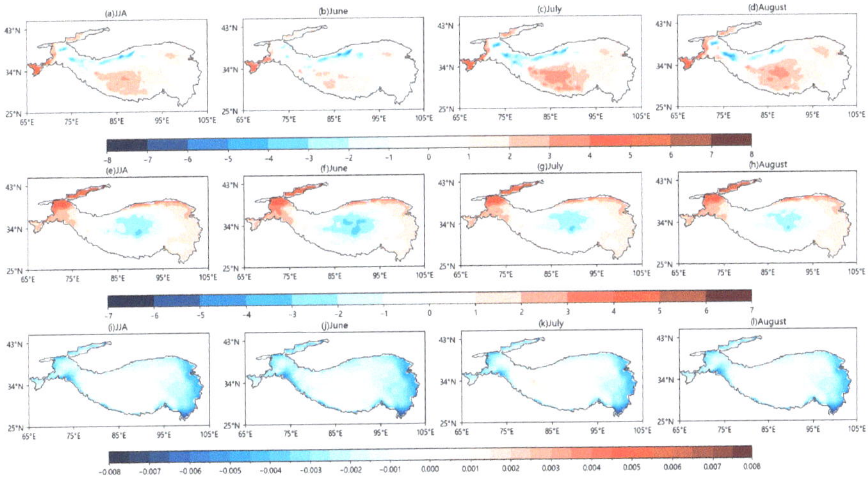

Figure 10. Spatial distribution of the differences in ground-air temperature (**top**), surface wind speed (**middle**), and specific humidity (**bottom**) on the TP as predicted by CMA−CPS and reanalyzed by CRA for different periods from 2001 to 2021 (units: °C, m/s, kg/kg, respectively). (**a**,**e**,**i**) Summer, (**b**,**f**,**j**) June, (**c**,**g**,**k**) July, (**d**,**h**,**l**) August.

Although the predictive capability of the CMA-CPS for SHS and their components in the summer over the TP has been systematically evaluated from the perspectives of bias, spatial similarity, and trend variation, the complex geographical environment of the TP prevents the use of observational station data for evaluation. We look forward to the establishment of more observational stations in the future to improve the model's predictive capability.

5. Conclusions

SHS data from eleven observation stations on the Tibetan Plateau TP and CRA reanalysis data were utilized to systematically evaluate the CMA-CPS's ability to predict summer SHS on the TP. The following main conclusions were drawn.

Both CRA and ERA5 reanalysis data exhibited similar correlations with observational data; however, CRA showed less dispersion. Therefore, CRA was selected as the comparative dataset.

The CMA-CPS predicted summer SH distribution on the TP revealed a pattern of high values in the west and lower values in the east, contrasting with LH, which displayed higher values in the east and lower values in the west. The SHS closely resembled LHt and aligned well with CRA results, indicating that CMA-CPS effectively captured the spatial distribution characteristics of SHS and its components on the TP during summer, albeit with predicted values approximately 25 W/m^2 lower.

Overall trends for SH and LH predicted by CMA-CPS were found to be −10 to 10 $W/m^2 \cdot (10a^{-1})$ and −5 to 15 $W/m^2 \cdot (10a^{-1})$, respectively. In comparison, CRA indicated a more pronounced weakening trend in SH > 15 $W/m^2 \cdot (10a^{-1})$ and a broader area of significant enhancement in LH. This suggests that trends predicted by CMA-CPS are smaller than those from CRA, which may account for the lesser interannual variability observed. Minimal interannual variations in SHS on the TP were shown by both CMA-CPS and CRA, likely due to the direct calculation of SHS leading to offsetting trends between SH and LH.

Among the predictions, LH demonstrated the highest overall skill, followed by SHS, while SH exhibited the lowest skill. Significant deviations in SH were observed in June and minimal deviations in August, especially in northern TP regions. Spatial similarity was highest during summer and in June, decreasing in August. For LH, the largest prediction deviations occurred in July, with significant discrepancies noted in northern and eastern TP. Approximately two-thirds of the years showed good *ACC* scores for LH, predominantly in the eastern and western TP, while *TCC* scores were consistently high, except in the northern TP. These results indicated that LH contributed more substantially to summer TP SHS than SH itself.

The prediction biases for SH and LH by CMA-CPS arose from different causes. The high bias in SH in the eastern TP was attributed to overestimated surface wind speed and ground-air temperature differences, while the low bias in the central TP resulted from underestimated surface wind speed. In the western TP, low ground-air temperature differences led to a low bias in SH. The spatial distribution of specific humidity discrepancies aligned with that of LH, suggesting that the prediction bias in specific humidity by CMA-CPS indirectly affected its LH prediction capability.

After a systematic evaluation, it has been determined that the predictive capability of the CMA-CPS for SHS and its components during the summer on the TP requires further improvement. The potential causes for this bias have been preliminarily discussed, highlighting the need for a more comprehensive evaluation of CMA-CPS's ability to predict the underlying surface factors of the TP. Such an assessment would provide a theoretical basis for enhancing the model's accuracy.

Author Contributions: Writing—original draft, X.C.; writing—review and editing, M.S., Y.W. and T.W. All authors have read and agreed to the published version of the manuscript.

Funding: This research was supported by the Second Tibetan Plateau Scientific Expedition and Research Program (STEP) (NO. 2019QZKK0103) and the National Natural Science Foundation of China, grant number: 42230608.

Data Availability Statement: The CRA40-land and ERA5-land datasets are available from the following sources: the CRA40-land dataset can be accessed from http://www.nmic.cn/ (accessed on 12 January 2024), and the ERA5-land dataset can be obtained from https://cds.climate.copernicus.eu/ (accessed on 12 January 2024).

Conflicts of Interest: The authors declare no conflict of interest.

References

1. Zhou, X.J.; Zhao, P.; Chen, J.M.; Chen, L.X.; Li, W.L. Impacts of thermodynamic processes over the Tibetan Plateau on the Northern Hemispheric climate. *Sci. China Ser. D Earth Sci.* **2009**, *52*, 1679–1693. [CrossRef]
2. Ma, Y.; Hu, Z.Y.; Tian, L.D.; Zhang, F.; Duan, A.; Yang, K.; Zhang, Y.; Yang, Y. Progress in research on climate system changes over the Tibetan Plateau and their impacts and mechanisms on the East Asian region. *Adv. Earth Sci.* **2014**, *29*, 207–215.
3. Duan, A.; Sun, R.; He, J. Impact of surface sensible heating over the Tibetan Plateau on the western Pacific subtropical high: A land–air–sea interaction perspective. *Adv. Atmos. Sci.* **2017**, *34*, 157–168. [CrossRef]
4. Ma, Y.; Ma, W.; Zhong, L.; Hu, Z.; Li, M.; Zhu, Z.K.; Han, C.; Wang, B.; Liu, X. Monitoring and Modeling the Tibetan Plateau's climate system and its impact on East Asia. *Sci. Rep.* **2017**, *7*, 44574. [CrossRef]
5. Duan, A.; Liu, Y.; Wu, G. The thermal conditions over the Tibetan Plateau from April to June and the anomalies of summer precipitation and atmospheric circulation in East Asia. *Earth Sci. Ser. D* **2003**, *33*, 997–1004.
6. Duan, A.; Wu, G. Weakening trend in the atmospheric heat source over the Tibetan Plateau during recent decades. Part I: Observations. *J. Clim.* **2008**, *21*, 3149–3164. [CrossRef]
7. Zhong, S.; He, J.; Guan, Z.; Wen, M. Climatic characteristics of atmospheric heat sources over the Tibetan Plateau from 1961 to 2001. *Acta Meteorol. Sin.* **2009**, *67*, 407–416.
8. Wang, M.; Guo, D.; Zhong, S. Applicability analysis of multi-source data in calculating atmospheric heat sources over the Tibetan Plateau. *Meteorology* **2019**, *45*, 1718–1726.
9. Ma, Y.; Hu, Z.; Wang, B.; Ma, W.; Chen, X.; Han, C.; Li, M.; Zhong, L.; Gu, L.; Sun, F.; et al. Progress and review of research on multi-sphere land-atmosphere interaction processes over the Tibetan Plateau. *Plateau Meteorol.* **2021**, *40*, 1241–1262.
10. Yu, W.; Liu, Y.; Yang, X.; Wu, G. Interannual and decadal variation characteristics and causes of surface sensible heat at different altitudes on the Tibetan Plateau. *Plateau Meteorol.* **2018**, *37*, 1161–1176.
11. Lei, R.; Yu, Y.; Zhou, G.; Li, J.; Liu, Y. Long-term trend analysis of sensible heat flux over the Tibetan Plateau from 1984 to 2020. *Plateau Meteorol.* **2023**, *42*, 833–847.
12. Wang, H.; Zhang, L.; Shi, X.; Li, D. Seasonal differences in the turning characteristics of surface sensible heat trends in central and eastern Tibetan Plateau. *Chin. J. Atmos. Sci.* **2022**, *46*, 133–150.

13. Yang, K.; Guo, X.F.; Wu, B.Y. Recent trends in surface sensible heat flux on the Tibetan Plateau. *Sci. China Earth Sci.* **2011**, *54*, 19–28. [CrossRef]
14. Yang, L.; Zhang, Q. Analysis of circulation and surface latent heat flux characteristics of summer precipitation anomalies in southern Xinjiang. *Plateau Meteorol.* **2007**, *26*, 435–441.
15. Yu, L.; Chen, H. Relationship between April land surface conditions and surface heating anomalies on the Tibetan Plateau and summer precipitation in China. *Plateau Meteorol.* **2012**, *31*, 1173–1182.
16. Qi, Y.; Yan, Y.; Li, J.; Chen, W. Relationship between surface latent heat flux from May to October on the Tibetan Plateau and precipitation in Qinghai during the same period. *Arid Zone Res.* **2019**, *36*, 529–536. [CrossRef]
17. Wang, Z.; Li, G. Diagnostic analysis of a nighttime rainstorm on the eastern slope of the Tibetan Plateau based on heat source effects. *Desert Oasis Meteorol.* **2023**, *17*, 96–103.
18. Li, Y.; Gao, X.; Ma, Y.; Hu, Z.; Li, Z.; Yang, L.; Jin, X.; Zhou, Y. Atmospheric heat engine efficiency and heat source characteristics over the Tibetan Plateau in summer. *Sci. China Earth Sci.* **2024**, *54*, 122–137.
19. Chen, L.; Pryor, S.C.; Wang, H.; Zhang, R. Distribution and variation of the surface sensible heat flux over the central and eastern Tibetan Plateau: Comparison of station observations and multireanalysis products. *J. Geophys. Res. Atmos.* **2019**, *124*, 6191–6206. [CrossRef]
20. Wu, Z.; Chen, Q. Comparative analysis of surface sensible heat on the Tibetan Plateau based on multiple data sources. *Plateau Mt. Meteorol. Res.* **2018**, *38*, 1–10.
21. Zhong, L.; Ma, Y.; Hu, Z.; Fu, Y.; Hu, Y.; Wang, X.; Cheng, M.; Ge, N. Estimation of hourly land surface heat fluxes over the Tibetan Plateau by the combined use of geostationary and polar-orbiting satellites. *Atmos. Chem. Phys.* **2019**, *19*, 5529–5541. [CrossRef]
22. Zhong, L.; Ge, N.; Ma, Y.; Fu, Y.; Ma, W.; Han, C.; Wang, X.; Cheng, M. Estimation of surface latent heat flux over the entire Tibetan Plateau using geostationary satellites. *Adv. Earth Sci.* **2021**, *36*, 773–784.
23. Ma, Y.; Zhong, L.; Wang, B.; Ma, W.; Chen, X.; Li, M. Determination of land surface heat fluxes over heterogeneous landscape of the Tibetan Plateau by using the MODIS and in situ data. *Atmos. Chem. Phys.* **2011**, *11*, 10461–10469. [CrossRef]
24. Hua, W.; Lin, Z.; Guo, D.; Fan, G.; Zhang, Y.; Yang, K.; Hu, Q.; Zhu, L. Simulated long-term vegetation–climate feedbacks in the Tibetan Plateau. *Asia-Pac. J. Atmos. Sci.* **2019**, *55*, 41–52. [CrossRef]
25. Zhang, S.; Chen, Y.; Lü, S.; Li, S.; Wang, S.Y. Simulation study on the impact of vegetation changes on the Tibetan Plateau on summer precipitation in eastern China. *Plateau Meteorol.* **2013**, *32*, 1236–1245.
26. Chen, J.; Wu, X.; Yin, Y.; Lu, C.; Xiao, H.; Huang, Q.; Deng, L.P. Thermal effects of the surface heat flux on cloud systems over the Tibetan Plateau in boreal summer. *J. Clim.* **2019**, *32*, 4699–4714. [CrossRef]
27. Li, M.; Yin, S.; Liu, X.; Liu, X.; Lü, Z.; Song, X.; Ma, Y.; Sun, F. Numerical simulation of turbulent flux changes over the Tibetan Plateau and its surroundings in the past decade. *Plateau Meteorol.* **2019**, *38*, 1140–1148.
28. Hu, W.; Ma, W.; Ma, Y.; Xie, Z. Evaluation of the simulation performance of the Noah-MP land surface model driven by GLDAS data on surface energy exchange over the Tibetan Plateau. *Plateau Meteorol.* **2020**, *39*, 486–498.
29. Xie, Z.; Hu, Z.; Liu, H.; Sun, G.; Yang, Y.; Lin, Y.; Huang, F. Evaluation of the simulation performance of the CLM4.5 land surface model on surface energy exchange over alpine meadows on the Tibetan Plateau. *Plateau Meteorol.* **2017**, *36*, 1–12.
30. Wang, S.; Ma, Y.; Wu, W. Distribution and variation characteristics of surface sensible and latent heat fluxes on the Tibetan Plateau based on the Noah-MP land surface model. *Plateau Meteorol.* **2023**, *42*, 25–34.
31. Wu, T. *Integrated Climate Model Prediction System on Sub-Seasonal to Seasonal to Interannual Scales (CMA-CPS)*; National Climate Center: Beijing, China, 29 August 2022.
32. Li, Y.; Zhu, R.; Liu, T.; Chang, Y.; Yin, Z. Analysis of future temperature and precipitation trends in the Shule River Basin based on the BCC-CSM2-MR model. *Plateau Meteorol.* **2021**, *40*, 535–546.
33. Zhai, Q.; Sun, F.; Ao, X.; Geng, S.; Li, C.; Li, Y.; Li, M. Simulation and prediction of temperature and precipitation in the Liao River Basin based on the BCC-CSM2-MR model CMIP6 experiment. *J. Meteorol. Environ.* **2022**, *38*, 27–36.
34. Zhang, Z.; Zhou, F.; Zhang, H.; Zhou, C.; Wang, X. Evaluation of the prediction of typical winter circulation systems by BCC_CSM1.1m. *J. Appl. Meteorol.* **2023**, *34*, 27–38.
35. Li, S.; Quan, W.; Wang, Z.; Chen, Y.; Su, T.; Yan, P. Evaluation of BCC-CSM2-MR Global Climate Model Simulation of Precipitation and Temperature in East Asia. *Arid Meteorol.* **2023**, *41*, 984–996.
36. Song, M.; Pei, Y.; Zhang, S.; Wu, T. Simulation Performance and Case Study of Extreme Events in Northwest China Using the BCC-CSM2 Model. *Remote Sens.* **2022**, *14*, 4922. [CrossRef]
37. Xue, J.; Shan, H.; Liang, J.H.; Dong, C.M. Assessment and Projections of Marine Heatwaves in the Northwest Pacific Based on CMIP6 Models. *Remote Sens.* **2023**, *15*, 2957. [CrossRef]
38. Liu, J.; Lu, Y. How well do CMIP6 models simulate the greening of the Tibetan Plateau? *Remote Sens.* **2022**, *14*, 4633. [CrossRef]
39. Zhang, H.; Liang, X.; Wang, F.; Xie, B.; Li, Q. Improvement of BCC_CSM1.1m's Cross-Seasonal Dynamic Prediction Capability for Summer Precipitation in China by CWRF Downscaling. *J. Atmos. Sci.* **2023**, *46*, 161–179. [CrossRef]
40. Ying, W.; Yan, H.; Luo, J.J. Seasonal predictions of summer precipitation in the middle-lower reaches of the Yangtze River with global and regional models based on NUIST-CFS1.0. *Adv. Atmos. Sci.* **2022**, *39*, 1561–1578. [CrossRef]
41. Gu, B.; Zheng, Z.; Feng, G.; Wang, Z. The capacity of seasonal forecast models for the forecast of the East Asian summer circulation and its response to tropical SST anomaly. *Chin. J. Atmos. Sci.* **2017**, *41*, 91–105. (In Chinese) [CrossRef]

42. Liang, P.; Lin, H. Sub-seasonal prediction over East Asia during boreal summer using the ECCC monthly forecasting system. *Clim. Dyn.* **2018**, *50*, 1007–1022. [CrossRef]
43. Li, J.; Zhu, Z.; Dong, W. Assessing the uncertainty of CESM-LE in simulating the trends of mean and extreme temperature and precipitation over China. *Int. J. Climatol.* **2017**, *37*, 2101–2110. [CrossRef]
44. Jie, W.; Vitart, F.; Wu, T.; Wu, T.; Liu, X.W. Simulations of the Asian summer monsoon in the sub-seasonal to seasonal prediction project (S2S) database. *Q. J. R. Meteorol. Soc.* **2017**, *143*, 2282–2295. [CrossRef]
45. Ma, Y.; Xie, Z.; Chen, Y.; Liu, S.; Che, T.; Xu, Z.; Shang, L.; He, X.; Meng, X.; Ma, W.; et al. Dataset of spatially extensive long-term quality-assured land–atmosphere interactions over the Tibetan Plateau. *Earth Syst. Sci. Data* **2024**, *16*, 3017–3043. [CrossRef]
46. Xin, X.; Wu, T.; Zhang, J.; Zhang, F.; Li, W.; Zhang, Y.; Lu, Y.; Fang, Y.; Jie, W.; Zhang, L.; et al. Introduction to BCC Model and Its CMIP6 Experiments. *Adv. Clim. Change Res.* **2019**, *15*, 533–539.
47. Li, G.; Lu, H.; Huang, C.; Fan, Y.; Zhang, B. Climatic Characteristics of Summer Surface Heat Source on the Tibetan Plateau and Its Impact on the Formation of Plateau Vortices. *J. Atmos. Sci.* **2016**, *40*, 131–141.
48. Wu, J.; Ren, H.; Zhang, S.; Liu, Y.; Liu, X. Seasonal Prediction Evaluation and Predictability Analysis of the Second Generation BCC Climate System Model. *J. Atmos. Sci.* **2017**, *41*, 1300–1315.
49. Abdolmanafi, A.; Saghafian, B.; Aminyavari, S. Evaluation of Global Ensemble Prediction Models for Forecasting Medium to Heavy Precipitations. *Meteorol. Atmos. Phys.* **2021**, *133*, 15–26. [CrossRef]
50. Li, N. Calculation and Variability Study of Surface Heat Source on the Tibetan Plateau. Ph.D. Thesis, Chinese Academy of Meteorological Sciences, Beijing, China, 2020. [CrossRef]
51. Wang, C.; Shi, R. Simulation Analysis of Land Surface Process Characteristics in Western Tibetan Plateau. *J. Glaciol. Geocryol.* **2007**, *29*, 73–81.
52. Tan, J.; Huang, A.; Shi, X.; Zhang, Y.; Zhang, Y.; Cao, L.; Wu, Y. Evaluation of BCC-CSM2-MR Model's Simulation Capability for Land Surface Processes in China. *Plateau Meteorol.* **2022**, *41*, 1335–1347.

remote sensing

MDPI

Article

Evaluation of Boundary Layer Characteristics at Mount Si'e Based on UAV and Lidar Data

Jiantao Dang [1], Xinrui Xie [2] and Xiaohang Wen [2,*]

[1] The Technical Department of Xichang Satellite, Xichang 615000, China; jiantao_dang@163.com
[2] Plateau Atmosphere and Environment Key Laboratory of Sichuan Province, College of Atmospheric Sciences, Chengdu University of Information Technology, Chengdu 610225, China; 3230101090@stu.cuit.edu.cn
* Correspondence: wxh@cuit.edu.cn

Abstract: The atmospheric boundary layer is a crucial transitional region connecting the surface with the free atmosphere, playing a bridging role in land-sea-air interactions and the interactions between different atmospheric layers. This study utilizes rotary-wing UAVs, high-resolution lidar, and WRF simulation data to analyze the vertical distribution characteristics of temperature, humidity, wind speed, and wind direction boundary layer over the Mount Si'e region in 4–6 April 2024. The results indicate that the boundary layer temperature decreases with increasing altitude, reaching up to 18°C, while humidity decreases with height, dropping to as low as 35%. Daytime wind speeds range from 4 to 8 m/s, decreasing to 2 to 4 m/s at night. The boundary layer height can reach up to 900 m during the day and drops to 100–200 m at night, showing distinct diurnal variation characteristics. UAV observations are in good agreement with lidar and WRF simulation results, highlighting the application value of UAVs in high temporal and spatial resolution boundary layer studies. The study also reveals the significant impact of complex terrain on boundary layer characteristics, providing scientific insights into the dynamic and thermal processes of the boundary layer and offering reference value for improving regional weather forecasting and numerical simulations.

Keywords: atmospheric boundary layer; LiDAR; UAV; wind profile; ABLH

Citation: Dang, J.; Xie, X.; Wen, X. Evaluation of Boundary Layer Characteristics at Mount Si'e Based on UAV and Lidar Data. *Remote Sens.* **2024**, *16*, 3816. https://doi.org/10.3390/rs16203816

Academic Editors: Tamas Várnai, Massimo Menenti, Yaoming Ma, Li Jia and Lei Zhong

Received: 29 August 2024
Revised: 9 October 2024
Accepted: 10 October 2024
Published: 14 October 2024

1. Introduction

The atmospheric boundary layer (ABL), also known as the planetary boundary layer, is the transitional zone between the surface-disturbed airflow and the free atmosphere where friction is negligible [1]. It is directly influenced by surface conditions and responds to these effects within an hour or less, with its height reaching up to several kilometers. It serves as a vital channel for the transport of momentum, heat, and various substances within the Earth's atmosphere, playing a significant role in the exchange of energy and matter between the Earth's surface and the free atmosphere [2–4]. A deeper understanding of the structure and characteristics of the boundary layer will help improve our understanding of energy and matter transfer and cycling within the Earth system [4–7]. The average elevation of the plateau exceeds 4 km, and it features a unique atmospheric boundary layer formed by intense solar radiation, resulting in significant temperature differences between the surface and the complex terrain [7,8]. The interaction between the intricate topography of the eastern plateau and diverse circulation systems creates the region's unique weather and climate characteristics. Therefore, studying the atmospheric boundary layer structure in this area is crucial for understanding the causes of weather and climate and their impacts on China's weather [9,10].

The atmospheric boundary layer (ABL) on the Tibetan Plateau has been a focal point of research due to its significant scientific value. Over the past decades, numerous studies have explored the ABL height (ABLH) and the vertical distribution of meteorological elements within the boundary layer across the plateau. The ABLH generally ranges from 2000 to 5000 m, with substantial regional and seasonal variations. For example, Duzheng Ye [11] reported an ABLH between 2000 and 3000 m, while other studies observed variations across different regions, such as around 2250 m in Dangxiong County [12], up to 3880 m in the Everest region [13], and as high as 5000 m in Gaize [14]. In northern areas of the plateau, such as Anduo and Naqu, the convective boundary layer can extend to around 3200 m, whereas the stable boundary layer typically remains much lower, around 1000 m [15].

The ABLH also exhibits significant diurnal and seasonal variations throughout the plateau. It generally increases rapidly from sunrise to noon, stabilizes in the afternoon, and decreases significantly at night [16,17]. During summer, the convective boundary layer can reach up to 4000 m, while the stable boundary layer dominates in the morning hours [18]. Seasonal variations show that the ABLH tends to be higher in spring and summer, especially in the eastern and western regions of the plateau, while it drops significantly during winter and rainy seasons in central and northern areas [19,20].

In summary, the ABLH generally peaks in the eastern plateau during spring, in the western plateau during summer, and in the central plateau during spring and autumn, with lower values observed during winter across most regions. Studies using COSMIC data have shown that, on average, the ABLH in winter exceeds 2000 m, decreases slightly in spring, rises again in summer, and drops in autumn [21]. Over the past 30 years, the ABLH on the plateau has shown a general increasing trend in winter, particularly since the early 21st century [22]. These variations in ABLH reflect the complex interplay between topography, weather systems, and surface conditions on the Tibetan Plateau, underscoring the need for high-resolution studies to better capture these dynamics.

Mount Si'e is located on the southeastern edge of the Tibetan Plateau, at the transition zone between the main body of the Tibetan Plateau and the Sichuan Basin. Although many studies have analyzed the variations of near-surface meteorological elements on the Tibetan Plateau, research on the eastern edge of the plateau remains limited, despite its significant impact on regional weather and climate patterns [23]. Moreover, existing studies on the atmospheric boundary layer (ABL) of the Tibetan Plateau are often constrained by low spatial and temporal resolution, which can result in certain limitations, such as an inability to capture fine-scale boundary layer processes and potential inaccuracies in representing local meteorological phenomena.

This study compares and analyzes UAV observations and WRF simulation data from 4–6 April 2024, at the Mount Si'e station on the southeastern edge of the plateau. Combined with lidar data from the Mount Si'e station, the study investigates the boundary layer characteristics in the eastern part of the plateau. The use of high-resolution UAV and lidar data allows for a detailed examination of the fine-scale boundary layer processes that were previously difficult to capture with conventional methods. By addressing the limitations of low spatial and temporal resolution found in earlier studies, this research aims to provide a more accurate and comprehensive understanding of the boundary layer structure and dynamics in this region. These advanced observational techniques help correct potential inaccuracies in previous findings and offer valuable insights for studying weather development and climate change in complex terrains like the southeastern Tibetan Plateau.

2. Materials and Methods

2.1. Data

The study area is located on Mount Si'e, where the atmospheric observations are conducted. Figure 1 illustrates the topography and vegetation fraction of the study area, providing context for the following sections.

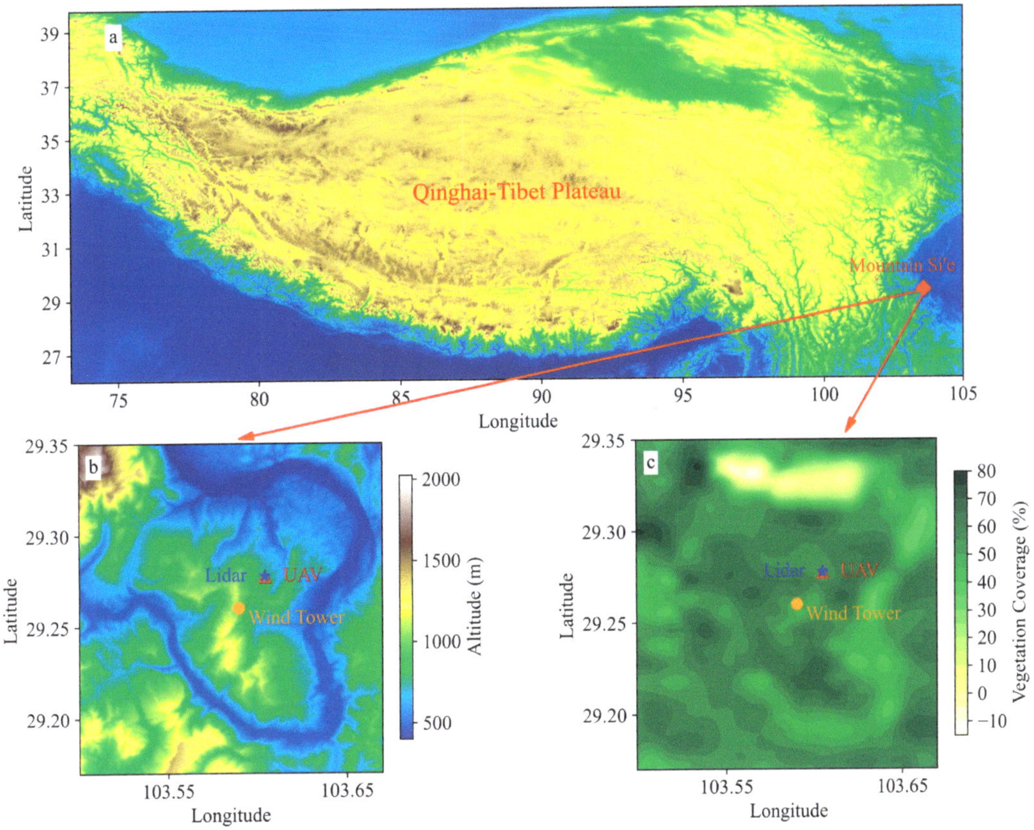

Figure 1. (**a**,**b**) The topography of Mount Si'e; (**c**) The vegetation fraction of Mount Si'e.

2.1.1. UAV Observation Data

The Unmanned Aerial Vehicle (UAV) observation data used in this study are derived from the Rotary Wing UAV Boundary Layer Meteorological Integrated Detection System (RWBMIS) by DJI Innovations M300 RTK Model, China. In this system, a rotary-wing UAV is used as a carrier to carry the Spirit Sniff Mini2 monitoring module, manufactured by Shenzhen Kefei Technology Co., Ltd., Shenzhen, China, to complete the observation of temperature, pressure, humidity, wind and other meteorological elements contours in a mobile observation mode. The UAV performs vertical ascents and descents during high-altitude detection experiments, with an ascent speed of 2 m/s. It offers a time resolution of 1 s and a vertical resolution of 1 m. The sensor is based on the ultrasonic measurement method without moving parts, with UAV translational motion compensation algorithm, UAV attitude compensation algorithm, and UAV rotational motion compensation algorithm, which can realize the measurement of real wind speed and direction information while moving, with fast response rate, small systematic error and high accuracy (Figure 2b). Due to the obvious time lag in the downlink data of the UAV, only 2 m/s uplink data is used in this study.

Figure 2. (**a**) Wind Tower at Mount Si'e; (**b**) UVA; (**c**) Lidar Wind3D 6000.

Table 1 lists the basic performance param of the rotary-wing UAV used in this study in detail.

Table 1. Measurement Param of UAV Meteorological Observation Systems.

Measuring Element	Measuring Range	Resolution	Accuracy
Temperature	−40~+80 °C	0.1 °C	±0.1 °C (+15~+25 °C); ±0.15 °C (0~+15 °C, +25~+40 °C); ±0.4 °C (−40~0 °C, +40~+80 °C)
Pressure	300~1100 hPa	0.1 hPa	±0.1 °C (+15~+25 °C); ±0.15 °C (0~+15 °C, +25~+40 °C); ±0.4 °C (−40~0 °C, +40~+80 °C)
Humidity	0~100% RH	0.1% RH	0~+40 °C: ±1.5% RH (0~90% RH); ±2.5% RH (90~100% RH) −40~0 °C, +40~+80 °C: ±3.0% RH (0~90% RH); ±4.0% RH (90~100% RH)
Wind Speed	0~50 m/s	0.1 m/s	±0.1 m/s (0–10 m/s); ±1% (11–30 m/s); ±2% (31–50 m/s)
Wind Direction	0~365°	±0.1°	1°

2.1.2. LiDAR Data

The wind speed and wind direction data used in this study were obtained from the Wind3D 6000, a three-dimensional scanning wind lidar manufactured by Qingdao Leice Transient Technology Co., Ltd., Qingdao, China (Figure 2c). Lidar (Light Detection and Ranging) operates on the principle of optical pulse coherent Doppler frequency shift detection, enabling fine-scale detection of three-dimensional wind fields in the lower and middle troposphere, including the atmospheric boundary layer. The Wind3D 6000 is equipped with a high-precision optical scanning mirror, supporting various three-dimensional scanning modes, such as Doppler Beam Swinging (DBS), Velocity Azimuth Display (VAD), Plan Position Indicator (PPI), Range Height Indicator (RHI), and Constant Altitude Plan Position Indicator (CAPPI). The maximum detection radius can reach up to 6 km, covering all the requirements of the study area. In this study, only data below 1500 m were used to facilitate comparative analysis with observational data. Table 2 details the basic performance param of the Wind3D 6000 lidar system used in this research.

Table 2. Technical Specifications of the Wind3D 6000 Lidar.

Indicator	Parameter
Operating Wavelength	1.5 μm
Detection Range	45–6000 m
Spatial Distance Resolution	45–6000 m
Temporal Resolution	4.5–150 m
Horizontal Wind Speed Measurement Error	Wind Speed Line ≤ 1 m/s
Horizontal Wind Direction Measurement Error	≤3°
Scanning Angle Accuracy	≤0.1°
Scanning Modes	Plan Position Indicator (PPI), Range Height Indicator (RHI), Doppler Beam Swinging (DBS), Sector Scanning, Volume Scanning, Custom Scanning Modes

2.1.3. Wind Tower Data

The wind speed and wind direction data used in this study, recorded at 1-min intervals over two days from 4–6 April 2024, were obtained from the Emei Mountain Atmospheric and Environmental Comprehensive Observation and Experiment Station of Chengdu University of Information Technology. This station (Figure 1a) is located on Mountain Si'e (29.26°N, 103.59°E) at an altitude of 970 m and will hereafter be referred to as the Mount Si'e Station. The Mount Si'e Station is situated in Mount Si'e Village, Hulu Ba Town, Shawan District, Leshan, Sichuan Province, southeast of the main peak of Mount Emei, Jin Ding, and 37.4 km away from it. Apart from higher elevation mountains to the northwest, the station is surrounded by the Dadu River on all other sides. The underlying surface is a complex mountainous forest with an average forest canopy height of 14 m. Mount Si'e falls within the subtropical montane evergreen broad-leaved forest climate zone, characterized by warm winters, hot summers, and abundant rainfall.

The meteorological gradient observation tower at the station is 60 m high and is equipped with an eddy covariance observation system, including a Campbell CSAT3A 3D ultrasonic anemometer, EC150 infrared CO_2 and H_2O gas analyzers, and a CNR-1 four-component radiometer (Figure 2a). There are five observation levels for wind speed, wind direction, air temperature, and relative humidity: two levels within the canopy at 2 m and 10 m, and three levels above the canopy at 20 m, 25 m, and 58 m. Rain gauges are installed at 38 m and 2 m to measure precipitation and canopy-intercepted rainfall. The flux observation system's ultrasonic probes are installed at 38 m. Data collection is performed using the Campbell CR6 Series data logger, with the eddy covariance system operating at a sampling frequency of 10 Hz.

2.2. WRF Simulation Design

The WRF is a next-generation mesoscale forecast model with a data-assimilation system that advances both the understanding and prediction of mesoscale weather and accelerates the transfer of research advances into operations. It was developed collaboratively by NCAR and the National Centers for Environmental Prediction (NCEP; http://www.mmm.ucar.edu/wrf/users/, accessed on 12 April 2024). The WRF version 3.7.1 modeling system, released in August 2015, is used in this study. The domain configuration features a 25-km resolution single grid system with the center located at 103.59°E, 29.26°N. The simulated area ranges from 70.47°E to 135.53°E longitude and 18.28°N to 53.57°N latitude, with a temporal resolution of 3 h.

Initial and boundary conditions for the large-scale atmospheric fields, as well as initial soil param (soil water, moisture, and temperature), are given by the 1° × 1° NCEP Global Final Analysis (FNL). The use of the FNL provides an upper bound on the skill that is achievable with regional climate prediction models. Boundary conditions at the specified zone are determined entirely by temporal interpolation from the 6-hourly FNL data [24]. Since the domain boundaries are located at the edge of the Tibet Plateau, there are high mountain areas with big topographic relief. In order to avoid vertical interpolation problems caused by the differences in topography between the forcing data in WRF model, 35 unevenly spaced full Eta levels were defined in the vertical dimension (eta level = 1.000, 0.995, 0.983, 0.970, 0.954, 0.934, 0.909, 0.880, 0.845, 0.807, 0.520, 0.468, 0.420, 0.376, 0.335, 0.298, 0.263, 0.231, 0.202, 0.175, 0.150, 0.127, 0.106, 0.088, 0.070, 0.055, 0.040, 0.026, 0.013, 0.000). The vertical coordinate η is defined as:

$$\eta = \frac{p_h - p_{ht}}{p_{hs} - p_{ht}} \tag{1}$$

where p_h is the hydrostatic component of the pressure, p_{hs} is surface pressure, and p_{ht} is a constant pressure at the model top (50 hPa) from the surface are used [24].

The details of the physical parameterizations are summarized in Table 3.

Table 3. Summary of the WRF configuration.

Physical Parameterizations	Configuration
Microphysics	Lin scheme
Cumulus parameterization	Tiedtke scheme
Longwave radiation	RRTM scheme
Shortwave radiation	Dudhia scheme
Cumulus clouds	Kain-Fritsch convection scheme
Planetary boundary layer	YSU PBL scheme
Surface layer	Noah land surface model

2.3. Calculation Method of ABLH

To facilitate analysis and comparison, this study divides the boundary layer into the convective boundary layer (CBL) and the stable boundary layer (SBL) for analysis. This classification is based on the potential temperature profile and the Richardson number (*Ri*). The boundary layer height is estimated using the parcel method and the Ri method, respectively. The calculation formula for *Ri* is as follows [25]:

$$Ri(z) = \frac{g(z-z_0)}{\theta_v(z)} \frac{\theta_v(z) - \theta_v(z_0)}{u(z)^2 + v(z)^2} \tag{2}$$

$$u(z) = V(z) \sin\alpha(z) \tag{3}$$

$$v(z) = V(z) cos\alpha(z) \tag{4}$$

where g is the acceleration due to gravity (9.81 m/s^2), z_0 is the ground elevation, θ_v is the imaginary temperature, u and v are the zonal and meridional wind components, respectively, V is the wind speed, and α is the wind direction.

The calculation of the imaginary temperature (θ_v) is as follows:

$$r_s = \frac{\varepsilon e_s}{P} \tag{5}$$

$$\begin{cases} t \geq 0, e_s = 6.112e^{\left(\frac{17.67t}{t+243.5}\right)} \\ t < 0, e_s = 6.1078 \times 10^{\left(\frac{7.5t}{t+237.3}\right)} \end{cases} \tag{6}$$

$$U_w = \frac{r_s}{r} \tag{7}$$

$$\theta_v = \theta(1 + 0.608r) \tag{8}$$

where r_s is the saturation mixing ratio, with $\varepsilon \approx 0.622$, r is the mixing ratio, e_s is the saturation vapor pressure, t is the temperature in degrees Celsius, U_w is the relative humidity, P is the atmospheric pressure, θ is the potential temperature.

The diagnostic methods for boundary layer height are as follows [26]:

$$\begin{cases} \theta_{200} - \theta_0 < 0, R_i(0 \sim 100) < 0 & \text{CBL} \\ else & \text{SBL} \end{cases} \tag{9}$$

here, θ_{200} and θ_0 represent the potential temperature at 200 m and at the surface.

The parcel method defines the boundary layer height as the altitude where a parcel of air, with an initial virtual potential temperature (θ_{v0}) equal to the maximum at the surface, reaches the same virtual potential temperature again upon vertical ascent. This method is used when the atmospheric boundary layer (ABL) is a convective boundary layer (CBL). When the ABL is a stable boundary layer (SBL), the *Ri* method is typically used to estimate the boundary layer height. The height of the SBL is defined as the level where the Richardson number equals or exceeds a specified critical value. In this study, the critical Richardson number (*Ric*) is set to *Ric* = 0.25.

3. Results

3.1. Vertical Variation Characteristics of Meteorological Elements in ABL

3.1.1. Vertical Profiles of Temperature, Pressure, and Humidity

Figure 3 illustrates the vertical distribution characteristics of temperature, pressure, and humidity at different times as observed by UAV soundings over the Mount Si'e Station. From the figure, it can be observed that temperature, pressure, and humidity exhibit regular variations, generally showing a decreasing trend with increasing altitude. Figure 1 also reveals distinct diurnal variation patterns: during the day, temperatures rise, humidity decreases, and pressure remains stable; at night, temperatures fall, humidity increases and pressure changes are minimal.

During the daytime (4 April at 16:00 and 17:00; CST, the same below), surface temperatures are relatively high, with a rapid decrease in temperature near the surface. The temperature is around 18°C near the ground and gradually decreases to about 8 °C with increasing altitude. This indicates that solar radiation during the day significantly increases the surface temperature, creating a pronounced temperature gradient. At night (4 April at 23:00 and 5 April at 09:00), temperatures are generally lower, especially near the surface, where the temperature gradient is smaller, indicating a relatively uniform temperature profile. This may be due to nighttime radiative cooling at the surface, significantly lowering the near-surface temperature and increasing the stability of the ABL, thereby suppressing vertical heat mixing. In the morning (5 April at 10:00 and 6 April at 09:00), solar radiation gradually intensifies, leading to a rise in temperature (Figure 3).

Pressure variations with height are minimal, and the pressure curves at different times nearly overlap, indicating that the vertical distribution of pressure is relatively stable during these observation periods. This stability in pressure suggests that no significant weather systems were interfering during the observations, and the pressure field remained relatively consistent over these periods.

The lowest humidity during the daytime is observed at 14:00 on 5 April, with minimum humidity reaching about 35%. This indicates that enhanced solar radiation during the day intensifies the evaporation at the surface and in the lower atmosphere, making the air drier [1]. At night, humidity is higher, particularly near the surface, exceeding 50%. This phenomenon can be explained by the drop in temperature at night, causing water vapor in the air to condense more easily, thereby increasing near-surface humidity [27,28]. Additionally, humidity is higher in the lower layers (from the surface to around 200 m) and decreases rapidly with height, which may be related to variations in atmospheric mixing processes and turbulence activity [29].

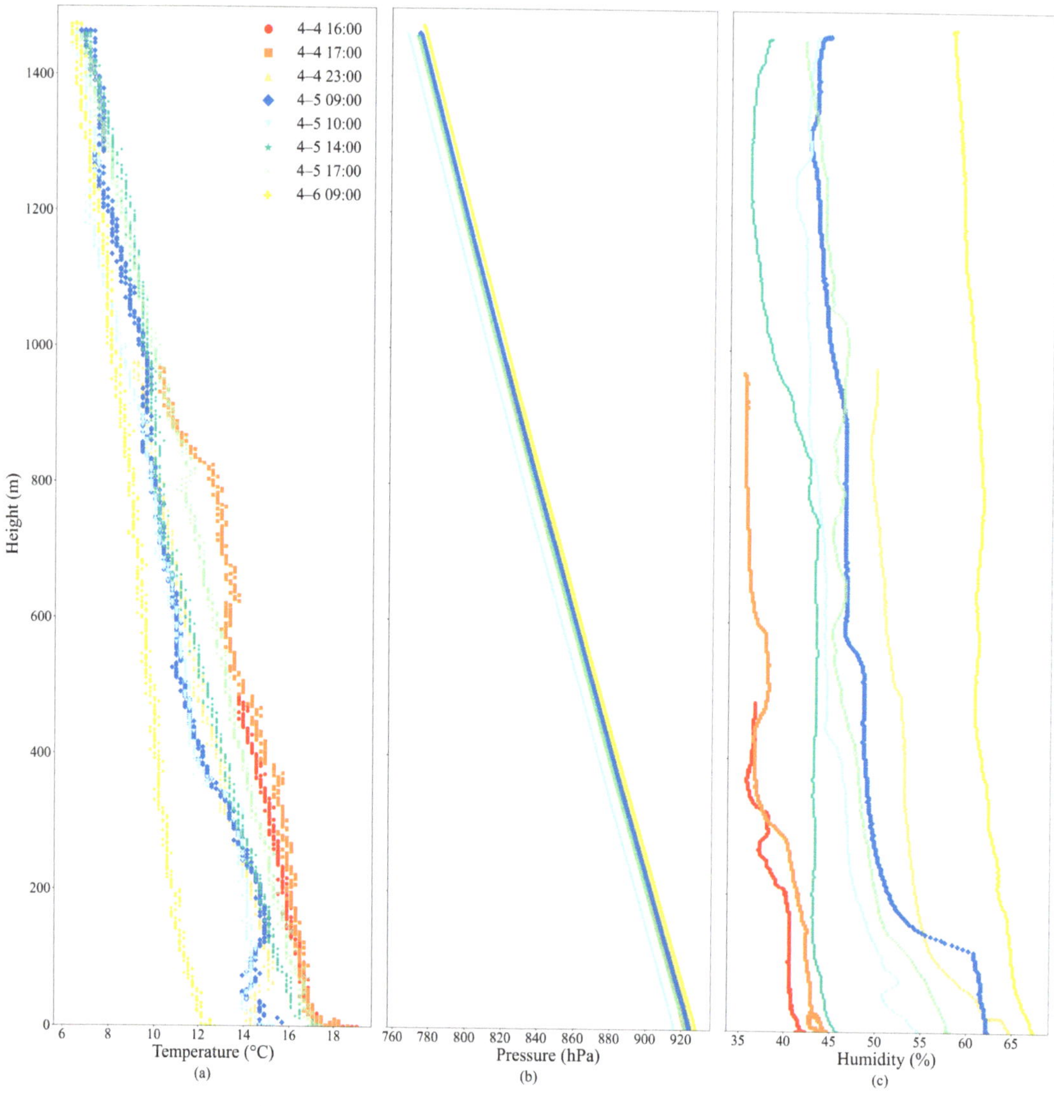

Figure 3. Vertical Distribution of (**a**) Temperature, (**b**) Pressure, and (**c**) Humidity Observed by UAV at Different Times.

3.1.2. Vertical Variation Characteristics of Wind Speed and Wind Direction

Figure 1b,c show the topography and vegetation fraction of the Mount Si'e region. The complex mountainous terrain and dense forest cover significantly impact the local wind field, creating a unique pattern of wind speed and direction distribution. Figure 4 shows that the main wind directions during April at the Mount Si'e area are concentrated in the south-southwest (SSW), south (S), and southwest (SW) directions. These directions not only exhibit a higher frequency of occurrence but also have relatively strong wind speeds. Particularly in the southwest and south-southwest directions, wind speeds often range between 4–8 m/s and sometimes even exceed 8 m/s. The figure also indicates a noticeable presence of northerly (N) winds. Although less frequent than the south-southwest and southwest winds, the northerly winds still have some influence. The wind speeds of the northerly winds are mostly in the low to moderate range (0–4 m/s), indicating that while northerly winds occur during April, their intensity is relatively weak. Overall, the April wind field characteristics at Mount Si'e are distinct, with a clear dominance of south-southwest and southwest winds, which are accompanied by higher wind speeds, whereas the occurrence of northerly winds is relatively infrequent and less intense.

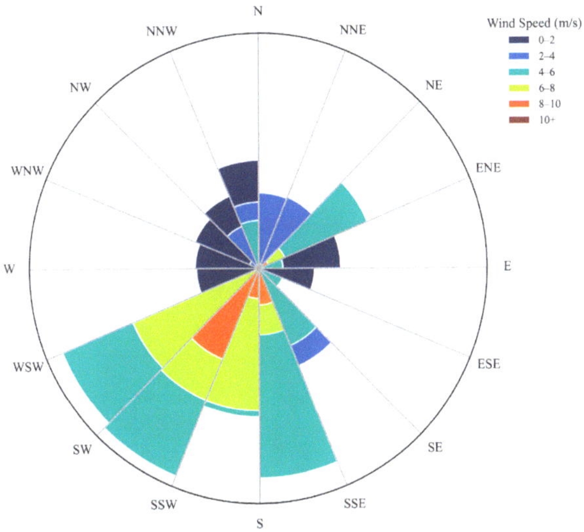

Figure 4. Wind rose diagram based on cumulative UAV observation data over the study period.

These wind direction and wind speed distribution characteristics are closely related to the complex mountainous forest's terrain of the Mount Si'e region. Complex mountainous terrain and forest cover typically significantly alter the local wind speed and direction distribution. The variations in elevation and vegetation density create wind corridors that channel the winds in specific directions, particularly from the southwest and south-southwest, enhancing wind speeds in these directions. The higher wind speeds in the south-southwest and southwest directions may be associated with the wind corridor effect created by the terrain undulations and variations in forest density in these directions, where the guidance of the terrain enhances the wind speed. The presence of forest vegetation can increase surface roughness, thereby reducing low-level wind speeds. However, in a complex mountainous environment, the interaction between terrain and vegetation can create zones of enhanced wind speed. In regions with dense forest cover, wind speeds tend to be lower near the ground due to increased surface roughness, while in open areas or along wind corridors, wind speeds can increase significantly. Changes in wind speed are also related to the microclimatic effects of the forest. Forest cover can alter thermal conditions at the surface, influencing vertical air movement and wind speed distribution. Forests can release moisture through transpiration, creating localized moist air regions near the ground. These areas tend to have relatively lower wind speeds, while higher wind speeds may occur in more open areas.

The influence of the mountainous terrain is also evident in the vertical distribution of wind characteristics, as shown in Figure 5. Figure 5, showing wind rose diagrams at different altitude levels, further reveals the vertical distribution characteristics of wind speed and direction. In the lower layer (0–200 m), wind speeds are primarily concentrated between 0–2 m/s and 2–4 m/s, with more dispersed wind directions. This aligns with the complex mountainous forest environment at the surface, where the forest canopy increases surface friction and local terrain induces changes in wind direction. As altitude increases, the influence of surface features diminishes, and wind directions become more aligned with larger-scale atmospheric flows. At the 200–600 m altitude layer, wind speeds increase, mainly ranging between 2–4 m/s and 4–6 m/s, and wind directions become more focused, showing enhancement from the southwest and southeast directions. This suggests that as altitude increases, the influence of terrain on the wind field gradually diminishes, and the impact of larger-scale weather systems becomes more apparent. At even higher altitudes, in the 600–1000 m and 1000–1500 m layers, wind speeds increase significantly, concentrating between 4–6 m/s and 8–10 m/s, with wind directions mainly from the south. This indicates more pronounced characteristics of large-scale atmospheric circulation, with the direct influence of the mountainous forest underlying surface substantially reduced.

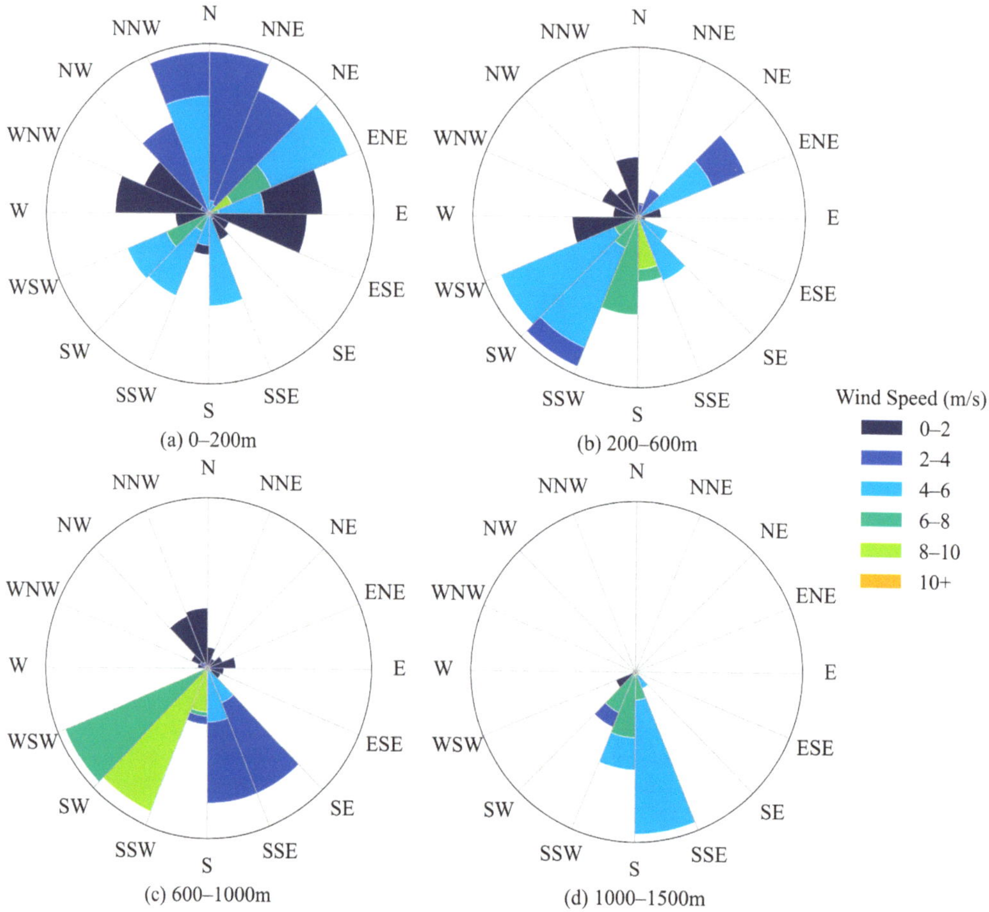

Figure 5. Wind Rose Diagrams at Different Altitude Levels (based on cumulative UAV observation data over the study period).

These changes in wind speed and direction are closely related to the presence of large-scale weather systems, such as troughs of low pressure or ridges of high pressure, and the diurnal variation processes induced by surface heating and cooling. The development of the atmospheric boundary layer increases in thickness due to surface heating during the day, enhancing turbulence and enabling the upward transfer of momentum, resulting in a significant increase in wind speed at higher levels. At night or in the early morning, surface cooling causes the boundary layer to thin, leading to a reduction in wind speed.

Moreover, convective diffusion processes also transfer momentum vertically from lower to higher levels, which is particularly evident during the day.

In summary, the complex mountainous forest underlying surface conditions result in lower wind speeds and more dispersed wind directions at lower levels. In contrast, at higher levels, large-scale atmospheric circulation begins to dominate, leading to increased wind speeds and more concentrated wind directions. This vertical variation pattern is a result of the interaction between atmospheric dynamic and thermal processes, particularly under the influence of atmospheric boundary layer development and convective diffusion.

Figure 4 presents the vertical profiles of boundary layer wind speed at different times over the Mount Si'e region from 4–5 April 2024. The wind speed distribution varies at different times. Overall, at lower altitudes (typically below 400 m), the wind speed fluctuates more significantly, while with increasing altitude, the wind speed tends to stabilize, and the profiles become smoother above 600 m. The wind speed profiles also show distinct diurnal variation characteristics. During the day (Figure 6f,g), wind speeds are relatively high between the surface and 400 m, ranging from 4–6 m/s, and at higher altitudes (Figure 6f), they can reach close to 8 m/s. This phenomenon is related to the enhancement of thermal turbulence caused by solar radiation heating the surface during the day, leading to increased wind speeds in the lower layers. At night (Figure 6c,e), the lower layer wind speeds are lower, usually around 2–4 m/s, and calm regions with near-zero wind speed appear at some altitudes. This could be due to the presence of an inversion layer or the formation of a stable boundary layer structure, which weakens turbulence and reduces wind speed.

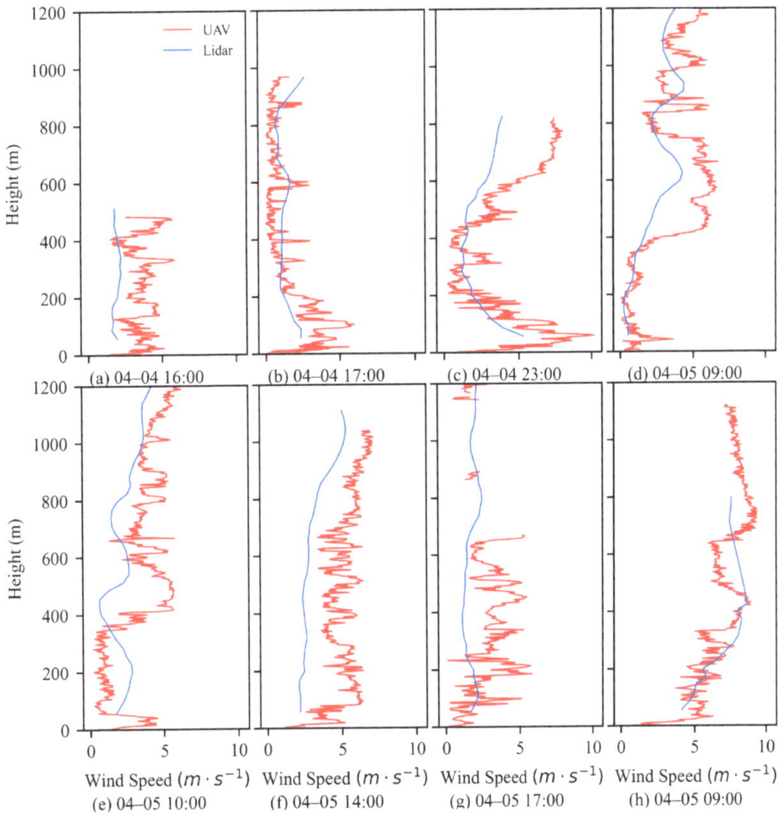

Figure 6. Comparison of April Boundary Layer Wind Speed Vertical Profiles at Mount Si'e.

The diurnal variation in wind speed may be influenced by changes in the local pressure gradient, terrain effects, and temperature differences between day and night [30]. During the day, solar radiation heats the surface, enhancing turbulence mixing in the boundary layer and leading to increased wind speeds [31]. At night, surface cooling due to heat loss stabilizes the boundary layer, suppressing turbulence and reducing wind speeds. The variations in the strength of turbulent diffusion also contribute to the day-night differences in wind speed distribution. During the day, strong turbulence results in more uniform wind speed distribution, while at night, reduced turbulence increases the wind speed gradient, creating noticeable vertical differences in wind speed distribution.

The trends observed from UAV and lidar data are generally similar, showing strong consistency in wind speed variations at larger scales. However, UAV observations provide higher-resolution wind speed data, revealing more detailed fluctuations, which is a significant advantage over lidar data.

Figure 7 shows the vertical profiles of wind direction at different times in the Mount Si'e region. The figure indicates significant changes in wind direction at different times and altitude levels. At 16:00 on 4 April, the wind direction gradually shifts from southerly near the surface to southwesterly at higher altitudes, possibly related to thermal differences induced by daytime solar radiation, which promote changes in the local wind field. At 17:00, there are significant fluctuations in wind direction across different altitude levels, with southerly winds in the lower layers and easterly winds in the higher layers, suggesting the possible presence of an inversion layer or turbulence activities within the boundary layer causing vertical differences in wind direction. At 23:00, the wind direction is easterly in the lower layers and southerly in the higher layers, indicating characteristics associated with the formation of a stable stratification at night.

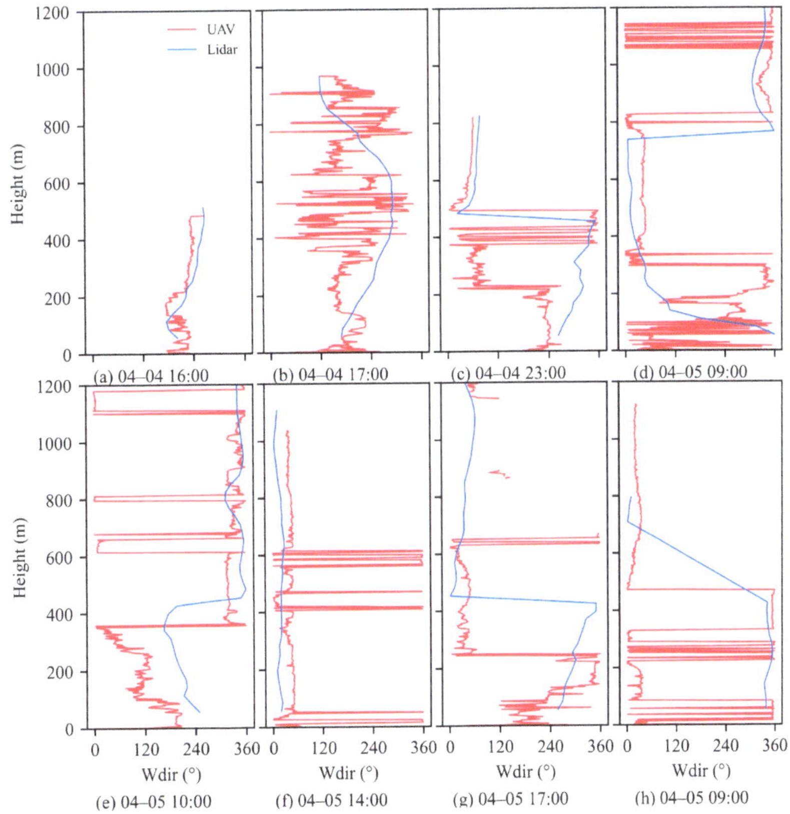

Figure 7. Comparison of April Boundary Layer Wind Direction Vertical Profiles at Mount Si'e.

In the morning of 5 April (09:00 and 10:00), the lower layer wind direction is northerly and northwesterly, gradually shifting to southerly and southwesterly in the mid-upper layers. This transition in wind direction may be related to the dissipation of the nocturnal radiative inversion layer and the redevelopment of the boundary layer in the morning. At 14:00, there is a significant change in wind direction, with easterly winds in the lower layers and southerly winds in the higher layers, possibly due to local thermal and momentum exchange processes. At 17:00, the wind direction in the lower layers is southerly, gradually shifting to easterly at higher altitudes, indicating that during the afternoon development of the boundary layer, wind direction is significantly influenced by turbulence mixing. At 09:00 on 5 April, the wind direction again transitions from northerly near the surface to southwesterly in the mid-upper layers, consistent with the characteristics of the developing boundary layer in the morning.

The trends observed from UAV and lidar data remain largely consistent, showing strong agreement. However, UAV observations can capture more detailed fluctuations in wind direction, demonstrating an advantage over lidar observations.

Figure 8 shows the temporal variation of wind speed and wind direction in the low levels of the boundary layer over the Mount Si'e region. The figure reveals distinct diurnal variation characteristics in both wind speed and wind direction. During the daytime, especially at higher levels (38 m and 58 m), wind speed significantly increases, reaching its maximum. This could be attributed to enhanced atmospheric turbulence caused by increased solar radiation during the day, which raises wind speeds. In contrast, wind speeds decrease significantly at night, particularly at lower levels (2 m and 10 m), a phenomenon that may be associated with stable stratification due to nighttime surface cooling.

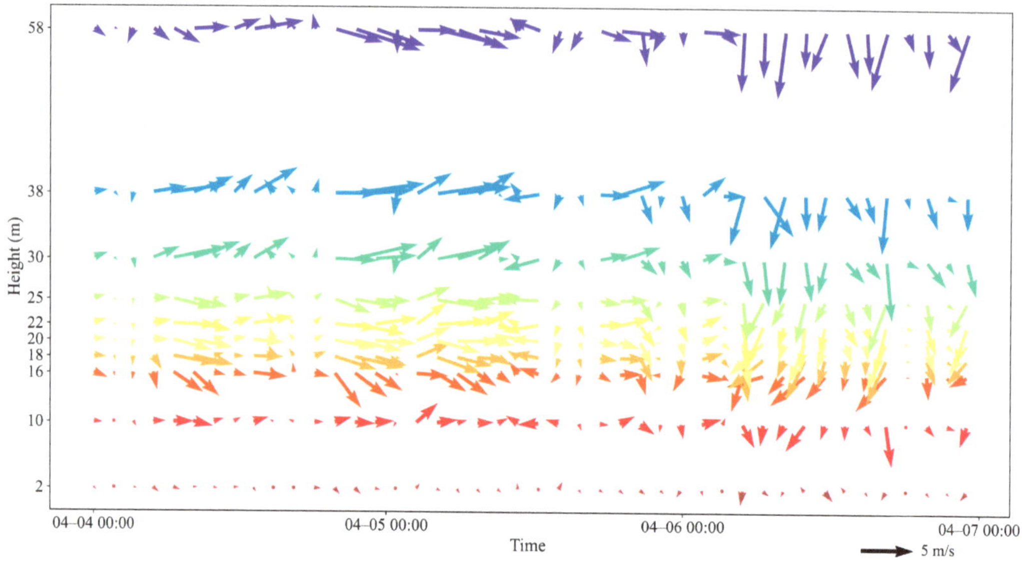

Figure 8. Temporal Variation of Wind Observed by the Boundary Layer Low-Level Wind Tower.

During the day, the wind direction tends to be more consistent across different altitude levels due to enhanced turbulence mixing, which leads to a unification of wind directions throughout the boundary layer. In contrast, at night, especially in the lower layers, wind direction varies frequently, showing multi-directionality. This reflects the instability of wind direction within the boundary layer, which is strongly influenced by local terrain, radiative cooling, and the formation of inversion layers. At lower levels (2 m and 10 m), wind speeds are lower, and wind directions fluctuate significantly due to the influence of surface roughness and complex terrain. This suggests that wind at these levels is more susceptible to local topographic and thermal effects. As the altitude increases to the middle levels (16 m and 30 m), wind speed begins to increase, and wind direction becomes more stable compared to lower levels but still shows some variability. This indicates a transitional zone where both local influences and larger-scale dynamics are at play. At the higher levels (38 m and 58 m), wind speeds reach their maximum and wind direction becomes more consistent. This suggests that at these levels, the influence of local terrain diminishes, and the wind is primarily controlled by larger-scale atmospheric processes. This pattern highlights that lower-level winds are more affected by surface conditions, resulting in greater variability in both wind speed and direction, while airflow at higher levels is more stable, with stronger wind speeds and more uniform wind directions.

Overall, the variation in wind speed and direction within the boundary layer reflects the combined impact of surface heating and cooling, local topography, and atmospheric boundary layer processes. Since lower-level air is closer to the surface, it is more vulnerable to temperature changes and topographic influences, resulting in lower wind speeds and more variable wind directions. In contrast, higher-level air is less affected by these surface influences, leading to higher wind speeds and more stable wind directions.

3.2. *The ABLH Characteristics*

3.2.1. Temporal Distribution Characteristics of ABLH

Figure 9 shows the diurnal variation of boundary layer height as observed by UAVs and simulated by the WRF model. It can be seen that the trends are generally consistent between the two, reflecting significant changes in boundary layer height over time. The boundary layer height begins to increase in the early morning, typically rising gradually after 08:00, reaching its peak around noon, with a maximum height of up to 900 m. This phenomenon is primarily due to solar radiation heating the surface during the day, which raises ground temperatures, enhances thermal convection, and leads to the thickening of the boundary layer, forming a substantial mixed layer. As solar radiation decreases, the boundary layer height starts to decline. At night, radiative cooling causes the air stratification to stabilize, significantly reducing turbulence activity and resulting in a notable decrease in boundary layer height, typically only reaching 100–200 m, forming a thin stable layer. This diurnal variation pattern indicates that the development of the boundary layer during the day is dominated by solar radiation, while at night, surface cooling and stable stratification cause the boundary layer to contract.

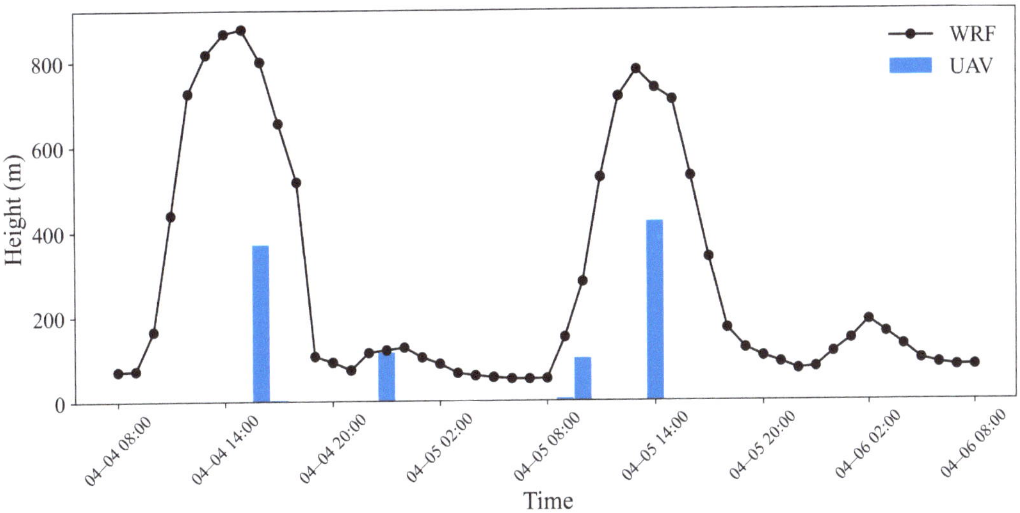

Figure 9. Comparison of Boundary Layer Heights from UAV Observations and WRF Simulations.

3.2.2. Spatial Distribution Characteristics of ABLH

Figure 10 shows the spatial distribution of ABLH at four different times of the day. At 02:00, the boundary layer height is generally low, with most areas below 300 m, indicating a stable stratification characteristic of nighttime conditions. By 08:00, the boundary layer height has slightly increased, yet remains relatively low, with most areas below 600 m. This suggests limited surface heating and weak turbulence in the early morning. At 14:00, the boundary layer height rises significantly, reaching between 600 and 1200 m, with some localized areas even higher. This increase is attributed to strong daytime heating and enhanced convective activity, which thickens the boundary layer. By 20:00, the boundary layer height decreases again, dropping back to around 300 m, reflecting the cooling effect as solar radiation diminishes.

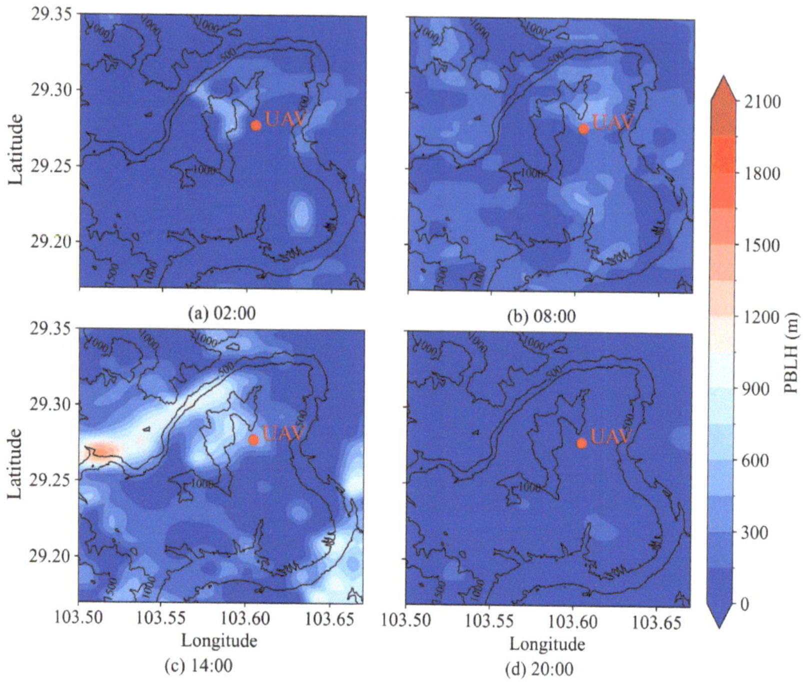

Figure 10. Spatial distribution of ABLH at different times based on daily averages from WRF simulations. (The solid lines represent the topographic contour lines at intervals of 500 m. The red dot indicates the location of the UAV observation station.).

These results are consistent with the previous analysis, showing that the boundary layer height increases significantly during the day, particularly in the afternoon, mainly due to the development of turbulence and the mixed layer triggered by surface solar heating, making the boundary layer thicker. At night, as solar radiation weakens and surface cooling occurs, the boundary layer becomes more stable, and its height decreases significantly. This stable boundary layer limits vertical turbulence mixing and reduces the exchange of heat and momentum in the near-surface atmospheric layer.

It is evident that changes in boundary layer height are directly influenced by surface heating and cooling processes, playing a crucial role in controlling the exchange of matter and energy in the atmosphere.

4. Discussions

This study provides crucial insights into the characteristics of the atmospheric boundary layer (ABL) at Mount Si'e, located on the southeastern edge of the Tibetan Plateau. During the observation period, the boundary layer height (ABLH) can reach up to 900 m during the day and drop to between 100 and 200 m at night. This diurnal variation emphasizes the strong influence of solar radiation and surface temperature changes on boundary layer dynamics, reinforcing previous findings [1,2]. This observation is consistent with the findings of Wen Xiaohang et al., who reported that ABLH is generally lower in forested areas [22].

The significant increase in ABLH during the day is primarily attributed to enhanced thermal convection resulting from solar heating of the surface. This aligns with studies conducted in other mountainous regions, where daytime surface heating plays a critical role in boundary layer formation. At night, surface radiative cooling leads to a decrease in ABLH, creating a stable nocturnal layer. Additionally, the unique microclimatic effects introduced by the complex topography and forest coverage in the Mount Si'e area notably influence wind speed and direction. Higher wind speeds from the southwest may arise from topographical wind corridor effects, while forest coverage increases surface roughness, thereby reducing near-surface wind speeds. The diurnal changes in wind speed and direction further highlight the profound impact of surface heating and cooling. During the day, enhanced solar radiation boosts turbulent mixing, resulting in increased wind speeds, particularly in the lower layers. Conversely, nighttime cooling stabilizes the boundary layer and reduces wind speeds. This pattern is consistent with observations in other mountainous regions, where complex terrain significantly affects the vertical distribution of wind speed and direction compared to flat areas [32].

Discrepancies between our findings and previous research warrant further exploration. While previous studies have documented various ABL heights across the Tibetan Plateau, the specific mechanisms leading to the pronounced nighttime reduction in ABLH observed in this study remain unclear. Factors such as radiative cooling and atmospheric stability at night may contribute to this phenomenon, as highlighted in studies of land-atmosphere interactions [8]. Clarifying these mechanisms is essential for a deeper understanding of ABL processes in this region.

Moreover, unexpected results, such as the pronounced differences in diurnal wind speeds, illustrate the inherent complexity of boundary layer dynamics. Our findings indicate that daytime wind speeds range from 4 to 8 m/s, decreasing to 2 to 4 m per second at night. This diurnal pattern may reflect the typical thermal stratification effects in mountainous regions, where daytime heating enhances turbulence and mixing, while nighttime cooling stabilizes the atmosphere.

However, limitations in this study primarily stem from the temporal and spatial extent of the observations. Although UAV technology allows for high-resolution data collection, the need for manual operation restricts the number and duration of observation points, potentially missing the full variability present across the Tibetan Plateau. Previous research has noted significant variations in boundary layer characteristics over short distances due to topographical changes [7,29]. Future research should consider extending observation durations and expanding the network to encompass a broader range of elevations and geographical features, enhancing the representativeness of the findings. Exploring the potential impacts of climate change on boundary layer characteristics should also be prioritized. Given the region's sensitivity to climatic variations, understanding how ABL dynamics may shift in response to global warming could have significant implications for regional hydrology, ecosystem stability, and climate feedback mechanisms.

5. Conclusions

The boundary layer plays a critical role in the atmosphere for the transport of momentum, heat, and matter. The variations of meteorological elements within it contain important information about the occurrence and development of convection. This study provides a detailed analysis of the April boundary layer characteristics in the Mount Si'e region of Sichuan, based on data from rotary-wing UAVs, lidar, and WRF simulations. Compared to traditional observation methods, UAVs offer high temporal resolution boundary layer profile data, facilitating a more in-depth analysis of the dynamic and thermal conditions of the boundary layer. The main findings of this study are as follows:

1. During April in the Mount Si'e region, the boundary layer height shows a notable diurnal variation. Daytime heating drives intense thermal convection, raising the boundary layer to a peak of around 900 m, generally ranging between 600 to 1200 m, with localized areas occasionally exceeding this. At night, radiative cooling causes the boundary layer to drop to 100–200 m, forming a stable layer, typically staying below 300 m.
2. Daytime temperatures reach up to 18°C, while humidity drops to around 35%, reflecting active vertical mixing. At night, temperatures drop significantly, reducing the near-surface temperature gradient, while humidity rises above 50%.

3. Prevailing winds in the Mount Si'e region mainly come from the south-southwest, south, and southwest. Daytime wind speeds range from 4 to 8 m/s, occasionally exceeding 8 m/s due to enhanced turbulence. At night, wind speeds decrease to 2–4 m/s, influenced by local topography, vegetation, and stable atmospheric conditions.

In conclusion, this study, through the combined use of rotary-wing UAVs, high-resolution soundings, and the WRF model, reveals the typical April boundary layer characteristics in the Mount Si'e region of Sichuan, providing valuable scientific data for understanding boundary layer processes in the eastern part of the plateau. The findings not only offer important insights for improving boundary layer parameterization schemes and numerical weather prediction models but also lay the foundation for further research into the complex dynamic and thermal processes within the boundary layer. This study demonstrates the potential of UAV observations in enhancing boundary layer parameterization schemes. High-resolution UAV data on boundary layer height, wind speed, and thermal conditions can be used to calibrate and validate existing parameterization schemes, particularly in complex terrain and local thermal environments, improving the accuracy of turbulent mixing and thermal exchange modeling. Furthermore, integrating UAV data into numerical weather prediction models via data assimilation can significantly improve the representation of boundary layer structure in models, thus enhancing forecasting accuracy for local convective weather and wind field variations. Future work will continue to explore the potential application of UAV data in numerical model data assimilation to improve the forecasting capabilities for local convective weather.

Author Contributions: Conceptualization, J.D., methodology and writing, X.X. and X.W. All authors have read and agreed to the published version of the manuscript.

Funding: The Second Tibetan Plateau Scientific Expedition and Research Program (STEP) (grant number 2019QZKK010304). The National Natural Science Foundation of China (grant number 41975096) and the Technological Innovation Capacity Enhancement Program of the Chengdu University of Information Technology (grant number KYQN202203) provided financial support for this study.

Data Availability Statement: The original contributions presented in the study are included in the article, further inquiries can be directed to the corresponding author.

Conflicts of Interest: The authors declare no conflicts of interest.

References

1. *An Introduction to Boundary Layer Meteorology*; Stull, R.B. (Ed.) Springer: Dordrecht, The Netherlands, 1988; ISBN 978-90-277-2769-5.
2. Su, Y.; Lü, S.; Fan, G. The Characteristics Analysis on the Summer Atmospheric Boundary Layer Height and Surface Heat Fluxes over the Qinghai-Tibetan Plateau. *Plateau Meteorol.* **2018**, *37*, 1470–1485.
3. Li, Y.; Hu, Z.; Zhao, H. Overview on the Characteristic of Boundary Layer Structure in Tibetan Plateau. *Plateau Mt. Meteorol. Res.* **2012**, *32*, 91–96.
4. Baklanov, A.A.; Grisogono, B.; Bornstein, R.; Mahrt, L.; Zilitinkevich, S.S.; Taylor, P.; Larsen, S.E.; Rotach, M.W.; Fernando, H.J.S. The Nature, Theory, and Modeling of Atmospheric Planetary Boundary Layers. *Bull. Amer. Meteor. Soc.* **2011**, *92*, 123–128. [CrossRef]
5. Zhang, Q.; Zhang, J.; Qiao, J.; Wang, S. Relationship of Atmospheric Boundary Layer Depth with Thermodynamic Processes at the Land Surface in Arid Regions of China. *Sci. China Earth Sci.* **2011**, *54*, 1586–1594. [CrossRef]
6. Zhao, M. A Review of the Research on the Effects of Boundary Layer and Land Surface Process on Heavy Rain in China. *Torrential Rain Disasters* **2008**, *2*, 186–190.
7. Ma, Y.; Hu, Z.; Wang, B.; Ma, W.; Chen, X.; Han, C.; Li, M.; Zhong, L.; Gu, L.; Sun, F.; et al. The Review of the Observation Experiments on Land-Atmosphere Interaction Progress on the Qinghai-Xizang(Tibetan)Plateau. *Plateau Meteorol.* **2021**, *40*, 1241–1262.
8. Fu, W.; Li, M.; Yin, S.; Lü, Z.; Wang, L.; Shu, L. Study on the Atmospheric Boundary Layer Structure of the Qinghai-Xizang Plateau under the South Branch of the Westerly Wind and the Plateau Monsoon Circulation Field. *Plateau Meteorol.* **2022**, *41*, 190–203.
9. Song, X.; Zhang, H.; Liu, X.; Fan, S.; Liu, H.; Hu, F.; Li, S.; Zhou, M.; Bian, L.; Xu, X. Determination of Atmospheric Boundary Layer Height in Unstable Conditions over the Middle Tibetan Plateau. *Acta Sci. Nat. Univ. Pekin.* **2006**, *42*, 328–333.
10. Jiang, X.; Li, Y.; He, G. The Climatic Characteristics Analysis of Radiosonde Data over East Part of Tibetan Plateau. *Plateau Mt. Meteorol. Res.* **2008**, *28*, 1–9.

11. Ye, D. *Meteorology of the Tibetan Plateau*, 1st ed.; Science Press: Beijing, China, 1979.
12. Xu, X.; Zhou, M.; Chen, J.; Bian, L.; Zhang, G.; Liu, H.; Li, S.; Zhang, H.; Zhao, J.; Suolang, D.; et al. Comprehensive Physical Representation of the Dynamic and Thermal Structure of the Land-Atmosphere Processes on the Tibetan Plateau. *Sci. Sin.* **2001**, *31*, 428–441.
13. Li, M.; Dai, Y.; Ma, Y.; Zhong, L.; Lü, S. Analysis on Structure of Atmospheric Boundary Layer and Energy Exchange of Surface Layer over Mount Qomolangma Region. *Plateau Meteorol.* **2006**, *25*, 807–813.
14. Chen, X.; Añel, J.A.; Su, Z.; De La Torre, L.; Kelder, H.; Van Peet, J.; Ma, Y. The Deep Atmospheric Boundary Layer and Its Significance to the Stratosphere and Troposphere Exchange over the Tibetan Plateau. *PLoS ONE* **2013**, *8*, e56909. [CrossRef] [PubMed]
15. Gu, L.; Yao, J.; Hu, Z.; Ma, Y.; Sun, F.; Yu, H.; Wang, S.; Yang, Y.; Guo, R.; Qin, Y. Characteristics of the Atmospheric Boundary Layer's Structure and Heating (Cooling) Rate in Summer over the Northern Tibetan Plateau. *Atmos. Res.* **2022**, *269*, 106045. [CrossRef]
16. Yang, K.; Koike, T.; Fujii, H.; Tamura, T.; Xu, X.; Bian, L.; Zhou, M. The Daytime Evolution of the Atmospheric Boundary Layer and Convection over the Tibetan Plateau: Observations and Simulations. *J. Meteorol. Soc. Jpn. Ser. II* **2004**, *82*, 1777–1792. [CrossRef]
17. Li, F.; Zou, H.; Zhou, L.; Ma, S.; Li, P.; Zhu, J.; Zhang, Y. Study of Boundary Layer Parameterization Schemes'Applicability of WRF Model over Complex Underlying Surfaces in Southeast Tibet. *Plateau Meteorol.* **2017**, *36*, 340–357.
18. Che, J.; Zhao, P. Characteristics of the Summer Atmospheric Boundary Layer Height over the Tibetan Plateau and Influential Factors. *Atmos. Chem. Phys.* **2021**, *21*, 5253–5268. [CrossRef]
19. Zuo, H.; Hu, Y.; Li, D.; Lü, S.; Ma, Y. Seasonal Transition and Its Boundary Layer Characteristics in Anduo Area of Tibetan Plateau. *Prog. Nat. Sci.* **2005**, *15*, 239–245.
20. Li, M.; Ma, Y.; Ma, W.; Ishikawa, H.; Sun, F.; Shin-Ya, O. Structural Difference of Atmospheric Boundary Layer between Dry and Rainy Seasons over the Central Tibetan Plateau. *J. Glaciol. Geocryol.* **2011**, *33*, 72–79.
21. Zhou, W.; Yang, S.; Jiang, X.; Guo, Q. Study of the Atmospheric Boundary Layer Height over the Tibetan Plateau Using COSMIC Radio Occultation Data. *Acta Meteorol. Sin.* **2018**, *76*, 117–133.
22. Wen, X.; Ma, J.; Chen, M. Analysis of Atmospheric Boundary Layer Characteristics on Different Underlying Surfaces of the Eastern Tibetan Plateau in Summer. *Remote Sens.* **2024**, *16*, 1645. [CrossRef]
23. Li, Y.; Li, Y.; Zhao, X. The Comparison and Analysis of ABL observational data between the East of Tibetan Plateau and Chengdu Plain (I). *Plateau Mt. Meteorol. Res.* **2008**, *28*, 30–35.
24. Lo, J.C.-F.; Yang, Z.-L.; Pielke, R.A., Sr. Assessment of Three Dynamical Climate Downscaling Methods Using the Weather Research and Forecasting (WRF) Model. *J. Geophys. Res.* **2008**, *113*, 2007JD009216. [CrossRef]
25. Seibert, P.; Beyrich, F.; Gryning, S.-E.; Joffre, S.; Rasmussen, A.; Tercier, P. Review and Intercomparison of Operational Methods for the Determination of the Mixing Height. *Atmos. Environ.* **2000**, *34*, 1001–1027. [CrossRef]
26. Yang, F.Y. Comparison of Determination Methods and Characteristics of Boundary Layer Height in Semi-Arid Area. PhD Thesis, Lanzhou University, Lanzhou, China, 2018.
27. Garratt, J.R. The Atmospheric Boundary Layer. *Earth-Sci. Rev.* **1994**, *37*, 89–134. [CrossRef]
28. Culf, A.D.; Fisch, G.; Malhi, Y.; Nobre, C.A. The Influence of the Atmospheric Boundary Layer on Carbon Dioxide Concentrations over a Tropical Forest. *Agric. For. Meteorol.* **1997**, *85*, 149–158. [CrossRef]
29. Mahrt, L. Stratified Atmospheric Boundary Layers. *Bound.-Layer Meteorol.* **1999**, *90*, 375–396. [CrossRef]
30. He, Y.; Zhu, X.; Sheng, Z.; He, M. Identification of Stratospheric Disturbance Information in China Based on the Round-Trip Intelligent Sounding System. *Atmos. Chem. Phys.* **2024**, *24*, 3839–3856. [CrossRef]

31. Ayet, A.; Chapron, B. The Dynamical Coupling of Wind-Waves and Atmospheric Turbulence: A Review of Theoretical and Phenomenological Models. *Bound.-Layer Meteorol.* **2022**, *183*, 1–33. [CrossRef]
32. Gohm, A.; Harnisch, F.; Vergeiner, J.; Obleitner, F.; Schnitzhofer, R.; Hansel, A.; Fix, A.; Neininger, B.; Emeis, S.; Schäfer, K. Air Pollution Transport in an Alpine Valley: Results from Airborne and Ground-Based Observations. *Bound.-Layer Meteorol.* **2009**, *131*, 441–463. [CrossRef]

remote sensing

MDPI

Article

Relationship between Tibetan Plateau Surface Heat Fluxes and Daily Heavy Precipitation in the Middle and Lower Yangtze River Basins (1980–2022)

Lu Li [1,2], Xiaohua Dong [1,2,*,†], Yaoming Ma [1,3,4,5,6,7,8,†], Hanyu Jin [9,10], Chong Wei [1,2] and Bob Su [11]

1 College of Hydraulic & Environmental Engineering, China Three Gorges University, Yichang 443002, China;
2 Engineering Research Center for the Ecological Environment of the Three Gorges Reservoir Area, Ministry of Education, Yichang 430072, China
3 Land-Atmosphere Interaction and Its Climatic Effects Group, State Key Laboratory of Tibetan Plateau Earth System, Resources and Environment (TPESRE), Institute of Tibetan Plateau Research, Chinese Academy of Sciences, Beijing 100101, China
4 College of Earth and Planetary Sciences, University of Chinese Academy of Sciences, Beijing 100049, China
5 College of Atmospheric Science, Lanzhou University, Lanzhou 730000, China
6 National Observation and Research Station for Qomolongma Special Atmospheric Processes and Environmental Changes, Dingri 858200, China
7 Kathmandu Center of Research and Education, Chinese Academy of Sciences, Beijing 100101, China
8 China-Pakistan Joint Research Center on Earth Sciences, Chinese Academy of Sciences, Islamabad 45320, Pakistan
9 Institute of International Rivers and Eco-Security, Yunnan University, Kunming 650504, China; jinhanyu@stu.ynu.edu.cn
10 Yunnan Key Laboratory of International Rivers and Transboundary Eco-Security, Yunnan University, Kunming 650504, China
11 Faculty of Geo-Information Science and Earth Observation, University of Twente, 7500 AE Enschede, The Netherlands
* Correspondence: xhdong@ctgu.edu.cn
† These authors contributed equally to this work.

Citation: Li, L.; Dong, X.; Ma, Y.; Jin, H.; Wei, C.; Su, B. Relationship between Tibetan Plateau Surface Heat Fluxes and Daily Heavy Precipitation in the Middle and Lower Yangtze River Basins (1980–2022). *Remote Sens.* **2024**, *16*, 3779. https://doi.org/10.3390/rs16203779

Academic Editor: Ismail Gultepe

Received: 23 August 2024
Revised: 8 October 2024
Accepted: 10 October 2024
Published: 11 October 2024

Abstract: Variable heat fluxes over the Tibetan Plateau (TP) interact thermally with the atmosphere, affecting weather in surrounding areas, particularly in the Middle and Lower Yangtze River (MLYR). However, the circulation patterns and time-lag effects between TP heat fluxes and MLYR precipitation remain unclear. This study identified 577 large-scale daily heavy precipitation events (LSDHPEs) in the MLYR from 1980 to 2022. We analyzed the weather causation and spatiotemporal correlations between the TP surface heat fluxes and MLYR LSDHPEs using self-organizing map clustering, singular value decomposition, and harmonic analysis of time series. The results found two dominant synoptic patterns of LSDHPEs at 500 hPa: one, driven by anticyclonic and cyclonic circulations coinciding with shifts in the West Pacific subtropical high and South Asian high, increased from 2000 to 2022; the other, influenced by MLYR cyclonic circulation, showed a significant decrease. For the first time, we revealed lagged effects of the latent heat anomalies (with a lag time of 1–10 d and 130–200 d) and sensible heat anomalies (with a lag time of 2–4 months) over the TP during LSDHPEs in the MLYR. The results may enhance our understanding of TP heat flux anomalies as precursor signals for early warning of heavy rainfall and flooding in the MLYR.

Keywords: large-scale daily heavy rainfall event; surface heat flux; self-organizing map; singular value decomposition; Tibetan Plateau; Middle and Lower Yangtze River

1. Introduction

Surface heat flux is the most important variable in land–atmosphere interactions on the Tibetan Plateau (TP) [1,2]. It is crucial in influencing the climate anomalies on the TP and in neighboring regions [3–6], which include the Middle and Lower Yangtze River

(MLYR) [7–10]. The MLYR is one of the most important economic development zones in China, with a dense population, and contributes 35% of the total GDP of China; however, it is influenced significantly by frequent heavy precipitation [11–14]. Previous studies demonstrated the existence of a linkage between precipitation in the MLYR and surface heat fluxes over the TP [7,15–17]. In recent decades, the amount of extreme precipitation events has increased in the MLYR [18,19], and the TP has become progressively warmer and wetter [20,21]. Correctly recognizing the relationship between the two under the background of climate change is essential because it is critical in developing early warning systems for rainstorms and floods in the MLYR.

Various researchers have studied the formation mechanism of heavy precipitation in the MLYR, aiming to understand the specific influence of the TP in this process [8,22–24]. The process by which the TP surface heat flux affects the precipitation in the MLYR can be summarized as follows. First, the unique TP topography reinforces mesoscale disturbances, indirectly increasing the surface heat flux over the TP [25,26]. Subsequently, surface heat changes of the plateau and its differences from the surrounding region can stimulate or hinder the development of atmospheric (e.g., vortices, clouds, and high potential vorticity systems) or monsoon systems in the plateau [27,28]. Furthermore, weather systems (e.g., vortices and jets) carrying large amounts of water vapor move eastward from the plateau and its surroundings to the MLYR to produce rainfall [29,30]. Chen [9] found that an eastward-moving cloud from the TP encountered a significant water vapor enhancement accompanied by a low-level vortex and jet streams up into the troposphere at an altitude of 5.5 to 7 km, a strong precursor of heavy rainfall in the MLYR in 2016. The surface diabatic heating of the TP can stimulate the vigorous vertical motion and mutual proximity of the South Asian High (SAH) and Western North Pacific Subtropical High (WNPSH) [31,32]. Moreover, Ma [33] also illustrated that intense latent heating resulted in a cyclonic anomaly in the middle and lower levels of the troposphere, leading to wind anomalies that shifted extreme precipitation southward over the Yangtze River from 1961 to 2013. However, current studies on the correlation between precipitation in the MLYR and the TP always focus on individual storm events or local weather systems (the Yangtze River region or the plateau). Few studies have categorized weather patterns covering the two regions from the perspective of daily heavy rainfall events.

Quantifying the temporal connection between surface heat flux over the TP and precipitation in the MLYR faces various challenges. Previous studies have established that TP influences MLYR precipitation [15,34,35]. Wan [36] noted that surface air temperatures over the TP rose 3 to 4 d before extreme precipitation occurred in southeastern China, while Zhao [8] found that vortices associated with 51 heavy rainfall events took 3-5 d to travel from the TP to the MLYR. Despite these insights into the movement of convective weather systems, the exchange of matter and energy between the land and atmosphere over the TP remains inadequately addressed. Specifically, spring sensible heat (SH) flux from the TP may signal abnormal summer precipitation in the MLYR [35,37,38]. Some studies have examined factors affecting surface heat flux (e.g., soil moisture content and snow cover) in relation to precipitation in the MLYR [31,39–41]. It is worth noticing that in these studies, the rainfall mainly refers to cumulative rainfall on a monthly or seasonal scale, and the lag time analyzed is also on a seasonal scale. However, research on the time-lag correlation between daily surface heat flux variations on the TP and extreme precipitation events in the MLYR is still lacking.

To address the identified research gaps, the objectives of this paper are to (1) analyze the synoptic patterns corresponding to Large-Scale Daily Heavy Precipitation events (LSDHPEs) in the MLYR and variation trends so that upon the appearance of a certain synoptic pattern, the LSDHPEs in the MLYR are to be expected and (2) quantify the lag time between heat flux changes over the TP and heavy precipitation in the MLYR so as to enable a certain prediction of the LSDHPEs in the MLYR. The remainder of this paper is organized as follows: Sections 2 and 3 describe our data sources and methods. Section 4.1 presents the spatio-temporal characteristics of precipitation in the MLYR, and Section 4.2

further displays the dominant synoptic patterns of LSDHPEs among the six categories and the dynamic and thermodynamic contributions to the trends of the LSDHPEs. Section 4.3 describes the time-lagged effect of the heat flux anomaly over the TP on LSDHPEs in the MLYR using the singular value decomposition (SVD) and harmonic analysis of time series (HANTS) methods. Sections 5 and 6 present the discussion and conclusions, respectively.

2. Study Regions and Data Acquisition

The study regions were the TP and MLYR in China (Figure 1) during the period from 1980 to 2022. The boundary of the MLYR was obtained from the National Earth System Science Data Center, with a specific range of 106–122°E, 24–36°N. This study used the TP boundary proposed by Zhang Yili in 2021, with a specific range of 25°59′30″–40°1′0″N, 67°40′37″–104°40′57″E.

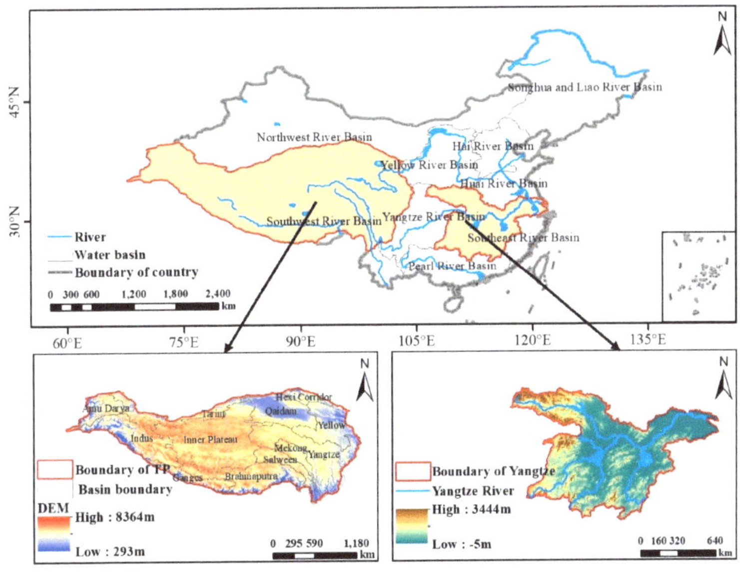

Figure 1. Location of the study area.

To accurately screen heavy precipitation events in the MLYR, three precipitation datasets were used: daily Multi-Source Weighted-Ensemble Precipitation (MSWEP V2.2) data at a spatial resolution of 0.1° × 0.1°, Climate Prediction Center (CPC) global precipitation data released by the National Oceanic and Atmospheric Administration at a spatial resolution of 0.5° × 0.5°, and the CN05.1 daily precipitation dataset, which employs an anomaly approach to process observed data from 2400 stations in China at a resolution of 0.25° × 0.25° [42]. Furthermore, daily geopotential height fields, u wind fields, and v wind fields at 500 hPa from ERA5-Land reanalysis data were used to examine the mesoscale generation mechanisms of LSDHPEs. SH and latent heat (LH) flux data were collected from ERA5 daily datasets.

3. Methodologies

To reach the objectives of this research, the following methodologies were employed: (1) screening for the LSDHPEs in the MLYR, that is, selecting the LSDHPEs from daily precipitation series from 1980 to 2022; (2) classification of synoptic patterns, that is, clustering the synoptic patterns from plum rain period (from May to August), finding the typical synoptic patterns leading to the LSDHPEs, and analyzing the synoptic pattern trends; (3) identification of the lagged correlation between heat fluxes and precipitation to calculate the spatial correlation between the precipitation anomaly fields of the LSDHPEs and the heat anomaly fields of the TP at different lead times and determining the lag times and sensitive areas in which the correlation coefficients are at high values. An overview of the study framework is provided in Figure 2. The terms in Figure 2 are organized from left to right as steps 2, 1, and 3.

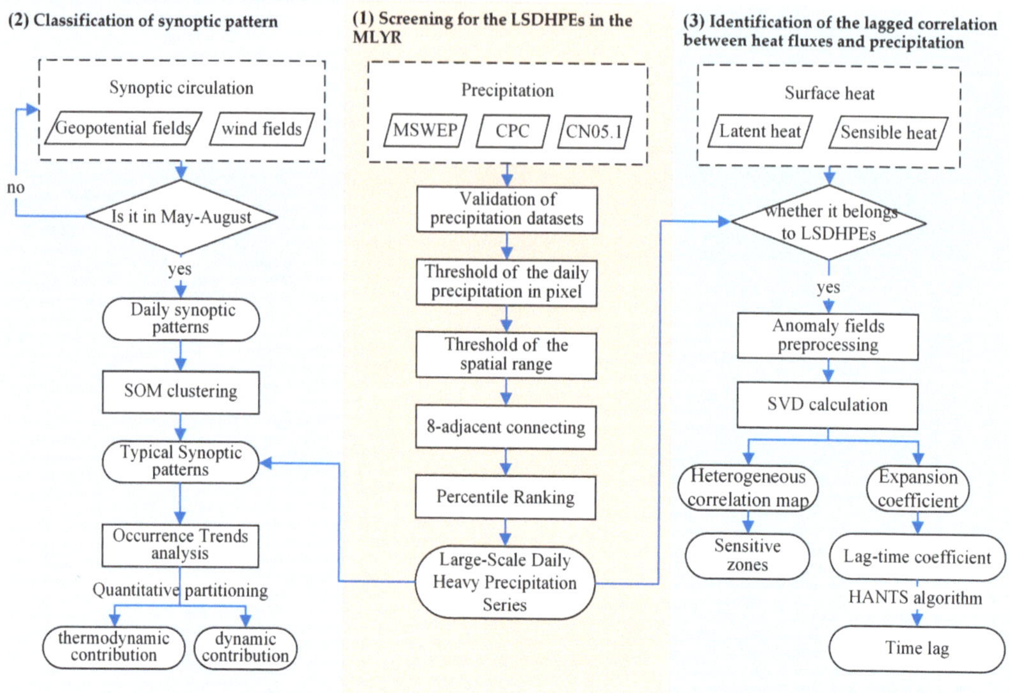

Figure 2. Research framework of this study.

3.1. Screening for the LSDHPEs in the MLYR

By comparing of the accuracy of three rainfall products, the one with the highest accuracy was selected to identify regional heavy precipitation events. LSDHPEs were derived by defining extreme hourly precipitation events [43] with some modifications. First, the number of pixels in the MLYR with precipitation exceeding 50 mm·d^{-1} were counted. Second, using the eight-connected-domain method in graph theory, areas of heavy rainfall were identified, and their areas were calculated; the pixels connected and the precipitation exceeding 50 mm were recognized as a daily heavy precipitation events (DHPEs). A DHPE must satisfy the spatial range and precipitation intensity thresholds to be classified as an LSDHPE. The 75th percentile of the DHPE area is the first-range threshold. Given that two recognized DHPEs may be spatially disjointed yet in proximity, the second regional threshold was the 85th percentile of the sequence of daily heavy precipitation grid point counts. Finally, once the daily precipitation reached either of the two thresholds, an LSDHPE was considered to occur.

3.2. Classification of Synoptic Patterns

LSDHPE circulation patterns are classified using the SOM cluster analysis method. As an unsupervised clustering method proposed by Kohonen [44], the SOM captures more robust and distinctive topology information compared to traditional clustering methods (such as k-means and Ward's method) [45] and has been widely applied in atmospheric and meteorological sciences [46,47]. In the MLYR, LSDHPEs are concentrated from May to August (Figure 3b). Incorporating the weather system development process into the classification samples indirectly adds time weights to typical weather events, improving the representativeness of the classification results. Therefore, continuous daily weather from May to August is more reasonable for clustering than LSDHPE weather. Consequently, daily geopotential height anomaly fields, u wind anomaly fields, and v wind anomaly fields from May to August at 500 hPa during 1980–2022 were used as inputs for the SOM to categorize weather regimes over the TP and the MLYR. Using two criteria from previous studies to quantify the accuracy of the SOM node numbers [48,49], we determined the optimal number of nodes to be 6 (Figure A2 in the Appendix A). All input data were standardized to eliminate the influence of time trends and various magnitudes of values for each parameter.

Under the background of global warming, dynamic and thermodynamic changes are closely associated with extreme events [1,50]. Referring to the methods used by Cassano et al. [51] to assess thermodynamic- and circulation-related changes in precipitation events, the line trends ($d \cdot y^{-1}\ y^{-1}$) of annual occurrence values of LSDHPEs in each SOM pattern were calculated. The equation was used to partition the contributions of thermodynamic and dynamic changes to the LSDHPE trend [49,51].

$$\frac{dE}{dt} = \sum_{i=1}^{K} \left(\overline{f_i}\frac{dE_i'}{dt} + \overline{E_i}\frac{df_i'}{dt} + \frac{d\left(E_i'f_i'\right)}{dt} \right) \tag{1}$$

where E represents the frequency of LSDHPE occurrences; f_i represents the frequency of the pattern i occurrences; K indicates the total number of SOM nodes; $\overline{f_i}$ and $\overline{E_i}$ indicate the mean values of E_i and f_i, respectively; and E_i' and f_i' indicate the deviation values of E_i and f_i, respectively. The terms on the right-hand side of the equation indicate, from left to right, the thermodynamic, dynamic and interaction contributions. Thermodynamic contributions are associated with circulation-related influences, such as radiation, green gas, and surface fluxes. Dynamic contribution refers to the influence of circulation, and interaction contribution represents the dynamic changes acting on thermodynamic changes, for example, the effect of surface-air interactions on the thermal conditions (soil heat flux) of the land surface [49].

3.3. Identifying the Lagged Correlation between Heat Fluxes and Precipitation

This study used a combination of singular value decomposition (SVD) and HANTS to identify lag correlations between the surface heat over the TP and precipitation in the MLYR. SVD is commonly used to determine the correlation between two meteorological fields [52–54]. This method allows the weather field to be divided into multiple linear combinations (also known as modes), with the new variables of these linear combinations being pairwise correlated [55]. The heterogeneous correlation map is an output of SVD that depicts the correlation between the grid pixel values in the k-th mode of one variable and k-th time coefficient series of another variable. In this study, precipitation and surface heat anomalies were normalized and weighted using the cosine of latitude in advance. Using all surface heat flux and precipitation anomaly fields corresponding to LSDHPEs as SVD inputs, we calculated the correlation coefficients (CCs) of the expansion coefficients of the two variables in the first mode (which had the largest covariance and was tested by the Monte Carlo method) [53]. Key zones can be selected from the pixels that pass significance tests and have high correlation values in heterogeneous maps.

Considering the time lag between heat flux and precipitation events, SVD calculations were performed on the rainfall fields corresponding to LSDHPEs and the heat fields of the preceding k-th day (with k ranging from 0 to 366). The resulting sequence of CCs served as the input for the HANTS calculation. Referring to the previous literature on parameter values in the HANTS algorithm [56,57], the smoothed curves obtained after excluding outliers can be used to identify the lag time corresponding to the interval with the maximum correlation between the heat flux field over the TP and the precipitation field in the MLYR.

4. Results

4.1. Features of Precipitation and LSDHPEs in the MLYR

Figure 3 shows the temporal distribution of the LSDHPEs selected from the three datasets in the MLYR during 1980–2022 and the spatial distribution of precipitation derived from the CN05.1 dataset. When comparing the distribution of LSDHPEs across the three precipitation datasets, the ability to identify LSDHPEs was closely related to the spatial resolution of the datasets (Figure 3a). Before 2016, a high spatial resolution corresponded to many LSDHPEs. The ranking of the LSDHPEs was MSWEP > CN05.1 > CPC. After 2016, this phenomenon was alleviated and the difference in LSDHPE amounts was reduced owing to improvements in the number and spatial coverage of surface stations [58]. The common denominator for the LSDHPEs of the three products was more than 60% of the total LSDHPEs in the MLYR occurred in June and July (Figure 3b).

The accuracies of the three precipitation datasets were evaluated using daily precipitation observations from surface stations in the MLYR (Figure A1 in the Appendix A). Notably, CN05.1 performed better than the others, capturing spatiotemporal features consistent with those of the other two datasets, MSWEP and CPC, indicating its robustness. Overall, the frequency and intensity of precipitation in the MLYR were high in the southeast and low in the northwest (Figure 3c); the high-value areas were located on the eastern border. The frequency of heavy precipitation events exhibited a similar distribution (Figure 3d). Moreover, the magnitude of precipitation was higher at lower latitudes than those at higher latitudes. In total, 577 LSDHPEs presented in CN05.1 were selected for further analysis.

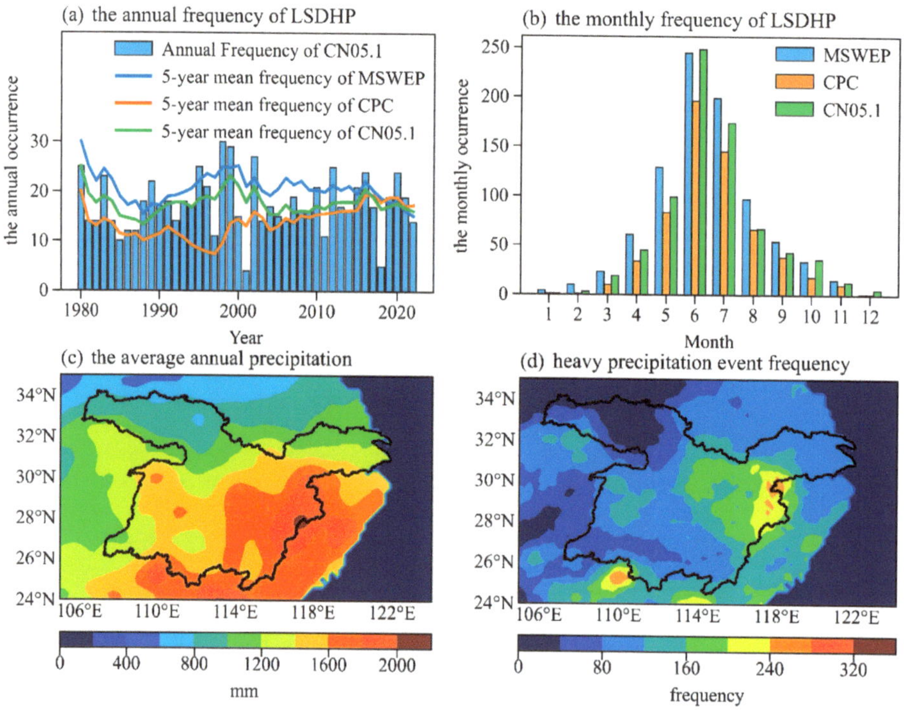

Figure 3. (**a**,**b**) Temporal distribution of LSDHPEs selected from MSWEP, CPC, and CN05.1 in MLYR. (**c**) The average annual precipitation of MLYR from CN05.1 during 1980–2022. (**d**) Spatial distribution of heavy precipitation events selected from CN05.1 in MLYR.

4.2. Typical Synoptic Pattern of LSDHPEs in the MLYR

The synoptic anomalous patterns classified by the SOM are presented in Figure 4. Overall, the synoptic patterns from May to August of 1980–2022 were categorized into six clusters based on the location of high (low) pressure anomaly systems and vortex anomalies. Notably, geopotential height anomalies were calculated considering the 1981–2022 period as the climatological mean benchmark and then normalized.

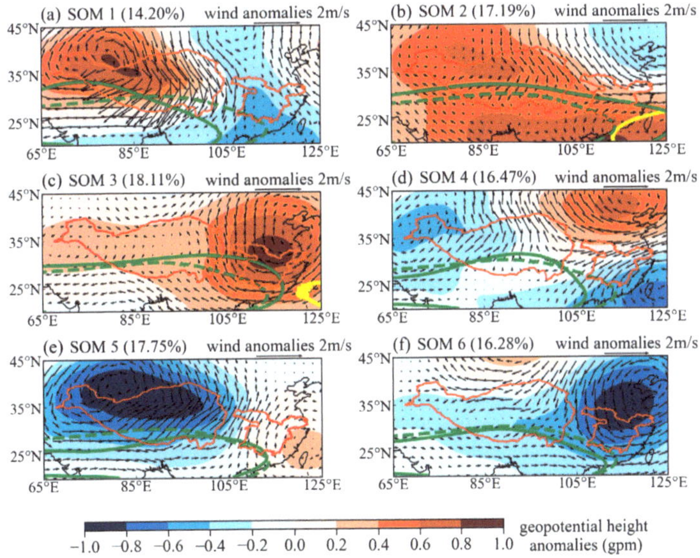

Figure 4. Synoptic patterns of SOM correspond to the May to August period during 1980–2022 at 500 hPa. (Contour represents mean geopotential height anomalies. Arrows represent the wind field anomalies. The yellow line represents the WNPSH outlined by 5880 gpm contour at 500 hPa. The green line represents the SAH outlined by a 12500 gpm contour at 200 hPa. The green dashed line indicates the SAH at 200 hPa relative to the climatic mean from May to August 1980–2022.)

The first SOM pattern (SOM 1) exhibited a typical anomalously high–low–high pattern (Figure 4a), with a north–south-oriented trough extending from the center of the MLYR and an east–west-oriented high-pressure ridge developing from the TP center. The strong anticyclonic anomaly in the west triggered northeasterly wind convergence at the eastern margin of the TP. In the synoptic patterns described in SOM 2 (Figure 4b), the high-pressure anomaly system strongly affected the TP and MLYR, with a low-pressure system over the northeast. This scheme was affected by a strong eastward-extended SAH at 200 hPa and a westward-extended WNPSH at 500 hPa. Additionally, the MLYR lies in the junction area between the cyclone and anticyclone, with dense isobars and westerly winds. SOM 3 represents a single east-high pattern over the MLYR, with a distinct westward-extended WNPSH and a slight eastward-extended SAH. Moreover, the commonality between SOM 3 and 4 is that the wind is easterly in the MLYR. The synoptic pattern of SOM 4 displayed a configuration opposite that of SOM 2. SOM 5 and 6 were affected by typical cyclone anomalies. SOM 5 was affected by a cyclone anomaly over the TP and an anticyclone in the southeastern portion of the study area, associated with a strong southwesterly airflow anomaly at the border of the TP and the MLYR. Furthermore, SOM 6 was controlled by a low-pressure anomaly system with a deep trough extending along an east–west axis from the MLYR to the TP, leading to airflow convergence south of the MLYR. The wind field represents the direction of water vapor movement. Overall, the main sources of water vapor for the MLYR were the South China Sea, East China Sea, and Bay of Bengal, consistent with the results of previous studies [9,30,59].

To further analyze the circulation pattern of heavy precipitation events, the composite mean synoptic fields of the LSDHPEs in each SOM Cluster (hereafter, LSC) are depicted in Figure 5. LSC 1(Figure 5a) accounted for 14.26% of the LSDHPEs. In addition to an additional north–south trough in the MLYR, the synoptic schemes of LSC 1 and SOM 1 were similar. Compared with SOM 1, the position of the SAH in LSC 1 has a significant eastward shift, which coincided with its position in the climatology of 1980–2022. The total column water vapor was concentrated in the eastern part of the MLYR, along with the anomalous cyclone. Furthermore, a mass of anomalous water vapor shifting in a northeasterly direction occurred behind a low-pressure trough in the lower troposphere above the MLYR (Figure A3 in the Appendix A). LSC 2 had the second-high frequency of LSDHPEs (21.73%). Its circulation pattern (Figure 5b) highlighted the south-high system and a deep trough of anomalously low pressure propagating from north to south in the MLYR. Moisture convergence and strong eastward water vapor were observed over the MLYR. In LSC 3 (Figure 5c), the eastern region was controlled by two anomalous high-pressure systems and an anomalous anticyclone located north of the MLYR, leading to a southerly anomalous wind. With the lowest frequency of LSDHPEs (8.83%), the water vapor in LSC 3 originated from the South China Sea and East China Sea. Moreover, the mean convective potential energy values in the MLYR were the lowest among all the patterns (Figure 6). Compared with SOM 4, the anticyclonic center above the MLYR in LSC 4 (Figure 5d) exhibited a significant westward shift. In addition to the joint influence of the WNPSH and SAH, the cyclone over the TP evoked the northerly wind blowing over the eastern margin in LSC 5, jointly leading to a reduction in the intensity and extension of positive precipitation anomalies in the MLYR (Figure 5e). Bearing the highest frequency among the LSDHPEs (23.43%), the circulation pattern of LSC 6 was consistent with that of SOM 6. Under the influence of a strong cyclone, westerly winds from the Bay of Bengal and India converged with northerly winds from North China. It blew over the MLYR, resulting in large amounts of water vapor southeast of the MLYR.

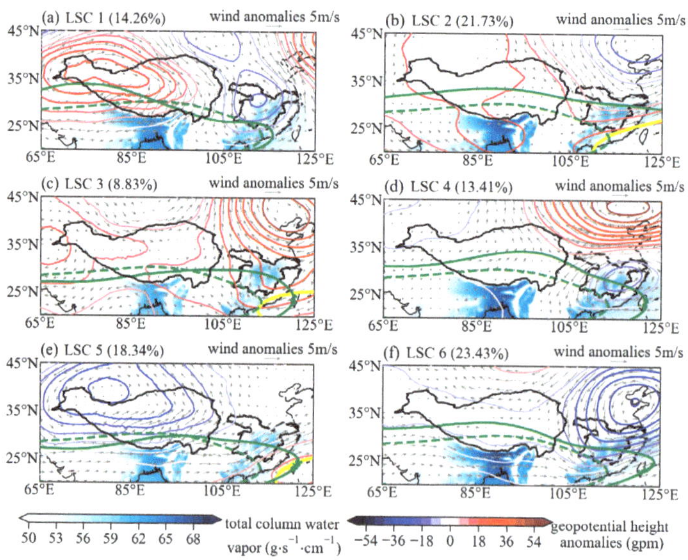

Figure 5. Composite mean synoptic fields of LSDHPEs in SOM patterns at 500 hPa. (The contours represent geopotential height anomalies. The left color bar represents the water vapor flux at 500 hPa. The yellow line represents the WNPSH outlined by 5880 gpm contour at 500 hPa. The green line represents the SAH outlined by the 12500 gpm contour at 200 hPa. The green dashed line indicates the SAH at 200 hPa relative to the climatic mean from May to August during 1980–2022.)

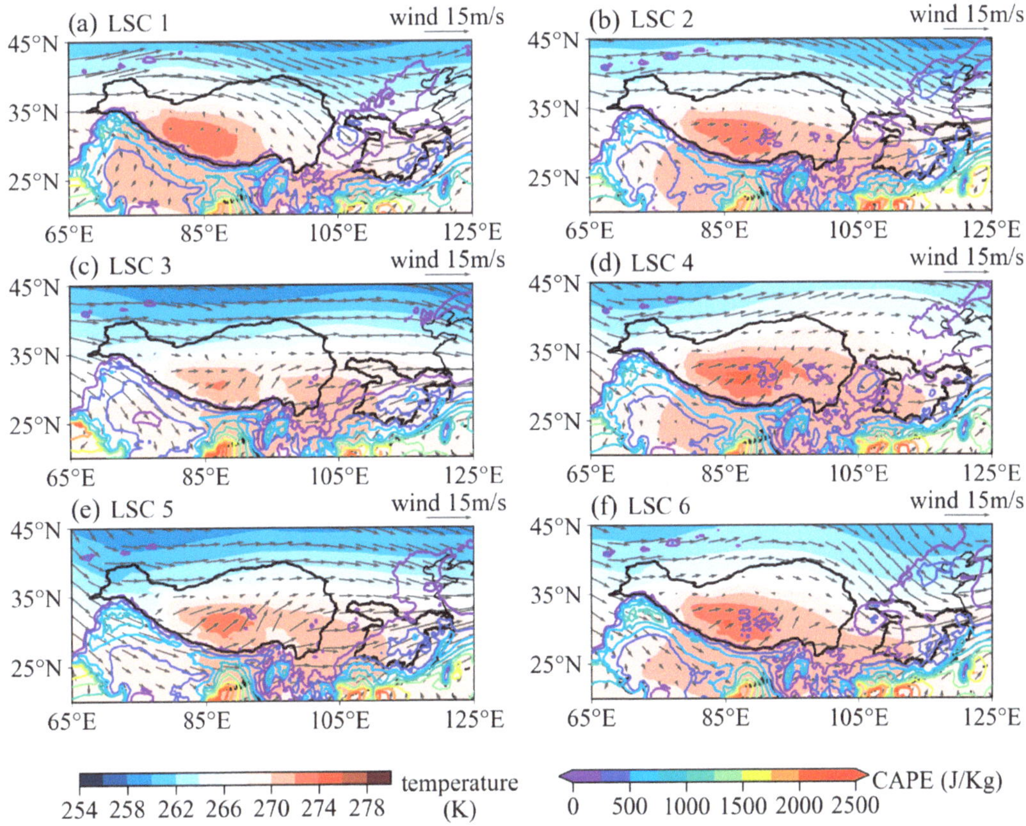

Figure 6. Mean air temperature at 500hPa and the total CAPE field of the LSC patterns. (The contours represent the CAPE. The left color bar represents the temperature at 500 hPa).

To determine the potential linkages between flux variation over the TP and precipitation in the MLYR, LSCs 1, 2, 5, and 6 were selected as the core representative types from the six patterns, based on their occurrence frequency. The typical synoptic pattern accounted for 65.42% of the daily circulation from May to August during 1980–2022 (Figure 4) and 77.76% of the LSDHPEs (Figure 5). All six synoptic patterns represented profound weather systems, with the distribution of geopotential height anomalies from the 850 hPa (Figure A4 in the Appendix A) to the 200 hPa (Figure A5 in the Appendix A).

Heavy precipitation events are often associated with high atmospheric Convective Available Potential Energy (CAPE), wind shear, and water vapor accumulation. As seen in the total water vapor distribution in Figure 5, the MLYR had high water vapor values. Combined with the distribution of air temperature at 500 hPa, wind fields, and the CAPE distribution in Figure 6, the water vapor in the MLYR was carried by westerly winds originating from the Bay of Bengal and the Arabian Sea. When these winds encountered the topography along the southern edge of the TP, they were blocked and diverted eastward, eventually reaching the MLYR. Additionally, a relatively high-temperature area extended from the east to southwest of the MLYR influenced by the temperature in the middle and lower troposphere (Figure A4 in the Appendix A). Influenced by temperature troughs, the abundant water vapor and high CAPE in the MLYR could trigger intense convective activity, making this region prone to heavy precipitation.

Figure 7 shows the linear trends in the annual occurrence of each SOM and LSC pattern. The significance of trends was evaluated using a t-test. SOM patterns 2 and 3 showed significantly increasing trends during 1980–2022, whereas the other SOM patterns showed decreasing or non-significant trends in annual occurrence. SOM 2 and LSC 2 notably showed significant and stable upward trends over the last 22 years. The comparison in Figure 5b shows that the synoptic pattern of LSC 2 is dominated by the synaptic forcing of the SAH and WNPSH, which account for 21.73% of LSDHPEs. As shown in Figure 7, LSDHPEs show a prominent decreasing trend in LSCs 1, 5, and 6, whereas a notable increasing trend is observed in LSCs 2 and 3. In LSC 1, 5, and 6 patterns, the SAH reached and covered the MLYR, whereas the WNPSH was not observed in the MLYR or nearby areas. However, in LSCs 2 and 3, the WNPSH was located in the coastal region near the MLYR. This indicated that the influence of WNPSH on the LSDHPEs in the MLYR gradually increased and the influence of SAH gradually decreased.

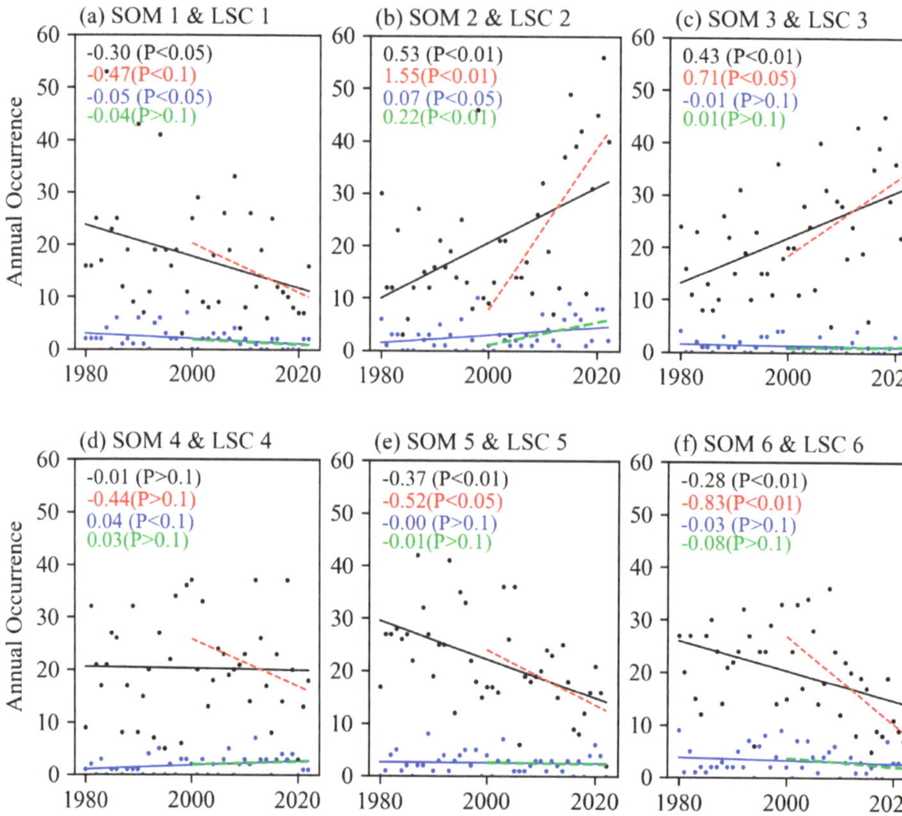

Figure 7. The annual occurrence of SOM and LSC patterns. (The blue dots and black dots indicate the occurrence of the SOM and LSC patterns, respectively. Solid lines represent the trend line from 1980 to 2022. Dashed lines represent the trend line of the corresponding dots from 2000 to 2022. The slope and P values (parentheses) are displayed in the top left corner.)

Table 1 shows that the total trend of LSDHPEs in the MLYR increased and was dominated by circulation pattern 2 (266.63%). Based on the explanation of the three contributions by Daniel [49], the increase has resulted from thermodynamic (36.34%) and dynamic changes (37.27%). In patterns 5 and 6, the occurrence of the LSDHPEs displayed a decreasing trend (-0.006 d·y^{-1} y^{-1} and 0.008 d·y^{-1} y^{-1}, respectively), and the dynamic contribution motivated the decrease. For all the SOM patterns, the interaction change term (0.009) controlled for the total trend of LSDHPEs occurrence.

Table 1. Quantitative partitioning of the LSDHPEs trend in each SOM pattern.

Items	SOM 1	SOM 2	SOM 3	SOM 4	SOM 5	SOM 6	Total
Trend in the thermodynamic influence	−0.006	0.010	−0.002	0.005	−0.001	−0.004	0.004
Trend in the dynamic influences	−0.004	0.010	0.003	0.000	−0.006	−0.006	−0.002
Trend in the interaction influences	−0.002	0.007	−0.001	0.001	0.001	0.002	0.009
Total trend	−0.012	0.028	0.001	0.007	−0.006	−0.008	0.011
Percent of total trend	−109.59	266.63	7.59	62.36	−53.74	−73.24	
Percent of thermodynamic	48.24	36.34	−245.07	80.75	9.64	47.47	
Percent of dynamic	33.66	37.27	433.84	−2.74	107.67	77.83	
Percent of the interaction	18.11	26.40	−88.77	21.99	−17.31	−25.30	

4.3. Time-Lagged Effect of TP Anomalous Surface Heat Flux on LSDHPEs

As shown in Figure 8a, the time-lagged CC curves between the anomalous LH and precipitation of the LSDHPEs exhibit three distinct peaks. The peak values of time-lagged CCs ranged from 0 to 10 d (mean value of 0.3178), 130 to 200 d (mean value of 0.352), and 330 to 365 d (mean value of 0.368). In contrast, the time-lagged CC curves between the anomalous SH and anomalous precipitation of the LSDHPEs showed two peaks, with the peak ranges varying between 60 and 135 d (mean value of 0.327) and 300 and 350 d (mean value of 0.371).

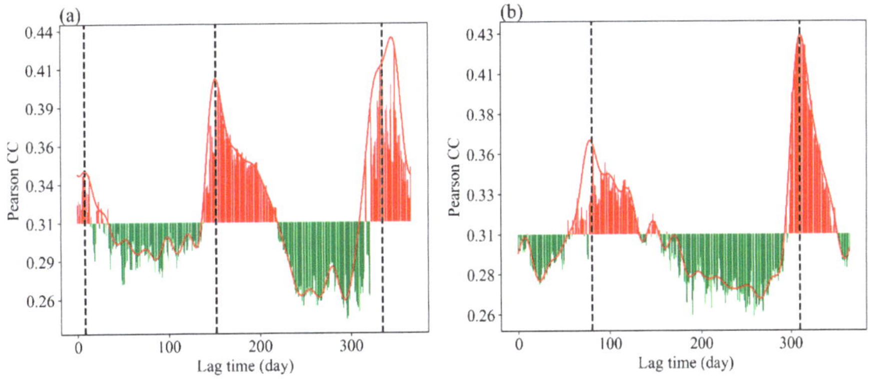

Figure 8. Time-lagged correlation coefficients between the surface heat flux anomalies over the TP and the precipitation anomalies in LSDHPEs. ((**a**), the correlation between the anomalous LH on the TP and the anomalous precipitation over the MLYR. (**b**), the correlation between the anomalous SH on the TP and the anomalous precipitation over the MLYR. All values (bar) passed the 95% significance level test; the baseline indicates the mean value of the correlation coefficient. Red line represents the fitted curve obtained by using the HANTS algorithm.).

The first peak in Figure 8a (the lag time range of 0–10 d, and maximum CC value of 0.34) corroborates the previous research findings of Dong [10], indicating that the anomalous LH over the southeastern margin of the TP can be a strong signal of precipitation in the MLYR and has a 3–5 d lead-time before the plum rain period(known as the Mei-yu period in China). In addition, the second peak range was roughly equal to 4–6 months, which may have resulted from the teleconnection between winter snow cover on the TP and summer rainfall in the MLYR [39].

In addition, the first peak (60–135 d) in Figure 8b corroborates previous research findings, indicating that variations in springtime SH over the TP can influence precipitation patterns in eastern China. Previous studies ascribed this lag effect to soil moisture dynamics [6]. The last peak in Figure 8a,b, which has exceeded 10 months to almost a year, is likely to be interpreted as an effect of the interannual variability of large-scale circulation systems, for example, El Niño-Southern Oscillation and summer monsoon [60–62].

To further identify the key zone of surface heat flux over the TP, the spatial distribution of the heterogeneous CC between the first mode of the anomalous heat flux and anomalous precipitation at the first peak is presented in Figure 9. The time corresponding to the first maximum peak in Figure 8 was selected. Specifically, the lag time is 90 d for SH and 7 d for LH. The heterogeneous CC between the anomalous LH and anomalous precipitation of the LSDHPEs was significantly negative at the eastern margin of the TP (Figure 9a). The heterogeneous CC between the anomalous SH and anomalous precipitation of the LSDHPEs was significantly negative west of the TP, whereas it was significantly positive in the center of the TP (Figure 9c). In addition, the sensitive areas of SH and LH in the MLYR were essentially the same (Figure 9b,d). The results indicate that when the anomalous LH decreases in the center and east of the TP and increases in the western regions, the anomalous precipitation manifests as a decrease over the northwestern and southeastern parts of the MLYR, with a time lag of 7 d (as seen in Figure 9a,b). When the anomalous SH increased in the center of the TP, the anomalous precipitation increased over the northwestern and southeastern parts of the MLYR, with a time lag of 90 d (Figure 9c,d). The distribution of the heterogeneous CC on the TP, as shown in the statistical histograms of Figure 9a,c, was almost evenly split between the positive and negative values. However, in the MLYR, the CC values exhibited a completely different distribution, being either positive or negative.

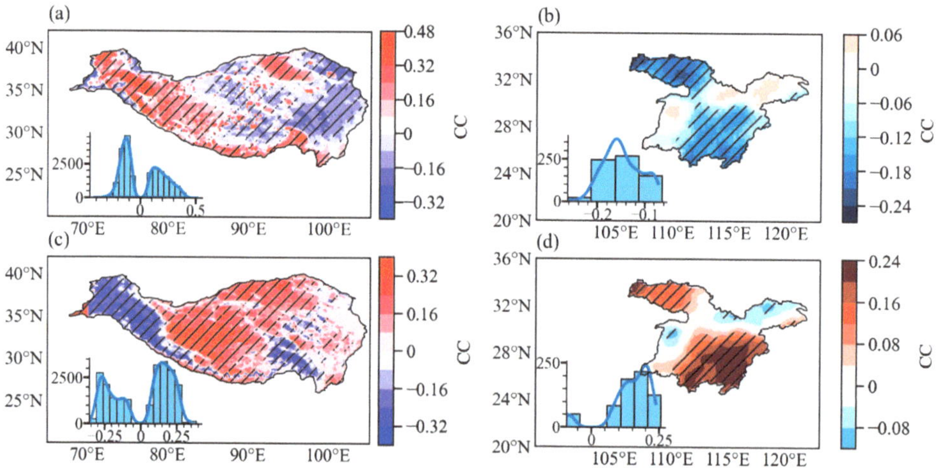

Figure 9. The spatial distribution of heterogeneous CCs between the anomalous LH and anomalous precipitation with a time lag of 7 days (**a**,**b**). The distribution of heterogeneous CCs between the anomalous SH and anomalous precipitation with a time lag of 90 days (**c**,**d**). (The slash lines are statistically significant at 5% level. The statistical histograms in each subfigure are the distribution of the CCs that passed the significance test.)

5. Discussion

This study investigates the correlation between surface heat fluxes over the TP and daily heavy precipitation events in the MLYR, revealing the synoptic patterns associated with LSDHPEs and the time-lagged relationship between heat fluxes on the TP and LSDHPEs. The data used in the study are publicly available remote sensing and reanalysis data, characterized by wide coverage, continuous observation, and high resolution. This remote sensing-based approach can be applied not only to the attribution analysis of precipitation events but also to studying the relationship between extreme events, such as heatwaves and compound droughts, and the TP. Furthermore, the study region is not limited to the TP and Yangtze River basin and can be extended to eastern China and even further regions.

Previous studies have contended that TP thermal forcing, specifically diabatic heating anomalies, showed a vital impact on the Asian summer monsoon and summer precipitation in Eastern China [6,15,16,31]. Our study investigated the time-lagged effects of the LH and SH anomalies over TP on precipitation anomalies in the MLYR from May to August from 1980 to 2022. We observed that SH exhibited a more positive correlation with precipitation than LH (Figure 9). The main reason for this difference may be the composition of diabatic heating during the development of the TP vortex [32,63]. Dong [64] noted that the TP vortex, which moved off the eastern slope of the TP and brought abundant water vapor to East China, was mainly affected by the SH during the developing stage. Once it leaves the TP, the apparent heat source of the atmosphere, sourced from condensation, will be beneficial to the eastward movement of the TP vortex [65].

Mesoscale air masses or weather systems, such as vortices, typically travel from the TP to the MLYR within approximately 3–5 d [22,66]. However, our lag-correlation analysis revealed a notably high CC between the anomalous adiabatic heating over the plateau and precipitation in the MLYR after 3 months (Figure 8). Delayed teleconnection could be explained by the memory of subsurface soil temperature and soil moisture over the TP [31]. These diabatic heating anomalies persist in deep soil layers for several months, altering the land-atmosphere thermal contrast, which in turn influences the planetary atmosphere, and triggering mesoscale waves to generate precipitation anomalies [6,7].

To identify the sensitivity areas over the TP for forecasting precipitation events in the MLYR, the dynamic and thermodynamic characteristics of the land-atmosphere interactions over the TP have been investigated. Topographic heating in the Himalayas [67] and the convection of air masses [68] in the central and eastern parts of the TP are considered precursor signals for anomalous precipitation in the MLYR. The latter is located on the water vapor transport path and is in line with the southwesterly moving monsoon flows over the MLYR [69,70]. Moreover, Dong [10] depicted the southeastern TP as the sensitivity area, in which the average MLYR precipitation in July from 1979 to 2014 was significantly negatively correlated with the regional latent heat fluxes over the TP. However, this study was not adopted this definition because it ignores the time lag between TP heat fluxes and MLYR precipitation. In addition, the lagged CCs of the mean heat flux in the sensitive area and the MLYR mean precipitation did not outperform the results of the mean TP heat flux and the MLYR mean precipitation (Figure A6 in the Appendix A). The interpretation of Figure 9 leads to the conclusion that SH affects atmospheric circulation, promoting the development of convective cloud activity over the TP. This, in turn, enhances westerly belt long waves and causes vortex systems to shift from the plateau to eastern or northern China.

A limitation of this study is that only two variables, the surface heat flux and the daily precipitation, were used to establish a linkage between the land–atmosphere interactions over the TP and the heavy precipitation events in the MLYR. Actually, various other variables could be especially influential, such as vegetation cover, soil moisture and temperature, snow depth over the TP, etc. These data can be obtained through remote sensing satellites, indicating the possibility to establish a linkage between the meteorological indicators and climatic events in two big geographic units that are separating afar. The potential implication of this research is that by including more remote sensing data, the prediction of the heavy precipitation in the MLYR can be extended in the lead time and the accuracy can also be improved.

6. Conclusions

To gain insight into the linkages between the precipitation in the MLYR and the surface heat fluxes of the TP, LSDHPEs in the MLYR from 1980 to 2022 were selected to analyze their synoptic circulation patterns and identify the dominant ones. Using statistical analysis and time-lag correlation analysis methods, this study quantified the undetermined spatiotemporal relationships between changes in the surface heat fluxes of the TP and the heavy rainfall events in the MLYR. The conclusions of this paper are summarized as follows:

(1) In the MLYR, the frequency of heavy rainfall events exhibited a unimodal distribution throughout the year, with June standing out as the peak period, characterized by a high concentration of intense precipitation. Depending on the frequency of occurrence, 77.76% of the LSDHPEs were categorized into four core types, from which two dominant circulation schemes were identified. One dominant weather pattern of the LSDHPEs during the plum rain period was an anomalous eastern low-level vortex pattern (SOM 6). The second was the southern anomalously high-pressure weather pattern (SOM 2), under the WNPSH and SAH combined effect. Under the two circulation patterns, anomalous airflow exhibits eastward movement in the eastern margin of the TP, and the topography of the TP hampers the convergence of water vapor toward the MLYR. In particular, the increasing trend in SOM 2 over the last 20 years resulted from dynamic and thermodynamic changes. Dynamic changes induced the decreasing trend in SOM 6.

(2) For all LSDHPEs, the surface SH anomalies were significantly correlated with precipitation in the MLYR, with time lags of 2–4 months and 300–350 d. Meanwhile, anomalous LH was correlated with anomalous precipitations, with time lags of 0–10 d, 130–200 d, and 330–365 d. The phenomenon associated with a time lag of more than two months was widely recognized as a memory effect of soil moisture. Therefore, a more detailed investigation should be conducted considering the soil moisture therein.

Author Contributions: Conceptualization, L.L. and X.D.; methodology, L.L. and H.J.; software, L.L. and H.J.; validation, L.L., X.D. and Y.M.; formal analysis, X.D.; investigation, L.L.; resources, Y.M.; data curation, L.L.; writing—original draft preparation, L.L.; writing—review and editing, X.D., C.W. and B.S.; visualization, L.L. and H.J.; supervision, X.D.; project administration, X.D.; funding acquisition, Y.M. All authors have read and agreed to the published version of the manuscript.

Funding: This research was funded by the Second Tibetan Plateau Scientific Expedition and Research Program (STEP, Grant No. 2019QZKK0103), the European Space Agency (ESA), and the National Remote Sensing Center of China (Grant No. 58516), the National Natural Science Foundation of China (Grant No. 42401030).

Data Availability Statement: The boundary of the MLYR can be obtained from the website http://www.geodata.cn/, accessed on 22 August 2024. The TP boundary can be obtained from the website https://data.tpdc.ac.cn/, accessed on 22 August 2024. MSWEP V2.2 data can be obtained from the website https://www.gloh2o.org/mswep/, accessed on 22 August 2024, Climate Prediction Center (CPC) global precipitation data can be obtained from the website https://psl.noaa.gov/data/gridded/data.cpc.globalprecip.html, accessed on 22 August 2024. ERA5 data are at https://cds.climate.copernicus.eu/, accessed on 22 August 2024. The CN05.1 daily precipitation data and observed data from 2400 stations in China are obtained by request and not publicly available.

Acknowledgments: We thank the Three Gorges University Advanced Computing Center for their assistance.

Conflicts of Interest: The authors declare no conflicts of interest.

Appendix A

Appendix A.1 Precision Validation of Three Precipitation Datasets

To validate the accuracy of the three precipitation products, the spatial resolution of all three products was resampled to 0.25° using linear interpolation, taking the observed daily precipitation data from 133 hydrological stations from 1980 to 2019 to validate the accuracy of three precipitation products and using the Pearson correlation coefficient (CC), the root mean square error (RMSE), and the relative bias (RB) to quantitatively evaluate the difference in site scale (Figure A1). The RMSE for the three rainfall products, MSWEP, CPC, and CN05.1 dataset, at the 133 stations are 13.844, 14.323, and 8.755, respectively. The corresponding CCs are 0.518, 0.498, and 0.83. Additionally, RBs are 390.559%, 370.773%, and 171.406%, respectively. Overall, the order of accuracy of the three rainfall products at the daily site scale is CN05.1 > MSWEP > CPC.

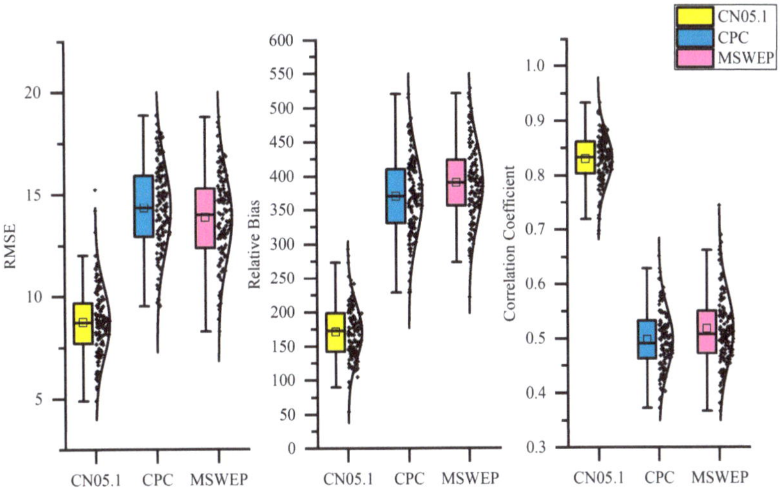

Figure A1. Box plots of RMSE, CC, and RB for three rainfall products at rainfall observation sites in the MLYR.

Appendix A.2 Accuracy Comparison of SOM Classification with Different Number of Nodes

Exploring a reasonable number of SOM nodes helps to accurately classify the synoptic circulation during the heavy rainfall concentration period in the MLYR. There are two principles for determining this, one is to capture a sufficiently large number of representative weather patterns, and the other is that the overlap between the categories should also be small. Referring to the previous method of determining the number of nodes and slightly modifying it [48,51], we calculate the topological error (mean Euclidean distance between daily original fields and the specific SOM pattern to which it belongs), quantization error (mean Euclidean distance between each mode of the SOM) [49], and correlation coefficient between daily original fields and the specific SOM pattern to which it belongs (Figure A2). Once the number of nodes exceeds six, the topological error and quantization error increase rapidly and stay at high values thereafter. By plotting the SOM weather type graph, it is found that the classification below six cannot completely segment all types of weather types. Therefore, the number of SOM classification nodes is determined to be six.

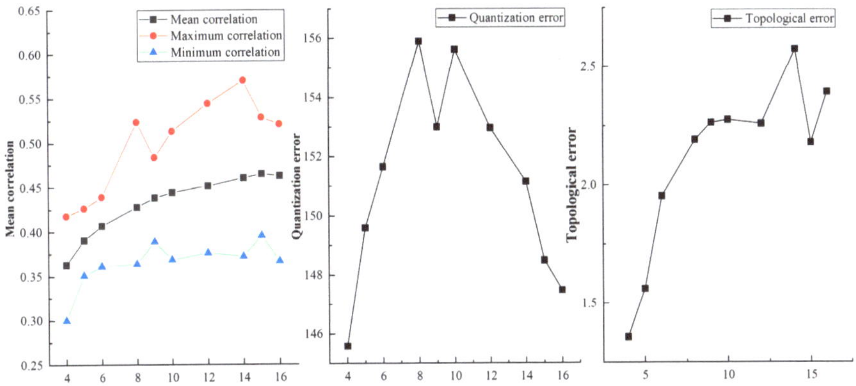

Figure A2. Evaluating the accuracy of SOM for different numbers of nodes.

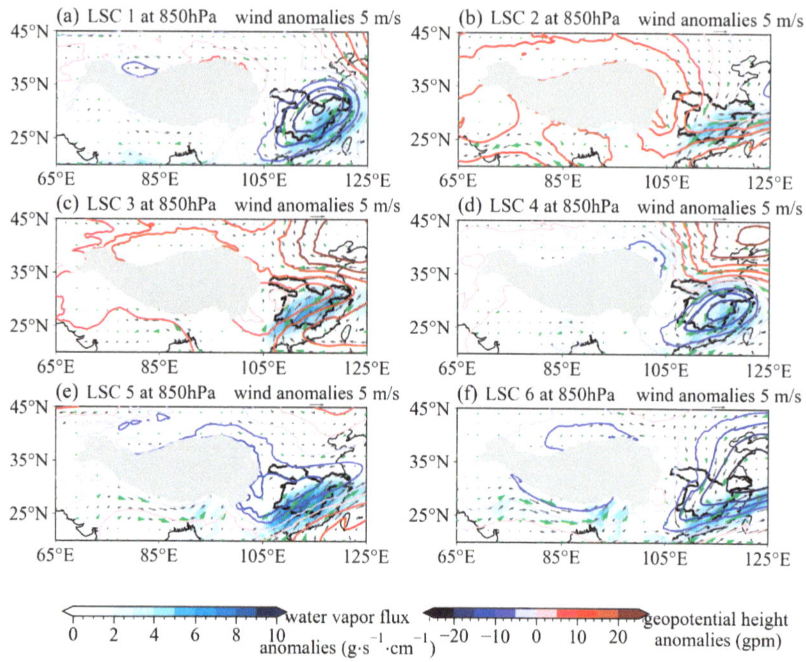

Figure A3. Water vapor flux anomalies of the LSDHPEs in SOM patterns at 850 hPa.

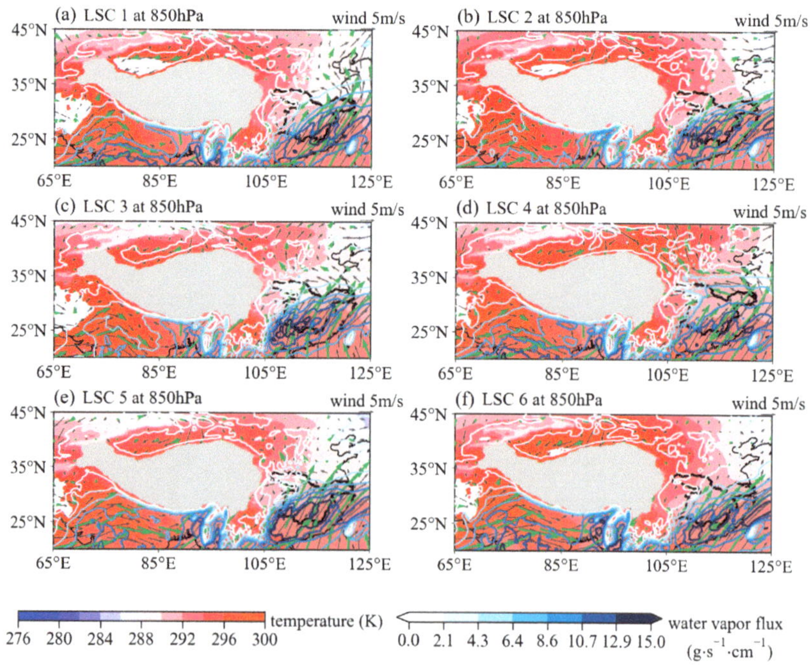

Figure A4. Temperature and water vapor flux of the LSC patterns at 850 hPa.

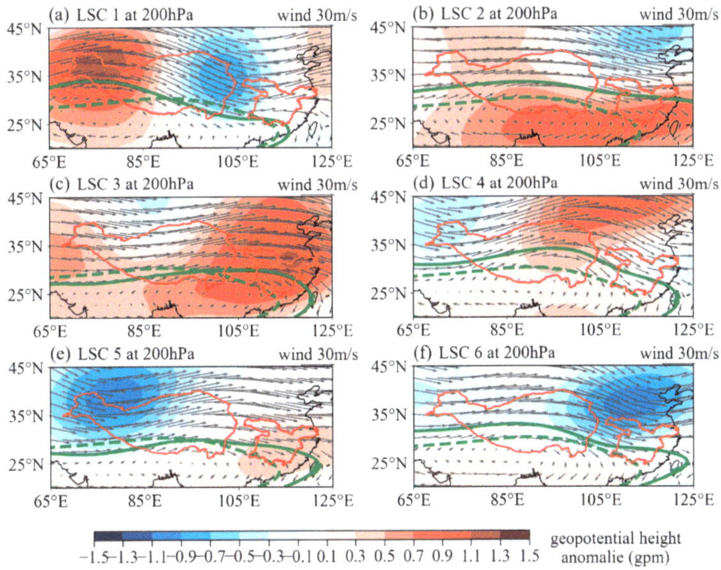

Figure A5. The synoptic patterns of the LSC at 200 hPa.

Appendix A.3 Lagged Correlation Coefficient Curves in Various Locations and SOM Modes

Zone A (98°–105°E, 23°–32.5°N) is proposed by Dong [10] as the location of high correlation with the precipitation in the MLYR.

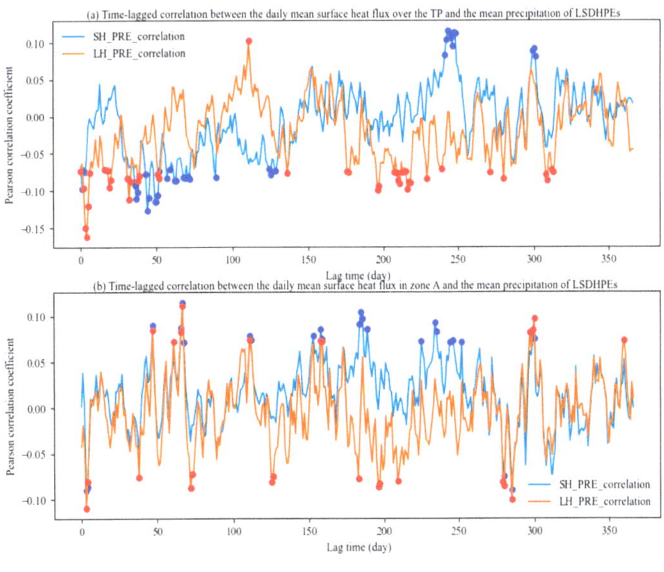

Figure A6. (**a**) Lagged correlation coefficient curves of the mean TP heat flux and MLYR mean precipitation. (**b**) Lagged correlation coefficient curves of the mean heat flux in the critical area A and the MLYR mean precipitation. Dots represent significance at the 95% confidence level.

References

1. Huang, J.; Zhou, X.; Wu, G.; Xu, X.; Zhao, Q. Global Climate Impacts of Land-Surface and Atmospheric Processes over the Tibetan Plateau. *Rev. Geophys.* **2023**, *61*, e2022RG000771. [CrossRef]
2. Yao, T.; Thompson, L.G.; Mosbrugger, V.; Zhang, F.; Ma, Y.; Luo, T.; Xu, B.; Yang, X.; Joswiak, D.R.; Wang, W.; et al. Third Pole Environment (TPE). *Environ. Dev.* **2012**, *3*, 52–64. [CrossRef]
3. Dong, W.; Lin, Y.; Wright, J.S.; Ming, Y.; Xie, Y.; Wang, B.; Luo, Y.; Huang, W.; Huang, J.; Wang, L.; et al. Summer rainfall over the southwestern Tibetan Plateau controlled by deep convection over the Indian subcontinent. *Nat. Commun.* **2016**, *7*, 10925. [CrossRef] [PubMed]
4. Xu, X.; Zhao, T.; Lu, C.; Guo, Y.; Chen, B.; Liu, R.; Li, Y.; Shi, X. An important mechanism sustaining the atmospheric "water tower" over the Tibetan Plateau. *Atmos. Chem. Phys.* **2014**, *14*, 11287–11295. [CrossRef]
5. Ye, Y.; Tian, R.; Jin, Z. Dynamic Effects of Atmosphere over and around the Tibetan Plateau on the Sustained Drought in Southwest China from 2009 to 2014. *Remote Sens.* **2023**, *15*, 2198. [CrossRef]
6. Han, Y.; Jiang, D.; Si, D.; Ma, Y.; Ma, W. Time-lagged Effects of the Spring Atmospheric Heat Source over the Tibetan Plateau on Summer Precipitation in Northeast China during 1961–2020: Role of Soil Moisture. *Adv. Atmos. Sci.* **2024**, *41*, 1527–1538. [CrossRef]
7. Xu, H.; Liang, X.-Z.; Xue, Y. Regional climate modeling to understand Tibetan heating remote impacts on East China precipitation. *Clim. Dyn.* **2022**, *62*, 2683–2701. [CrossRef]
8. Zhao, Y.; Chen, D.; Deng, Y.; Son, S.-W.; Wang, X. How Were the Eastward-Moving Heavy Rainfall Events from the Tibetan Plateau to the Lower Reaches of the Yangtze River Enhanced? *J. Clim.* **2021**, *34*, 607–620. [CrossRef]
9. Chen, Y.; Zhang, A.; Wang, B.; Fu, Y. A Heavy Precipitation Event in the Yangtze River Basin Led by an Eastward Moving Tibetan Plateau Cloud System in the Summer of 2016. *J. Geophys. Res.-Atmos.* **2020**, *125*, e2020JD032429. [CrossRef]
10. Dong, L.; Xu, X.; Zhao, T.; Ren, H. Linkage between moisture transport over the Yangtze River Basin and a critical area of the Tibetan Plateau during the Meiyu. *Clim. Dyn.* **2019**, *53*, 2643–2662. [CrossRef]
11. Xu, C.; Yuan, C.; Li, X.; Lin, Y.; Fan, H. Projection of disaster-causing risk of extreme precipitation in the Yangtze River Basin based on CMIP6. *Hydrol. Res.* **2023**, *54*, 401–407. [CrossRef]
12. Sun, S.; Zhao, Y.; He, Y.; Xia, Z.; Chen, S.; Zhang, Y.; Sun, Q. Exacerbated climate risks induced by precipitation extremes in the Yangtze River basin under warming scenarios. *Front. Ecol. Evol.* **2023**, *11*, 1127875. [CrossRef]
13. Bi, Y.; Zheng, L.; Wang, Y.; Li, J.; Yang, H. Coupling relationship between urbanization and water-related ecosystem services in China's Yangtze River economic Belt and its socio-ecological driving forces: A county-level perspective. *Ecol. Indic.* **2023**, *146*, 109871. [CrossRef]
14. Wang, Y.; Xu, Y.; Tabari, H.; Wang, J.; Wang, Q.; Song, S.; Hue, Z. Innovative trend analysis of annual and seasonal rainfall in the Yangtze River Delta, eastern China. *Atmos. Res.* **2020**, *231*, 104673. [CrossRef]
15. Xu, X.; Lu, C.; Ding, Y.; Shi, X.; Guo, Y.; Zhu, W. What is the relationship between China summer precipitation and the change of apparent heat source over the Tibetan Plateau? *Atmos. Sci. Lett.* **2013**, *14*, 227–234. [CrossRef]
16. Ge, J.; Youa, Q.; Zhang, Y. Effect of Tibetan Plateau heating on summer extreme precipitation in eastern China. *Atmos. Res.* **2019**, *218*, 364–371. [CrossRef]
17. Wang, Z.; Duan, A.; Wu, G. Time-lagged impact of spring sensible heat over the Tibetan Plateau on the summer rainfall anomaly in East China: Case studies using the WRF model. *Clim. Dyn.* **2014**, *42*, 2885–2898. [CrossRef]
18. Li, X.; Zhang, K.; Gu, P.; Feng, H.; Yin, Y.; Chen, W.; Cheng, B. Changes in precipitation extremes in the Yangtze River Basin during 1960–2019 and the association with global warming, ENSO, and local effects. *Sci. Total Environ.* **2021**, *760*, 144244. [CrossRef]
19. Zhang, Q.; Xu, C.-Y.; Zhang, Z.; Chen, Y.D.; Chun-ling Liu, H.L. Spatial and temporal variability of precipitation maxima during 1960-2005 in the Yangtze River basin and possible association with large-scale circulation. *J. Hydrol.* **2008**, *353*, 215–227. [CrossRef]
20. Xie, Y.; Huang, J.; Wu, G.; Liu, Y.; Dong, W.; Lu, M. Oceanic repeaters boost the global climatic impact of the Tibetan Plateau. *Sci. Bull.* **2023**, *68*, 2225–2235. [CrossRef]
21. Kuang, X.; Jiao, J.J. Review on climate change on the Tibetan Plateau during the last half century. *J. Geophys. Res. Atmos.* **2016**, *121*, 3979–4007. [CrossRef]
22. Yang, H.; Deng, Y.; Cui, C.; Wang, X.; Dong, X. Dynamic Trigger and Moisture Source of Two Typical Meiyu Front Rainstorms Associated with Eastward-Moving Cloud Clusters from the Tibetan Plateau. *J. Meteorol. Res.* **2022**, *36*, 478–499. [CrossRef]
23. Wang, M.; Wang, J.; Duan, A.; Yang, J.; Liu, Y. Quasi-biweekly impact of the atmospheric heat source over the Tibetan Plateau on summer rainfall in Eastern China. *Clim. Dyn.* **2019**, *53*, 4489–4504. [CrossRef]
24. Li, J.; Chen, Y.; Mao, J.; Zhai, P. Increase in optimal configuration of 25–60-day atmospheric circulations for Yangtze heavy rainfall under global warming background. *Weather. Clim. Extrem.* **2023**, *42*, 100630. [CrossRef]
25. Xu, X.; Lu, C.; Shi, X.; Gao, S. World water tower: An atmospheric perspective. *Geophys. Res. Lett.* **2008**, *35*, L20815. [CrossRef]
26. Su, Z.; Ma, Y.; Chen, X.; Dong, X.; Du, J.; Han, C.; He, Y.; Hofste, J.G.; Li, M.; Li, M.; et al. Monitoring Water and Energy Cycles at Climate Scale in the Third Pole Environment (CLIMATE-TPE). *Remote Sens.* **2021**, *13*, 3661. [CrossRef]
27. Zhao, Y.; Xu, X.; Ruan, Z.; Chen, B.; Wang, F. Precursory strong-signal characteristics of the convective clouds of the Central Tibetan Plateau detected by radar echoes with respect to the evolutionary processes of an eastward-moving heavy rainstorm belt in the Yangtze River Basin. *Meteorol. Atmos. Phys.* **2019**, *131*, 697–712. [CrossRef]

28. Zhao, Y.; Sun, R.; Xie, Z.; Duan, A. Upper-Troposphere Saddle-Like Response to Springtime Surface Sensible Heating Over the Tibetan Plateau: Combined Effect From Baroclinic and Barotropic Process. *J. Geophys. Res.-Atmos.* **2022**, *127*, e2021JD036002. [CrossRef]
29. Zeng, J.; Huang, A.; Wu, P.; Huang, D.; Zhang, Y.; Tang, J.; Zhao, D.; Yang, B.; Chen, S. Typical Synoptic Patterns Responsible for Summer Regional Hourly Extreme Precipitation Events Over the Middle and Lower Yangtze River Basin, China. *Geophys. Res. Lett.* **2023**, *50*, e2023GL104829. [CrossRef]
30. Chen, J.; Li, Y.; Xiong, B.; Wang, Y.; Zhou, S.; Huang, Y. Comparison of Moisture Sources of Summer Precipitation in 1998 and 2020 in the Middle and Lower Reaches of Yangtze River Basin. *Int. J. Climatol.* **2023**, *43*, 3493–3505. [CrossRef]
31. Zhu, C.; Ullah, W.; Wang, G.; Lu, J.; Li, S. Diagnosing Potential Impacts of Tibetan Plateau Spring Soil Moisture Anomalies on Summer Precipitation and Floods in the Yangtze River Basin. *J. Geophys. Res. Atmos.* **2023**, *128*, e2022JD037671. [CrossRef]
32. Wu, D.; Zhang, F.; Wang, C. Impacts of Diabatic Heating on the Genesis and Development of an Inner Tibetan Plateau Vortex. *J. Geophys. Res.-Atmos.* **2018**, *123*, 11691–11704. [CrossRef]
33. Ma, J.; Wei, K.; Chen, W.; Wang, T.; Shi, D. The southward movement preference of large-scale persistent extreme precipitation events over the Yangtze River Valley during Mei-yu period. *Clim. Dyn.* **2023**, *60*, 3729–3747. [CrossRef]
34. Tao, S.; Ding, Y. Observational Evidence of the Influence of the Qinghai-Xizang (Tibet) Plateau on the Occurrence of Heavy Rain and Severe Convective Storms in China. *Bull. Am. Meteorol. Soc.* **1981**, *62*, 23–30. Available online: https://www.jstor.org/stable/26221519 (accessed on 20 August 2024). [CrossRef]
35. Zhan, C.; Shi, Q.; Liang, S. Assessing the impacts of the spring sensible heat flux over the Tibetan Plateau on Asian summer monsoon rainfall using observational and reanalysis data. *Int. J. Climatol.* **2019**, *40*, 2342–2358. [CrossRef]
36. Wan, B.; Gao, Z.; Chen, F.; Lu, C. Impact of Tibetan Plateau Surface Heating on Persistent Extreme Precipitation Events in Southeastern China. *Mon. Weather. Rev.* **2017**, *145*, 3485–3505. [CrossRef]
37. Sun, Y.; Shan, X.; Zhou, S.; Wang, M.; Wang, C.; Deng, Z. Impacts of Tibetan Plateau sensible heat and El Nino-Southern Oscillation on precipitation over South China under the background of the PDO. *Front. Environ. Sci.* **2023**, *11*, 1156206. [CrossRef]
38. Huang, W.; Zhao, Y.; Sun, C.; Wang, H.; Wang, X. Climate modulation of summer rainstorm activity in eastern China based on the Tibetan Plateau spring heating. *Arab. J. Geosci.* **2020**, *13*, 126. [CrossRef]
39. Yang, K.; Qi, Q.; Wang, C. Possible impacts of vegetation cover increment on the relationship between winter snow cover anomalies over the Third Pole and summer precipitation in East Asia. *NPJ Clim. Atmos. Sci.* **2023**, *6*, 140. [CrossRef]
40. Qi, W.; He, L.; Hai, J.; Wei, W. Relationship between spring soil moisture in the Tibetan Plateau and summer precipitation in the Yangtze river basin and its possible mechanism. *Chin. J. Geophys.-Chin. Ed.* **2016**, *59*, 3985–3995. [CrossRef]
41. Zhang, C.; Jia, X.; Wen, Z. Increased impact of the Tibetan Plateau Spring Snow Cover to the Meiyu Rainfall over the Yangtze River Valley after 1990s. *J. Clim.* **2021**, *34*, 5985–5997. [CrossRef]
42. Xu, Y.; Gao, X.; Shen, Y.; Xu, C.; Shi, Y.; Giorgi, F. A Daily Temperature Dataset over China and Its Application in Validating a RCM Simulation. *Adv. Atmos. Sci.* **2009**, *26*, 763–772. [CrossRef]
43. Xu, X.; Huang, A.; Huang, D.; Gu, C. What Are the Dominant Synoptic Patterns Leading to the Summer Regional Hourly Extreme Precipitation Events Over Central-Eastern Tibetan Plateau and Sichuan Basin. *Geophys. Res. Lett.* **2023**, *50*, e2022GL102342. [CrossRef]
44. Kohonen, T. Self-organized formation of topologically correct feature maps. *Biol. Cybern.* **1982**, *43*, 59–69. [CrossRef]
45. Ng, C.P.; Zhang, Q.; Li, W. Changes in Hourly Extreme Precipitation Over Eastern China From 1970 to 2019 Dominated by Synoptic-Scale Precipitation. *Geophys. Res. Lett.* **2021**, *48*, e2020GL090620. [CrossRef]
46. Shin, Y.; Kim, J.-H.; Chun, H.-Y.; Jang, W.; Son, S.-W. Classification of Synoptic Patterns With Mesoscale Mechanisms for Downslope Windstorms in Korea Using a Self-Organizing Map. *J. Geophys. Res. Atmos.* **2022**, *127*, e2021JD035867. [CrossRef]
47. Loikith, P.C.; Lintner, B.R.; Sweeney, A. Characterizing Large-Scale Meteorological Patterns and Associated Temperature and Precipitation Extremes over the Northwestern United States Using Self-Organizing Maps. *J. Clim.* **2017**, *30*, 2829–2847. [CrossRef]
48. Lee, S.; Feldstein, S.B. Detecting Ozone- and Greenhouse Gas–Driven Wind Trends with Observational Data. *Science* **2013**, *339*, 563–567. [CrossRef]
49. Horton, D.E.; Johnson, N.C.; Singh, D.; Swain, D.L.; Rajaratnam, B.; Diffenbaugh, N.S. Contribution of changes in atmospheric circulation patterns to extreme temperature trends. *Nature* **2015**, *522*, 465–469. [CrossRef]
50. Cohen, J.; Coumou, D.; Screen, J.A.; Furtado, J.C.; Barlow, M.; Francis, J.; Dethloff, K.; Whittleston, D. Recent Arctic amplification and extreme mid-latitude weather. *Nat. Geosci.* **2014**, *7*, 627–637. [CrossRef]
51. Cassano, J.J.; Uotila, P.; Lynch, A.H.; Cassano, E.N. Predicted changes in synoptic forcing of net precipitation in large Arctic river basins during the 21st century. *J. Geophys. Res.* **2007**, *112*, G04S49. [CrossRef]
52. Gong, H.; Wang, L.; Zhou, W.; Chen, W.; Wu, R.; Liu, L. Revisiting the Northern Mode of East Asian Winter Monsoon Variation and Its Response to Global Warming. *J. Clim.* **2018**, *31*, 9001–9014. [CrossRef]
53. Chen, Y.; Wang, A.; Feng, G. The relative contribution of large-scale circulation and local soil moisture to summer precipitation over Asian mid-low latitudes. *Clim. Dyn.* **2022**, *60*, 2097–2112. [CrossRef]
54. Lin, R.; Zhu, J.; Zheng, F. The Application of the SVD Method to Reduce Coupled Model Biases in Seasonal Predictions of Rainfall. *J. Geophys. Res.-Atmos.* **2019**, *124*, 11837–11849. [CrossRef]
55. Wallace, J.M.; Smith, C.; Bretherton, C.S. Singular Value Decomposition of Wintertime Sea Surface Temperature and 500-mb Height Anomalies. *J. Clim.* **1992**, *5*, 561–576. [CrossRef]

56. Zhong, L.; MA, Y.; Salama, M.S.; Su, Z. Assessment of vegetation dynamics and their response to variations in precipitation and temperature in the Tibetan Plateau. *Clim. Chang.* **2010**, *103*, 519–535. [CrossRef]
57. Roerink, G.J.; Menenti, M.; Verhoef, W. Reconstructing cloudfree NDVI composites using Fourier analysis of time series. *Int. J. Remote Sens.* **2000**, *21*, 1911–1917. [CrossRef]
58. Sun, Q.; Miao, C.; Duan, Q.; Sorooshian, S.; Hsu, K.-L. A Review of Global Precipitation Data Sets: Data Sources, Estimation, and Intercomparisons. *Rev. Geophys.* **2017**, *56*, 79–107. [CrossRef]
59. Chen, B.; Xu, X.-D.; Zhao, T. Main moisture sources affecting lower Yangtze River Basin in boreal summers during 2004–2009. *Int. J. Climatol.* **2012**, *33*, 1035–1046. [CrossRef]
60. Xia, Y.; Yao, S.; Sun, T.; Guo, Z. Role of the Low-Latitude Quasi-Biweekly Oscillation in the Extreme Persistent Heavy Rainfall in the Mei-Yu Season over the Middle and Lower Reaches of the Yangtze River. *J. Clim.* **2023**, *36*, 3817–3832. [CrossRef]
61. Zhang, J.; Zhao, K.; Wang, Y.; Kong, X.; Shao, X.; Liang, Y. Modulation of centennial-scale hydroclimate variations in the middle Yangtze River Valley by the East Asian-Pacific pattern and ENSO over the past two millennia. *Earth Planet. Sci. Lett.* **2021**, *576*, 117220. [CrossRef]
62. Xiao, M.; Zhang, Q.; Singh, V.P. Influences of ENSO, NAO, IOD and PDO on seasonal precipitation regimes in the Yangtze River basin, China. *Int. J. Climatol.* **2015**, *35*, 3385–3720. [CrossRef]
63. Deng, M.; Meng, X.; Sheng, D.; Niu, H.; Wu, P.; Li, Z.; Zhao, L.; Chen, H. Observed surface heat fluxes partitioning during the local growing season over the Tibetan Plateau. *Agric. For. Meteorol.* **2024**, *356*, 110186. [CrossRef]
64. Dong, Y.; Li, G.; Xie, X.; Yang, L.; Zhang, P.; Zeng, B. Mechanism of Diabatic Heating on Precipitation and the Track of a Tibetan Plateau Vortex over the Eastern Slope of the Tibetan Plateau. *Adv. Atmos. Sci.* **2024**, *41*, 155–172. [CrossRef]
65. Li, L.; Zhang, R.; Wen, M.; Duan, J. Development and eastward movement mechanisms of the Tibetan Plateau vortices moving off the Tibetan Plateau. *Clim. Dyn.* **2019**, *52*, 4849–4859. [CrossRef]
66. Xu, H.; Zhao, D.; Yin, J.; Duan, Y.; Gao, W.; Li, Y.; Zhou, L. Indirect Effects of Binary Typhoons on an Extreme Rainfall Event in Henan Province, China From 19 to 21 July 2021. 3. Sensitivities to Microphysics Schemes. *J. Geophys. Res.-Atmos.* **2023**, *128*, e2022JD037936. [CrossRef]
67. Wu, G.; Liu, Y.; He, B.; Bao, Q.; Duan, A.; Jin, F.-F. Thermal Controls on the Asian Summer Monsoon. *Sci. Rep.* **2012**, *2*, 404. [CrossRef]
68. Hu, L.; Deng, D.; Gao, S.; Xu, X. The seasonal variation of Tibetan Convective Systems: Satellite observation. *J. Geophys. Res. Atmos.* **2016**, *121*, 5512–5525. [CrossRef]
69. Xu, X.D.; Shi, X.Y.; Wang, Y.Q.; Peng, S.Q.; Shi, X.H. Data analysis and numerical simulation of moisture source and transport associated with summer precipitation in the Yangtze River Valley over China. *Meteorol. Atmos. Phys.* **2008**, *100*, 217–231. [CrossRef]
70. Zhao, Y.; Xu, X.; Liu, L.; Zhang, R. Effects of convection over the Tibetan Plateau on rainstorms downstream of the Yangtze River Basin. *Atmos. Res.* **2019**, *219*, 24–35. [CrossRef]

remote sensing

MDPI

Communication

Characterizing the Supercooled Cloud over the TP Eastern Slope in 2016 via Himawari-8 Products

Qiuyu Wu [1], Jinghua Chen [1,2,*] and Yan Yin [1]

[1] Collaborative Innovation Center on Forecast and Evaluation of Meteorological Disasters, China Meteorological Administration Aerosol-Cloud and Precipitation Key Laboratory, Nanjing University of Information Science & Technology, Nanjing 210044, China; 202212030010@nuist.edu.cn (Q.W.); yinyan@nuist.edu.cn (Y.Y.)

[2] State Key Laboratory of Severe Weather, Chinese Academy of Meteorological Sciences, Beijing 100081, China

* Correspondence: jhchen@nuist.edu.cn; Tel.: +86-15951847906

Abstract: Supercooled liquid water (SLW) refers to droplets in clouds that remain unfrozen at temperatures below 0 °C. SLW is an important intermediate hydrometeor in the processes of snowfall and rainfall that can modulate the radiation budget. This study investigates the distribution of supercooled cloud water over mainland China using the East Asia–Pacific cloud macro- and microphysical properties dataset (2016), derived from Himawari-8 observations. The results show that the highest frequency of SLW in liquid-phase stratus clouds occur at the eastern slope of the Tibetan Plateau, the western side of the Sichuan Basin. Additional SLW is mostly found in liquid-phase clouds over the Sichuan Basin and its adjacent areas in southern China. In the region with the highest frequency of SLW, the mechanical forcing of the Tibetan Plateau causes the convergence of low-level airflow within the basin, which also carries moisture that is forced to ascend stably, creating a favorable condition for the formation of supercooled clouds. As the airflow continues to ascend, it encounters the mid-to-upper-level westerlies and temperature inversion. At the mid-to-upper level, the westerlies exhibit stronger wind speeds, directing flow towards the basin. Concurrently, the temperature inversion stabilizes the atmospheric stratification, limiting the further ascent of airflow. This inversion can also restrain convection and upward motion within the clouds, allowing for SLW to exist and persist for an extended period.

Keywords: supercooled liquid water; topographic effects; the eastern margin of the Tibetan Plateau

Citation: Wu, Q.; Chen, J.; Yin, Y. Characterizing the Supercooled Cloud over the TP Eastern Slope in 2016 via Himawari-8 Products. *Remote Sens.* **2024**, *16*, 3643. https://doi.org/10.3390/rs16193643

Academic Editor: Carmine Serio

Received: 17 August 2024
Revised: 14 September 2024
Accepted: 26 September 2024
Published: 29 September 2024

1. Introduction

Supercooled water clouds (SWCs) are composed of liquid-phase particles or ice-phase particles and liquid particles with temperatures below 0 °C. The cloud top temperature of clouds containing SWC is generally 0 °C~−40 °C, with significant amounts of supercooled water observed between 0 °C and −20 °C [1]. In SWCs, cloud droplets below 0 °C are supercooled cloud water or supercooled liquid water (SLW).

SLW plays a crucial role in artificial weather modification. It impacts cloud precipitation, snowfall, and cloud lifetime by affecting ice particles in clouds [2]. Tabazadeh (2002) proposed a theoretical model, and based on it, they found that the surface crystallization rate of ice in SWC is much higher than the volume-based freezing rate and that the surface crystallization rate depends on the droplet size, temperature, and ambient phase [3]. Based on airborne in situ observations, Maciel and Diao (2022) considered that ice crystals are initially surrounded by supercooled liquid water droplets, and then, as they grow, pure ice segments are formed [4]. Furthermore, the concentration and location of SLW determine the location and timing of cloud seeding [5]. Morrison (2013) utilizes satellite observations to show that differences in the SLW between two regions lead to distinct changes in cloud microphysics after cloud seeding [6]. Li (2022) uses both in situ aircraft and satellite remote sensing observations to find that in regions seeded with supercooled water, droplets quickly transform into ice crystals, causing cloud-top sinking and precipitation [7]. Yang (2022) demonstrated that seeding at the height where SLW and ice crystals coexist showed obvious responses in the cloud microphysical properties [8].

In light of the mentioned factors, the unique characteristics of supercooled water and its role in aviation safety are critical. In aviation, when the aircraft flies through clouds, the SLW suspended in clouds induces the formation of flight ice [9]. Aircraft icing leads to engine failure and subsequent aviation accidents. Ice accretion on aircraft wings and the deposition or coating of ice on an airplane can damage the aircraft components [10].

The distribution of SLW has impacts on radiative energy balance, and SLW near the cloud top increases the cloud-top radiative cooling [11]. Between 40° and 70°S, clouds with supercooled liquid tops dominate the population of liquid clouds, and clouds with supercooled liquid tops contribute between 27% and 38% of the total reflected solar radiation [12]. In addition, SLW is important for the phase change process in clouds at different stages, and cloud phases regulate climate by affecting radiation [13]. However, in global climate models, the radiation in supercooled clouds cannot be simulated well, causing deviations in the simulation results of climate models [14]. For all of the above reasons, SLW strongly influences the weather and climate.

So far, much attention has been paid to the microphysics and radiation equilibrium of SLW, but the impacts regarding the distribution of SWC have received little attention. Hou (2010), based on CALIOP and MODIS global measurement data, found that the highest amount of SWCs is over the land of Europe and East Asia (mainly in China) [15]. In this study, the author investigates the occurrence and distribution of SLW in mainland China to explore the possible causes of the eastern boundary of the Tibetan Plateau (TP) becoming a high-frequency region for SLW. The motivation of the present study concerns providing references for the scientific implementation of artificial rain, the formulation of aircraft flight routes, and the effect of SWCs on the climate in China.

After analyzing the distribution characteristics of SLW, it was found that SLW is mainly concentrated in the part where the Sichuan Basin borders the TP and SLW. Notably, the transition zone between the eastern boundary of the Qinghai–Tibetan Plateau and the Sichuan Basin has always been a key focus of studies. The eastern edge of the Qinghai–Tibetan Plateau features complex terrain and is influenced by various weather systems, leading to frequent occurrences of extreme summer precipitation [16]. In winter, the leeward slope effect along the eastern edge of the Tibetan Plateau exacerbates air pollution in the Sichuan Basin [17]. The complex terrain between the eastern slope of the Qinghai–Tibetan Plateau and the Sichuan Basin creates a topographically dependent atmospheric environment, which is analyzed in this paper to explain the high frequency of SLW on the eastern slope of the TP.

2. Data and Methods

The East Asia–Pacific cloud macro- and microphysical properties dataset was developed using the next-generation geostationary satellite Himawari-8, the dataset includes parameters such as cloud cover (CF), cloud detection (CM), cloud phase detection (CP), cloud type (CT), and supercooled water detection (SWC), with temporal resolution of 1 h and spatial resolutions of $0.1° \times 0.1°$ [18]. Compared to CALIOP and MODIS, this dataset is derived from the Himawari-8 satellite and has been processed using a novel algorithm, resulting in cloud products with enhanced temporal and spatial precision [19].

The fifth generation atmospheric reanalysis data ERA5 provided by the ECMWF (European Centre for Medium-Range Weather Forecasts) on pressure levels, includes 37 layers of temperature, u-component of wind, v-component of wind, vertical velocity, specific humidity and divergence with pressure levels 1000 hPa–0.01 hPa, and surface pressure (temporal resolution of 1 h and spatial resolutions of $0.25° \times 0.25°$) [20].

The MODIS Level-2 Cloud product (MOD06_L2) consists of cloud optical and physical parameters; it is derived using remotely sensed infrared and visible and near-infrared solar reflected radiances, and the infrared channel radiances are used to derive cloud top temperature and cloud top height [21]. MOD06_L2 consists of parameters at a spatial resolution of either 1 km or 5 km (at nadir), and each MOD06_L2 product file covers a 5 min time interval [22].

The method for calculating the frequency of SLW involves assigning a value of 1 to each $0.1^\circ \times 0.1^\circ$ grid point that contains SLW per hour and a value of 0 where it is absent. The frequency of SLW is obtained by summing the value of 1 for each occurrence (as Equation (1)). The statistics of CT and CP with SWC are similar. (Similar to the statistics for SLW, the frequency of various CT and CP is denoted by 1 when supercooled water is present and then cumulatively tallied.)

$$F = \sum_{i,j}^{n} grid_{i,j}\ (grid_{i,j} = 1\ where\ SLW\ at\ grid_{i,j}\ is\ greater\ than\ 0) \quad (1)$$

When selecting specific areas, due to the varying spatial resolutions of the utilized data, the chosen regions correspond to irregular high frequency zones of SLW. These are then interpolated to different resolutions, and the approximate region is shown in Figure 1. Due to differences in the temporal precision of the data, the cloud top temperature (CTT) and cloud top pressure (CTP) of non-supercooled clouds are also included when discussing CTT and CTP. Normalization was employed in the statistical analysis of CTP and CTT (as Equation (2)); (x) is the original data, (x_{min}) is the minimum value in the dataset, and (x_{max}) is the maximum value. This formula subtracts the minimum value from the original data and then divides by the range (maximum value minus minimum value), and the resulting values will be scaled between 0 and 1.

$$x = \frac{x - x_{min}}{x_{max} - x_{min}} \quad (2)$$

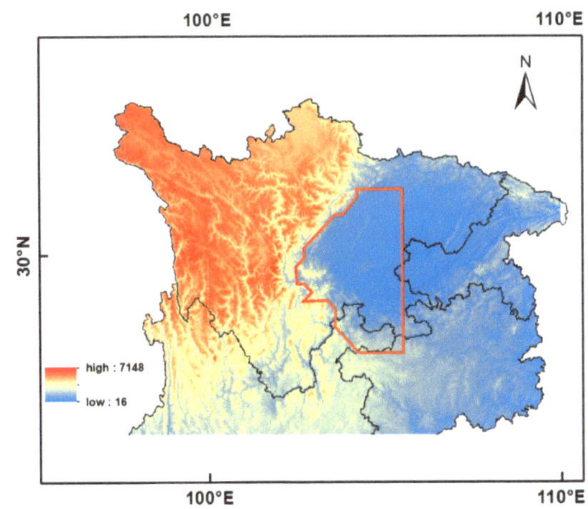

Figure 1. Typical area geographical location (the area within the red box is a typical region).

3. Results

Based on the East Asia–Pacific cloud macro- and microphysical properties dataset, the frequency of SLW and the probability of SLW occurring within clouds are shown in Figure 2, respectively. The maximum SLW frequency, exceeding 1000 occurrences, is observed at the junction of the Sichuan Basin and the eastern slope of the Tibetan Plateau. In the Sichuan Basin and its adjacent areas, the frequency of SLW ranges from 600 to 1000. A smaller frequency of SLW is distributed predominantly in Southeast China. Overall, the distribution of SLW decreases progressively from the confluence of the Sichuan Basin and the Tibetan

Plateau, extending towards the southeastern regions of China. The probability distribution of SLW occurrence in clouds mirrors the frequency distribution of SLW. In the eastern slope of the Tibetan Plateau and within the Sichuan Basin, the probability surpassing of SLW in clouds is significantly higher, exceeding a probability of 42%. In the southern part of the Sichuan Basin and the transitional region between the Sichuan Basin and the Yunnan–Guizhou Plateau, the probability of encountering supercooled water in clouds is above 36%. Previous studies have demonstrated that SWCs are mostly observed over the ocean near the storm-track regions and high-latitude regions [23]. However, in China, SLW is mainly distributed in mid-latitude clouds. The Sichuan Basin and its surrounding areas exhibit a significantly higher incidence of SLW in clouds than the precipitation-rich Southeast China coast.

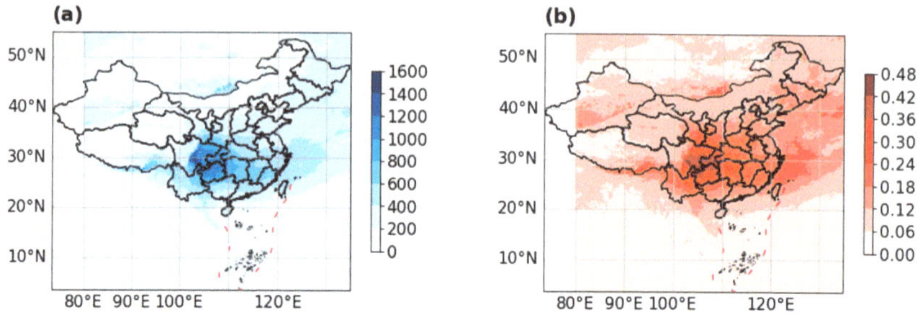

Figure 2. The distribution of supercooled cloud water in 2016: (**a**) the frequency of supercooled cloud water occurrences; (**b**) the probability of supercooled cloud water occurring within clouds.

Figure 3 further counts the frequency of the occurrence of SLW in different CTs and CP in 2016. It found that SLW almost appears in liquid water clouds (Figure 3a), and only a small portion was found in mixed-phase clouds in North China (Figure 3b). According to the statistical analysis of CTs, the frequency of SLW in China mainly occurs in nimbostratus clouds (Figure 3d). The highest frequency, exceeding 900 occurrences, is found in the elongated belt-like region at the transition between the Sichuan Basin and the eastern slope of the Tibetan Plateau. In the junction of Sichuan, Chongqing, and Guizhou (the transitional zone bridging the Sichuan Basin and the Yunnan–Guizhou Plateau), the frequency of nimbostratus clouds containing SLW exceeds 600. In addition to nimbostratus, SLW is also prevalent in altostratus clouds (Figure 3c). The highest frequency of altostratus clouds containing SLW, exceeding 400 occurrences, is observed along where the eastern slope of the Tibetan Platea meets the Sichuan Basin. Subsequently, within the Sichuan Basin and the southern part of Guizhou, the frequency hovers around 300.

In general, SLW is predominantly found within the liquid-phase stratiform clouds surrounding the Sichuan Basin. The stratiform clouds are primarily nimbostratus and altostratus. Yu discovered that downstream on the eastern side of the Tibetan Plateau, the dominance of nimbostratus and altostratus clouds resulted in extremely large cloud optical thickness and attributed this to the topography of the Tibetan Plateau [24]. The work described here suggests the transitional region between the Sichuan Basin and the Tibetan Plateau is a critical area for the distribution of SLW. Therefore, this article proposes that the SLW in China is mainly distributed in liquid-phase stratus clouds that may be impacted by the topographic factors of the Tibetan Plateau. Even a liquid cloud is likely to occur preferentially in the storm region [14], and though with abundant clouds and many storms passing through southeastern China [25], the frequency of SLW is far less than that in the Sichuan Basin, which has long been affected by the Tibetan Plateau.

To further verify the above inference, the study selected regions within the inner sections of the Sichuan Basin and along the eastern slopes of the Tibetan Plateau, which are characterized by a higher frequency of SLW, for a detailed analysis as representative areas (Figure 1). Due to variations in the temporal and spatial resolution of different datasets, the analysis was only including January, February, November, and December, as these months have a higher frequency of SLW and the greatest probability of SLW occurrence in clouds. Concurrently, the months of January, February, November, and December, which are the winter season in the northern hemisphere, provide a suitable environment for SLW due to the lower temperatures relative to other seasons.

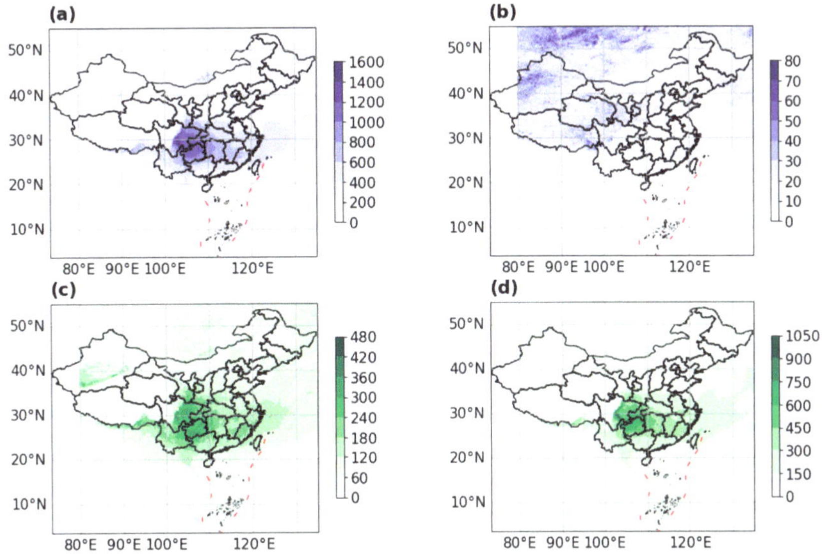

Figure 3. Distribution of supercooled water in different cloud types and cloud phases. (**a**) Distribution of supercooled water in liquid cloud. (**b**) Distribution of supercooled water in mixed-phase clouds. (**c**) Distribution of supercooled water in altostratus clouds. (**d**) Distribution of supercooled water in nimbostratus clouds.

In the above conclusion, SLW mainly appears in nimbostratus clouds and altostratus clouds, which basically are liquid-phase clouds. To explore the reason, use MODIS satellite data to statistically measure CTP and CTT for the months of January, February, November, and December. The statistical results are presented in Figure 4. The distribution of CTP exhibits a multi-peak pattern, with the main peaks occurring between 450 hPa and 700 hPa (Figure 4a). The temperature distribution at the cloud tops is characterized by a unimodal pattern, predominantly spanning from −20 °C to 10 °C (Figure 4b). When the cloud temperature is between −20 °C and 0 °C, a large amount of SLW exists [26]. The predominant range of CTT lies between −20 °C and 0 °C, fostering an environment conducive to the presence of SLW. Furthermore, the CTP primarily resides within the mid-tropospheric layer, signifying that in typical areas during the winter season, mid-level and low-level clouds account for the majority of cloud layers. Mid-level or low-level nimbostratus and altostratus clouds with cloud top temperatures between −20 °C and 0 °C create conditions conducive to the presence of SLW.

Downstream of the Tibetan Plateau, nimbostratus and altostratus clouds often develop from the gradual ascent of moist air, resulting in relatively weak turbulence within the cloud [25,27]. In the typical area, SWCs form in an environment with weak ascending air currents. We conducted an analysis of vertical divergence for different months (January, February, November, and December), as shown in Figure 5. The trend in all four months is relatively consistent. The low-level (from the surface to 800 hPa) divergence is less than 0. The mid–high-level (700 hPa~500 hPa) divergence is greater than 0. It indicates convergence in the low-level and mid-level divergence in the region. This may be related to the Tibetan Plateau acting as a barrier to the westerly winds, with notable effects during specific months (January, February, November, and December) [28].

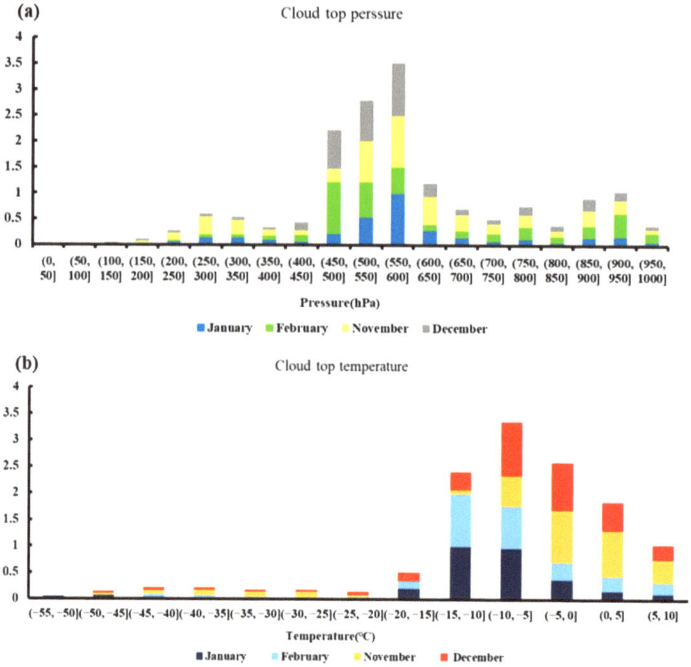

Figure 4. Cloud top pressure and cloud top temperature statistics. (**a**) Statistics of cloud top pressure in January, February, November, and December. (**b**) Statistics of cloud top temperature in January, February, November, and December (the values on the Y-axis are normalized and then aggregated).

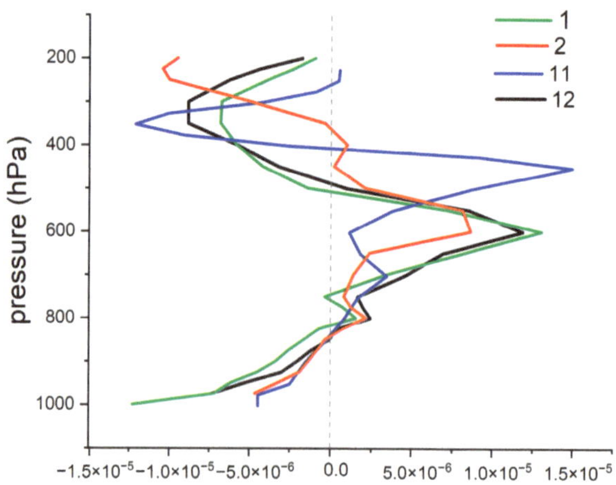

Figure 5. Average vertical divergence.

Based on ERA5 reanalysis data, Figure 6 shows the larger-scale annual mean wind and water vapor flux at 850 hPa. It is clear that the blocking of the terrain causes low-altitude westerly airflow upstream of the TP divided into two branches, and then, two air flows bypassing the Qinghai–Tibetan Plateau and the northwestern part of the Yunnan–

Guizhou Plateau converge toward the basin this is consistent with previous research results [29]. Meanwhile, the southern branch of the low-level westerly carries water vapor from the Indian subcontinent and the Bay of Bengal into the Sichuan Basin [30]. Low-level convergence maintains large-scale stable uplift. However, the moist airflow converging in the basin is obstructed by topography again. We used an ERA5 reanalysis data analysis of integrated water vapor transport (IVT) and wind in the vertical direction at the link between the Sichuan Basin and the Tibetan Plateau (Figure 7). The low-level airflow in the basin climbs toward the leeward slope of the Tibetan Plateau. The westerly flow, upon traversing the Tibetan Plateau, is considerably slowed down, which leads to pronounced divergence within the mid-tropospheric levels on the Tibetan Plateau's leeward side (Figures 5 and 7). The ascending low-level airflow encounters middle-layer divergence corresponding to the variation in the westerly, which suppresses its further ascent. Due to the westerly airflow blowing toward the basin, the climbing airflow, with a much lower wind speed than the westerlies, subsequently alters its upward direction to follow the westerly flow. From the perspective of IVT distribution (Figure 7, between longitude 105° E and 104° E), the transmission of water vapor to the Tibetan Plateau is blocked by terrain [31]. This work posits that within a large-scale environmental context, the topographical influence of the Tibetan Plateau causes low-level moist air currents to ascend from the Sichuan Basin towards the Tibetan Plateau, furnishing favorable conditions for cloud formation. However, divergence in the mid-latitude westerlies curtails the vertical development of clouds, confining them to the mid-troposphere. This also corresponds to the cloud top pressure statistics in Figure 4, which means that the obstruction of the Qinghai–Tibetan Plateau and the topographical variations between the Sichuan Basin and the eastern boundary of the Tibetan Plateau provides good conditions for the formation and maintenance of SLW.

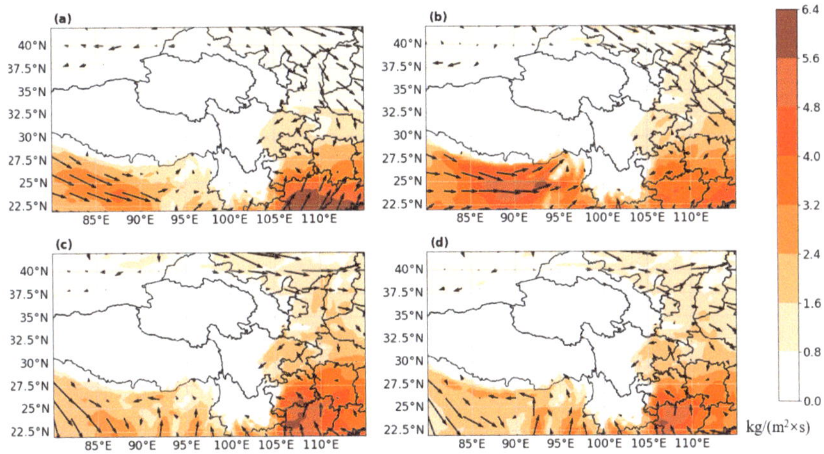

Figure 6. The average water vapor flux and the average winds at 850 hPa in (**a**) January, (**b**) February, (**c**) November, and (**d**) December.

The frequency of temperature inversions at different heights when SWCs exist in the region is analyzed, as shown in Table 1. In January, February, November, and December, the height with the highest temperature inversion frequency is 650 hPa, and the temperature inversion mainly occurs between 775 hPa and 550 hPa. Temperature inversion will make the upper air temperature higher than the lower air, causing the upper air to be less dense, hindering the upward movement of the lower air and inhibiting the convective movement of the air [32]. Temperature inversion not only makes the atmosphere stratification more stable but also reduces the vertical transport of water vapor and lowers the relative humidity of the atmosphere [33]. The inversion layer also weakens the radiative cooling of the ground so that the lower part of the stratus clouds will not be overcooled, maintaining the stability of the stratus clouds [34]. As a result, under inversion conditions, stratus clouds are more plausible to form and persist without being destroyed by convective motion and radiative cooling so that SLW can exist for a long time.

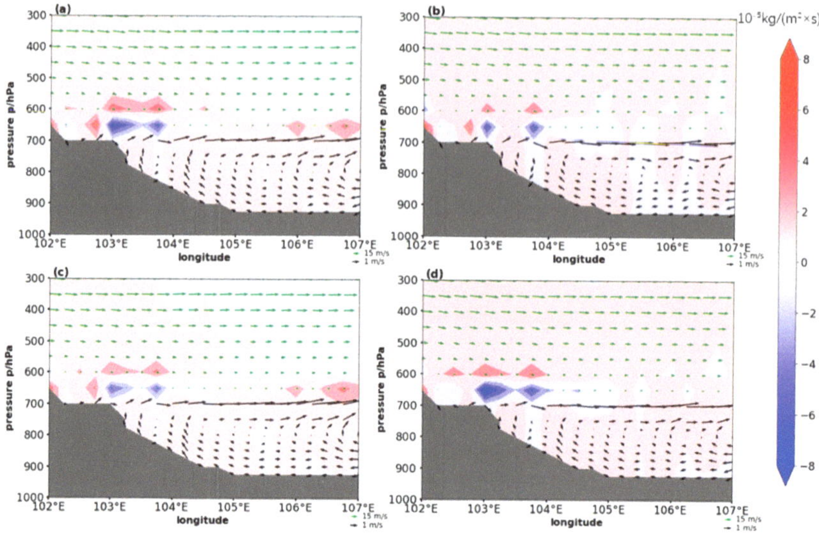

Figure 7. The average IVT (integrated water vapor transport) and the average winds in (**a**) January, (**b**) February, (**c**) November, and (**d**) December (the values > 8 or <−8, treated as the same value for interpolation, and positive values show that water vapor is transported in the direction of the wind, while negative values indicate the opposite).

Table 1. Statistical results of temperature inversion.

	Frequency of Temperature Inversion	**The Height and Frequency of Temperature Inversions**				
		600 hPa	**650 hPa**	**700 hPa**	**750 hPa**	**775 hPa**
January	0.795	7817	6007	4661	3288	1908
February	0.819	2566	3102	1934	779	289
November	0.687	5042	5432	5204	3005	1571
December	0.685	7169	8275	6138	4239	1939

4. Discussion

Yu proposed that the cloud amount and surface correlate negatively over the Yangtze River valley and Sichuan basin, a coupling between the surface temperature and the cloud radiative forcing, and the cloud can affect the surface temperature through changing cloud radiative forcing [24]. Although this study briefly analyzes the causes of supercooled water in typical areas in the winter of 2016, the frequency of SLW was also higher in September, October, and March in 2016. That is, in spring and summer, the supercooled water frequency is relatively low, which is likely to be related to surface temperature and the cloud–radiation feedback mechanism. Rising surface temperature leads to a decreasing static stability and relative humidity at low levels [24], which suppress the formation of supercooled stratus clouds. Meanwhile, the reduction of the supercooled stratus clouds would lead to cloud radiative cooling and favor further surface warming. In addition, the above-mentioned cloud feedback mechanism may also be related to the presence of more SLW at the top of the cloud. Supercooled cloud droplets can have a higher albedo and a lower absorption coefficient than warm cloud droplets, meaning that they reflect more SWR (shortwave radiation) and emit more LWR (longwave radiation) [35].

In the statistical results of this study, SLW are almost found in liquid-phase clouds. In addition to the considered inversion temperature and low surface temperature, which are beneficial to the maintenance of stratus clouds, the lack of ice phase particles should also be considered. The lack of ice particles in stratus clouds cannot consume SLW, let alone form precipitation, which may cause stratus clouds to maintain for a longer period of time under favorable weather conditions, resulting in SLW also persisting in the clouds for a long time [13].

5. Conclusions

Using Himawari-8 product, it was found that the high value area of SLW is located in the part where the Sichuan Basin borders the TP and that SLW occurs mainly in nimbostratus and altostratus clouds in these regions. This work analyzes the relevant causes from two perspectives: large-scale weather conditions and local dynamic conditions.

When obstructed by the TP, the low-level westerly airflow splits into southern and northern branches, with the southern branch carrying moisture and converging with the northern branch in the Sichuan Basin. In the Sichuan Basin, the converging airflow encounters the leeward slope of the eastern edge of the Qinghai–Tibetan Plateau, causing the moisture-laden air to ascend the Tibetan Plateau. That provides a favorable large-scale environment for the formation of nimbostratus and altostratus clouds. Simultaneously, the presence of localized temperature inversions contributes to atmospheric stability. Inversions occur at heights concentrated around 650 hPa, while the cloud-top pressure peaks are between 550 hPa and 600 hPa. This indicates that when ascending air forms clouds, it is hampered by inversions, resulting in weaker convection. During this ascent, the airflow encounters the stable mid-level westerlies. The combined effects of westerly-induced divergence, the guidance of ascending airflow by the westerlies toward the basin, and the presence of temperature inversions typically results in the formation of mid-level or low-level stratiform clouds and a few high-level clouds rather than convective clouds. In the winter, temperatures within the mid-to-low troposphere (specifically, cloud interior temperatures) typically range from approximately -20 °C to 0 °C. At these temperatures and altitudes, cloud particles have difficulty spontaneously freezing. As a result, the majority of SWCs remain in a liquid phase that allows for the substantial presence of SLW.

In studies on pollutants in the Sichuan Basin, weather patterns influenced by topography have been categorized into several main types [29]. Different weather patterns exhibit variations in cloud types and cloud heights as well as differences in the occurrence of SLW. This paper does not analyze the differences in cloud types and supercooled water under different weather conditions. Future research should employ multiple methods to conduct more detailed analyses of the occurrence of SLW occurrence under various weather conditions and topographical influences.

Author Contributions: Data curation, writing—original draft preparation, Q.W.; methodology, review, and editing, J.C.; conceptualization, funding acquisition, Y.Y. All authors have read and agreed to the published version of the manuscript.

Funding: This work is supported by National Science Foundation of China (42230604, 42075067), and the Open Research Program of the State Key Laboratory of Severe Weather (Grant No. 2023LASW-B25).

Data Availability Statement: The data presented in this study is available on request from the cor-responding author.

Acknowledgments: The datasets are provided by National Tibetan Plateau/Third Pole Environment Data Center (http://data.tpdc.ac.cn). (https://cstr.cn/18406.11.Atmos.tpdc.272822) (accessed on 8 September 2022). This study utilized ERA5 reanalysis data from the Copernicus Climate Change Service (C3S) at ECMWF. We thank the European Centre for Medium-Range Weather Forecasts (ECMWF) for providing the data. The views expressed are those of the authors and do not necessarily reflect those of the European Commission or ECMWF. This study also utilized MODIS Cloud (06_L2) data from the NASA MODIS Adaptive Processing System (MODAPS). We thank the MODIS Atmosphere Science Team for providing the data. The data were processed using the MODIS instrument on the Terra and Aqua platforms. The authors are solely responsible for the content and conclusions of this study. The data storage and analysis in this paper were completed on the supercomputing system at the Supercomputing Center of Nanjing University of Information Science & Technology.

Conflicts of Interest: The authors declare no conflicts of interest.

References

1. Wang, Z.; Letu, H.; Shang, H.; Zhao, C.; Li, J.; Ma, R. A Supercooled Water Cloud Detection Algorithm Using Himawari-8 Satellite Measurements. *J. Geophys. Res. Atmos.* **2019**, *124*, 2724–2738. [CrossRef]
2. Jiang, H.; Cotton, W.R.; Pinto, J.O.; Curry, J.A.; Weissbluth, M.J. Cloud Resolving Simulations of Mixed-Phase Arctic Stratus Observed during BASE: Sensitivity to Concentration of Ice Crystals and Large-Scale Heat and Moisture Advection. *J. Atmos. Sci.* **2000**, *57*, 2105–2117. [CrossRef]
3. Tabazadeh, A.; Djikaev, Y.S.; Reiss, H. Surface Crystallization of Supercooled Water in Clouds. *Proc. Natl. Acad. Sci. USA* **2002**, *99*, 15873–15878. [CrossRef] [PubMed]
4. Maciel, F.V.; Diao, M. The Transition from Supercooled Liquid Water to Ice Crystals in Mixed-Phase Clouds Based on Airborne In-Situ Observations. *Atmos. Meas. Tech. Discuss.* **2022**, *17*, 4843–4861. [CrossRef]
5. Bruintjes, R.T. A Review of Cloud Seeding Experiments to Enhance Precipitation and Some New Prospects. *Bull. Am. Meteorol. Soc.* **1999**, *80*, 805–820. [CrossRef]
6. Morrison, A.E.; Siems, S.T.; Manton, M.J. On a Natural Environment for Glaciogenic Cloud Seeding. *J. Appl. Meteorol. Climatol.* **2013**, *52*, 1097–1104. [CrossRef]
7. Li, D.; Zhao, C.; Yue, Z.; Liu, C.; Sun, Y.; Cohen, J.B. Response of Cloud and Precipitation Properties to Seeding at a Supercooled Cloud-Top Layer. *Earth Space Sci.* **2022**, *9*, e2021EA001791. [CrossRef]
8. Yang, Y.; Zhao, C.; Fu, J.; Cui, Y.; Dong, X.; Mai, R.; Xu, F. Response of Mixed-Phase Cloud Microphysical Properties to Cloud-Seeding Near Cloud Top Over Hebei, China. *Front. Environ. Sci.* **2022**, *10*, 865966. [CrossRef]
9. Brown, S.; Lengaigne, J.; Sharifi, N.; Pugh, M.; Moreau, C.; Dolatabadi, A.; Martinu, L.; Klemberg-Sapieha, J.E. Durability of Superhydrophobic Duplex Coating Systems for Aerospace Applications. *Surf. Coat. Technol.* **2020**, *401*, 126249. [CrossRef]
10. Li, S.; Paoli, R. Modeling of Ice Accretion over Aircraft Wings Using a Compressible OpenFOAM Solver. *Int. J. Aerosp. Eng.* **2019**, *2019*, e4864927. [CrossRef]
11. Chen, C.; Cotton, W.R. The Physics of the Marine Stratocumulus-Capped Mixed Layer. *J. Atmos. Sci.* **1987**, *44*, 2951–2977. [CrossRef]
12. Bodas-Salcedo, A.; Hill, P.G.; Furtado, K.; Williams, K.D.; Field, P.R.; Manners, J.C.; Hyder, P.; Kato, S. Large Contribution of Supercooled Liquid Clouds to the Solar Radiation Budget of the Southern Ocean. *J. Clim.* **2016**, *29*, 4213–4228. [CrossRef]
13. Murray, B.J.; Carslaw, K.S.; Field, P.R. Opinion: Cloud-Phase Climate Feedback and the Importance of Ice-Nucleating Particles. *Atmos. Chem. Phys.* **2021**, *21*, 665–679. [CrossRef]
14. Li, J.; Lv, Q.; Zhang, M.; Wang, T.; Kawamoto, K.; Chen, S.; Zhang, B. Effects of Atmospheric Dynamics and Aerosols on the Fraction of Supercooled Water Clouds. *Atmos. Chem. Phys.* **2017**, *17*, 1847–1863. [CrossRef]
15. Hou, T.; Lei, H.; Hu, Z. A Comparative Study of the Microstructure and Precipitation Mechanisms for Two Stratiform Clouds in China. *Atmos. Res.* **2010**, *96*, 447–460. [CrossRef]
16. Zhang, L.; Guo, X.; Zhao, T.; Xu, X.; Zheng, X.; Li, Y.; Luo, L.; Gui, K.; Zheng, Y.; Shu, Z. Effect of Large Topography on Atmospheric Environment in Sichuan Basin: A Climate Analysis Based on Changes in Atmospheric Visibility. *Front. Earth Sci.* **2022**, *10*, 997586. [CrossRef]
17. Zhao, R.; Chen, B.; Zhang, W.; Yang, S.; Xu, X. Formation Mechanisms of Persistent Extreme Precipitation Events over the Eastern Periphery of the Tibetan Plateau: Synoptic Conditions, Moisture Transport and the Effect of Steep Terrain. *Atmos. Res.* **2024**, *304*, 107341. [CrossRef]
18. National Tibetan Plateau Data Center. HUSI Letu East Asia-Pacific Cloud Macro- and Microphysical Properties Data Set (2016). 2022. Available online: https://data.tpdc.ac.cn/en/data/70c753d2-0471-4689-abab-798e2270c8b2/ (accessed on 8 September 2022).
19. Letu, H.; Yang, K.; Nakajima, T.Y.; Ishimoto, H.; Nagao, T.M.; Riedi, J.; Baran, A.J.; Ma, R.; Wang, T.; Shang, H.; et al. High-Resolution Retrieval of Cloud Microphysical Properties and Surface Solar Radiation Using Himawari-8/AHI next-Generation Geostationary Satellite. *Remote Sens. Environ.* **2020**, *239*, 111583. [CrossRef]
20. Hersbach, H.; Bell, B.; Berrisford, P.; Hirahara, S.; Horányi, A.; Muñoz-Sabater, J.; Nicolas, J.; Peubey, C.; Radu, R.; Schepers, D.; et al. The ERA5 Global Reanalysis. *Q. J. R. Meteorol. Soc.* **2020**, *146*, 1999–2049. [CrossRef]
21. Baum, B.A.; Menzel, W.P.; Frey, R.A.; Tobin, D.C.; Holz, R.E.; Ackerman, S.A.; Heidinger, A.K.; Yang, P. MODIS Cloud-Top Property Refinements for Collection 6. *J. Appl. Meteorol. Climatol.* **2012**, *51*, 1145–1163. [CrossRef]
22. Platnick, S.; King, M.D.; Meyer, K.G.; Wind, G.; Amarasinghe, N.; Marchant, B.; Arnold, G.T.; Zhang, Z.; Hubanks, P.A.; Ridgway, B.; et al. MODIS Cloud Optical Properties: User Guide for the Collection 6 Level-2 MOD06/MYD06 Product and Associated Level-3 Datasets. *Version* **2015**, *1*, 145.

23. Hu, Y.; Rodier, S.; Xu, K.; Sun, W.; Huang, J.; Lin, B.; Zhai, P.; Josset, D. Occurrence, Liquid Water Content, and Fraction of Supercooled Water Clouds from Combined CALIOP/IIR/MODIS Measurements. *J. Geophys. Res.* **2010**, *115*, 2009JD012384. [CrossRef]
24. Yu, R.; Wang, B.; Zhou, T. Climate Effects of the Deep Continental Stratus Clouds Generated by the Tibetan Plateau. *J. Clim.* **2004**, *17*, 2702–2713. [CrossRef]
25. Qi, H.; Cai, F.; Lei, G.; Cao, H.; Shi, F. The Response of Three Main Beach Types to Tropical Storms in South China. *Mar. Geol.* **2010**, *275*, 244–254. [CrossRef]
26. Yang, C.A.; Diao, M.; Gettelman, A.; Zhang, K.; Sun, J.; McFarquhar, G.; Wu, W. Ice and Supercooled Liquid Water Distributions Over the Southern Ocean Based on In Situ Observations and Climate Model Simulations. *J. Geophys. Res. Atmos.* **2021**, *126*, e2021JD036045. [CrossRef]
27. Yang, S.; Zou, X. Temperature Profiles and Lapse Rate Climatology in Altostratus and Nimbostratus Clouds Derived from GPS RO Data. *J. Clim.* **2013**, *26*, 6000–6014. [CrossRef]
28. Schiemann, R.; Lüthi, D.; Schär, C. Seasonality and Interannual Variability of the Westerly Jet in the Tibetan Plateau Region. *J. Clim.* **2009**, *22*, 2940–2957. [CrossRef]
29. Zhao, S.; Feng, T.; Tie, X.; Wang, Z. The Warming Tibetan Plateau Improves Winter Air Quality in the Sichuan Basin, China. *Atmos. Chem. Phys.* **2020**, *20*, 14873–14887. [CrossRef]
30. Liu, B.; Li, Y. Southwesterly Water Vapor Transport Induced by Tropical Cyclones over the Bay of Bengal during the South Asian Monsoon Transition Period. *J. Meteorol. Res.* **2022**, *36*, 140–153. [CrossRef]
31. Ning, G.; Yim, S.H.L.; Wang, S.; Duan, B.; Nie, C.; Yang, X.; Wang, J.; Shang, K. Synergistic Effects of Synoptic Weather Patterns and Topography on Air Quality: A Case of the Sichuan Basin of China. *Clim Dyn* **2019**, *53*, 6729–6744. [CrossRef]
32. Zhu, W.; Xu, X.; Zheng, J.; Yan, P.; Wang, Y.; Cai, W. The Characteristics of Abnormal Wintertime Pollution Events in the Jing-Jin-Ji Region and Its Relationships with Meteorological Factors. *Sci. Total Environ.* **2018**, *626*, 887–898. [CrossRef] [PubMed]
33. Enz, J.W.; Hofman, V.; Thostenson, A. Air Temperature Inversions. *NDSU Ext.* **2019**, *1705*, 1–16.
34. Zhang, J.; Zheng, Y.; Li, Z.; Xia, X.; Chen, H. A 17-Year Climatology of Temperature Inversions above Clouds over the ARM SGP Site: The Roles of Cloud Radiative Effects. *Atmos. Res.* **2020**, *237*, 104810. [CrossRef]
35. Ricaud, P.; Del Guasta, M.; Lupi, A.; Roehrig, R.; Bazile, E.; Durand, P.; Attié, J.-L.; Nicosia, A.; Grigioni, P. Supercooled Liquid Water Clouds Observed over Dome C, Antarctica: Temperature Sensitivity and Surface Radiation Impact. *Atmos. Chem. Phys. Discuss.* **2022**, *24*, 613–630. [CrossRef]

remote sensing

MDPI

Article

Vertical Distribution of Optical Turbulence at the Peak Terskol Observatory and Mount Kurapdag

Artem Y. Shikhovtsev [1,*], Chun Qing [2,3], Evgeniy A. Kopylov [4], Sergey A. Potanin [4,5,6] and Pavel G. Kovadlo [1]

1 Institute of Solar-Terrestrial Physics SB RAS, Lermontov St. 126a, 664033 Irkutsk, Russia; kovadlo2006@rambler.ru
2 Key Laboratory of Atmospheric Optics, Anhui Institute of Optics and Fine Mechanics, Chinese Academy of Sciences, Hefei 230031, China; chunqing@aiofm.ac.cn
3 Advanced Laser Technology Laboratory of Anhui Province, Hefei 230037, China
4 Institute of Astronomy, Russian Academy of Sciences, Leninsky Prospekt 14, 119017 Moscow, Russia; kopylov.phd@ya.ru (E.A.K.); sr.potanin@gmail.com (S.A.P.)
5 Faculty of Physics, Lomonosov Moscow State University, Leninskie Gory 1, 119991 Moscow, Russia
6 Sternberg Astronomical Institute, Lomonosov Moscow State University, Universitetskii Ave. 13, 119234 Moscow, Russia
* Correspondence: ashikhovtsev@iszf.irk.ru; Tel.: +7-(3952)-428265

Abstract: Atmospheric turbulence characteristics are essential in determining the quality of astronomical images and implementing adaptive optics systems. In this study, the vertical distributions of optical turbulence at the Peak Terskol observatory (43.27472°N 42.50083°E, 3127 m a.s.l.) using the Era-5 reanalysis and scintillation measurements are investigated. For the closest reanalysis grid node to the observatory, vertical profiles of the structural constant of the air refractive index turbulent fluctuations C_n^2 were obtained. The calculated $C_n^2(z)$ vertical profiles are compared with the vertical distribution of turbulence intensity obtained from tomographic measurements with a Shack–Hartmann sensor. The atmospheric coherence length at the location of Terskol Peak was estimated. Using a combination of atmospheric models and paramaterization schemes of turbulence, $C_n^2(z)$ profiles at Mt. Kurapdag were obtained. The values of atmospheric coherence length at Peak Terskol are compared with estimated values of this length at the ten astronomical sites, including Ali, Lenghu and Daocheng.

Keywords: optical turbulence; profiles; Fried parameter; air refractive index

Citation: Shikhovtsev, A.Y.; Qing, C.; Kopylov, E.A.; Potanin, S.A.; Kovadlo, P.G. Vertical Distribution of Optical Turbulence at the Peak Terskol Observatory and Mount Kurapdag. *Remote Sens.* **2024**, *16*, 2102. https://doi.org/10.3390/rs16122102

Academic Editors: Alexander Kokhanovsky and Riccardo Buccolieri

Received: 22 March 2024
Revised: 4 June 2024
Accepted: 7 June 2024
Published: 10 June 2024

1. Introduction

Adaptive optics systems that correct optical distortions in a wide field of view require information about the vertical profiles of optical turbulence and wind speed. The key characteristics of optical turbulence are the structural constant of the air refractive index turbulent fluctuations (C_n^2) and the atmospheric coherence length, which are often associated with the Fried parameter (r_0) and seeing [1,2]. These quantities determine the potential for achieving high angular resolution in astronomical observations.

There are several techniques for measuring the vertical profiles of optical turbulence. In astronomical observations, the vertical profiles above a telescope can be determined using a multidirectional Shack–Hartmann wavefront sensor [3]. Specifically, the vertical profiles of C_n^2 are calculated using the Solar Differential Image Motion Monitor (S-DIMM+) or Slope Detection and Ranging (SLODAR) techniques [4–12]. A comprehensive review of methods for measuring optical turbulence vertical profiles is provided in a recent paper [13]. Griffiths R. et al., present the results of an intercomparison of turbulence profiling instruments, including Stereo Scintillation Detection and Ranging (S-SCIDAR), DIMM, 24-hour Shack–Hartmann Image Motion Monitor (24hSHIMM), Ring-Image Next Generation Scintillation Sensor (RINGSS), and Multi-Aperture Scintillation Sensor-Differential Image Motion Monitor (MASS-DIMM) [13]. It is important to note that the comparison of these turbulence profiling instruments revealed that the largest deviations correspond to estimates of the isoplanatic angle (θ_0), the values of which are highly sensitive to turbulence in the upper atmospheric layers. These deviations, shown as scattered points, are due to limited accuracy in determining the strength of optical turbulence at different atmospheric levels and, presumably, differences in radiation propagation paths during measurements.

Measurement of optical turbulence vertical profiles is a difficult procedure [14]. Most often, the profile measurements are carried out only for a limited number of nights or days. In order to estimate the optical turbulence characteristics over long time intervals or, in other words, for a statistical ensemble, special atmospheric models are widely used [15–20]. The atmospheric models are applied to estimate the characteristics of mesoscale, micrometeorological and optical turbulence, on spatial scales comparable to the aperture diameters of ground-based telescopes. Below, a few recent papers demonstrating interesting results in this direction are discussed. A very interesting approach to predict optical turbulence was demonstrated by Cuevas et al. [21], who applied a combination of models to improve the forecasts of C_n^2 profiles and seeing at Paranal using the Weather Research and Forecasting model. Also, Macatangay et al. [22] utilized advanced numerical simulations with the Weather Research and Forecasting model to predict astronomical seeing. The meteorological model used by the authors reproduced variations very well, especially in comparison with smoothed DIMM data. Using the 21-yr European Centre for Medium-Range Weather Forecasts' fifth set of reanalysis (Era-5) data, the optical turbulence and the wind speed characteristics were simulated at the Lenghu site [23]. Analysis of the optical turbulence statistics shows that the calculated median value of seeing is slightly lower than measured seeing. In particular, the Era-5 derived value is $0.72''$ and the measured seeing is $0.75''$. Also, in our opinion, significant fundamental results in the optical turbulence diagnostics and forecasting with atmospheric models were achieved by English scientific teams, resulting in the development of new methods and tools for studying optical turbulence [24,25].

When estimating optical turbulence characteristics using atmospheric models, a crucial step is the selection of a small-scale turbulence parameterization scheme [26]. The correct selection and adjustment of the parameterization scheme to the atmospheric conditions of a particular site are necessary to obtain representative estimates of the key characteristics of optical turbulence. In mountainous regions like Tibet and the Caucasus, turbulence behavior is complex. The choice of parameterization scheme for mountain regions should be based on an analysis of preliminary measurement data. It is important to note that further research is needed to understand the dependencies in turbulence evolution and determine the extent to which these dependencies, specific to a given mountainous region, can be applied to other regions.

The present study focuses on the simulation of optical turbulence above the Peak Terskol Observatory (43.27472°N 42.50083°E, 3127 m a.s.l.) and Mt. Kurapdag (41.79609°N, 47.37428°E, 3553 m a.s.l.). These sites were chosen as reference locations due to the availability of special atmospheric measurements data [27]. Figure 1 illustrates the locations of the Terskol Peak Observatory, known for its high astroclimatic characteristics, Mount Kurapdag, characterized by low water vapor content [28], and the Big Telescope Alt-Azimuthal Telescope (BTA), the largest optical telescope with a non-segmented mirror in Eurasia [29]. Recent studies show that new suitable locations for ground-based telescopes include mountain regions of Dagestan. In this regard, the figure shows the location of the mountain village of Gunib as a place with potentially high characteristics.

In March 2023, the special experiments at the Peak Terskol Observatory were carried out to characterize the optical turbulence. Together with the measurements of optical turbulence profiles and meteorological characteristics in the surface layer, variations of $C_n^2(z)$ were simulated using a gradient method. Below, we discuss the results of estimating optical turbulence profiles using vertical shears of wind speed, mast meteorological measurements and measured height distributions of optical turbulence.

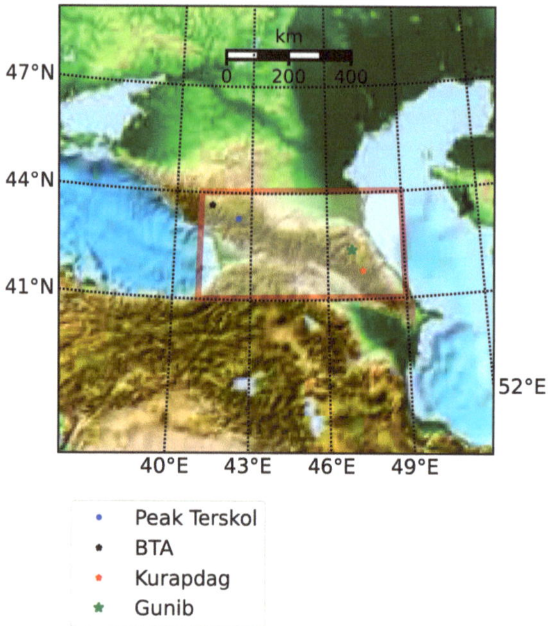

Figure 1. Sites of the Terskol Peak Observatory, Big Telescope Alt-Azimuthal Telescope (BTA), Gunib with high astroclimatic indicators and Mount Kurapdag with low water vapor content. Red rectangle shows the BTA region.

2. Data

This study aims to estimate the image quality at nighttime at the Terskol Peak Observatory, where optical turbulence monitoring is carried out, and at a new location, on the slope of Mt. Kurapdag, which is considered one of the reference sites for methodological tests. In this study, we solved a number of tasks. Firstly, using remote telescopic measurements, we determined vertical distributions of the optical turbulence intensity over the Peak Terskol observatory. Vertical distributions of turbulence intensity were obtained under conditions of good and mediocre image quality. Secondly, in order to estimate the vertical profiles of the structural constant of air refractive index turbulent fluctuations C_n^2, we used a gradient method, the foundations of which have been created by Tatarsky. Using the gradient method, the vertical profiles of optical turbulence over the Terskol Peak Observatory and Mt. Kurapdag were determined.

For the simulation of $C_n(z)$ vertical profiles, we used the European Centre for Medium-Range Weather Forecasts' fifth set of reanalysis (Era-5) data [30]. The Era-5 reanalysis contains observational data from around the world from 1940 to the present. Compared to previous databases, Era-5 demonstrates significant improvements in both spatial resolution and accuracy of atmospheric characteristics [31,32]. The reanalysis contains assimilated meteorological characteristics, including air temperature and wind speed components at various pressure levels. The meteorological characteristics are available with a flat horizontal resolution of about 30 km and a continuous 1-hour temporal resolution. In particular, for the grid node closest to the observatory, hourly values of air temperature and horizontal components of wind speed at various pressure surfaces were used.

3. Results

3.1. Vertical Distributions of Optical Turbulence from Shack–Hartmann Sensor Measurements at the Peak Terskol Observatory

In solar astronomy, Shack–Hartmann wavefront sensors with different fields of views are typically employed for estimated wavefront distortions [33]. The parameters of standard sensor depend on the optical turbulence characteristics. In this paper, we present the results of an observing campaign of the Shack–Hartmann sensor at the Peak Terskol Observatory in March 2023. We used the Shack–Hartmann sensor placed in the focal plane of a MEADE LX200 Telescope (Meade Instruments Corporation, Guangdong, China), which is installed inside the Schmidt–Cassegrain Telescope (Figure 2). The sensor is equipped with a high frame-rate camera. The subaperture size of the Shack–Hartmann sensor is 6×6 mm. The exposure time is 5 ms. The sampling is 65 frame/s. In the experiments, we analyzed 20 full time series and 10,000 hartmannograms were obtained during the night of 19 March 2023. Analysis of these measurement data made it possible to estimate and analyze the character of optical turbulence above the telescope.

Figure 2. The dome of the Schmidt–Cassegrain Telescope at the Peak Terskol Observatory.

Obtained data sets correspond to average seeing conditions suitable for astronomical observations and thus were selected for processing. Assuming that the turbulence structure obeys the Kolmogorov model, we can estimate the height distribution of optical turbulence strength. For these calculations, we used an expansion of the spatial spectrum of amplitude perturbations into modes. The model spectrum of the scintillation is related to the height distribution of optical turbulence strength:

$$s_k = \sum_{i=1}^{Nu} W_k(z_i) C_n^2(z_i) \Delta z_i, \quad (1)$$

where $W_k(z_i)$ is the weight function, which depends on the turbulent layer height and subaperture geometry, Δz_i is the turbulent layer thickness, Nu is the number of turbulent layers. Details of the method are described in the paper [34].

By solving this equation in terms of $C_n^2(z_i)\Delta z_i$, we obtain the vertical distributions of optical turbulence strength above the Peak Terskol Observatory. Night-time vertical distributions of optical turbulence strength $C_n^2 \Delta z$ at the Peak Terskol Observatory for different image qualities are shown in Figures 3 and 4. Image quality is estimated through the β parameter:

$$\beta = \frac{0.98\lambda}{r_0}, \quad (2)$$

where variance of the differential displacements of the solar subimages σ_α^2 is related to the Fried parameter r_0 by the following equation:

$$\sigma_\alpha^2 = K_i \lambda^2 r_0^{-5/3} D^{-1/3}, \quad (3)$$

where K_i is the numerical constant, λ is the light wavelength and D is the telescope

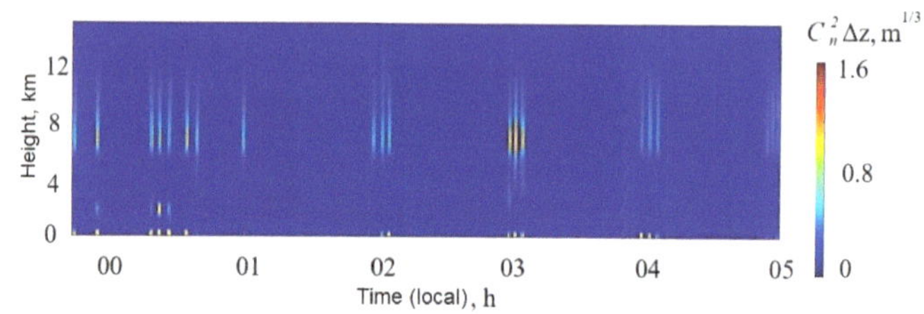

Figure 3. Night-time vertical distributions of optical turbulence strength $C_n^2\Delta z$ at the Peak Terskol Observatory for the average image quality, 19 March 2023. In the figure, we used a local time.

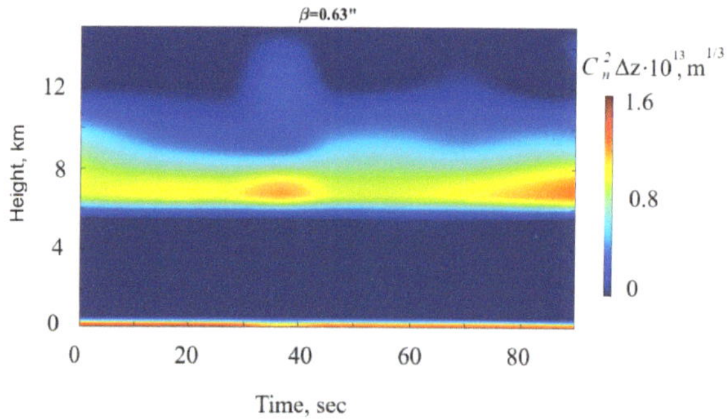

(**a**)

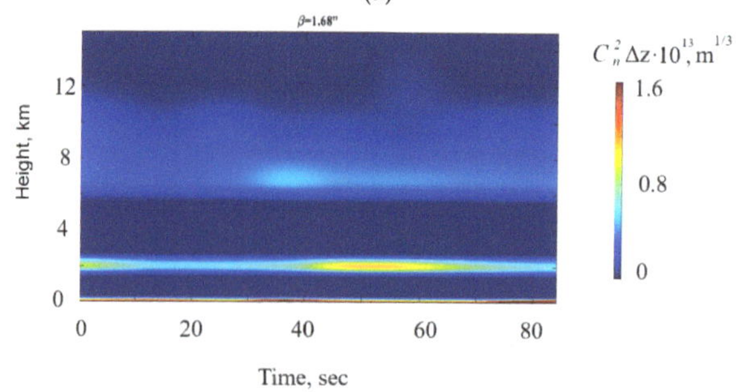

(**b**)

Figure 4. Night-time vertical distributions of optical turbulence strength $C_n^2\Delta z$ at the Peak Terskol Observatory for (**a**) the best (β = 0.63″) and (**b**) the worst atmospheric conditions (β = 1.68″), 19 March 2023. In the figure, we used a local time.

Analyzing these figures, we can see that pronounced turbulent layers are formed at heights of 7–8 km and 2–2.5 km above the ground. The turbulent layer at these heights leads to a significant decrease in image quality. For example, the parameter β determined by the integral C_n^2 over height increases from 0.63 to 1.68″. These changes are mainly due to generation of intensive turbulent fluctuations of air refractive index within these layers.

It is necessary to emphasize that, conditionally, we consider three different cases, including good seeing (β ranges from 0.5 to 1.0″), average seeing ($\beta \sim 1.2''$) and bad seeing (β ranges from 1.5 to 2.0″). Good seeing is associated with a thin lower turbulent layer (∼50–100 m) and the absence of a layer at heights of 2–2.5 km. With a thickness of the night boundary layer of 150–250 m, the seeing increases to ∼1.2″. Values higher than 1.5″ are often associated with the development of optical turbulence within the surface layer and an additional turbulent layer at heights of 2–2.5 km.

3.2. *Analysis of Sonic Measurements within the Atmospheric Surface Layer–Estimation of Optical Turbulence Strength* C_n^2

The lower part of the atmospheric boundary layer is known to make a major contribution to the formation of integral optical turbulence (along line of sight) and, as a consequence, to the resolving power of ground-based telescope [35,36]. Within this layer, variations in the intensity of optical turbulence are most poorly described, both in time and in height. At the very least, the accuracy of estimating and forecasting optical turbulence based on standard atmospheric models or limited optical remote measurement data is insufficient for planning observation times. In order to obtain the correct shape of optical turbulence profiles within the lowest atmospheric layers under the different atmospheric conditions, we analyzed measurement data carried out with a sonic anemometer. The sonic anemometer "Meteo-2" was placed on a 7-m meteorological mast at the Peak Terskol Observatory site (Figure 5). The observatory is located in the vicinity of the Elbrus, on a flat platform at an altitude of 3100 m above sea level.

(a)

Figure 5. *Cont.*

(b)

Figure 5. (a) The dome of Schmidt–Cassegrain Telescope and (b) Meteo-2 sonic anemometer.

Below, for calculations of atmospheric turbulence characteristics, we used long-period quasi-continuous measurement data of the sonic anemometer (from January 2023 to December 2023). Thanks to the high frequency measurements (10 Hz) of the wind field and the sonic temperature, we calculated a number of the turbulence characteristics, including surface values of C_n^2.

Based on the 1-year measurements with the sonic anemometer, the 3-min average values of C_n^2 were obtained [37]. In Figure 6, we present the number of nights of sonic measurements for each month. The fewest number of nights analyzed occurs in February–April. The largest number corresponds to January as well as the warm period of observations.

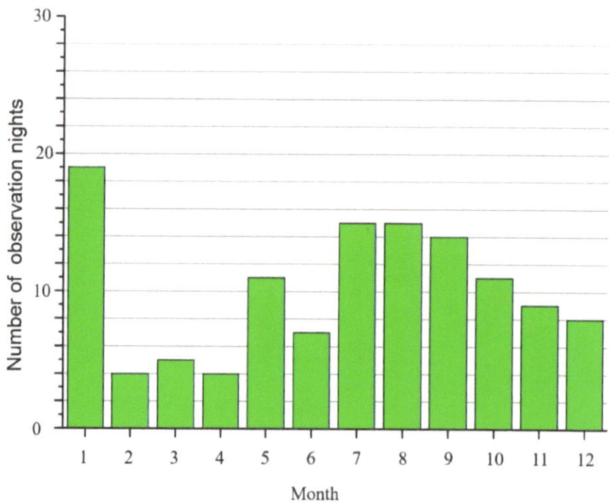

Figure 6. The number of nights of sonic measurements by month, 2023.

Distributions of measured values of C_n^2 within the surface layer of the atmosphere for different seasons at the Peak Terskol Observatory site are depicted in Figures 7 and 8.

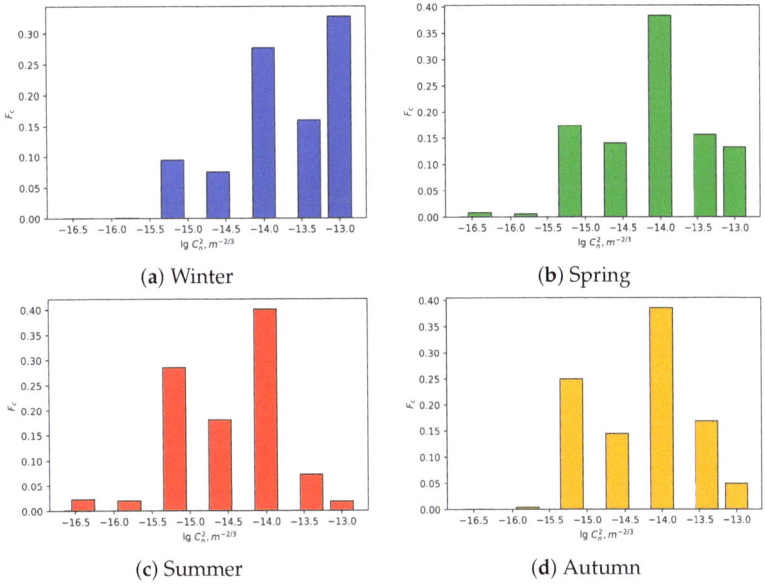

Figure 7. Distributions of measured values of C_n^2 within the surface layer of the atmosphere at the Peak Terskol Observatory site under average atmospheric conditions, 2023.

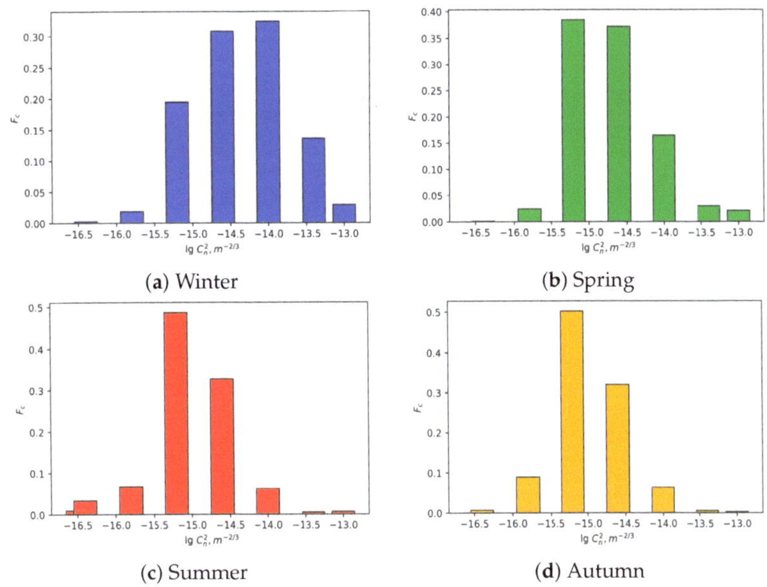

Figure 8. Distributions of measured values of C_n^2 within the surface layer of the atmosphere at the Peak Terskol Observatory site under clear sky and low intensity of optical turbulence, 2023.

Table 1 contains the calculated percentiles of C_n^2 in the surface layer of the atmosphere. Calculated values of C_n^2 obtained from the results of processing sonic measurement data are higher in comparison with estimations derived from optical measurement data under clear sky conditions. This fact is confirmed by previous studies [38]. Higher amplitudes of C_n^2 are largely determined by the influence of dynamic processes on the formation of small-scale turbulence in the surface layer of the atmosphere. Due to the fact that astronomical observations are carried out under good astro-optical conditions, when the intensity of optical turbulence along the line of sight is minimal, calculated values in the upper part of Table 1 differ from the statistics corresponding only to clear sky conditions. In this regard, we evaluated the statistics of surface turbulence for medium atmospheric conditions and low intensity of optical turbulence.

Table 1. Percentiles of C_n^2 in the surface layer of the atmosphere at the Peak Terskol Observatory, 2023.

Season	10%, $m^{-2/3}$	25%, $m^{-2/3}$	50%, $m^{-2/3}$	75%, $m^{-2/3}$
		Average atmospheric conditions		
Winter	$2.7 \cdot 10^{-15}$	$7.0 \cdot 10^{-15}$	$8.6 \cdot 10^{-15}$	$8.0 \cdot 10^{-14}$
Spring	$4.0 \cdot 10^{-16}$	$6.8 \cdot 10^{-16}$	$4.4 \cdot 10^{-15}$	$1.5 \cdot 10^{-14}$
Summer	$3.0 \cdot 10^{-16}$	$5.0 \cdot 10^{-16}$	$1.1 \cdot 10^{-15}$	$1.2 \cdot 10^{-14}$
Autumn	$3.4 \cdot 10^{-16}$	$6.0 \cdot 10^{-16}$	$3.8 \cdot 10^{-15}$	$3.5 \cdot 10^{-14}$
		Clear sky, light surface wind $V \leq 2.5$ m/s		
Winter	$5.0 \cdot 10^{-16}$	$1.1 \cdot 10^{-15}$	$2.6 \cdot 10^{-15}$	$1.1 \cdot 10^{-14}$
Spring	$4.8 \cdot 10^{-16}$	$5.7 \cdot 10^{-16}$	$1.1 \cdot 10^{-15}$	$3.1 \cdot 10^{-15}$
Summer	$1.4 \cdot 10^{-16}$	$4.2 \cdot 10^{-16}$	$5.3 \cdot 10^{-16}$	$2.2 \cdot 10^{-15}$
Autumn	$1.6 \cdot 10^{-16}$	$3.5 \cdot 10^{-16}$	$6.0 \cdot 10^{-16}$	$3.0 \cdot 10^{-15}$

Analysis of estimations shows that the intensity of turbulent fluctuations of the air refractive index has a pronounced seasonal dependence. The highest intensity of optical turbulence in the surface layer of the atmosphere is observed in winter. During this period, the median values of C_n^2, estimated for nighttime (from 0:00 to 6:00 local time), are equal to $8.6 \cdot 10^{-15}$ $m^{-2/3}$ and $2.6 \cdot 10^{-15}$ $m^{-2/3}$ for average atmospheric condition and under clear sky, respectively. In spring, we observe an increase in the non-dimensional number of atmospheric situations F_c with low intensity optical turbulence. At this time, during periods of high image quality, the median of C_n^2 is about $8.0 \cdot 10^{-16}$ $m^{-2/3}$.

In comparison with winter and spring, in summer and autumn, turbulence in the surface layer is significantly suppressed: the number of atmospheric situations with strong turbulence $C_n^2 \geq 10^{-14}$ $m^{-2/3}$ does not exceed 7%. We should emphasize that the largest number of cases with the best atmospheric conditions ($C_n^2 \geq 10^{-16}$ $m^{-2/3}$), occurs in summer and autumn. In this period, about 7–10% of the observing time corresponds to weak optical turbulence.

3.3. Vertical Distribution of Optical Turbulence Strength Estimated from Era-5 Reanalysis

For the simulation of the vertical profiles of optical turbulence, the gradient method was used. Correction of the vertical profiles was performed using sonic measurements and optical observations of image scintillation. The calculations were based on the following equation:

$$C_n^2(z) = \alpha L_0(z)^{4/3} M^2(z), \tag{4}$$

where α is the numerical constant, which is equal to 2.8. The parameter $M(z)$ is the vertical gradient of the air refractive index, and L_0 is the outer scale of turbulence.

The calculation of the air refractive index gradients $M(z)$ at different heights in the atmosphere is performed using the vertical gradients of the natural logarithm of the air potential temperature $\frac{\partial ln\theta}{\partial z}$:

$$M = \left(\frac{-79 \cdot 10^{-6} P}{T}\right) \frac{\partial ln\theta}{\partial z}, \tag{5}$$

where P is the atmospheric pressure, T is the air temperature, θ is the potential air temperature. The potential air temperature is related to the air temperature by the following formula:

$$\theta = T\left(\frac{1000}{P}\right)^{0.286}. \quad (6)$$

The key parameter is the outer scale of atmospheric turbulence L_0 [39]. In general, vertical changes of the outer scale L_0 can be determined through vertical gradients of the horizontal component of the wind velocity S:

$$L_0^{4/3} = \begin{cases} 0.1^{4/3} \cdot 10^{a+b\cdot S}, & \text{troposphere} \\ 0.1^{4/3} \cdot 10^{c+d\cdot S}, & \text{stratosphere} \end{cases}, \quad (7)$$

where

$$S = \left[\left(\frac{\partial u}{\partial z}\right)^2 + \left(\frac{\partial v}{\partial z}\right)^2\right]^{0.5}, \quad (8)$$

u and v are the horizontal components of wind velocity. Under conditions of weak turbulence, the coefficient a = 1.64, b = 42, c = 0.506 and d = 50.

Use of standard coefficients a = 1.64, b = 42, c = 0.506 and d = 50 gives mediocre results. In particular, we calculated a vertical profile $C_n^2(z)$ for 18–19 March 2023 (Figure 9a). By integrating this profile using Formula (9), we obtained the average estimation of the β parameter:

$$\beta_m = \frac{0.98\lambda}{\left(0.423 sec\alpha_s \left(\frac{2\pi}{\lambda}\right)^2 \int_0^H C_n^2(z)dz\right)^{-3/5}}, \quad (9)$$

where H is the height of the optically active atmosphere, α_s is the zenith angle, and λ is the wavelength of light. On average, the model value of β_m is higher than the measured β and the estimated C_n^2 in the upper atmospheric layers are higher than the expected values. In order to correct the vertical profile of $C_n^2(z)$, we calculated new parameterization coefficients a_{new} and b_{new} by finding the best-fitting model linking measured and model near-surface values of C_n^2.

Specifically, new coefficients were calculated by minimizing the standard deviation between the model values C_n^2 and the measured, sonic, values of $C_n^2(m)$:

$$\sum_i^{T_L} (C_{ni}^2(m) - C_{ni}^2)^2 \rightarrow min, \quad (10)$$

where $C_{ni}^2(m)$ is the measured value of C_n^2 in the surface layer, C_{ni}^2 is the model (calculated) value of $C_n^2(surf)$. Summation is performed over the entire time implementation, the length of which is T_L. Using this minimization condition, the weight coefficients a_{new} and b_{new} are estimated. New parametrization coefficients at the Terskol Peak Observatory differ significantly from standard values: a_{new} = 0.95 and b_{new} = 10. Figure 9b shows the corrected vertical profile of C_n^2 with the new parametrization coefficients. This profile corresponds to 19 March 2023.

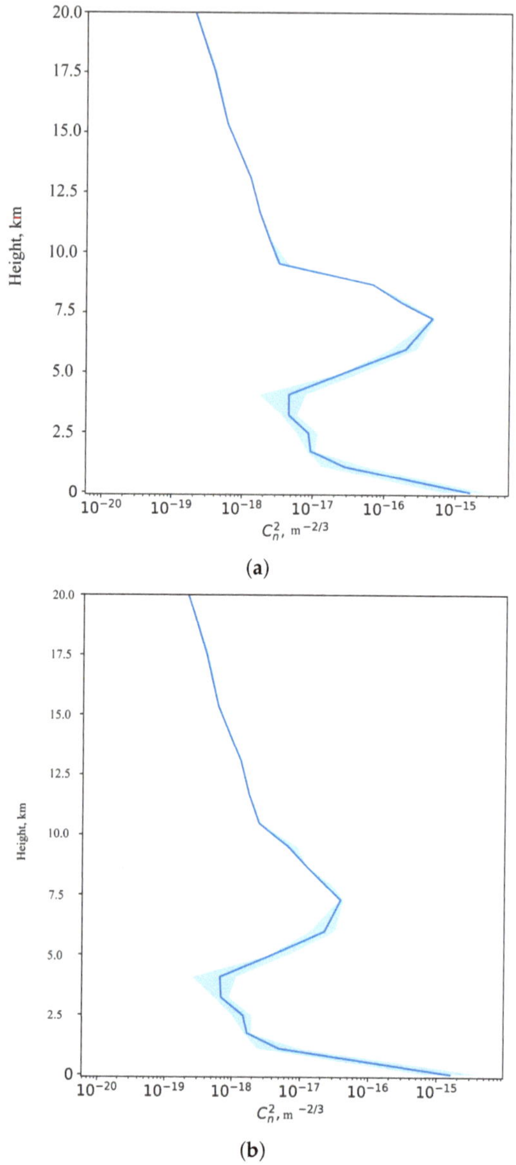

Figure 9. Calculated vertical profile of optical turbulence at the Peak Terskol Observatory, 18–19 March 2023. Shading corresponds to the interval between the first and the third quartiles estimated for all nights in March 2023. (**a**) $a = 1.64$, $b = 42$, $r_0 = 5.66$ cm, $\beta = 1.79$ arc sec; (**b**) $a_{new} = 0.95$, $b_{new} = 10$, $r_0 = 8.20$ cm, $\beta = 1.23$ arc sec.

One of the most important characteristics of optical turbulence is the height of the atmospheric boundary layer (ABL). In order to determine the ABL height, we used a well-known method based on the profile of the bulk Richardson number. The ABL height is defined as the height where the bulk Richardson number exceeds a critical threshold. From Figure 9, we can see that the new coefficients used to parameterize the night variations of C_n^2 made it possible to reduce the height of ABL. Obtained night-time heights of the atmospheric boundary layer range from 100 to 300 m. As a result, the estimated strength of optical turbulent fluctuations decreases more rapidly with height at night above Terskol Peak. We associate this decrease of turbulent fluctuations strength within the lower atmospheric layer above the Terskol Peak with the low heights of the atmospheric boundary layer at night.

Considering a large region, the minimum height of the atmospheric boundary layer is observed above Terskol Peak not only on individual nights, but also on average, over a long time period (Figure 10). The high quality of astronomical images at Terskol Peak is related to the low height of the atmospheric boundary layer. On some nights, its height is limited by 50–100 m. Mt Kurapdag and Big Telescope Alt-Azimuthal are located in worse atmospheric conditions, the mean night-time heights of the atmospheric boundary layer are ∼250–300 m and 250–350 m, respectively (Figure 10a). We believe that the observed character of vertical changes is in good agreement with theoretical and experimental data on the evolution of the atmospheric boundary layer.

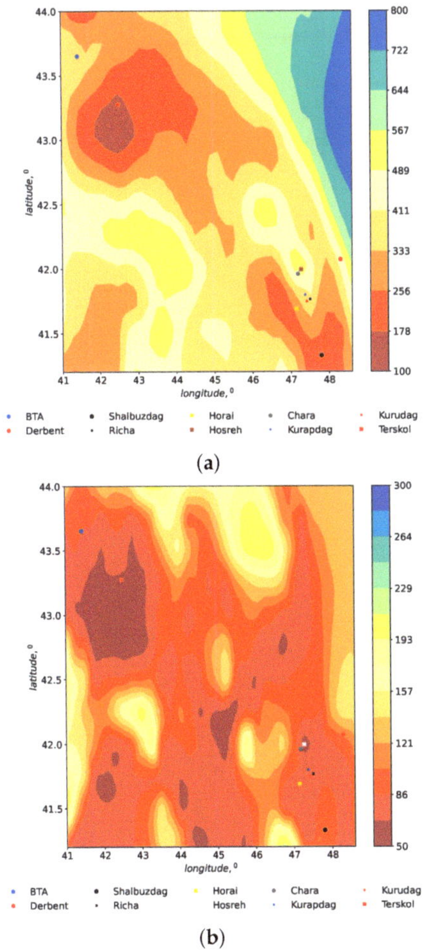

Figure 10. Distributions of atmospheric boundary layer height (in meters) within the Peak Terskol Observatory region. (**a**) Average atmospheric conditions, 2013–2023; (**b**) 19 March 2023.

The data of optical and meteorological measurements as well as the fields of wind speed and air temperature extracted from the reanalysis Era-5, served as the basis for calculation of the average profiles of optical turbulence above the observatory and Mt. Kurapdag. In Figure 11a we demonstrate calculated vertical profiles of optical turbulence above the Peak Terskol Observatory averaged over January 2023 and June 2023.

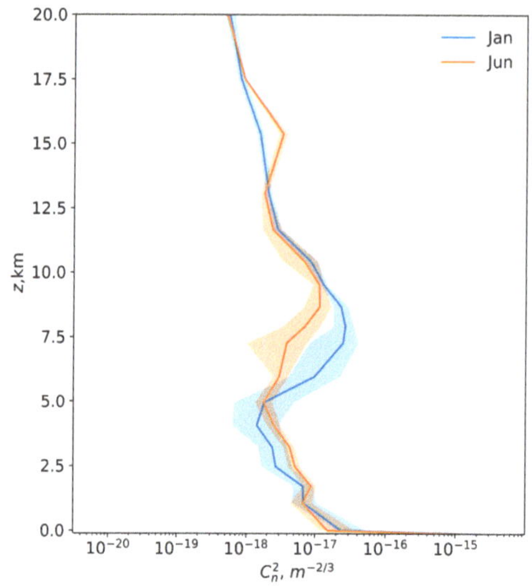

(**a**)

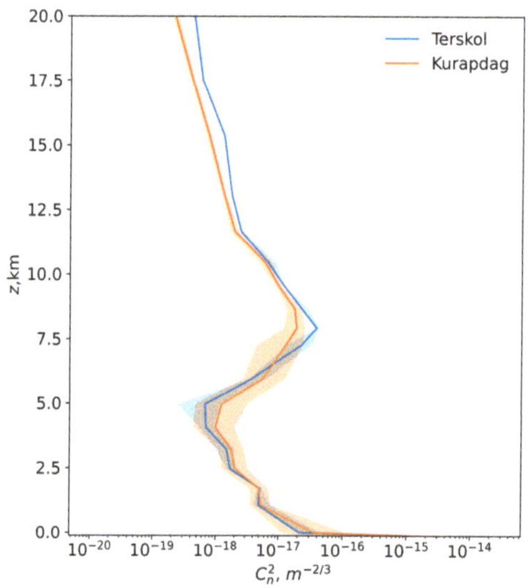

(**b**)

Figure 11. Night-time vertical profiles of optical turbulence above the Peak Terskol Observatory and Mt. Kurapdag. (**a**) Peak Terskol Observatory, 2023. The blue line corresponds to January. The orange line corresponds to June. The shading corresponds to the interval between the first and the third quartiles; (**b**) Mt. Kurapdag, 2013–2023. The blue line corresponds to the Peak Terskol Observatory. The orange line corresponds to Mt. Kurapdag. The shading corresponds to the interval between the first and the third quartiles.

Analyzing these profiles, it can be noted that the structure of turbulence in the upper atmospheric layers changes significantly in summer. This fact is indicated by deformations of the vertical profiles of optical turbulence. On the one hand, these deformations manifest in a greater vertical uniformity of optical turbulence and suppression of turbulent fluctuations in the surface layer of the atmosphere. On the other hand, the turbulence strength in the upper atmospheric layers decreases in summer.

Also, using Era-5 reanalysis data and identified patterns in optical turbulence behaviour at the Peak Terskol Observatory site, we calculated averaged vertical profiles of $C_n^2(z)$ above the Mt. Kurapdag. For comparison, Figure 11b shows vertical profiles above Terskol Peak and Kurapdag Mountain over the period of 2013–2023.

Analysis of these vertical profiles indicates the presence of an intense lower turbulence layer. It should be noted that within the surface layer, the intensity of turbulence, in terms of C_n^2, at the Mt. Kurapdag site exceeds by more than an order of magnitude the values of C_n^2 observed at the Peak Terskol Observatory. Moreover, at the Mt. Kurapdag site, turbulence in this atmospheric layer is characterized by high intensity regardless of the parameterization coefficients. The nature of the turbulence is largely due to strong vertical wind speed shears near the ground.

4. Estimation of the Fried Parameter at the Sites with Good Astroclimatic Conditions

Figure 12 displays the monthly averaged Fried parameter estimated using Era-5 reanalysis data at the ten astronomical sites. European Southern Observatory, Tibetan Plateau, and the Ethiopian Plateau are some of the most favorable sites. In contrast to the Fried parameter of the European Southern Observatory and Tibetan Plateau (TP) sites, the Ethiopian Plateau has a different pattern, possibly owing to different climate patterns. Throughout the year, the Ethiopian Plateau atmospheric coherence length is high in winter and low in summer. In winter, atmospheric coherence length at the Ethiopian Plateau is better than that at the ESO and TP sites, indicating that the Ethiopian Plateau has high potential conditions that are comparable to some of the best astronomical observatories in the world.

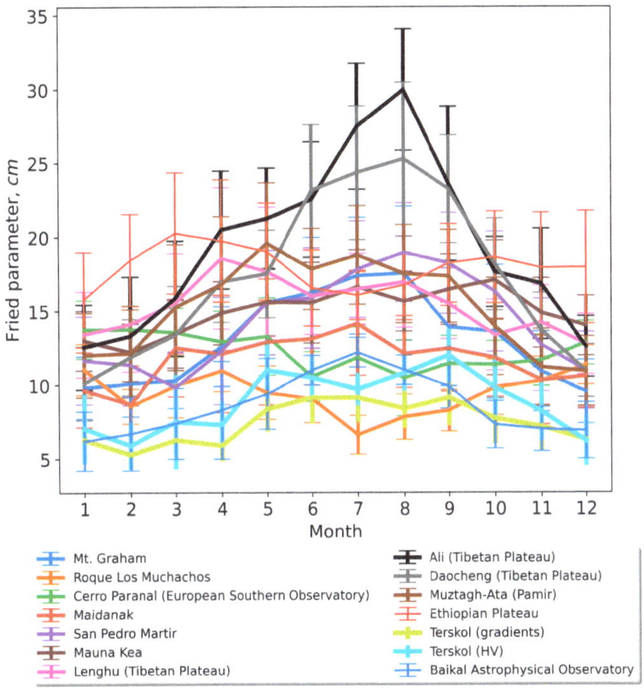

Figure 12. Variations of the Fried parameter $r_0 = 0.98\lambda/\beta$ calculated for the period 2018–2022.

The Estimated Fried parameter at the Ali site located in western China and the Tibetan plateau, one of the best astronomical sites, is about 20 cm in the best months. These sites are presently under study for astronomy [5]. The Era-5 derived values of the Fried parameter corresponding to spatial coherence length are consistent with measured values at the Maidanak Observatory [40].

We also compare the variation of the Fried parameter at the site of Peak Terskol Observatory. The figure also shows variations of the Fried parameter r_0 at the Terskol Peak Observatory. As demonstrated in the figure, the site of Terskol Peak has a high image quality, close to that of the best places in the world. In particular, in summer, β is about 1.0–1.2″. In terms of r_0, these values correspond to 8.3–10.0 cm. In addition to the average monthly values of the Fried parameter obtained using the gradient method, the figure shows the results of applying the Hufnagel–Vally (HV) climate model to describe vertical changes in C_n^2 [41]. The HV model is based on the processing of experimental data and, on average, describes the structure of optical turbulence well. Nevertheless, detailed information about the vertical distribution of optical turbulence cannot be reflected using this model. Analysis of the figure shows that the gradient model and the model give similar results in terms of r_0.

5. Discussion

In this paper, we present the results of studies on optical turbulence in the atmospheric boundary layer and free atmosphere, up to a height of 20 km. To determine the structural constant of turbulent fluctuations of the air refractive index, we refined the parameterization scheme for calculating this quantity. In the region of interest, this turbulent parameter was calculated by comparing reanalysis data, mast meteorological measurements, and observations of image scintillation.

Specifically, the processing of mast measurement data allowed us to estimate not only the statistics of C_n^2 in the atmospheric surface layer but also to select optimal parameterization coefficients that determine the contribution of wind speed shear instability to turbulence formation. By using these coefficients, we obtained more accurate estimates of C_n^2 values at different heights in the atmospheric boundary layer, enabling us to refine the shapes of the vertical profiles of optical turbulence. The validity of our approach, which utilizes constant parameterization coefficients within the atmospheric boundary layer, is supported by the fact that the structure of turbulence in this layer is primarily influenced by processes in the surface layer of the atmosphere, at least for long time periods.

The accuracy of our calculations is further confirmed by the results of estimating the height of the atmospheric boundary layer. Figure 10 shows the distributions of the atmospheric boundary layer height (in meters) within the Peak Terskol Observatory region. Analysis of these distributions reveals the formation of a relatively thin boundary layer above the Observatory, indicating that turbulence predominantly develops near the ground. During the measurements, the thickness of the atmospheric boundary layer was minimal compared to average atmospheric conditions (Figure 10a).

It can be noted that the gradient method outputs are in good agreement with the data of optical measurements of turbulence at different height levels in the atmosphere, at least for the Terskol Peak Observatory. Based on the selected profile calculation scheme for the Terskol Peak Observatory, we estimated the optical turbulence characteristics for other sites as well. Analysis of spatial distribution of the atmospheric boundary layer height, as well as calculated profiles of optical turbulence, can serve as the basis for choosing optimal site for a ground-based astronomical telescope. Taking into account the recent data on the spatial distribution of astroclimatic characteristics within this macroregion, the following areas suitable for optical telescopes are located: in mountain Dagestan, including Gunib, as well as Shalbuzdag and Peak Terskol Observatory. The location of Mt. Kurapdag is not good due to strong surface winds and strong surface optical turbulence, which is the main reason of high values of β (2–3″).

6. Conclusions

In this work we demonstrated that, using the gradient method and additional measurement data, we can estimate the vertical profiles of optical turbulence more accurately. The increase of accuracy of optical turbulence profiles within the lower atmosphere is due to using both measured values of C_n^2 and applying new parametrization coefficients between vertical shears of wind speed and model values of C_n^2. Moreover, the gradient method takes into account variations in the height of the atmospheric boundary layer estimated from reanalysis data. Estimations of the atmospheric boundary layer height and the vertical profiles of wind shears are important for obtaining smooth vertical profiles of C_n^2 and L_0. Above the atmospheric boundary layer, vertical profiles of optical turbulence are reconstructed only on the basis of vertical profiles of wind speed shears with parameterization coefficients $a = 1.64$ and $b = 42$, that describe C_n^2 variations well for weak turbulence. Thus, we can conclude that the hypotheses of optical turbulence embedded in the gradient method are valid as, in average, used atmospheric models converge to the measurements: the measured and model values of β are practically identical.

For the first time, by linking simulation outputs to measurement data, we have obtained representative optical turbulence vertical profiles for the sites of Peak Terskol Observatory and the Mt. Kurapdag. We have shown that the Kurapdag mountain should not be considered as a potential site for a new optical telescope.

Author Contributions: A.Y.S. and P.G.K. developed the methodology and performed the investigation; E.A.K. conducted the formal analysis; C.Q. and S.A.P. were engaged in measurements and data analysis. All authors have read and agreed to the published version of the manuscript.

Funding: Section 3 was supported by the Ministry of Science and Higher Education of the Russian Federation. The optical turbulence profiling was funded by RSF grant No. 23-72-00041.

Data Availability Statement: Data used are available on request from the corresponding author.

Acknowledgments: Measurements were verified using the Unique Research Facility Large Solar Vacuum Telescope http://ckp-rf.ru/usu/200615/ accessed on 1 February 2024.

Conflicts of Interest: The author declares no conflict of interest.

References

1. Coulman, C.E. Fundamental and applied aspects of astronomical "seeing". *Annu. Rev. Astron. Astrophys.* **1985**, *23*, 19–57. [CrossRef]
2. Tokovinin, A. The Elusive Nature of "Seeing". *Atmosphere* **2023**, *14*, 1694. [CrossRef]
3. Ran, X.; Zhang, L.; Rao, C. The AC-SLODAR: Measuring daytime normalized optical turbulence intensity distribution based on slope autocorrelation. *Mon. Not. R. Astron. Soc.* **2024**, *528*, 3981–3991. [CrossRef]
4. Hickson, P.; Ma, B.; Shang, Z.; Xue, S. Multistar turbulence monitor: A new technique to measure optical turbulence profiles. *Mon. Not. R. Astron. Soc.* **2019**, *485*, 2532–2545. [CrossRef]
5. Hickson, P.; Feng, L.; Hellemeier, J.A.; Shen, Z.; Xue, S.; Yao, Y.; Ma, B.; Chen, H.; Yang, R. Optical turbulence at Ali, China—Results from the first year of lunar scintillometer observation. *Mon. Not. R. Astron. Soc.* **2020**, *494*, 5992–6000. [CrossRef]
6. Wilson, R. SLODAR: Measuring optical turbulence altitude with a Shack–Hartmann wavefront sensor. *Mon. Not. R. Astron. Soc.* **2002**, *337*, 103–108. [CrossRef]
7. Deng, J.; Song, T.-F.; Liu, Y. A Review of Daytime Atmospheric Optical Turbulence Profile Detection Technology. *Chin. Astron. Astrophys.* **2023**, *47*, 257–284. [CrossRef]
8. Wang, Z.; Zhang, L.; Kong, L.; Bao, H.; Guo, Y.; Rao, X.; Zhong, L.; Zhu, L.; Rao, C. A modified S-DIMM+: Applying additional height grids for characterizing daytime seeing profiles. *Mon. Not. R. Astron. Soc.* **2018**, *478*, 1459–1467. [CrossRef]
9. Subramanian, S.K.; Rengaswamy, S. Forward modelling of turbulence strength profile estimation using S-DIMM+. *Proc. SPIE* **2023**, *12638*, 1263812. [CrossRef]
10. Shikhovtsev, A.Y. A Method of Determining Optical Turbulence Characteristics by the Line of Sight of an Astronomical Telescope. *Atmos. Ocean. Opt.* **2022**, *35*, 303–309. [CrossRef]
11. Joo, J.Y.; Ha, H.S.; Lee, J.H.; Kim, Y.S.; Butterley, T. SLODAR System Development for Vertical Atmospheric Disturbance Profiling at Geochang Observatory. *Curr. Opt. Photonics* **2024**, *8*, 30–37. [CrossRef]
12. Bennoui, F.; Bahloul, D. Estimation of the Atmospheric Turbulence Parameters Using the Angle-of-Arrival Covariance Function. *Atmos. Ocean. Opt.* **2023**, *36*, 569–577. [CrossRef]

13. Griffiths, R.; Bardou, L.; Butterley, T.; Osborn, J.; Wilson, R.; Bustos, E.; Tokovinin, A.; Le Louarn, M.; Otarola, A. A comparison of next-generation turbulence profiling instruments at Paranal. *Mon. Not. R. Astron. Soc.* **2024**, *529*, 320–330. [CrossRef]
14. Zhang, K.; Luo, T.; Wang, F.-F.; Sun, G.; Liu, Q.; Qing, C.; Li, X.-B.; Weng, N.-Q.; Zhu, W.-Y. Influence of low clouds on atmospheric refractive index structure constant based on radiosonde data. *Acta Phys. Sin.* **2022**, *71*, 089202. [CrossRef]
15. Bolbasova, L.A.; Shikhovtsev, A.Y.; Kopylov, E.A.; Selin, A.A.; Lukin, V.P.; Kovadlo, P.G. Daytime optical turbulence and wind speed distributions at the Baikal Astrophysical Observatory. *Mon. Not. R. Astron. Soc.* **2019**, *482*, 2619–2626. [CrossRef]
16. Qing, C.; Wu, X.; Li, X.; Luo, T.; Su, C.; Zhu, W. Mesoscale optical turbulence simulations above Tibetan Plateau: First attempt. *Opt. Express* **2020**, *28*, 4571–4586. [CrossRef]
17. Bi, C.; Qing, C.; Qian, X.; Luo, T.; Zhu, W.; Weng, N. Investigation of the Global Spatio-Temporal Characteristics of Astronomical Seeing. *Remote Sens.* **2023**, *15*, 2225. [CrossRef]
18. Yang, Q.; Wu, X.; Han, Y.; Chun, Q.; Wu, S.; Su, C.; Wu, P.; Luo, T.; Zhang, S. Estimating the astronomical seeing above Dome A using Polar WRF based on the Tatarskii equation. *Opt. Express* **2021**, *29*, 44000–44011. [CrossRef]
19. Quatresooz, F.; Griffiths, R.; Bardou, L.; Wilson, R.; Osborn, J.; Vanhoenacker-Janvier, D.; Claude, O. Continuous daytime and nighttime forecast of atmospheric optical turbulence from numerical weather prediction models. *Opt. Express* **2023**, *31*, 33850–33872. [CrossRef]
20. Wu, X.-Q.; Xiao, C.-Y.; Esamdin, A.; Xu, J.; Wang, Z.-W.; Xiao, L. Quantitative Analysis of Seeing with Height and Time at Muztagh-Ata Site Based on ERA5 Database. *Res. Astron. Astrophys.* **2024**, *24*, *1*, 015006. [CrossRef]
21. Cuevas, O.; Marin, J.C.; Blazquez, J.; Meyer, C. Combining C_n^2 models to forecast the optical turbulence at Paranal. *Mon. Not. R. Astron. Soc.* **2024**, *529*, 2208–2219. [CrossRef]
22. Macatangay, R.; Rattanasoon, S.; Butterley, T.; Bran, S.H.; Sonkaew, T.; Sukaum, B.; Sookjai, D.; Panya, M.; Supasri, T. Seeing and turbulence profile simulations over complex terrain at the Thai National Observatory using a chemistry-coupled regional forecasting model. *Mon. Not. R. Astron. Soc.* **2024**, *530*, 1414–1423. [CrossRef]
23. Bi, C.; Qing, C.; Qian, X.; Zhu, W.; Luo, T.; Li, X.; Cui, S.; Weng, N. Astroclimatic parameters characterization at lenghu site with ERA5 products. *Mon. Not. R. Astron. Soc.* **2024**, *527*, 4616–4631. [CrossRef]
24. Masciadri, E.; Jabouille, P. Improvements in the optical turbulence parameterization for 3D simulations in a region around a telescope. *Astron. Astrophys.* **2001**, *376*, 727–734. [CrossRef]
25. Turchi, A.; Masciadri, E.; Fini, L. Forecasting surface-layer atmospheric parameters at the Large Binocular Telescope site. *Mon. Not. R. Astron. Soc.* **2017**, *466*, 1925–1943. [CrossRef]
26. Shikhovtsev, A. Reference optical turbulence characteristics at the Large Solar Vacuum Telescope site. *Publ. Astron. Soc. Jpn.* **2024**, psae031. [CrossRef]
27. Bolbasova, L.A.; Kopylov, E.A. Long-Term Trends of Astroclimatic Parameters above the Terskol Observatory. *Atmosphere* **2023**, *14*, 1264. [CrossRef]
28. Khaikin, V.B.; Shikhovtsev, A.Y.; Shmagin, V.E.; Lebedev, M.K.; Kopylov, E.A.; Lukin, V.P.; Kovadlo, P.G. Eurasian Submillimeter Telescopes (ESMT) project. Possibility of submm image quality improvement using adaptive optics. *Zhurnal Radioelektroniki [J. Radio Electron.]* **2022**, *7*. (In Russian) [CrossRef]
29. Shikhovtsev, A.Y.; Kovadlo, P.G. Statistical estimations of the vapor content and optical thickness of the atmosphere using reanalysis and radiosonding data as applied to millimeter telescopes. *Optika Atmosfery i Okeana* **2024**, *37*, 169–175.
30. Hersbach, H.; Bell, B.; Berrisford, P.; Hirahara, S.; Horányi, A.; Muñoz-Sabater, J.; Nicolas, J.; Peubey, C.; Radu, R.; Schepers, D.; et al. The ERA5 global reanalysis. *Q. J. R. Meteorol. Soc.* **2020**, *146*, 1999–2049. [CrossRef]
31. Huang, J.; Yin, J.; Wang, M.; He, Q.; Guo, J.; Zhang, J.; Liang, X.; Xie, Y. Evaluation of Five Reanalysis Products with Radiosonde Observations Over the Central Taklimakan Desert During Summer. *Earth Space Sci.* **2021**, *8*, 2021EA001707. [CrossRef]
32. Rao, P.; Wang, F.; Yuah, X.; Liu, Y.; Jiao, Y. Evaluation and comparison of 11 sets of gridded precipitation products over the Qinghai-Tibet Plateau. *Atmos. Res.* **2024**, *302*, 107315. [CrossRef]
33. Townson, M.J.; Kellerer, A.; Saunter, C.D. Improved shift estimates on extended Shack–Hartmann wavefront sensor images. *Mon. Not. R. Astron. Soc.* **2015**, *452*, 4022–4028. [CrossRef]
34. Potanin, S.A.; Kornilov, M.-V.; Savvin, A.D.; Safonov, B.S.; Ibragimov, M.A.; Kopylov, E.A.; Nalivkin, M.A.; Shmagin, V.E.; Huy, L.X.; Thao, N.T. A Facility for the Study of Atmospheric Parameters Based on the Shack–Hartmann Sensor. *Astrophys. Bull.* **2022**, *77*, 214–221. [CrossRef]
35. Kornilov, V.; Safonov, B.; Kornilov, M.; Shatsky, N.; Voziakova, O.; Potanin, S.; Gorbunov, I.; Senik, V.; Cheryasov, D. Study on atmospheric optical turbulence above mount Shatdzhatmaz in 2007–2013. *Publ. Astron. Soc. Pac.* **2014**, *126*, 482–495. [CrossRef]
36. Panchuk, V.E.; Afanas'ev, V.L. Astroclimate of Northern Caucasus-myths and reality. *Astrophys. Bull.* **2011**, *66*, 233–254. [CrossRef]
37. Odintsov, S.L.; Gladkikh, V.A.; Kamardin, A.P.; Nevzorova, I.V. Determination of the Structural Characteristic of the Refractive Index of Optical Waves in the Atmospheric Boundary Layer with Remote Acoustic Sounding Facilities. *Atmosphere* **2019**, *10*, 711. [CrossRef]
38. Lukin, V.P.; Botygina, N.N.; Gladkikh, V.A.; Emaleev, O.N.; Konyaev, P.A.; Odintsov, S.L.; Torgaev, A.V. Joint measurements of atmospheric turbulence level with optical and acoustic meters. *Atmos. Ocean. Opt.* **2015**, *28*, 254–257. [CrossRef]
39. Rao, R. Effect of Outer Scale of Atmospheric Turbulence on Imaging Resolution of Large Telescopes. *Guangxue Xuebao/Acta Opt. Sin.* **2023**, *43*, 2400001.

40. Tillayev, Y.; Azimov, A.; Ehgamberdiev, S.; Ilyasov, S. Astronomical Seeing and Meteorological Parameters at Maidanak Observatory. *Atmosphere* **2023**, *14*, 199. [CrossRef]
41. Mahmood, D.A.; Naif, S.S.; Al-Jiboori, M.H.; Al-Rbayee, T. Improving Hufnagel-Andrews-Phillips model for prediction C_n^2 using empirical wind speed profiles. *J. Atmos. Sol. Terr. Phys.* **2022**, *240*, 105952. [CrossRef]

remote sensing

MDPI

Article

Retrieval of At-Surface Upwelling Radiance and Albedo by Parameterizing Cloud Scattering and Transmittance over Rugged Terrain

Junru Jia [1,2], Massimo Menenti [1,3], Li Jia [1,*], Qiting Chen [1] and Anlun Xu [4]

1 Key Laboratory of Remote Sensing and Digital Earth, Aerospace Information Research Institute, Chinese Academy of Sciences, Beijing 100101, China; jiajr@radi.ac.cn (J.J.); m.menenti@radi.ac.cn (M.M.); chenqt@aircas.ac.cn (Q.C.)
2 University of Chinese Academy of Sciences, Beijing 100049, China
3 Faculty of Civil Engineering and Earth Sciences, Delft University of Technology, 2628 CN Delft, The Netherlands
4 Dali National Climate Observatory, Dali 671003, China; xualun@126.com
* Correspondence: jiali@aircas.ac.cn; Tel.: +86-10-6480-7982

Abstract: Accurate and continuous estimation of surface albedo is vital for assessing and understanding land–surface–atmosphere interactions. We developed a method for estimating instantaneous all-sky at-surface shortwave upwelling radiance and albedo over the Tibetan Plateau. The method accounts for the complex interplay of topography and atmospheric interactions and aims to mitigate the occurrence of data gaps. Employing an RTLSR-kernel-driven model, we retrieved surface shortwave albedo with a 1 km resolution, incorporating direct, isotropic diffuse; circumsolar diffuse; and surrounding terrain irradiance into the all-sky solar surface irradiance. The at-surface upwelling radiance and surface shortwave albedo estimates were in satisfactory agreement with ground observations at four stations in the Tibetan Plateau, with RMSE values of 56.5 W/m^2 and 0.0422, 67.6 W/m^2 and 0.0545, 98.6 W/m^2 and 0.0992, and 78.0 98.6 W/m^2 and 0.639. This comparison indicated an improved accuracy of at-surface upwelling radiance and surface albedo and significantly reduced data gaps. Valid observations increased substantially in comparison to the MCD43A2 data product, with the new method achieving an increase ranging from 40% to 200% at the four stations. Our study demonstrates that by integrating terrain, cloud properties, and radiative transfer modeling, the accuracy and completeness of retrieved surface albedo and radiance in complex terrains can be effectively improved.

Keywords: land surface albedo; at-surface upwelling radiance; Tibetan Plateau; rugged terrain

Citation: Jia, J.; Menenti, M.; Jia, L.; Chen, Q.; Xu, A. Retrieval of At-Surface Upwelling Radiance and Albedo by Parameterizing Cloud Scattering and Transmittance over Rugged Terrain. *Remote Sens.* **2024**, *16*, 1723. https://doi.org/10.3390/rs16101723

Academic Editor: Yi Luo

Received: 4 March 2024
Revised: 15 April 2024
Accepted: 16 April 2024
Published: 13 May 2024

1. Introduction

The land surface albedo is defined as the proportion of the total incident solar irradiance reflected in the 0.2–5 μm wavelength range in the viewing hemisphere [1–4]. It is a primary factor controlling the land surface radiation balance and significantly influences water and heat balances at the interface of the Earth's surface and the atmosphere. The spatial and temporal variation of land surface albedo introduces perturbations in radiative forcing, which drives land cover conditions, regional to global climate and environmental changes [3]. Accordingly, it is a crucial land surface property in both climate and land surface processes models [5]. The literature documents the accuracy of albedo required for climate modeling, e.g., ±0.05 according to [3] and ±0.02 according to [6]. The Tibetan Plateau is the world's highest plateau and mountain area, with a mean elevation of approximately 4500 m.a.s.l. (meters above sea level). It is the third-largest reservoir of water ice in the world and is often referred to as the Water Tower of Asia. As a high-elevation heat and water source, the energy balance of the Tibetan Plateau significantly influences both regional and global weather and climate [7]. Investigating the spatial and temporal variations of land surface albedo and better understanding the intricate climate–surface albedo feedbacks in the Tibetan Plateau is of great importance in the context of regional and global climate [8].

(a) Current approaches to retrieving land surface albedo and limitations due to cloud cover

Currently, satellite data offer a practical technique to estimate albedo at larger scales, and numerous algorithms and products on surface albedo have been developed over several decades. Since the launch of the Terra (1999) and Aqua (2002) satellites carrying the Moderate Resolution Imaging Spectroradiometer (MODIS), the estimation of land–surface albedo using remote sensing data has advanced significantly [9]. The MODIS retrieval system of the MCD43 product generates multiple BRDF data products [10]. The most accurate and reliable retrieval description of the BRDF is obtained by fitting the Ross Thick-Li Sparse (RTLSR) BRDF model to available cloud-free observations within each 16 day window (referred to as "full model inversion").

The retrieval of land surface albedo from polar-orbiting satellite data includes three main steps: estimation of the at-surface reflectance, modeling of the bidirectional reflectance distribution function (BRDF), and narrow-to-broadband conversion of albedo [9,11]. MCD43A Version 6 (V006.1) is the newest daily MODIS surface albedo product. The MODIS albedo algorithm uses cloudless atmospherically corrected bidirectional reflectance data (BRF, i.e., the MODIS level-2 surface reflectance product) as input and takes into account the quality flag to select observations in the spectral bands 1 through 7 (i.e., from 620 nm to 2155 nm) of MODIS level 1B data, which has been calibrated and geolocated using at-aperture radiances [9] The MODIS surface reflectance product is corrected for the effects of atmospheric gases and aerosol only, i.e., it applies to clear sky conditions over flat terrain.

A semiempirical kernel-driven model, i.e., the RossThick-LiSparseR (RTLSR) kernel-driven BRDF model, is used in the MODIS BRDF/albedo product algorithm to describe the BRDF as a linear combination of three kernels: isotropic scattering, volume scattering, and geometric optics scattering [10]. This requires three kernel-weighting coefficients, which are least-squares estimates using clear-sky at-surface reflectance. At least seven clear-sky observations during a 16-day period are required in the full model inversion, in which a linear RTLSR kernel-driven bidirectional reflectance distribution function (BRDF) model is fitted using least squares to obtain the BRDF parameters [11,12]. The semiempirical kernel-driven model reconstructs the BRDF for any illumination and view geometry. If the results of the full model inversion are unsatisfactory, or if the number of clear-sky observations are insufficient ($4 \leq$ number of observations < 7 in the MODIS Collection V004 and V005 BRDF/albedo product, and $2 \leq$ number of observations < 7 in the MODIS Collection V006 BRDF/albedo product), the magnitude inversion method is used. This magnitude inversion uses the available multi-angle clear-sky observations to adjust the a priori BRDF parameters to update the RTSLR parameters to be consistent with the actual surface conditions [11,12]. If the results of the magnitude inversion are unsatisfactory, or if the number of observations are less than the minimum number of observations required as mentioned above, no retrieval results are available, and the pixel is assigned a value using a global BRDF database.

Despite using 16 days of accumulated surface reflectance data, it is at times not possible to gather the sufficient clear-sky multi-angular observations of surface reflectance to fulfill the minimum requirements of the BRDF model due to cloud coverage [13]. From 2000 to 2009, annually, approximately 20% to 40% of global land pixels lack daily valid retrievals sufficient to meet the requirements of the BRDF model [14]. Particularly in the Tibetan Plateau region, the MODIS albedo product struggles to gather sufficient valid retrievals. Due to cloud cover, atmospheric contamination, sensor malfunction and variations in surface topography, there are many gaps in the land surface albedo data set [9,15].

Cloud cover is the primary factor that impacts the retrieval of albedo, as it limits the number of clear-sky observations. This scarcity of clear-sky multi-angular observations affects the quality of albedo retrieval, sometimes even preventing successful retrieval.

Given the importance of spatial and temporal continuity of the observations required for various applications, algorithms designed for gap-filling are essential to filling in missing data. Enhancing input data quality for surface albedo estimation and filling gaps in existing albedo products are key strategies to improve current albedo datasets [14,16–19].

(b) Terrain effects on irradiance and reflected radiance

Terrain effects have a significant impact on both incoming solar radiance and outgoing reflected radiance, ultimately playing a substantial role in the estimation of surface albedo. Surface albedo changes with the illumination and viewing geometries, particularly in mountain areas, due to the combination of effects at the observed target and on the surrounding terrain. The change in surface albedo with the direct and diffuse irradiance, the evolution of the surface cover type, especially rapid and short-lived changes due to snowfall, and the topographic effect regulating the direct, diffuse and terrain-reflected radiation [20] need to be accounted for in the retrieval of land surface albedo.

Land surface albedo can then be retrieved once the BRDF is established. The semiempirical kernel-driven model is used to describe how reflectance varies with illumination and view angles, which is applied to each spectral channel of the instrument. There has been growing consideration of terrain effects in the remote-sensing-based surface albedo algorithms in recent years. Gao et al. [21] proposed an improved retrieval method by adding a topographic factor to the MODIS Bidirectional Reflectance Anisotropy of Land Surface model. However, this method did not address the correction of land surface reflectance in obstructed pixels in rugged areas and ignored the reflected sky diffuse irradiance. Wen et al. [22] proposed an extended multi-sensor combined BRDF inversion model to characterize subpixel-scale effects, which performed well compared to the ground measurement.

(c) All-sky retrieval of surface albedo

In some studies [23–25], researchers delved deeper into estimating downwelling solar radiation in alpine regions under all sky conditions and worked towards improving the accuracy of retrieved surface albedo. Efficiently and accurately estimating radiation under cloudy skies is crucial due to the substantial influence of cloud cover on the data acquired by optical remote sensors. Previous studies have used numerous satellite-based parameters as inputs for complex radiative transfer models [26,27]. These models incorporate detailed descriptions of physical processes, but they are resource-intensive due to numerous variables and the high computational burden limits the spatial resolution of the outcomes [28]. A possible solution is to use lookup tables generated offline. LUTs are created through the application of atmospheric radiative transfer models across numerous scenarios. These model outputs are then resampled to develop LUTs for downwelling solar radiation estimation. The advantage of LUT-based methods is their avoidance of complex input parameters at the retrieval stage [29]. As regards the treatment of radiative transfer under cloudy conditions, parameterization methods have been established [30–32]. Many parameterization schemes are spectrum-dependent [33,34]. Van Laake and Sanchez-Azofeifa et al. [33] applied the cloud parameterization model of Stephens et al. [30] to estimate photosynthetically active radiation (PAR) using MODIS atmospheric products. Huang et al. [35] used a broadband parameterization to estimate solar surface irradiance (SSI) inspired by the cloud parameterization model of Fu et al. [32].

Modern techniques and methodologies, like those used in the Satellite Application Facility on Climate Monitoring (CM-SAF) all-sky data, effectively simulate cloud reflectance and attenuation effects on radiative flux, despite cloud complex microphysical and optical properties. Several estimation approaches exist, including empirical relationships, physical principles, optimization techniques, and machine learning models [36–41].

(d) Overview of the new approach

Recognizing the limitations of current approaches, especially under all-sky and rugged terrain conditions, this study aims to introduce an innovative method for estimating surface albedo and at-surface upwelling radiance over the Tibetan Plateau using MODIS data. We have been able to estimate the components of surface solar irradiance and of upwelling radiance: direct, isotropic and circumsolar diffuse, and terrain irradiance over rugged terrain under cloudy sky [42]. The approach proposed in this paper aimed to substantially reduce the occurrence of data gaps and improve the quality of the estimated albedo. Notably, in an earlier study, we focused on the estimation of the surface solar irradiance (SSI) and its four components, i.e., direct, isotropic diffuse, circumsolar diffuse, and surrounding terrain irradiance, factoring in the actual terrain geometry derived from DEM data [42]. Under cloudy conditions, our model integrates cloud properties, such as cloud cover fraction and cloud reflectance and transmittance, to improve the estimation of SSI and at-surface upwelling radiance. This method allows for an estimation of bidirectional surface reflectance even under cloudy conditions, improving the continuity of input data for BRDF modelling and the retrieval of surface albedo. As in the MODIS algorithm, the anisotropy of the bidirectional reflectance of the land surface is described using the RTLSR-kernel-driven BRDF model in our study. The semiempirical kernel-driven model is applied to reconstruct the BRDF, but the reflectances are calculated for the four components of surface solar irradiance under all sky conditions. Ultimately, the surface albedo is calculated as the ratio of the total reflected radiance to the total incident solar irradiance, aligning with the fundamental definition of surface albedo. Our method enables daily albedo estimation in a region critically important for global climate dynamics.

2. Description of Study Area and Data

2.1. Study Area and In Situ Radiation Observations

The Tibetan Plateau is encompassed by formidable mountain ranges: the Inner Himalayas to the south, the Arkin Qilian Mountains to the north, the Karakorum Mountains to the west, and the Hengduan Mountains to the east, as illustrated in Figure 1. This plateau, the largest in China and the highest globally, spans an approximate area of 2.05 million km^2, accounting for a quarter of China's landmass. The average elevation of the plateau exceeds 4500 m above sea level (m.a.s.l.).

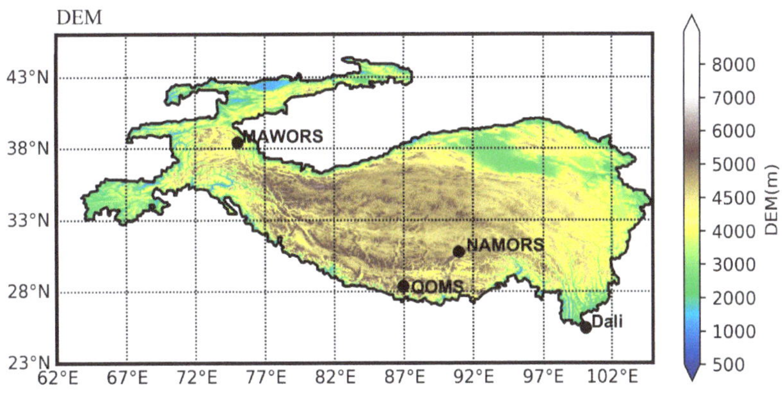

Figure 1. A DEM map of the third pole showing the study area and locations of the ground stations (black dots).

The ground measurements were obtained from radiation balance stations located on the Tibetan Plateau (Table 1 and Figure 1), namely the NamCo Monitoring and Research for Multisphere Interactions Station (NAMORS), the Dali Station, Muztagh Ata for Westerly Environment Observation and Research Station (MWWORS), and Qomolangma Atmospheric and Environmental Observation and Research Station (QOMS). The four stations—with MAWORS in the westerly zone and QOMS, NAMORS, and Dali in the monsoon zone—are geographically and climatically positioned to represent different terrain conditions, ecosystem types (land cover types, ranging from desert, alpine steppe to cropland), and climate zones on the Tibetan Plateau (Table 1). Each station provides unique insights into different atmospheric and environmental aspects, enhancing the overall representativeness of the study. MAWORS is located in the westerly climate zone and has a dry and cold climate. This site provides essential data on the impact of westerly driven weather [43]. The QOMS is located in a semi-arid plateau monsoon climate [44] at 4298 m.a.s.l. and 30 km from the northern boundary of Mt. Everest. It is dominated by alpine Gobi desert, relatively flat and open terrain characterized mainly by bare ground and sparse and short vegetation. MAWORS is located close to Muztagh Mountain and Karakuli Lake and lies at 3668 m.a.s.l. The NAMORS is situated on the southeast bank of Namco Lake, surrounded by the Nyenchen Tanglha Mountains and has an elevation of 4730 m.a.s.l. It is classified as an alpine meadow and belongs to the semi-arid plateau monsoon climate zone. The Dali Station, situated at 1999 m.a.s.l., nestles between the Cangshan Mountains to the west and Erhai Lake to the east. This location allows sampling a crucial water vapor transport channel in Hengduan Moutains on the southeastern edge of the Tibetan Plateau. Dali station is located in the windward zone of the Southwest Monsoon of the Bay of Bengal, belonging to the subtropical highland monsoon climate type. Detailed information about the locations and the instruments for Dali NAMORS, MWWORS, and QOMS stations can be found in [45,46].

Table 1. Geographic characteristics of the four stations with radiation measurements over Tibetan Plateau (30 m resolution slope and aspect derived from 30 m resolution ASTER-GDEM2; 1 km slope and aspect resampled from 30 m resolution slope and aspect).

Station Name	Latitude	Longitude	Elevation (m.a.s.l.)	Slope (30 m/1 km)	Aspect (30 m/1 km)	Land Cover	Soil Type
MAWORS	38.42°N	75.03°E	3668	5°/19°	119°/106°	Alpine desert	Sand and gravel
NAMORS	30.77°N	90.96°E	4730	3°/5°	256°/155°	Alpine steppe	Sandy and loam
QOMS	28.36°N	86.95°E	4298	7°/8°	45°/172°	Alpine desert	Sand and gravel
Dali	25.72°N	100.19°E	1990	3°/3°	309°/109°	Farmland	Paddy soil

We used time-averaged observations with a temporal resolution of 30 min at QOMS, MAWORS and NAMORS station and or 1 h at Dali station. Every station provides distinct observations on various atmospheric and land surface processes. The surface radiation components include the incoming and outgoing shortwave and longwave radiations. The ground measurements of surface shortwave albedo were obtained as the ratio of shortwave upwelling radiation to shortwave downwelling radiation.

Although the four stations are sparsely distributed, their strategic locations in diverse ecological and climatic settings across the Tibetan Plateau, coupled with the comprehensive analysis of daily atmospheric conditions over a year at the MODIS Aqua and Terra overpass time, provide a robust and representative dataset for our study. We believe that our evaluation covered the solar irradiance variability across this unique and complex region.

2.2. Satellite Data

The Terra and Aqua satellites utilize MODIS sensors to capture images in 36 spectral bands ranging from 0.62 μm to 14.385 μm. These images have a broad ground swath of approximately 2330 km and different spatial resolutions of 250 m, 500 m, and 1 km depending on the spectral band. The MODIS level 1B data undergo radiometric calibration, while the level 2 data are atmospherically corrected and formatted in an ungridded orbital swath, resulting in a product that spans five minutes. The Terra and Aqua products will be referred to collectively with the identifiers "MOD" and "MYD", respectively. For this study, MODIS Aqua and Terra data were employed, and the atmospheric products (e.g., MOD04/MYD04 L2, MOD05/MYD05 L2, MOD06/MYD06 L2, and MOD07/MYD07 L2) and the calibrated radiance product (MOD021KM/MYD021KM) from Table 2 were used as input data. The first seven spectral bands of MOD021KM/MYD021KM are used to estimate shortwave TOA radiance in this study. The instantaneous MODIS TOA spectral radiance was converted to TOA total shortwave radiance according to narrowband-to-broadband conversion method [47]. The study utilized eight variables—namely, aerosol optical depth (AOD), precipitable water content, ozone optical depth, surface pressure, cloud top pressure, cloud fraction, and cloud optical depth, all obtained from MODIS atmospheric data products—to estimate atmospheric transmittance. The MOD03/MYD03 provided the geolocation of the observations. The MODIS BRDF/Albedo MCD43A3 C6.1 dataset, available from the NASA Land Processes Distributed Active Archive Center (LP DAAC), produces albedo quantities at a resolution of 500 m in a sinusoidal projection. The MODIS MCD43A2 C6.1dataset is BRDF/Albedo) quality dataset is produced daily using 16 days of Terra and Aqua MODIS data at 500 m resolution for the corresponding 16-day MCD43A3 BRDF/Albedo dataset. The MCD43A2 contains individual band quality and observation information for the MODIS land bands 1–7, along with the overall BRDF/Albedo quality information. These quantities have been produced daily since 2000 with a synthesis period of 16 days, using data from both the Terra and Aqua satellites.

Table 2. MODIS level 2 data products and DEM data used in this study.

Product Short Name	Product Name	Spatial Resolution	Temporal Resolution	Data Type
MOD021km/MYD021km	Calibrated Radiance	1 km	5 min	EV_250_Aggr1km_RefSB EV_500_Aggr1km_RefSB
MOD03/MYD03 L2	Geolocation dataset	1 km	5 min	Longitude/latitude
MOD04/MYD04 L2	Aerosol optical thickness	10 km	5 min	Corrected optical depth land
MOD05/MYD05 L2	Total precipitable water vapor	5 km	5 min	Water vapor NIR retrieval
MOD06/MYD06 L2	Cloud	1 km or 5 km	5 min	Surface pressure Cloud top pressure Cloud fraction Cloud optical thickness
MOD07/MYD07 L2	Atmosphere profile	5 km	5 min	Total ozone
ASTER-GDEM2	DEM	30 m	-	DEM

The ASTER-GDEM2 dataset was used for calculating topographic parameters, i.e., slope, aspect, SVF (sky-view factor) and TVF (terrain-view factor) [48,49], which is one of the most widely used high-resolution digital topographic datasets to date. The 30 m resolution ASTER-GDEM2 data were averaged to estimate the mean elevation of each 1 km $\times$ 1 km grid, i.e., the same spatial resolution as the MODIS surface reflectance data. The slope, aspect, SVF and TVF of the terrain were retrieved at a 30 m resolution from the DEM generated using the data acquired by ASTER-GDEM2 data and then averaged to 1 km resolution.

3. Method

In this study, we retrieved surface albedo by calculating first all the relevant radiative fluxes under all sky conditions, taking into account the terrain effects on both the incoming and the reflected radiant energy. The procedure includes four steps: (1) estimating instantaneous surface solar irradiance (SSI) and its four components under all sky conditions over rugged terrain using a physically-based model [42,50] (see Section 3.1); (2) estimating the instantaneous at-surface upwelling directional radiance under all sky conditions in rugged terrain [24,51] (see Section 3.2); (3) estimating land surface reflectance under clear and cloudy sky on a rugged surface (see Section 3.3); and (4) estimating the surface albedo of rugged terrain [52] (see Section 3.4). The workflow used to retrieve the surface albedo is illustrated schematically in Figure 2. The detailed methods are described in the following sections.

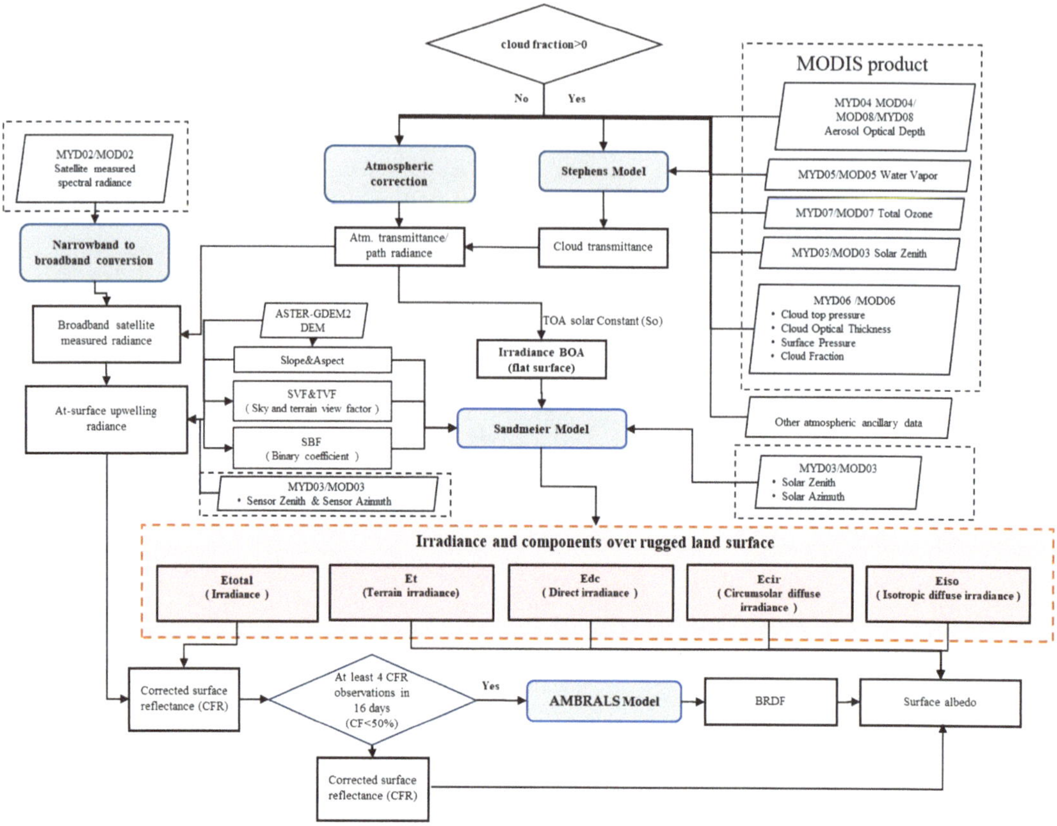

Figure 2. Schematic diagram of the workflow for the retrieval of the surface albedo in rugged terrain.

3.1. Estimation of Surface Solar Irradiance under Full Sky Conditions over Rugged Terrain

Over rugged terrain, the temporal and spatial distribution of surface solar irradiance (SSI) is the result of complex interactions among the incoming solar beam, the atmosphere, and the surface. At the bottom of the atmosphere (BOA), the SSI on a rugged terrain element includes four components, i.e., direct irradiance (E_{dc}, W/m^2), isotropic diffuse irradiance (E_{iso}, W/m^2), circumsolar diffuse irradiance (E_{cir}, W/m^2) and terrain irradiance (E_t, W/m^2) [42,50] and can be described as:

$$E_{total} = E_{dc} + E_{cir} + E_{iso} + E_t \quad (1)$$

where E_{total} (W/m^2) is the at-surface solar irradiance (SSI); E_{dc} (W/m^2) is the solar radiation directly reaching the surface through the atmosphere in rugged terrain; E_{cir} (W/m^2) is the scattering of direct sunlight in the atmosphere, and the effect of topography can be modeled using the same method used for direct irradiance; E_{iso} (W/m^2) is a function of the proportion of the sky hemisphere not obstructed by the topography; E_t (W/m^2) is the irradiance reflected by the terrain surrounding the observed facet. We have proposed a method of estimating the four components in Equation (1) under full sky conditions over rugged terrain [51]; a brief description is given below.

3.1.1. SSI under Clear Sky Conditions over Rugged Terrain

Under clear sky conditions, the direct solar irradiance is influenced by the slope and azimuth of the observed terrain facet and of the surrounding terrain. Solar radiation is absorbed by water vapor, ozone and other uniformly mixed gases, while aerosols either scatter or absorb radiation [53]. In mountainous regions, two primary factors influence the irradiance onto the observed facet: the orientation of the observed facet relative to the Sun beam and the impact of the facet orientation combined with the surrounding terrain. Part of the radiation scattered in the atmosphere can reach the rugged surface as diffuse irradiance. The direct irradiance (E_{dc}, W/m^2), anisotropic circumsolar diffuse irradiance (E_{cir}, W/m^2), isotropic diffuse irradiance (E_{iso}, W/m^2), a function of the proportion of the sky hemisphere not obstructed by the topography) and terrain irradiance for a tilted facet under clear sky conditions (E_t, W/m^2), a complex term, as it consists of both the direct and diffuse irradiance reflected from surrounding terrain [54,55]) can be described by the following equations:

$$E_{dc} = \Theta E_d \frac{cos\theta_i}{cos\theta_s} \quad (2)$$

$$E_{cir} = E_f k \frac{cos\theta_i}{cos\theta_s} \quad (3)$$

$$E_{iso} = E_f(1-k)V_d \quad (4)$$

$$E_t = \left(E_d + E_f\right)V_t\rho_{adj} \quad (5)$$

where Θ is a binary coefficient introduced by [45] that controls the cast shadows and is set to zero for a facet in the shadow and to one for a sunlit facet. θ_i is the solar incident angle between the normal to the facet and the Sun line of illumination (LoI). Both θ_i and Θ are derived from ASTER-GDEM2. θ_s is the solar zenith from MOD03/MYD03 products. k is the ratio of direct irradiance on a flat surface (E_d, W/m^2) to the irradiance at the top of the atmosphere, which is also called Hay's anisotropy index [56]; ρ_{adj} is the surface reflectance of the adjacent objects. V_d is the sky view factor (SVF), which is defined as the ratio of the sky portion seen from a specific facet to that on an unobstructed horizontal facet. V_t is the terrain view factor (TVF), defined as the portion of hemisphere overlying the observed facet and obstructed by the surrounding terrain. The sky view and terrain view factors are calculated using the analytical algorithm developed by [48]. The sky view and terrain view factors are also derived from the ASTER-GDEM2.

In Equations (2)–(5), E_d (W/m^2) is the direct irradiance on flat surface and E_f (W/m^2) is the diffuse irradiance on flat surface, and they are calculated as:

$$E_d = E_0 \cdot cos\ (\theta_s) \cdot T_B \quad (6)$$

$$E_f = E_0 \cdot cos\theta_s \cdot T_D \quad (7)$$

where E_0 (W/m^2) is the extraterrestrial solar irradiance, T_B is the direct beam transmittance, and T_D is the diffuse transmittance under clear sky introduced by [57]. They are calculated as:

$$E_0 = S_0 \cdot SE_d \quad (8)$$

$$T_B = \tau_o \cdot \tau_w \cdot \tau_g \cdot \tau_r \cdot \tau_a \quad (9)$$

$$T_D = 0.5\tau_o \cdot \tau_w \cdot \tau_g \cdot (1 - \tau_a \cdot \tau_r) \quad (10)$$

where S_0 (W/m^2) is the solar constant (1367 W/m^2). SE_d is Sun–Earth distance. Transmittance factors (τ_r, τ_a, τ_o, τ_w, τ_g) introduced by [57] are influenced by five radiation-damping processes under clear sky conditions: (1) Rayleigh scattering, (2) aerosol extinction, (3) ozone absorption, (4) water vapor absorption, and (5) permanent gas absorption.

The input variables for calculation of transmittance factor ($\tau_r, \tau_a, \tau_o, \tau_w, \tau_g$) are the surface pressure, precipitable water, optical thickness of the ozone layer, ozone thickness, and AOD, which can be obtained from MODIS data products (Table 2).

3.1.2. SSI under Cloudy Sky Conditions over Rugged Terrain

Under cloudy conditions, cloud tops reflect a fraction of the solar radiation back into space while allowing the rest to pass through. In this study, the influence of clouds on radiation is modeled using a parameterization method described in detail in [30,58]. This method categorizes the solar shortwave range into two broad bands: 0.30–0.75 μm and 0.75–4 μm, and is designed to estimate the reflection, transmission, and absorption of radiation by clouds within these bands. The direct irradiance ($E_{dc}{}'$, W/m^2) is direct irradiance in the cloud-free fraction of a pixel. The isotropic diffuse irradiance ($E_{iso}{}'$, W/m^2) in a pixel is composed of three components: the isotropic diffuse irradiance in the cloudless fraction, the isotropic diffuse irradiance transmitted by the clouds, and the isotropic diffuse irradiance scattered by the clouds. The circumsolar diffuse irradiance ($E_{cir}{}'$, W/m^2) in a pixel is composed of the circumsolar diffuse irradiance in the cloudless part and the circumsolar diffuse irradiance scattered by the clouds.

Under cloudy conditions, the direct irradiance ($E_{dc}{}'$, W/m^2), anisotropic circumsolar diffuse irradiance ($E_{cir}{}'$, W/m^2), isotropic diffuse irradiance ($E_{iso}{}'$, W/m^2), and terrain irradiance for a tilted facet under cloudy sky conditions can be described by the following equations [43]:

$$E_{dc}{}' = (1 - cf) \times E_{dc} \tag{11}$$

$$E_{iso}{}' = E_f(1 - cf)(1 - k)V_d + T_B{}'E_0 cf cos\theta_s V_d + (1 - T_B)E_{iso}cf \tag{12}$$

$$E_{cir}{}' = E_f(1 - cf)k\frac{cos\theta_i}{cos\theta_s} + (1 - T_B{}')\ E_{cir}cf \tag{13}$$

$$E_t{}' = \left(E_{dc}{}' + E_f{}'\right)V_t\rho_{adj} \tag{14}$$

$$E_f{}' = T_B{}'E_0\ cos\theta_i cf + (1 - cf)E_f \tag{15}$$

$$T_B{}' = \tau_o \cdot \tau_w{}' \cdot \tau_g{}' \cdot \tau_r{}' \cdot \tau_a \cdot \tau_c \tag{16}$$

where τ_c is the cloud transmittance according to [30]. τ_c is estimated from cloud optical thickness from MOD06/MYD06; cf is the cloud fraction from MOD06/MYD06, defined as the fraction of each pixel covered by clouds and determines the impact of the clouds on the solar radiation. $E_{dc}{}'$ is estimated as a proportion of clear-sky direct irradiance under cloudy sky conditions. $E_{iso}{}'$ is the isotropic diffuse irradiance under cloudy sky conditions. $E_{cir}{}'$ is circumsolar diffuse irradiance under cloudy sky. $E_t{}'$ is the terrain irradiance under cloudy sky conditions. $E_f{}'$ is the diffuse component of irradiance on a horizontal surface under cloudy sky conditions. $T_B{}'$ is the total transmittance under cloudy conditions. Transmittance factors ($\tau_w{}'$, $\tau_g{}'$, $\tau_r{}'$) under cloudy conditions were introduced by [42].The input variables for estimation of transmittance factors ($\tau_w{}'$, $\tau_g{}'$, $\tau_r{}'$) under cloudy conditions include not only the surface pressure, precipitable water, optical thickness of the ozone layer, ozone thickness, and AOD from the MODIS data product but also cloud top pressure provided by MOD06/MYD06 product (Table 2).

3.2. *At-Surface Upwelling Radiance under Clear and Cloudy Sky in Rugged Terrain*

In this study, at-surface upwelling radiance was estimated and then used to obtain the bidirectional surface reflectance, which is the key radiative variable in the estimation of at-surface reflectance as Ross–Li model input data for retrieving surface albedo. The surface reflected radiant exitance was divided by the SSI to derive surface reflectance in a specific direction towards the sensor (see Section 3.1). The at-surface upwelling radiance L_{BOA} (W/m^2/sr) was estimated from the at-satellite TOA radiance (W/m^2/sr) by taking into account the effects of the atmosphere described above.

3.2.1. At-Surface Upwelling Radiance under Clear Sky in Rugged Terrain

Terrain characteristics not only affect SSI but also radiance reflected by the surface under different atmospheric conditions. We used a simple model to remove the reflected radiance of the surrounding terrain irradiance by surface following previous study [21]. This model estimates the at-surface upwelling radiance in rugged terrain, taking into account topographic influence. Under clear sky conditions, the at-sensor TOA radiance from a target pixel observed by MODIS (provided by the MYD02/MOD02 products) is composed of (1) the at-surface upwelling radiance of the pixel (L_{BOA}) attenuated by the atmosphere, (2) the irradiance from the surrounding terrain that is reflected by the observed facet and attenuated by the atmosphere (surrounding terrain contribution ($L_{TOA,\ topo\_neighbor}$), and (3) the atmospheric contribution (L_P) and is expressed as:

$$L_{TOA} = (T_B + T_D)\ cos\theta_v\ L_{BOA} + L_{TOA,\ topo\_neighbor} + L_P \tag{17}$$

where L_{TOA} (W/m^2/sr) is the observed at-sensor TOA radiance made available in the MYD02/MOD02 product, $L_{TOA,topo\_neighbor}$ (W/m^2/sr) is the TOA value of the irradiance from the surrounding topography that is reflected by the observed pixel surface, and L_p (W/m^2/sr) is the path radiance along the satellite observation direction under clear sky conditions as estimated using MODTRAN. The spectral TOA radiance (W/m^2/sr) in MOD02/MYD02 is converted into shortwave TOA radiance according to [47]. θ_v is the satellite observation zenith angle from the MOD03/MYD03 product. Assuming that at TOA, the proportion of the terrain irradiance to the total surface solar irradiance is the same as the proportion of terrain radiance to observed radiance at BOA, $L_{TOA,\ topo\_neighbor}$ can be estimated as:

$$L_{TOA,\ topo\_neighbor} = \frac{E_t}{E_{total}} \cdot (L_{TOA} - L_p) \tag{18}$$

Therefore, under clear sky conditions, the at-surface upwelling radiance of the pixel, L_{BOA}, is calculated by deducting the contributions from surrounding terrain and the atmosphere (combining Equations (17) and (18)) and expressed as:

$$L_{BOA} = \frac{(L_{TOA} - L_p)}{(T_B + T_D)cos\theta_v} - \frac{E_t}{E_{total}}\left[\frac{(L_{TOA} - L_p)}{(T_B + T_D)cos\theta_v}\right] \tag{19}$$

The first term of Equation (19) is the surface reflected radiance at observation angles at the bottom of atmosphere after atmospheric correction. The second term of Equation (19) is the irradiance from surrounding topography reflected by the observed surface facet.

3.2.2. At-Surface Upwelling Radiance under Cloudy Sky in Rugged Terrain

Under cloudy sky conditions, the at-sensor TOA radiance includes the radiation reflected by the land surface and transmitted by clouds, path radiance, the radiation scattered by the surroundings of the observed target and the radiation reflected by clouds. In order to increase the number of valid retrievals of surface reflectance, we estimated the at-surface shortwave upwelling radiance under cloudy sky conditions ($L_{BOA}{}'$, W/m^2/str) as:

$$L_{BOA}{}' = \frac{(L_{TOA} - L_p)}{T_{all}{}'} - \frac{E_t}{E_{total}}\left[\frac{\left(L_{TOA} - L_p - \frac{E_{cf}}{\pi}\right)}{T_{all}{}'}\right] \tag{20}$$

$$T_{all}{}' = \left[(T_B + T_D)(1 - cf) + T_B{}'cf\right]cos\theta_v \tag{21}$$

$$E_{cf} = E_0\rho_{cld} \times cf(\tau_{oz}\overline{\tau_r})^2\tau_c cos_{\theta_v} cos\theta_s \quad (22)$$

where T_{all}' is atmospheric transmittance under cloudy sky at observation angle θ_v. E_{cf} (W/m^2) is the cloud reflected radiative flux observed by a satellite at TOA. ρ_{cld} is the cloud reflectance from [30].

3.3. Estimating Land Surface Reflectance under Clear and Cloudy Sky Conditions on a Rugged Surface

At-surface reflectance is the ratio of at-surface upwelling radiant exitance at given incident and observation angles to SSI. The at-surface reflectances under clear and cloudy sky conditions are:

$$\rho_{BOA} = \frac{L_{BOA}\ \pi}{E_{dc} + E_{cir} + E_{iso} + E_t} \quad (23)$$

$$\rho_{BOA}' = \frac{L_{BOA}'\ \pi}{E_{dc}' + E_{cir}' + E_{iso}' + E_t'} \quad (24)$$

where ρ_{BOA} is the shortwave surface reflectance under clear sky and ρ_{BOA}' is the shortwave surface reflectance under cloudy sky.

3.4. Retrieval of Surface Albedo over Rugged Terrain

Rugged terrain alters the ratio of direct, diffuse, and terrain-reflected radiations [59]. The actual surface albedo quantifies the fraction of the sunlight energy reflected by the surface of the Earth.

We used two complementary methods to retrieve the surface albedo: (1) if there are at least three observations with cf $\leq$ 50% within a 16-day window, we construct the BRDF using the same method as in the MODIS MCD43 BRDF/albedo algorithm; (2) if there are fewer than three observations with cf $\leq$ 50% within a 16-day window (cloudy conditions), we estimate the surface albedo as the ratio of the at-surface upwelling radiant exitance to the SSI.

In the first method, the surface albedo (α) of rugged terrain can be written as:

$$\alpha = \frac{E_{dc}\rho_{dh}(\theta_i,\varphi_i;\theta_v,\theta\varphi_v,\ 2\pi) + (E_{cir}+E_{iso})\rho_{hh}(\theta_i,\varphi_i,2\pi;2\pi) + E_t\rho_{hh} + (E_{cir}+E_{iso})\rho_{hd}(\theta_i,\varphi_i,2\pi;\theta_v,\varphi_v)}{E_{total}} \quad (25)$$

where $\rho_{dh}(\theta_i,\varphi_i;\theta_v,\varphi_v,2\pi)$, also called the black-sky albedo (BSA), is the directional-hemispheric reflectance, which is the hemispheric integral of the radiance reflected by a surface into the view hemisphere when illuminated by a point-like direct illumination source. $\rho_{hd}(\theta_i,\varphi_i,2\pi;\theta_v,\varphi_v)$ is the reflectance for the hemispheric irradiance in a specific direction of observation (line of sight, LoS) and is an integral of the bidirectional reflectance over all incidence angles and is equal to the hemispherical directional reflectance factor (HDRF). $\rho_{hh}(\theta_i,\varphi_i,2\pi;2\pi)$ is the bi-hemispheric reflectance, also called the white-sky albedo (WSA). Although $\rho_{dh}(\theta_i,\varphi_i;\theta_v,\varphi_v,2\pi)$ and $\rho_{hd}(\theta_i,\varphi_i,2\pi;\theta_v,\varphi_v)$ are defined from different perspectives of reflectance measurement, their values can be considered approximately equivalent under many practical circumstances due to their similar roles in characterizing the reflective properties of the surface [60].

The $\rho_{dh}(\theta_i,\varphi_i;\theta_v,\varphi_v,2\pi)$ and $\rho_{hh}(\theta_i,\varphi_i,2\pi;2\pi)$ are as follows:

$$\begin{aligned}\rho_{dh}(\theta_i,\varphi_i;\theta_v,\varphi_v,2\pi) = &f_{iso}\left(g_{0iso} + g_{1iso}\theta^2 + g_{2iso}\theta^3\right)\\ &+f_{vol}\left(g_{0vol} + g_{1vol}\theta^2 + g_{2vol}\theta^3\right)\\ &+f_{geo}\left(g_{0geo} + g_{1geo}\theta^2 + g_{2geo}\theta^3\right)\end{aligned} \quad (26)$$

$$\rho_{hh}(\theta_i,\varphi_i,2\pi;2\pi) = f_{iso}g_{iso} + f_{vol}g_{vol} + f_{geo}g_{geo} \quad (27)$$

where f_k (k = iso, vol, geo) is BRDF model kernel weights in kernel-driven BRDF model. g_{jk} (j = 0, 1, 2; k = iso, vol, geo) and g_k coefficients are taken from [10].

The BRDF quantifies the angular distribution of radiance reflected by an illuminated surface [61]. We used an RTLSR-kernel-driven BRDF model that linearly combines three kernels: isotropic scattering, volume scattering, and geometric optics scattering according to the algorithm of the MCD43 BRDF/albedo product [62]. The Ross–Li kernel-driven BRDF model equation is:

$$R(\theta_i, \theta_v, \varphi_i, \varphi_v) = f_{iso} + f_{vol}k_{vol}(\theta_i, \theta_v, \varphi_i, \varphi_v) + f_{geo}k_{geo}(\theta_i, \theta_v, \varphi_i, \varphi_v) \quad (28)$$

where $R(\theta_i, \theta_v, \varphi_i, \varphi_v)$ is the at-surface bidirectional reflectance, f_{iso} is the coefficient of the isotropic term, k_{vol} is the volume scattering term from the Ross–Thick kernel, k_{geo} is the geometrical optical scattering term of the Li–Ross kernel, f_{vol} is the coefficient of k_{vol}, and f_{geo} is the coefficient of k_{geo}.

Given the at-surface reflectance ρ_{BOA} at specific incident and view angles obtained from Equations (23) and (24) under clear and cloudy sky conditions, we can derive analytical solutions for the model parameters f_k in Equation (28) by minimizing a least squares error function as applied in the MCD43 BRDF/albedo algorithm (when applying $\frac{\partial e^2}{\partial f_k} = 0$):

$$e^2 = \frac{1}{d}\sum_{i=1}^{n} \frac{[\rho_{BOA,i}(\theta_{in}, \theta_{vn}, \varphi_{in}, \varphi_{vn}) - R_i(\theta_{in}, \theta_{vn}, \varphi_{in}, \varphi_{vn})]^2}{\omega_i} \quad (29)$$

where n is the number of observations (the minimum number of observations is taken as 3 in our method), d is degree of freedom (number of observations minus number of unknown parameters), and ω_i is a weight given to each respective observation. ρ_{BOA} is estimated by Equations (23) and (24).

If the number of observations with cf $\leq 50\%$ within a 16-day window is insufficient to determine the parameters of the BRDF model, we apply the second method to estimate the surface albedo as the ratio of the at-surface upwelling radiance to the SSI from Equations (23) and (24).

Once all the components in Equation (25) are estimated through the above procedures, land surface albedo over rugged terrain under clear and cloudy sky conditions is obtained.

3.5. Validation Approach

We used four error metrics for the evaluation of our retrievals by comparison with in situ observations: the bias (Bias), root mean square error (RMSE), mean absolute percentage error (MAPE), and correlation coefficient (R), which are calculated as follows:

$$\text{Bias} = \sum_{i=1}^{n}(x_i - y_i) \quad (30)$$

$$\text{RMSE} = \sqrt{\frac{1}{n}\sum_{i=1}^{n}(x_i - y_i)^2} \quad (31)$$

$$\text{MAPE} = \frac{1}{n}\sum_{i=1}^{n}\left|\frac{x_i - y_i}{x_i}\right| \quad (32)$$

$$R = \frac{\sum_{i=1}^{n}(x_i - \overline{x}) \times (y_i - \overline{y})}{\sum_{i=1}^{n}(x_i - \overline{x})^2 \times (y_i - \overline{y})^2} \quad (33)$$

where x_i is the in situ observation value, y_i is the value estimated using the method developed in this study, $\overline{x}$ is the average observation value, $\overline{y}$ is the average estimated value, and n is the number of estimates and measurements.

4. Results

The method proposed in Section 3 and the data introduced in Section 2 were applied to estimate both instantaneous at-surface upwelling shortwave radiance and surface albedo taking into account the coupled effects of the topography and atmosphere at 1 km spatial resolution on the Tibetan Plateau. The method was evaluated against in situ daily measurements recorded daily at the Dali, NAMORS, MAWORS, and QOMS stations on the Tibetan Plateau in 2018. This validation revealed that at-surface upwelling estimates and the surface albedo retrieval are in satisfactory agreement with the ground observations (Sections 4.2 and 4.3) and our approach clearly improved the spatial and temporal coverage of the albedo retrievals (Section 4.4).

4.1. Number of Valid Observations in the Improved Method versus MODIS Product

Both our surface albedo retrieval method and the MODIS albedo algorithm (MCD43A3) require at least 4 valid surface reflectance observations every 16 days to construct the BRDF for surface albedo inversion. The MODIS MCD43A2 product provides the BRDF/Albedo band quality and information on the days of valid observation of surface reflectance within each 16-day period for MODIS bands 1 to 7. The advantage of our method over the MODIS surface albedo algorithm can then be best documented by comparing the number of valid surface reflectance observations. This comparison was made for the Dali, MAWORS, NAMORS, and QOMS stations (Figure 3). For the MODIS MCD43A2 product, the number of valid observations is the sum of the number of valid daily surface reflectance (clear sky) observations that can be used to successfully retrieve daily surface albedo within each 16-day moving window in each month. For our method, we calculated the number of valid observations in each month in the same way but used the conditions of the cloud fraction less than 50% (other than totally clear sky as required by the MODIS algorithm), taking into account the accuracy in the estimating cloud reflectance and transmittance (see Section 4.2). The number of valid observations with our method was, as expected, much higher than with the MODIS algorithm, i.e., 200% higher at Dali, 66% higher at QOMS, 100% higher at MAWORS, and 40% higher at NAMORS station. This difference is that this lacks the advantage of estimating the at-surface upwelling radiance under cloudy conditions as our method does. The performance of this advantage of our method is evaluated in the next Section 4.2. At Dali Station, the absence of days, with zero cloudy days in July and the presence of clear sky conditions only on certain days in August, highlighted the limitations of the MODIS algorithm in providing more effective albedo retrievals. The situation at QOMS Station mirrors that at Dali Station.

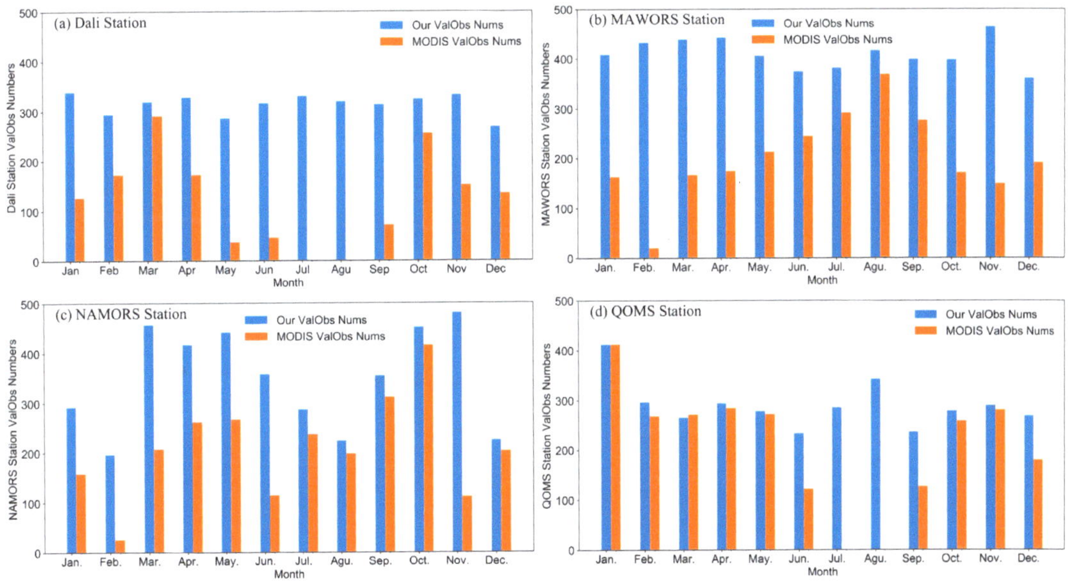

Figure 3. Comparison of the monthly distribution of the sum of the number of valid daily surface reflectance observations (ValObs) between our method and MODIS MCD43A2 that can be used for successful retrieval of daily surface albedo within each 16-day moving window at the four stations on the Tibetan Plateau in 2018: (**a**) Dali Station, (**b**) MAWORS Station, (**c**) NAMORS Station, and (**d**) QOMS Station.

4.2. Evaluation of Retrieved At-Surface Upwelling Radiance

Our method relies on the estimation of the at-surface upwelling radiance under all sky conditions, which requires the estimation of cloud reflectance and transmittance, in addition to atmospheric absorption and scattering and to terrain irradiance (see Section 3.2). We compared our estimation of at-surface upwelling radiance with a set of ground observations at the four stations on the Tibetan Plateau in 2018. We used time averaged observations of a temporal interval of 30 min (QOMS, MAWORS, and NAMORS) or 1 h (Dali). We applied four error metrics—BIAS, MAPE, RMSE, and R—calculated using Equations (30)–(33). Better estimates of the at-surface upwelling radiance were obtained for MAWORS, NAMORS, and QOMS than for Dali when the cloud fraction (CF) of the pixel at the ground site measured by MOD06/MYD06 was less than 50% (Figure 4).

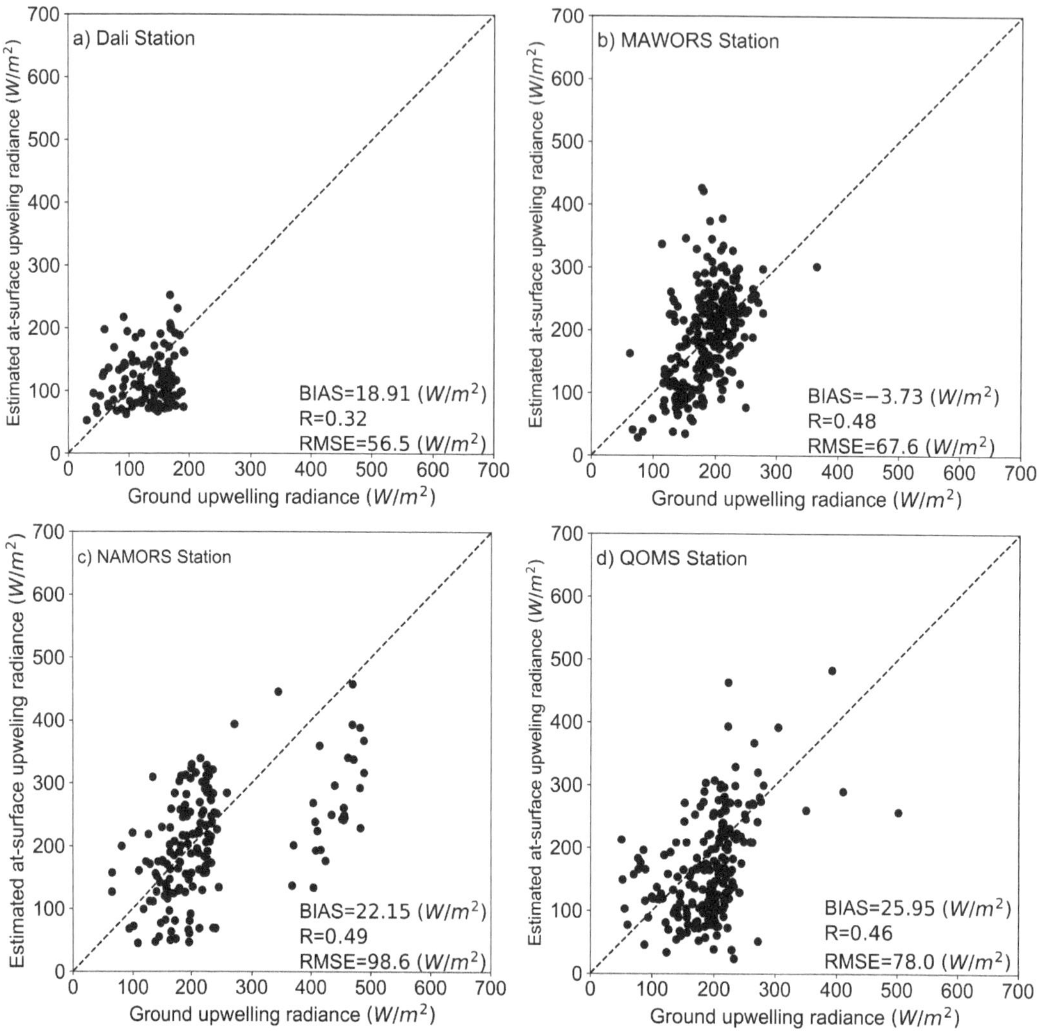

Figure 4. Validation of the instantaneous at-surface upwelling estimated using the method developed in this study when CF < 50% by comparison with the ground observations at Dali, NAMORS, MAWORS, and QOMS stations on the Tibetan Plateau in 2018 during the MODIS Terra and Aqua overpass time.

The estimates of at-surface upwelling radiance at MAWORS, NAMORS, QOMS, and Dali stations were also evaluated at different cloud fractions (Figure 5). Five cloud fraction bins were considered: CF = 0%, 0% < CF ≤ 20%, 20 < CF ≤ 50%, 50% < CF ≤ 75%, and 75% < CF ≤ 100%. Under clear sky conditions (CF = 0%) and low cloud fraction (0% < CF ≤ 20%), all stations have relatively low MAPE values, indicating the higher accuracy of the estimates. As the cloud fraction increases (20% < CF ≤ 50% and 50% < CF ≤ 75%), all stations show varying increases in MAPE, indicating that the presence of clouds makes the estimates of at-surface upwelling radiance more challenging. At the highest cloud fraction level (75% < CF ≤ 100%), there is a significant increase in MAPE for all stations, particularly for QOMS and NAMORS. This indicates that estimation errors increase significantly with high cloud coverage, suggesting that the current parameterization of cloud reflectance and transmittance should be improved to better deal with nearly complete cloud cover conditions.

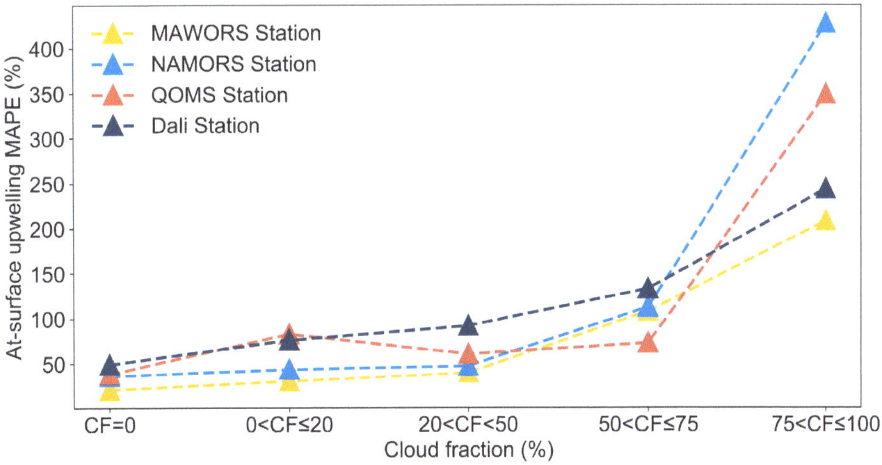

Figure 5. The MAPE for at-surface upwelling radiance estimates at different cloud fractions at Dali, MAWORS, NAMORS, and QOMS stations on the Tibetan Plateau in 2018.

Using the method described here, we estimated instantaneous at-surface upwelling radiance corresponding to the overpass time applied to the MODIS orbital products, i.e., a spatial map was produced every 5 min with a spatial resolution of 1 km. Figure 6 shows an example on 2 January 2018, with large cloud cover over the Tibetan Plateau, to demonstrate the ability of our method to provide more valid estimates of at-surface upwelling radiance.

4.3. Evaluation of Retrieved Surface Albedo

We compared our shortwave surface albedo with the shortwave WSA and BSA of MODIS (MCD43A3) and with ground measurements of surface albedo at four stations on the Tibetan Plateau: Dali, MAWORS, NAMORS, and QOMS (Figure 7). The MODIS retrievals apply to clear sky only, and the MODIS albedo products do not include blue-sky albedo retrievals. In general, the blue-sky albedo is computed by linear interpolation of WSA and BSA on the basis of the fraction of diffuse skylight [63], but as shown in Section 3.4, the blue-sky albedo depends nonlinearly on BSA and WSA in a rugged terrain. We compared our broadband albedo retrievals with both MODIS WSA and BSA (Figure 7) with the ground measurements. Our method led to a significant reduction in RMSE at the four observation sites when compared to the RMSE of MODIS WSA and BSA data (Figure 7). The best performance for the MODIS WSA, BSA, and our retrievals was achieved at the Dali station, with RMSE values of 0.0585, 0.0640, and 0.0422, respectively. At the MAWORS site, the RMSE values for our retrievals were 34% and 31% lower than for WSA and BSA, respectively (Figure 7). Our algorithm includes the calculation of the effect of terrain on at-surface irradiance and at-surface upwelling radiance and also retrievals albedo for all sky conditions. The average cloud coverage in 2018 at the four stations Dali, NAWORS, MAWORS, and QOMS was 53.9%, 55.7%, 61.0%, and 38.3%, respectively. The four error metrics (Table 3) provide further evidence on the satisfactory performance of our proposed method. The terrain slopes at the four stations—Dali, NAWORS, MAWORS, and QOMS—are 3°, 3°, 5°, and 7°, respectively. It should be noted that our estimates of albedo apply to a plane facet parallel to a 1 km × 1 km terrain facet. The retrievals of surface albedo are based on the mean slope and aspect of the terrain within each 1 km × 1 km grid, determined by averaging slope and aspect calculated at a 30 m × 30 m spatial resolution within each grid. If this facet has the same orientation as the terrain at the location of the radiometer used to measure radiance value, the terrain geometry of our retrievals and in situ measurements should be comparable. The footprint of an in situ radiometer, however, is much smaller than the 1 km × 1 km pixel applied in our retrieval. Differences in aspect have an impact when differences in slope are larger, as it is the case at MAWORS (Table 1), which may explain the lower accuracy of our retrievals at this site. In addition, we noted a few instances of less reliable in situ radiometric measurements, which led to a few outliers with an impact on error metrics difficult to evaluate. The MAWORS site has the highest cloud cover in 2018, with the slope being the second-largest of the four sites. Therefore, it seems to have the greatest atmospheric and topographical influence.

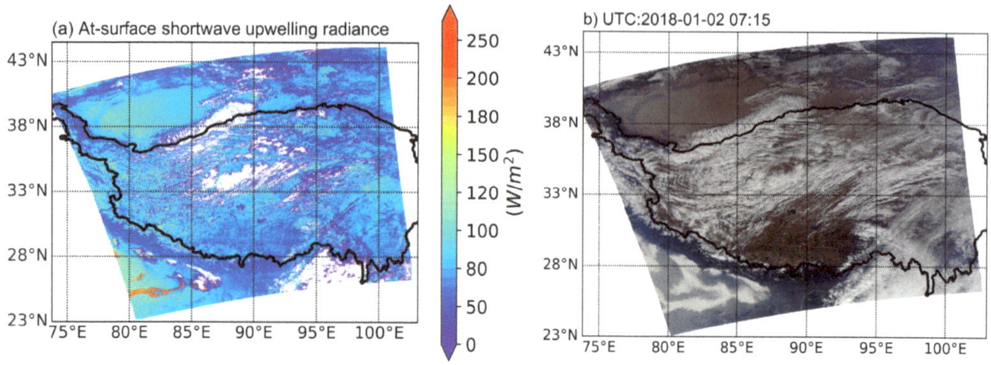

Figure 6. (**a**) The estimates of instantaneous at-surface upwelling radiance of the Tibetan Plateau (MYD Swath: UTC 2018-01-02-0715) and (**b**) MODIS true color composite images of bands 4, 3, and 1 at the same time.

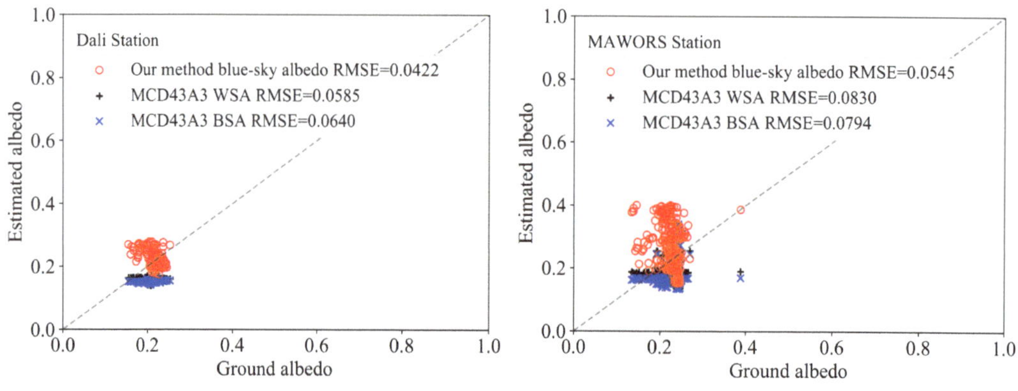

Figure 7. *Cont.*

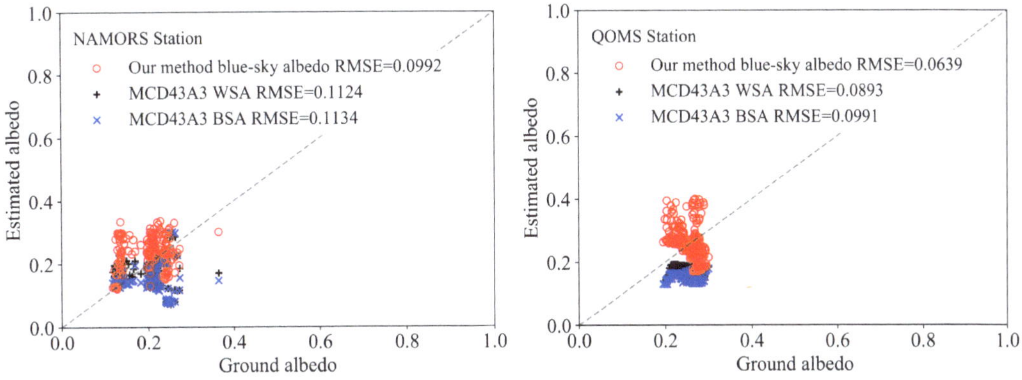

Figure 7. Comparison of clear day results of the estimated broadband surface albedo using the new method proposed in this paper, the MODIS shortwave WSA and BSA and the ground measured albedo at Dali, MAWORS, NAMORS, and QOMS stations on the Tibetan Plateau in 2018.

Table 3. Validation metrics of the estimated surface albedo and comparison with MODIS WSA and BSA under clear sky conditions.

Station	Method	R	RMSE	Bias	MAPE
Dali	Our method	0.44	0.0422	−0.0109	16%
	MODIS WSA	0.30	0.0585	0.054	25%
	MODIS BSA	0.31	0.0640	0.061	28%
NAMORS	Our method	0.26	0.0992	−0.036	20%
	MODIS WSA	0.21	0.1124	0.040	24%
	MODIS BSA	0.20	0.1134	0.043	22%
MAWORS	Our method	0.20	0.0545	−0.035	25%
	MODIS WSA	0.16	0.0830	0.041	23%
	MODIS BSA	0.14	0.0794	0.049	25%
QOMS	Our method	0.42	0.0639	0.012	20%
	MODIS WSA	0.16	0.0893	0.083	31%
	MODIS BSA	0.21	0.0991	0.095	36%

Figure 8 shows the performance of the surface broadband albedo retrievals using our method when compared with ground measurements under different cloud conditions. As expected, the performance was best when the cloud fraction was 0% at Dali, MAWORS, NAMORS, and QOMS. As the cloud fraction increased, the MAPE between the estimated and observed surface albedo increased slightly, i.e., the uncertainty of the estimates increased at Dali and NAWORS but not MAWORS and QOMS (Figure 8).

4.4. Spatial and Temporal Coverage of Retrieved Surface Albedo in Rugged Terrain

Our method improves significantly the spatial and temporal coverage of valid retrievals. The case of the MAWORS station, located in rugged terrain, documents this aspect effectively (Figure 9). Our estimates of surface albedo achieve nearly full spatial coverage (Figure 9f) and, in the cloud-covered areas, show texture characteristics consistent with the surrounding terrain. The high-surface-albedo regions in the MCD43A3 products and our retrievals appear to correlate with flatter areas. By examining areas of mismatch between our results and MCD43A3 albedos, we noted that topography influences surface albedo retrieval, where terrain features like ridges and valleys can cause significant variability (Figure 9). Such terrain effect under cloudy conditions can be captured well by our method, but retrieval with the MODIS albedo products failed (Figure 9d,e). The temporal coverage is also clearly improved by our method, as shown by a time series of the estimated broadband shortwave surface albedo by our method, the MODIS WSA and BSA, and the ground-measured albedo throughout the year 2018 at MAWORS Station (Figure 10). Gaps in MODIS retrievals from late January to early March were substantial (Figure 10), while there are very few gaps in our retrievals. Moreover, the retrievals on cloudy days (as shown by the gaps in the MODIS albedo time series) are consistent with retrievals on clear-sky days, notwithstanding the larger error in albedo retrieval on cloudy days.

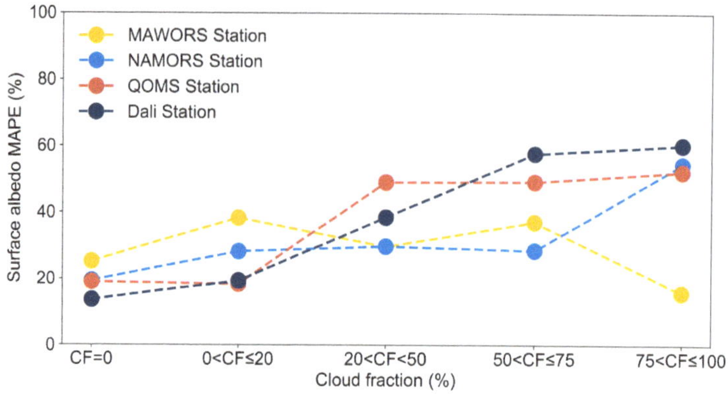

Figure 8. The MAPE between the estimated surface albedo by our method and the ground measurements of albedo at different cloud fractions at Dali, MAWORS, NAMORS, and QOMS stations in 2018.

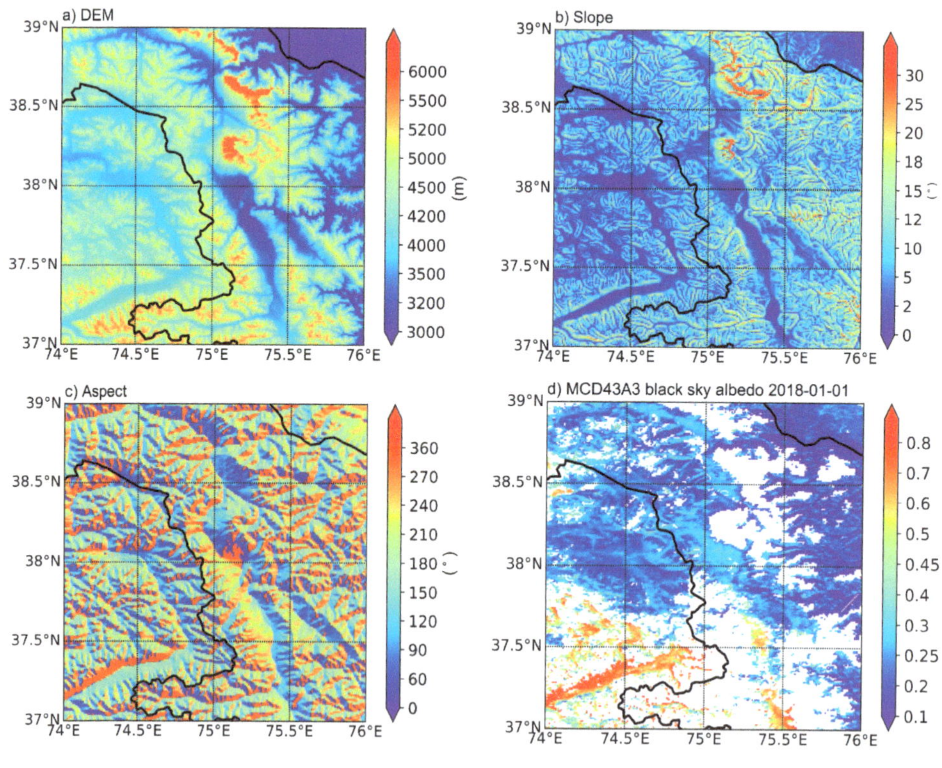

Figure 9. *Cont.*

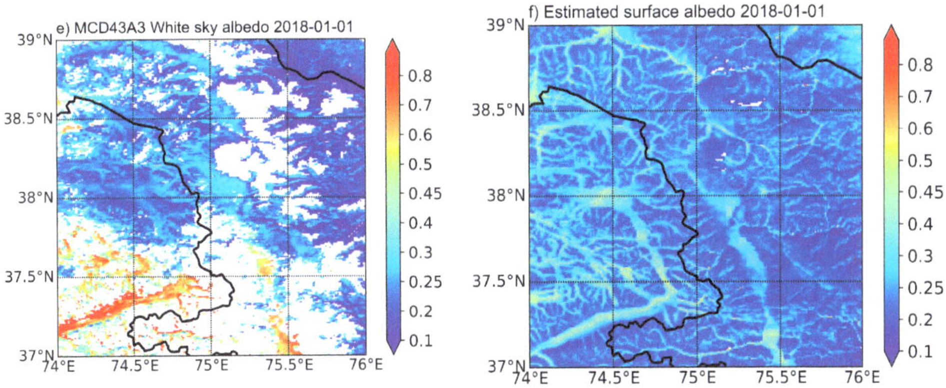

Figure 9. Spatial distribution over an area in the surrounding of the MAWORS station on the Tibetan Plateau: (**a**) elevation (m), (**b**) slope (°), (**c**) aspect (°), (**d**) MCD43A3 black-sky albedo, (**e**) MCD43A3 white-sky albedo, and (**f**) our retrievals of surface broadband albedo on 2018.01.01.

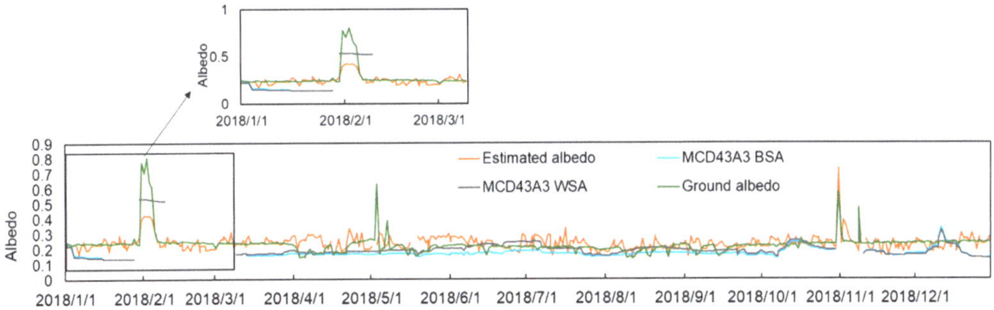

Figure 10. The time series of the estimated broadband surface albedo by our method, MODIS shortwave WSA and BSA, and ground-measured albedo at the MAWORS site on the Tibetan Plateau in 2018.

5. Discussion

5.1. Overall Evaluation of Results

Estimating the surface albedo is of great importance for the regional-scale surface radiation budget and the global climate monitoring of the land–surface–atmosphere system. Moderate-solution satellite data such as MODIS are valuable to generate global daily time series of albedo data products [10]. The limitation in the retrieval of surface albedo relates to a combination of limited available observations to evaluate and improve the retrieval accuracy and the heterogeneity of the land surface.

There are three key issues that warrant further attention. First, this estimation of broadband albedo from a set of narrowband reflectance with conversion coefficients does not necessarily comply with the definition of surface albedo as the ratio of hemispheric radiative fluxes [44,64,65]. This deviation arises because surface albedo, as implied by its definition, has a nonlinear relationship with the incident and reflected narrowband radiance. Moreover, the conversion coefficient set is sensor-specific and cannot be applied universally. Second, the interaction of atmospheric and terrain effects is often oversimplified in the process of estimating surface reflectance from at-satellite radiance at top of atmosphere (TOA) or when estimations are limited to clear sky conditions [66–69]. In reality, optical properties of clouds and aerosols and terrain effects jointly influence the solar radiation reaching the surface. Radiative transfer models that consider the coupled interactions between the rugged terrain and the atmosphere are helpful to improve the quality of both the derived surface properties (e.g., surface reflectance and albedo) and atmospheric properties (e.g., aerosol loading and size distribution) [70,71] Third, most available albedo products generated through this procedure also assume a flat and homogeneous land surface, as seen in products generated with the data acquired by sensors like MODIS, the multi-angle imaging spectroradiometer (MISR), and polarization and directionality of the earth reflectances (POLDER) [10,72,73].

In this study, we developed a method for estimating instantaneous at-surface shortwave upwelling radiance and surface shortwave albedo under all sky conditions, taking into account the complex interactions between topography and atmosphere on the Tibetan Plateau. This method aimed to reduce the occurrence of data gaps (e.g., due to clouds) and improve the quality of the estimated albedo. Our retrieval method considers all components of all-sky SSI, i.e., direct, isotropic diffuse, circumsolar diffuse, and irradiance from surrounding terrain. This method rests on retrieving at-surface upwelling radiance under all sky conditions to estimate BRF then using the RTLSR model to retrieve daily surface albedo on the Tibetan Plateau at 1 km resolution.

The results show, as expected, that the accuracy of estimated instantaneous at-surface upwelling radiance improves as the cloud fraction decreases, as illustrated in Figure 5. The results also show that the method not only largely reduces the frequency of gaps encountered during albedo retrieval but also improves the surface shortwave albedo retrieval accuracy under clear sky conditions. As expected, the number of valid observations with our method was higher than that with the MCD43A2 product data at four stations on the Tibetan Plateau (Figure 3). Although the R value of at-surface upwelling radiance at Dali Station is relatively low, the model demonstrates small bias and a low RMSE (Figure 4). This station is located in the southeastern Tibetan Plateau, i.e., within the transition region between the Indian and East Asian summer monsoon circulations [74]. The Dali station belongs to a frequently cloudy area, and the percentage of missing observations in the MCD43A3 albedo data is up to 39%. The main reason for low R can be that the temporal resolution of the ground observations is too coarse for the instantaneous at-surface upwelling retrievals at Dali station, where the temporal variability of clouds is very high.

The overall accuracy of estimating surface albedo also improves as the cloud cover decreases (Figure 3). The study highlights that the proposed method for shortwave surface albedo retrieval achieves a substantial increase in the number of valid observations compared to the MODIS MCD43 product. This increase, reaching up to 200% at certain locations, is due to the capability of the new method to estimate the at-surface upwelling radiance under continuous non-clear sky conditions. Our method also improved significantly the spatial and temporal coverage of retrievals, with a high level of consistency between retrievals under cloudy and cloud-free conditions (Figures 9 and 10).

The study also demonstrates that the MCD43A3 albedo model tends to underestimate surface shortwave albedo when compared with ground measurements at four stations on the Tibetan Plateau (alpine steppe and farmland) over sparsely vegetated areas, consistent with the literature [75–77]. Using the increased number of valid observations under all sky conditions and taking topography effect into account, our method shows a significant improvement in RMSE values at the observation sites compared to the MCD43A3 data, indicating a more accurate reflection of terrain and atmospheric influences on albedo.

5.2. Limitations of Our Method

Distinguishing between snow/ice and cloud cover in multispectral imagery can be challenging due to their similar spectral properties [78]. Clouds typically exhibit higher reflectance in the visible (VIS), near-infrared (NIR), and shortwave infrared (SWIR) bands along with lower temperatures compared to the underlying Earth surface. Cloud detection relies on well-defined characteristics, such as high reflectance in VIS and NIR/SWIR bands, coupled with lower temperatures in thermal bands. Establishing accurate reflectance and temperature thresholds facilitates the identification of diverse cloud types. Resolving spatial heterogeneity of snowfall over the Tibetan Plateau is a challenge, impacting the accuracy of retrieved albedo. The misclassification of snow as clouds in the MOD06/MYD06 cloud data product leads to inaccuracies in the retrieved albedo. The snow-covered surfaces have high albedo, and mistaking these areas for clouds can result in inaccurate estimations of the surface albedo.

In our calculation of at-surface upwelling radiance and surface albedo under cloudy conditions, we assume that the cloud distribution within a pixel is uniform, plane-parallel, and static. However, the validity of this assumption depends on the specific type of cloud and its actual spatial distribution within the pixel, which can result in significant discrepancies between our estimates and observations. In Stephens's model [30] used in this study, it is assumed that the clouds are liquid water clouds, and assuming the wrong cloud type can lead to less accurate estimates of cloud reflectance and transmittance.

Terrain has multiple impacts on albedo across spatial scales [16]. Hao et al. [20] focused on a quantitative exploration of the topographic effect, elucidating that the BSA exhibits a decreasing trend with an increase in mean slope. Moreover, an augmented solar zenith angle accentuates the diminishing trend in BSA, particularly as mean terrain slope increases. Gao et al. [21] proposed an improved method by removing the surrounding topographic effect on land surface reflectance and considering the topographic effect in retrieving land surface albedo. By considering the influence of the topography, our method incorporates various critical factors such as the illumination angle, observing angle, effects of shadowing, and the reflectance of radiation due to the surrounding terrain. Currently, we can retrieve surface albedo with a spatial resolution of 1 km $\times$ 1 km. However, this resolution restricts our ability to accurately capture the heterogeneity of land cover within each 1 km $\times$ 1 km pixel. Such pixels may contain diverse land cover types, each with distinct albedo properties, and this diversity can result in a generalized smoothing effect that conceals minor variations. Our method may still be inadequate to reveal complex terrain and land cover changes at the sub-pixel level. Wen et al. [79] illustrated the scaling effects and the necessary corrections for surface albedo assessments in rugged terrain, while Liu et al. [80] have provided insights into how sub-pixel details can refine our understanding of pixel-wide land surface albedo. Furthermore, Wen et al. [81] has introduced an advanced multi-sensor BRDF inversion model, designed specifically to account for sub-pixel topographic influences. These studies collectively underscore the value of integrating sub-pixel topography and land cover data to enhance the accuracy of surface BRDF and land surface albedo estimations, particularly in rugged terrain. Future advancements could stem from accounting for the spatial variability of terrain at the sub-pixel level, utilizing high-resolution DEM data and sophisticated downscaling methods. Such refinements are anticipated to bridge the gap between current capabilities and the nuanced realities of terrain effects on albedo measurements.

The tower-mounted radiometers used to measure radiative fluxes at the land-atmosphere interface are typically installed horizontally on flat terrain, i.e., the field-measured shortwave upward and downward radiation apply to a flat and horizontal plane. These conditions apply to the radiometers installed at our four stations (i.e., QOMS, MAWORS, NAMORS and Dali; see Table 1), where the slopes are about 5° or less (Table 1). On the other hand, our estimates of shortwave upward and downward radiation apply to the plane parallel to a 1 km $\times$ 1 km terrain facet. This means that the footprints captured by space-borne and in situ radiometers may be inherently different. Radiometers on the ground are usually designed to measure all-sky radiation from all directions.

In situ radiometers observe at-surface upwelling radiance from an entire hemisphere, but mostly originating from a small area on the surface, and atmospheric conditions have limited impact. However, satellite observations are more influenced by atmospheric conditions, as they record radiation from a larger area and are impacted by highly variable atmospheric factors, such as cloud cover and atmospheric aerosols. The accuracy of estimating at-surface upwelling radiance using satellite data is more complicated due to the limitations of satellite observation angles and distance. This makes it more difficult to match in situ measurements of at-surface upwelling radiance. The study estimated at-surface upwelling radiance and surface albedo by averaging the slope and aspect of the terrain within each 1 km × 1 km grid, calculated from 30 m × 30 m resolution data. This approach ensures that the terrain orientation in the retrieval process aligns with that at the radiometer location, providing comparable terrain geometry apply. However, the radiometer's footprint is significantly smaller than the 1 km × 1 km pixel used in the retrieval.

In summary, our method has made certain advancements in estimating all-sky surface albedo over rugged terrain on the Tibetan Plateau, but it also reveals several potential limitations and directions for improvement. Future research should focus on enhancing the model adaptability to complex terrains and varying climatic conditions, further improving estimation accuracy and reliability. Through continuous validation, calibration, and refinement, we anticipate further enhancing our understanding and representation of the complex interplay between surface characteristics and atmospheric conditions on the Tibetan Plateau.

6. Conclusions

The primary aim of this study was to introduce an innovative approach for the all-sky retrieval of surface albedo across the Tibetan Plateau with consideration of the terrain characteristics and atmospheric effects, particularly clouds. This method was designed to significantly reduce the occurrence of data gaps mainly due to clouds and enhance the accuracy of the estimated albedo. Notably, in a previous companion study, we addressed the retrieval of surface solar irradiance and its four components, including direct, isotropic diffuse, circumsolar diffuse, and surrounding terrain irradiance. Both surface solar irradiance and at-surface upwelling radiance have been estimated by accounting for the actual terrain geometry derived from DEM data. Under cloudy conditions, the radiative transfer model incorporates cloud properties such as cloud cover fraction, reflectance, and transmittance to estimate surface solar irradiance and at-surface upwelling radiance. The RTLSR kernel-driven model is utilized to describe the BRDF of observed targets and to estimate the reflectance for the four components of SSI in all sky conditions, which are subsequently used to determine the total reflected irradiance within the viewing hemisphere. Ultimately, the surface albedo is calculated as the ratio of the total reflected radiance to the total incident solar irradiance, in alignment with the fundamental concept of surface albedo.

Our study demonstrates that by integrating terrain information and cloud properties into radiative transfer modeling, the accuracy of estimating surface albedo and radiance in complex terrains can be effectively improved. This method is significant for understanding and monitoring the surface radiation energy balance in complex terrain areas like the Tibetan Plateau and provides more accurate input data for climate modeling and analysis. We anticipate that the outcomes of our research will offer a viable solution for accurate estimation of surface albedo on a global scale and will have a positive impact on global change research and environmental monitoring.

Author Contributions: Conceptualization, J.J., M.M. and L.J.; methodology, J.J., M.M. and L.J.; software, J.J.; validation, J.J.; formal analysis, J.J., M.M., L.J. and Q.C.; investigation, J.J., M.M. and L.J.; resources, L.J. and M.M.; data curation, J.J., L.J. and A.X.; writing—original draft preparation, J.J.; writing—review and editing, L.J., M.M. and Q.C.; visualization, J.J.; supervision, M.M. and L.J.; project administration, L.J. and M.M.; funding acquisition, L.J. and M.M. All authors have read and agreed to the published version of the manuscript.

Funding: This research was jointly funded by the Second Tibetan Plateau Scientific Expedition and Research Program (STEP) (Grant No. 2019QZKK0103) and the National Natural Science Foundation of China project (Grant No. 91737205). MM acknowledges the support of the Chinese Academy of Sciences President's International Fellowship Initiative (Grant No. 2020VTA0001) and the MOST High Level Foreign Expert Program (Grant No. G2022055010L).

Data Availability Statement: The datasets presented in this article are not readily available because the data are part of an ongoing study.

Acknowledgments: We thank the Institute of Tibetan Plateau Research of the Chinese Academy of Sciences and the Dali National Climate Observatory for providing valuable ground observation data.

Conflicts of Interest: The authors declare no conflicts of interest.

References

1. Zhang, X.; Jiao, Z.; Zhao, C.; Qu, Y.; Liu, Q.; Zhang, H.; Tong, Y.; Wang, C.; Li, S.; Guo, J.; et al. Review of Land Surface Albedo: Variance Characteristics, Climate Effect and Management Strategy. *Remote Sens.* **2022**, *14*, 1382. [CrossRef]
2. Dickinson, R.E. Land Surface Processes and Climate—Surface Albedos and Energy Balance. In *Theory of Climate, Proceedings of a Symposium Commemorating the Two-Hundredth Anniversary of the Academy of Sciences of Lisbon*; Saltzman, B., Ed.; Advances in Geophysics; Elsevier: Amsterdam, The Netherlands, 1983; Volume 25, pp. 305–353.
3. Henderson-Sellers, A.; Wilson, M.F. Surface albedo data for climatic modeling. *Rev. Geophys.* **1983**, *21*, 1743–1778. [CrossRef]
4. Liang, S.; Zhao, X.; Liu, S.; Yuan, W.; Cheng, X.; Xiao, Z.; Zhang, X.; Liu, Q.; Cheng, J.; Tang, H.; et al. A long-term Global LAnd Surface Satellite (GLASS) data-set for environmental studies. *Int. J. Digit. Earth* **2013**, *6*, 5–33. [CrossRef]
5. Dickinson, R.E. Land processes in climate models. *Remote Sens. Environ.* **1995**, *51*, 27–38. [CrossRef]
6. Sellers, P.J.; Meeson, B.W.; Hall, F.G.; Asrar, G.; Murphy, R.E.; Schiffer, R.A.; Bretherton, F.P.; Dickinson, R.E.; Ellingson, R.G.; Field, C.B.; et al. Remote sensing of the land surface for studies of global change: Models—Algorithms—Experiments. *Remote Sens. Environ.* **1995**, *51*, 3–26. [CrossRef]
7. Ma, Y.; Yao, T.; Zhong, L.; Wang, B.; Xu, X.; Hu, Z.; Ma, W.; Sun, F.; Han, C.; Li, M.; et al. Comprehensive study of energy and water exchange over the Tibetan Plateau: A review and perspective: From GAME/Tibet and CAMP/Tibet to TORP, TPEORP, and TPEITORP. *Earth-Sci. Rev.* **2023**, *237*, 104312. [CrossRef]
8. Lin, X.; Wen, J.; Liu, Q.; You, D.; Wu, S.; Hao, D.; Xiao, Q.; Zhang, Z.; Zhang, Z. Spatiotemporal Variability of Land Surface Albedo over the Tibet Plateau from 2001 to 2019. *Remote Sens.* **2020**, *12*, 1188. [CrossRef]
9. Qu, Y.; Liang, S.; Liu, Q.; He, T.; Liu, S.; Li, X. Mapping Surface Broadband Albedo from Satellite Observations: A Review of Literatures on Algorithms and Products. *Remote Sens.* **2015**, *7*, 990–1020. [CrossRef]
10. Schaaf, C.B.; Gao, F.; Strahler, A.H.; Lucht, W.; Li, X.; Tsang, T.; Strugnell, N.C.; Zhang, X.; Jin, Y.; Muller, J.-P.; et al. First operational BRDF, albedo nadir reflectance products from MODIS. *Remote Sens. Environ.* **2002**, *83*, 135–148. [CrossRef]
11. Strahler, A.H.; Muller, J.-P.; MODIS Science Team Members. MODIS BRDF/Albedo Product: Algorithm Theoretical Basis Document Version 5.0. *MODIS Doc.* **1999**, *23*, 42–47.
12. Jin, Y.; Schaaf, C.B.; Gao, F.; Li, X.; Strahler, A.H.; Lucht, W.; Liang, S. Consistency of MODIS surface bidirectional reflectance distribution function and albedo retrievals: 1. Algorithm performance. *J. Geophys. Res. Atmos.* **2003**, *108*. [CrossRef]
13. Aijun, C.; Lingen, B.; Yujie, L. Deriving albedo over cloudy areas with composite inversion. *Proc. SPIE* **2007**, *6753*, 75337.
14. Liu, N.F.; Liu, Q.; Wang, L.Z.; Liang, S.L.; Wen, J.G.; Qu, Y.; Liu, S.H. A statistics-based temporal filter algorithm to map spatiotemporally continuous shortwave albedo from MODIS data. *Hydrol. Earth Syst. Sci.* **2013**, *17*, 2121–2129. [CrossRef]
15. Vermote, E.F.; El Saleous, N.Z.; Justice, C.O. Atmospheric correction of MODIS data in the visible to middle infrared: First results. *Remote Sens. Environ.* **2002**, *83*, 97–111. [CrossRef]
16. Ju, J.; Roy, D.P.; Shuai, Y.; Schaaf, C. Development of an approach for generation of temporally complete daily nadir MODIS reflectance time series. *Remote Sens. Environ.* **2010**, *114*, 1–20. [CrossRef]
17. Samain, O.; Geiger, B.; Roujean, J.L. Spectral Normalization and Fusion of Optical Sensors for the Retrieval of BRDF and Albedo: Application to VEGETATION, MODIS, and MERIS Data Sets. *IEEE Trans. Geosci. Remote Sens.* **2006**, *44*, 3166–3179. [CrossRef]
18. Quaife, T.; Lewis, P. Temporal Constraints on Linear BRDF Model Parameters. *IEEE Trans. Geosci. Remote Sens.* **2010**, *48*, 2445–2450. [CrossRef]
19. Moody, E.G.; King, M.D.; Platnick, S.; Schaaf, C.B.; Feng, G. Spatially complete global spectral surface albedos: Value-added datasets derived from Terra MODIS land products. *IEEE Trans. Geosci. Remote Sens.* **2005**, *43*, 144–158. [CrossRef]
20. Hao, D.; Wen, J.; Xiao, Q.; Wu, S.; Lin, X.; Dou, B.; You, D.; Tang, Y. Simulation and Analysis of the Topographic Effects on Snow-Free Albedo over Rugged Terrain. *Remote Sens.* **2018**, *10*, 278. [CrossRef]
21. Gao, B.; Jia, L.; Menenti, M. An Improved Method for Retrieving Land Surface Albedo Over Rugged Terrain. *IEEE Geosci. Remote Sens. Lett.* **2014**, *11*, 554–558. [CrossRef]
22. Wen, J.; Lin, X.; Wu, X.; Bao, Y.; You, D.; Gong, B.; Tang, Y.; Wu, S.; Xiao, Q.; Liu, Q. Validation of the MCD43A3 Collection 6 and GLASS V04 Snow-Free Albedo Products Over Rugged Terrain. *IEEE Trans. Geosci. Remote Sens.* **2022**, *60*, 1–11. [CrossRef]
23. Yang, C.; Vidal, A. Combination of digital elevation models with SPOT-1 HRV multispectral imagery for reflectance factor mapping. *Remote Sens. Environ.* **1990**, *32*, 35–45. [CrossRef]
24. Roupioz, L.; Nerry, F.; Jia, L.; Menenti, M. Improved Surface Reflectance from Remote Sensing Data with Sub-Pixel Topographic Information. *Remote Sens.* **2014**, *6*, 10356–10374. [CrossRef]
25. Roupioz, L.; Jia, L.; Nerry, F.; Menenti, M. Correction of sub-pixel topographical effects on land surface albedo retrieved from geostationary satellite (FengYun-2D) observations. *IOP Conf. Ser. Earth Environ. Sci.* **2014**, *17*, 012270. [CrossRef]
26. Ma, Y.; Pinker, R.T. Modeling shortwave radiative fluxes from satellites. *J. Geophys. Res. Atmos.* **2012**, *117*. [CrossRef]

27. Pinker, R.T.; Laszlo, I. Modeling Surface Solar Irradiance for Satellite Applications on a Global Scale. *J. Appl. Meteorol. Climatol.* **1992**, *31*, 194–211. [CrossRef]
28. Tang, W.; Qin, J.; Yang, K.; Liu, S.; Lu, N.; Niu, X. Retrieving high-resolution surface solar radiation with cloud parameters derived by combining MODIS and MTSAT data. *Atmos. Chem. Phys.* **2016**, *16*, 2543–2557. [CrossRef]
29. Liang, S.; Zheng, T.; Wang, D.; Wang, K.; Liu, R.; Tsay, S.-C.; Running, S.; Townshend, J. Mapping High-Resolution Incident Photosynthetically Active Radiation over Land from Polar-Orbiting and Geostationary Satellite Data. *Photogramm. Eng. Remote Sens.* **2007**, *73*, 1085–1089.
30. Stephens, G.L.; Ackerman, S.; Smith, E.A. A Shortwave Parameterization Revised to Improve Cloud Absorption. *J. Atmos. Sci.* **1984**, *41*, 687–690. [CrossRef]
31. Slingo, A. A GCM parameterization for the shortwave radiative properties of water clouds. *J. Atmos. Sci.* **1989**, *46*, 1419–1427. [CrossRef]
32. Fu, Q.; Yang, P.; Sun, W.B. An Accurate Parameterization of the Infrared Radiative Properties of Cirrus Clouds for Climate Models. *J. Clim.* **1998**, *11*, 2223–2237. [CrossRef]
33. Van Laake, P.E.; Sanchez-Azofeifa, G.A. Simplified atmospheric radiative transfer modelling for estimating incident PAR using MODIS atmosphere products. *Remote Sens. Environ.* **2004**, *91*, 98–113. [CrossRef]
34. Huang, G.; Liu, S.; Liang, S. Estimation of net surface shortwave radiation from MODIS data. *Int. J. Remote Sens.* **2012**, *33*, 804–825. [CrossRef]
35. Huang, G.; Liang, S.; Lu, N.; Ma, M.; Wang, D. Toward a Broadband Parameterization Scheme for Estimating Surface Solar Irradiance: Development and Preliminary Results on MODIS Products. *J. Geophys. Res. Atmos.* **2018**, *123*, 12,180–12,193. [CrossRef]
36. Cano, D.; Monget, J.M.; Albuisson, M.; Guillard, H.; Regas, N.; Wald, L. A method for the determination of the global solar radiation from meteorological satellite data. *Solar Energy* **1986**, *37*, 31–39. [CrossRef]
37. Hammer, A.; Heinemann, D.; Hoyer, C.; Kuhlemann, R.; Lorenz, E.; Müller, R.; Beyer, H.G. Solar energy assessment using remote sensing technologies. *Remote Sens. Environ.* **2003**, *86*, 423–432. [CrossRef]
38. Zhang, Y.; He, T.; Liang, S.; Wang, D.; Yu, Y. Estimation of all-sky instantaneous surface incident shortwave radiation from Moderate Resolution Imaging Spectroradiometer data using optimization method. *Remote Sens. Environ.* **2018**, *209*, 468–479. [CrossRef]
39. Wang, T.; Yan, G.; Chen, L. Consistent retrieval methods to estimate land surface shortwave and longwave radiative flux components under clear-sky conditions. *Remote Sens. Environ.* **2012**, *124*, 61–71. [CrossRef]
40. Ryu, Y.; Jiang, C.; Kobayashi, H.; Detto, M. MODIS-derived global land products of shortwave radiation and diffuse and total photosynthetically active radiation at 5 km resolution from 2000. *Remote Sens. Environ.* **2018**, *204*, 812–825. [CrossRef]
41. Yang, L.; Zhang, X.; Liang, S.; Yao, Y.; Jia, K.; Jia, A. Estimating Surface Downward Shortwave Radiation over China Based on the Gradient Boosting Decision Tree Method. *Remote Sens.* **2018**, *10*, 185. [CrossRef]
42. Jia, J.; Menenti, M.; Jia, L.; Chen, Q.; Xu, A. Estimation of All-Sky Solar Irradiance Components over Rugged Terrain Using Satellite and Reanalysis Data: The Tibetan Plateau Experiment. *IEEE Trans. Geosci. Remote Sens.* **2024**.
43. Zhu, C.-S.; Cao, J.-J.; Xu, B.; Huang, R.-J.; Wang, P.; Ho, K.F.; Shen, Z.; Liu, S.-X.; Han, Y.; Tie, X.-X.; et al. Black Carbon Aerosols at Mt. Muztagh Ata, a High-Altitude Location in the Western Tibetan Plateau. *Aerosol Air Qual. Res.* **2016**, *16*, 752–763. [CrossRef]
44. Peng, S.; Wen, J.; Xiao, Q.; You, D.; Dou, B.; Liu, Q.; Tang, Y. Multi-Staged NDVI Dependent Snow-Free Land-Surface Shortwave Albedo Narrowband-to-Broadband (NTB) Coefficients and Their Sensitivity Analysis. *Remote Sens.* **2017**, *9*, 93. [CrossRef]
45. Ma, Y.; Hu, Z.; Xie, Z.; Ma, W.; Wang, B.; Chen, X.; Li, M.; Zhong, L.; Sun, F.; Gu, L.; et al. A long-term (2005–2016) dataset of hourly integrated land–atmosphere interaction observations on the Tibetan Plateau. Earth Syst. *Sci. Data* **2020**, *12*, 2937–2957. [CrossRef]
46. Liu, H.; Feng, J.; Sun, J.; Wang, L.; Xu, A. Eddy covariance measurements of water vapor and CO_2 fluxes above the Erhai Lake. *Sci. China Earth Sci.* **2015**, *58*, 317–328. [CrossRef]
47. Tang, B.; Li, Z.-L.; Zhang, R. A direct method for estimating net surface shortwave radiation from MODIS data. *Remote Sens. Environ.* **2006**, *103*, 115–126. [CrossRef]
48. Zaksek, K.; Oštir, K.; Kokalj, Ž. Sky-View Factor as a Relief Visualization Technique. *Remote Sens.* **2011**, *3*, 398. [CrossRef]
49. Allen, R.G.; Trezza, R.; Tasumi, M. Analytical integrated functions for daily solar radiation on slopes. *Agric. For. Meteorol.* **2006**, *139*, 55–73. [CrossRef]
50. Sandmeier, S.; Itten, K.I. A physically-based model to correct atmospheric and illumination effects in optical satellite data of rugged terrain. *IEEE Trans. Geosci. Remote Sens.* **1997**, *35*, 708–717. [CrossRef]
51. Dubayah, R. Estimating net solar radiation using Landsat Thematic Mapper and digital elevation data. *Water Resour. Res.* **1992**, *28*, 2469–2484. [CrossRef]
52. Lucht, W.; Schaaf, C.B.; Strahler, A.H. An algorithm for the retrieval of albedo from space using semiempirical BRDF models. *IEEE Trans. Geosci. Remote Sens.* **2000**, *38*, 977–998. [CrossRef]
53. Yang, K.; Huang, G.W.; Tamai, N. A hybrid model for estimating global solar radiation. *Sol. Energy* **2001**, *70*, 13–22. [CrossRef]
54. Duguay, C.R. An approach to the estimation of surface net radiation in mountain areas using remote sensing and digital terrain data. *Theor. Appl. Climatol.* **1995**, *52*, 55–68. [CrossRef]

55. Dozier, J.; Bruno, J.; Downey, P. Faster solution to the horizon problem. *Comput. Geosci.* **1981**, *7*, 145–151. [CrossRef]
56. Hay, J.E. Calculation of monthly mean solar radiation for horizontal and inclined surfaces. *Sol. Energy* **1979**, *23*, 301–307. [CrossRef]
57. Yang, K.; Koike, T.; Huang, G.; Tamai, N.J.R.d.i.s.e. Development and validation of an advanced model for estimating solar radiation from surface meteorological data. *Recent Dev. Sol. Energy* **2007**, *1*, 53.
58. Stephens, G.L. Radiation Profiles in Extended Water Clouds. II: Parameterization Schemes. *J. Atmos. Sci.* **1978**, *35*, 2123–2132. [CrossRef]
59. Lin, X.; Wu, S.; Hao, D.; Wen, J.; Xiao, Q.; Liu, Q. Sloping Surface Reflectance: The Best Option for Satellite-Based Albedo Retrieval Over Mountainous Areas. *IEEE Geosci. Remote Sens. Lett.* **2021**, *19*, 3002705. [CrossRef]
60. Schaepman-Strub, G.; Schaepman, M.E.; Painter, T.H.; Dangel, S.; Martonchik, J.V. Reflectance quantities in optical remote sensing—Definitions and case studies. *Remote Sens. Environ.* **2006**, *103*, 27–42. [CrossRef]
61. Huang, X.; Jiao, Z.; Dong, Y.; Zhang, H.; Li, X. Analysis of BRDF and Albedo Retrieved by Kernel-Driven Models Using Field Measurements. *IEEE J. Sel. Top. Appl. Earth Obs. Remote Sens.* **2013**, *6*, 149–161. [CrossRef]
62. Wanner, W.; Li, X.; Strahler, A.H. On the derivation of kernels for kernel-driven models of bidirectional reflectance. *J. Geophys. Res. Atmos.* **1995**, *100*, 21077–21089. [CrossRef]
63. Chrysoulakis, N.; Mitraka, Z.; Gorelick, N. Exploiting satellite observations for global surface albedo trends monitoring. *Theor. Appl. Climatol.* **2019**, *137*, 1171–1179. [CrossRef]
64. Liang, S. Narrowband to broadband conversions of land surface albedo I: Algorithms. *Remote Sens. Environ.* **2001**, *76*, 213–238. [CrossRef]
65. Carrer, D.; Pinault, F.; Lellouch, G.; Trigo, I.F.; Benhadj, I.; Camacho, F.; Ceamanos, X.; Moparthy, S.; Munoz-Sabater, J.; Schüller, L.; et al. Surface Albedo Retrieval from 40-Years of Earth Observations through the EUMETSAT/LSA SAF and EU/C3S Programmes: The Versatile Algorithm of PYALUS. *Remote Sens.* **2021**, *13*, 372. [CrossRef]
66. Chavez, P.S. An improved dark-object subtraction technique for atmospheric scattering correction of multispectral data. *Remote Sens. Environ.* **1988**, *24*, 459–479. [CrossRef]
67. Smith, G.M.; Milton, E.J. The use of the empirical line method to calibrate remotely sensed data to reflectance. *Int. J. Remote Sens.* **1999**, *20*, 2653–2662. [CrossRef]
68. Nazeer, M.; Nichol, J.E.; Yung, Y.-K. Evaluation of atmospheric correction models and Landsat surface reflectance product in an urban coastal environment. *Int. J. Remote Sens.* **2014**, *35*, 6271–6291. [CrossRef]
69. Bilal, M.; Nazeer, M.; Nichol, J.E.; Bleiweiss, M.P.; Qiu, Z.; Jäkel, E.; Campbell, J.R.; Atique, L.; Huang, X.; Lolli, S. A Simplified and Robust Surface Reflectance Estimation Method (SREM) for Use over Diverse Land Surfaces Using Multi-Sensor Data. *Remote Sens.* **2019**, *11*, 1344. [CrossRef]
70. Govaerts, Y.M.; Wagner, S.; Lattanzio, A.; Watts, P. Joint retrieval of surface reflectance and aerosol optical depth from MSG/SEVIRI observations with an optimal estimation approach: 1. Theory. *J. Geophys. Res. Atmos.* **2010**, *115*. [CrossRef]
71. Shi, H.; Xiao, Z.; Wen, J.; Wu, S. An Optical–Thermal Surface–Atmosphere Radiative Transfer Model Coupling Framework With Topographic Effects. *IEEE Trans. Geosci. Remote Sens.* **2022**, *60*, 4400312. [CrossRef]
72. Deschamps, P.; Breon, F.-M.; Leroy, M.; Podaire, A.; Bricaud, A.; Buriez, J.C.; Seze, G. The POLDER Mission: Instrument Characteristics and Scientific Objectives. *IEEE Trans. Geosci. Remote Sens.* **1994**, *32*, 598–615. [CrossRef]
73. Diner, D.J.; Beckert, J.C.; Reilly, T.H.; Bruegge, C.J.; Conel, J.E.; Kahn, R.A.; Martonchik, J.V.; Ackerman, T.P.; Davies, R.; Gerstl, S.A.W.; et al. Multi-angle Imaging SpectroRadiometer (MISR) instrument description and experiment overview. *IEEE Trans. Geosci. Remote Sens.* **1998**, *36*, 1072–1087. [CrossRef]
74. Shi, Z.; Sha, Y.; Liu, X. Effect of Yunnan–Guizhou Topography at the Southeastern Tibetan Plateau on the Indian Monsoon. *J. Clim.* **2017**, *30*, 1259–1272. [CrossRef]
75. Sobrino, J.A.; Franch, B.; Oltra-Carrió, R.; Vermote, E.F.; Fedele, E. Evaluation of the MODIS Albedo product over a heterogeneous agricultural area. *Int. J. Remote Sens.* **2013**, *34*, 5530–5540. [CrossRef]
76. Cescatti, A.; Marcolla, B.; Santhana Vannan, S.K.; Pan, J.Y.; Román, M.O.; Yang, X.; Ciais, P.; Cook, R.B.; Law, B.E.; Matteucci, G.; et al. Intercomparison of MODIS albedo retrievals and in situ measurements across the global FLUXNET network. *Remote Sens. Environ.* **2012**, *121*, 323–334. [CrossRef]
77. Zhang, H.; Zhang, X.; Cui, L.; Dong, Y.; Liu, Y.; Xi, Q.; Cao, H.; Chen, L.; Lian, Y. Enhancing Leaf Area Index Estimation with MODIS BRDF Data by Optimizing Directional Observations and Integrating PROSAIL and Ross–Li Models. *Remote Sens.* **2023**, *15*, 5609. [CrossRef]
78. Stillinger, T.; Roberts, D.A.; Collar, N.M.; Dozier, J. Cloud Masking for Landsat 8 and MODIS Terra Over Snow-Covered Terrain: Error Analysis and Spectral Similarity between Snow and Cloud. *Water Resour. Res.* **2019**, *55*, 6169–6184. [CrossRef] [PubMed]
79. Wen, J.; Liu, Q.; Liu, Q.; Xiao, Q.; Li, X. Scale effect and scale correction of land-surface albedo in rugged terrain. *Int. J. Remote Sens.* **2009**, *30*, 5397–5420. [CrossRef]

80. Liu, W.; Hu, B.; Wang, S. Improving Land Surface Pixel Level Albedo Characterization Using Sub-Pixel Information Retrieved from Remote Sensing. In Proceedings of the IGARSS 2008—2008 IEEE International Geoscience and Remote Sensing Symposium, Boston, MA, USA, 7–11 July 2008; Volume 2, pp. 801–804.
81. Wen, J.; You, D.; Han, Y.; Lin, X.; Wu, S.; Tang, Y.; Xiao, Q.; Liu, Q. Estimating Surface BRDF/Albedo over Rugged Terrain Using an Extended Multi-Sensor Combined BRDF Inversion (EMCBI) Model. *IEEE Geosci. Remote Sens. Lett.* **2022**, *19*, 2503505. [CrossRef]

remote sensing

MDPI

Article

Analysis of Atmospheric Boundary Layer Characteristics on Different Underlying Surfaces of the Eastern Tibetan Plateau in Summer

Xiaohang Wen [1,2], Jie Ma [3,*] and Mei Chen [1]

1 Plateau Atmosphere and Environment Key Laboratory of Sichuan Province, College of Atmospheric Sciences, Chengdu University of Information Technology, Chengdu 610225, China; wxh@cuit.edu.cn (X.W.); 3220101022@stu.cuit.edu.cn (M.C.)
2 State Key Laboratory of Severe Weather, Chinese Academy of Meteorological Sciences, Beijing 100081, China
3 Luzhou Meteorological Bureau, Luzhou 646099, China
* Correspondence: mjie9@hotmail.com

Abstract: The atmospheric boundary layer is a key region for human activities and the interaction of various layers and is an important channel for the transportation of momentum, heat, and various substances between the free atmosphere and the surface, which has a significant impact on the development of weather and climate change. During the Second Tibetan Plateau Scientific Expedition and Research Program (STEP) in June 2022, utilizing the comprehensive stereoscopic observation experiment of the "Plateau Low Vortex Network", this study analyzed the variation characteristics and influencing factors of the atmospheric boundary layer height (ABLH) at three stations with different underlying surface types on the Qinghai–Tibet Plateau (QTP): Qumalai Station (grassland), Southeast Tibet Observation and Research Station for the Alpine Environment (SETORS, forest), and Sieshan Station (cropland). The analysis utilized sounding observation data, microwave radiometer data, and ERA5 reanalysis data. The results revealed that the temperature differences between the sounding observation data and microwave radiometer data were minor at the three stations, with a notable temperature inversion phenomenon observed at Sieshan Station. Regarding water vapor density, the differences between the sounding observation data and microwave radiometer data were relatively small at Sieshan Station. The relative humidity increased with height at Sieshan Station, whereas it increased and then decreased with height at SETORS and Qumalai Station. The ABLH at all sites reached its maximum value around noon, approximately 1500 m, and exhibited mostly convective boundary layer (CBL) characteristics. During the night, the ABLH mostly showed a stable boundary layer (SBL) pattern, with heights around 250 m. In summer, latent heat flux (LE) and sensible heat flux (H) in the eastern plateau were generally lower than those in the western plateau except at 20:00, where they were higher. Vertical velocity (w) in the eastern plateau was greater than in the western plateau. Among Sieshan Station and SETORS, LE, and H had the most significant impact on ABLH, while at Qumalai Station, ABLH was more influenced by surface long-wave radiation (Rlu). These four influencing factors showed a positive correlation with ABLH. The impact of different underlying surface types on ABLH primarily manifests in surface temperature variations, solar radiation intensity, vegetation cover, and terrain. Grasslands typically exhibit a larger range of ABLH variations, while the ABLH in forests and mountainous cropland areas is relatively stable.

Keywords: Tibetan Plateau; atmospheric boundary layer; different underlying surfaces; ERA5; impact factors

Citation: Wen, X.; Ma, J.; Chen, M. Analysis of Atmospheric Boundary Layer Characteristics on Different Underlying Surfaces of the Eastern Tibetan Plateau in Summer. *Remote Sens.* **2024**, *16*, 1645. https://doi.org/10.3390/rs16091645

Academic Editor: Carmine Serio

Received: 2 April 2024
Revised: 24 April 2024
Accepted: 2 May 2024
Published: 5 May 2024

1. Introduction

The atmospheric boundary layer (ABL), or planetary boundary layer (PBL), is the part of the troposphere that is directly influenced by the ground, and it responds to the action of the ground on a time scale of 1 h or less, up to a height of several kilometers. It is an important channel for the transport of momentum, heat, and various materials in the Earth's atmosphere and has an important role in the exchange of energy and materials between the Earth's surface and the free atmosphere [1–4]. A deeper understanding of the structure and characteristics of the ABL will help to better understand the transfer and circulation of energy and matter in the Earth system [5–9]. The average altitude of the TP is above 4 km, and the plateau has a unique ABL formed by strong solar radiation, a significant temperature difference between the ground and the complex topography [8,10,11]. The interplay between the eastern plateau's intricate topography and diverse circulation systems contributes to unique weather and climate features in the region. Therefore, it is important to study the ABL structure in this region to understand the genesis of weather and climate and its influence on the weather in China [11,12].

Based on the important research value of the ABL on the TP, many scholars have conducted a lot of research on the Atmospheric Boundary Layer height (ABLH) and the vertical distribution of meteorological elements in the boundary layer on the TP in recent decades. Li et al. [13–15] found that the height of the convective mixed layer in the northern TP is higher in the dry season than in the rainy season, and the specific humidity in the dry season is smaller than that in the wet season. The convective ABL is formed later on rainy days than on cloudy and sunny days. And glacial winds affect the ABLH in the Everest region with a maximum height of 3888 m. Su et al. [1] found that the main influencing factors of ABLH in the western region are 0–10 cm soil water content and 10 m wind speed, while the main influencing factor of ABLH in the eastern region is cloudiness. It has been shown that the ABLH in the plateau under the south branch of the westerly wind field is higher than that under the summer wind field in the plateau [8]. Ma et al. [16] used radio sounding data to find that the temperature decreases at a rate of 0.74 K/100 m in the summer in the northern TP, and the relative humidity increases and then decreases with height. Gu et al. [17] found that inversion humidity occurred in the summer in both the Amdo and Nagqu regions, with strong inversion humidity mainly occurring at midnight or midday. Wang and Ma [18] studied the ABL structure in the Everest region and found that the wind speed in summer varies greatly from day to day, especially in the afternoon, and the wind speed increases with height at about 1800–2300 m and then decreases upward. height. Chen et al. [19] found that the winter ABL in the western TP can reach up to 9.2 km above sea level, which is close to the height of the top of the troposphere in the plateau region. Xu et al. [9] found that the mean ABLH in the eastern and downstream areas of the plateau is high during the day and low at night, while the mean ABLH in the central part of the plateau is lower around sunrise and then gradually increases with time.

The above studies have investigated the ABL of the TP from various aspects, including the ABLH, the vertical trends of meteorological elements within the boundary layer, and the factors affecting the development of ABLH, which are important for further understanding the structure and characteristics of ABL on the plateau. However, these studies mostly focus on a single underlying surface, and the comparative analysis of ABL characteristics in different underlying surfaces is less. This study compares and analyzes the sounding observations and ERA5 reanalysis data from the Qumalai and SETORS stations located in the eastern part of the plateau and the Sieshan station located at the southeastern edge of the plateau in June 2022 and combines them with the microwave radiometer data from the Sieshan station, to study the characteristics of ABL in the different underlying surfaces in the eastern part of the plateau and the factors influencing ABLH, which can help to understand the process and structure of the atmospheric boundary layer in the area and to provide some references for the study on the development of the weather and the climate change.

2. Data and Methods

2.1. Data

2.1.1. Observation Data

The three stations in this paper are as follows: Sieshan station is located on the southeastern edge of the TP (Figure 1). Its underlying surface is dominated by farmland and its altitude is 807 m (Table 1). SETORS is located in the southeast of the TP (Figure 1), and its underlying surface is dominated by evergreen needleleaf forests, with an altitude of 3327 m (Table 1). Qumalai station is located in the eastern TP. Its underlying surface is mainly grassland, and its altitude is 4199 m (Table 1).

Table 1. Description of geographic features of three sites.

Station	Latitude	Longitude	Elevation (m)	Land Cover
Qumalai	34.13°N	95.81°E	4199	grassland
SETORS	29.75°N	94.75°E	3327	evergreen needleleaf forests
Sieshan	29.27°N	103.60°N	807	farmland

In this study, sounding data from three stations (Qumalai station, SETORS, and Sieshan station) of the Second Tibetan Plateau Scientific Expedition and Research on the TP were used from June 10 to 18 and 26 to 30. The Qumalai station uses the RS92 sounding instrument of Vaisala company, which collects and sends a set of data of pressure, temperature, humidity, wind speed and wind direction per second, among which the range of atmospheric pressure measurement is 1080 to 3 hPa, the resolution is 0.4 hPa when it is larger than 100 hPa, and 0.3 hPa when it is smaller than 100 hPa, and the combined uncertainty is 1.0 hPa when it is larger than 100 hPa in the sounding. The measurement range of the temperature sensor is −90–60 °C, the resolution is 0.2 °C, and the combined uncertainty is 0.5 °C in the sounding. The measurement range of the humidity sensor is 0–100%RH with a resolution of 2%RH, and the combined uncertainty is 5%RH in the sounding. The wind speed is measured with a resolution of 1 m/s, and the combined uncertainty is 0.15 m/s in the sounding (Table 2) [20]. And the Sieshan station and SETORS use an RS41-SG type sounding instrument; it is the latest generation of Vaisala company full digital sounding instrument that collects and sends a set of pressure, temperature, humidity, wind speed, and wind direction every second, including atmospheric pressure measurement range for the surface pressure to 3 hPa, the resolution is 0.01 hPa, the combined uncertainty in the sounding is 1.0 hPa when it is greater than 100 hPa, 0.3 hPa when it is between 100 and 10 hPa, and 0.04 hPa when it is less than 10 hPa. The temperature sensor measurement range is −95–60 °C, the resolution is 0.01 °C, the combined uncertainty is 0.3 °C when the height is less than 16 km and 0.4 °C when the height is more than 16 km in the sounding. The measurement range of the humidity sensor is 0–100%RH, the resolution is 0.1%RH, and the combined uncertainty is 4%RH in the sounding. Wind speed resolution is 0.1 m/s, and the combined uncertainty is 0.15 m/s. The wind direction measurement range is 0–360°, the resolution is 0.1°, and the combined uncertainty is 2° (Table 3).

Table 2. RS92 sounding instrument main performance.

Projects	Atmospheric Pressure (hPa)	Temperature (°C)	Humidity (%RH)	Wind Speed (m/s)
Measurement Range	1080 to 3	−90–60	0–100	/
Resolution	0.4 (>100) 0.3 (<100)	0.2	2	1
combined uncertainty	1.0 (>100) 0.6 (<100) 0.6 (<10)	0.5	5	0.15

Table 3. RS41-SG sounding instrument main performance.

Projects	Atmospheric Pressure (hPa)	Temperature (°C)	Humidity (%RH)	Wind Speed (m/s)	Wind Direction (°)
Measurement Range	surface pressure-3	−95–60	0–100	/	0–360
Resolution	0.01	0.01	0.1	0.1	0.1
combined uncertainty	1.0 (>100) 0.3 (<100) 0.04 (<10)	0.3 (<16 km) 0.4 (>16 km)	4	0.15	2

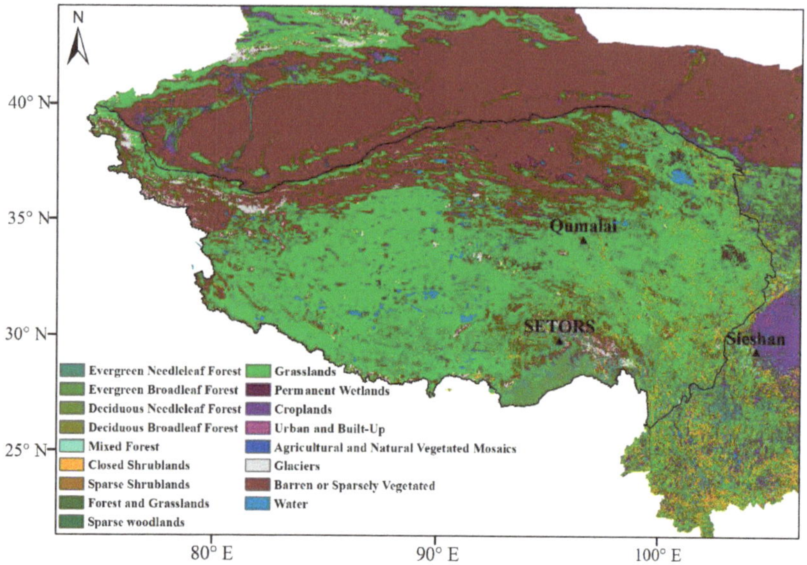

Figure 1. Spatial pattern of land use and station distribution [21].

2.1.2. Microwave Radiometer Data

In this study, the microwave radiometer model MWP967KV is used to automatically invert atmospheric temperature, humidity, cloud water distribution, water vapor, liquid water content, and other atmospheric parameters in the headspace in real time, and can continuously detect the temperature (°C), water vapor density (g/m^3), relative humidity (%) and liquid water content (g/m^3) from the ground to an altitude of 10,000 m. MWP967KV observation: one layer every 25 m below 500 m, one layer every 100 m from 500 m to 2000 m, and one layer every 250 m from 2000 m to 10,000 m, for a total of 83 layers. Each layer includes temperature, water vapor density, relative humidity, and liquid water content and returns a set of data every 2 min. Since the ABL is mainly studied in this paper, only data below 5000 m are used, and interpolation is carried out on the data to make the vertical resolution below 5000 m become 10 m to compare and analyze with the observation data.

2.1.3. ERA5 Reanalysis Data

The ERA5 [22] reanalysis data are the fifth generation of atmospheric reanalysis data for global climate from the ECMWF (European Centre for Medium-Range Weather Forecasts). The reanalysis data combine model data with observations from around the world to form a globally complete and consistent dataset. ERA5 is generated from 4D-Var data assimilation and model forecasts in CY41R2 of the ECMWF Integrated Forecast System (IFS) and also provides "surface or single level" data containing 2D parameters such as precipitation, top-of-atmosphere radiation, and vertical integrals over the entire depth of the atmosphere, while parameters such as 2 m temperature and soil temperatures are generated by coupling the atmospheric model in IFS with a land-surface model (HTESSEL). The time coverage ranges from 1950 to now. The time resolution is 1 h, and the spatial resolution is 0.25° × 0.25°. In this study, the data used are the hourly vertical velocity and hourly LE, H, Rlu, and ABLH data of ERA5 reanalysis data in June 2022.

2.2. Analysis Method

2.2.1. Calculation Method of ABLH

In this paper, for the convenience of analysis and comparison, the sounding data of the three sites are uniformly processed into a vertical resolution of 10 m. The ABL is divided into a convective boundary layer and a stable boundary layer for analysis, and this division is based on the potential temperature profile and Richardson number (Ri) profile to discriminate [23]. The ABLH was estimated by the gas block method and Ri method, respectively. The Ri can be calculated using the following formula [24]:

$$Ri(z) = \frac{g(z - z_0)}{\theta_v(z)} \frac{\theta_v(z) - \theta_v(z_0)}{u(z)^2 + v(z)^2} \tag{1}$$

$$u(z) = V(z)sin\alpha(z) \tag{2}$$

$$v(z) = V(z)cos\alpha(z) \tag{3}$$

where g is the acceleration of gravity, taken as 9.81 m/s^2, z_0 is the ground elevation, θ_v is the imaginary temperature, while u and v are the latitudinal and longitudinal wind components, respectively, V is the wind speed, and α is the wind direction.

The formula for calculating the virtual potential temperature (θ_v) is as follows:

$$r_s = \frac{\varepsilon e_s}{P} \tag{4}$$

$$t \geq 0, e_s = 6.112e^{\left(\frac{17.67t}{t+243.5}\right)}; t < 0, e_s = 6.1078 * 10^{\left(\frac{7.5t}{t+237.3}\right)} \tag{5}$$

$$U_w = \frac{r_s}{r} \tag{6}$$

$$\theta_v = \theta(1 + 0.608r) \tag{7}$$

In the above equation r_s is the saturation mixing ratio, $\varepsilon \approx 0.622$, e_s is the saturation vapor pressure, t is the Celsius temperature, U_w is the relative humidity, r is the mixing ratio, P is the atmospheric pressure, and θ is the potential temperature [25].

The boundary layer height is discriminated as follows:

$$\theta_{200} - \theta_0 < 0, R_i(0 \sim 100) < 0\ i.e.\theta_v\ (100) - \theta_v\ (0) < 0\ \ CBL;\ else\ SBL \tag{8}$$

where θ_{200} and θ_0 represent the potential temperature at 200 m and at ground level, respectively, and $\theta_v(100)$ and $\theta_v(0)$ represent the virtual potential temperature at 100 m and at ground level, respectively [25].

The relative maximum value θ_{v0} of the virtual potential temperature θ_v on the ground, rises vertically until the height of θ_{v0} is detected again, and we define this height as the ABLH and this method is referred to as the air-block method [26], as shown in Figure 2. When the ABL is a convective boundary layer (CBL) structure, the gas block method is used to determine the ABLH [9,23–25,27].

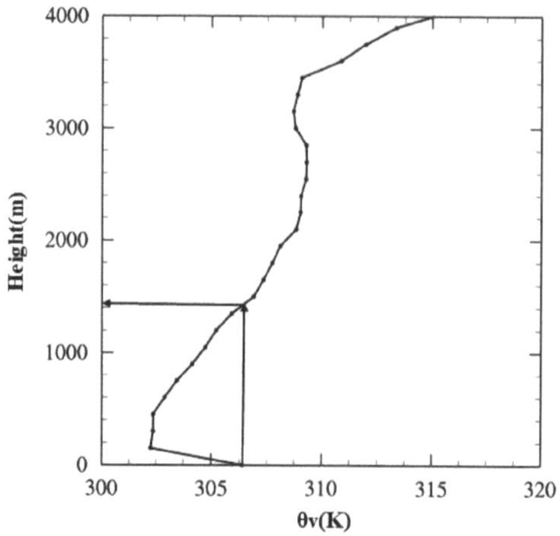

Figure 2. Schematic diagram of CBL height determination by the gas block method.

When the ABL is a stable boundary layer (SBL) structure, the Ri method is usually used to estimate the ABLH. The SBL height is defined as the height where the Ri is equal to or greater than a pre-given critical value [28]. In this study, the critical Richardson number (Ric) is used as a common value Ric = 0.25 [29].

2.2.2. Water Vapor Density (ρ_v) Calculation

$$q = \frac{\varepsilon e}{P} \tag{9}$$

$$\rho_v = \frac{\varepsilon e}{R_d T} \tag{10}$$

where e is the actual water vapor pressure, q is the specific humidity, ρ_v is the water vapor density, R_d is the specific gas constant of air, which is 287 J·K^{-1}·Kg^{-1}, and T is the Kelvin temperature.

2.2.3. Correlation Coefficient (R) Calculation

R is a statistical indicator that reflects the closeness of the correlation between the variables. The value of R is between −1 and 1. If the coefficient is positive then the two variables are positively correlated. If the coefficient is negative then the correlation is negative. The greater the absolute value, the stronger the correlation.

$$R = \frac{\sum_{i=1}^{n}(x_i - \overline{x})(y_i - \overline{y})}{\sqrt{\sum_{i=1}^{n}(x_i - \overline{x})^2 \sum_{i=1}^{n}(y_i - \overline{y})^2}} \tag{11}$$

where y_i is the ERA5 reanalysis data for each variable, x_i is the ABLH, and n is the number of samples.

3. Results and Analysis

3.1. Analysis of ABL Characteristics

3.1.1. Vertical Variation Characteristics of Meteorological Elements in the ABL

Figure 3 shows the comparison of average temperature and vertical distribution characteristics of water vapor density between the microwave radiometer and observation data at Sieshan station. It can be seen from the figure that the temperature changes regularly at four times, and the difference between the two data is not obvious. However, it can be seen from Figure 3a that the temperature at 02:00 at the lower level is significantly higher than that at the other three times. However, there is no significant difference between the four times in Figure 3b, and compared with the microwave radiometer data (Figure 3a), there is an obvious temperature inversion in the lower layer. As for the water vapor density, there is a large difference between the four times of the microwave radiometer data. It can be seen from Figure 3c that the water vapor density is the highest at 02:00 below 2800 m and the lowest at 08:00 within 1000–3600 m. Different from Figure 3c, there is no significant difference in the variation of water vapor density at the four times in Figure 3d, but the water vapor density at 14:00 within 1000 m to 3600 m is slightly lower than the other three times.

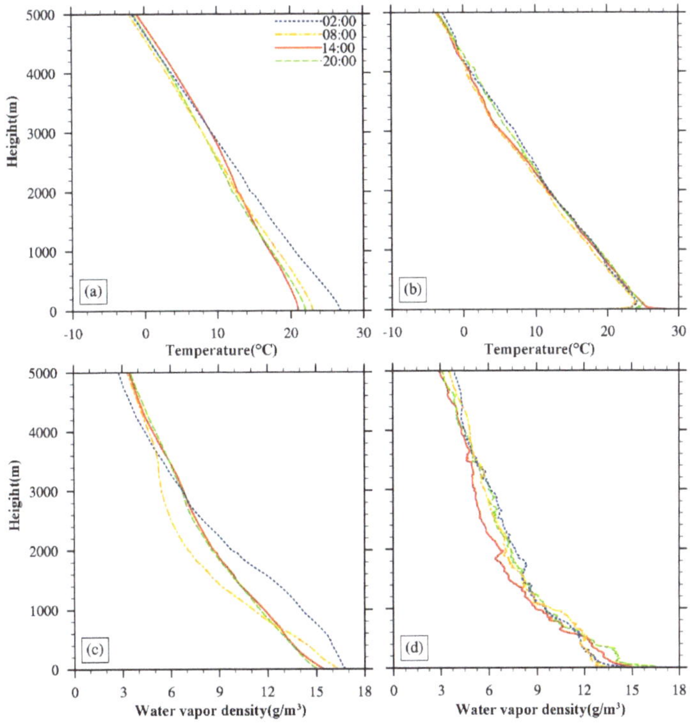

Figure 3. The vertical distribution characteristics of mean temperature and water vapor density of microwave radiometer data (**a**,**c**) and observation data (**b**,**d**) at Sieshan station (The 4 subfigures of **a**–**d** share a common legend).

The vertical structure of the thermal and dynamic properties of the ABL has a significant effect on the transport and distribution of material within it [30], and the distribution of water vapor in the vertical direction is then influenced by the turbulent action within the ABL. Figure 4 shows the characteristics of the vertical distribution of relative humidity and specific humidity at the three stations, from which it can be seen that the vertical variation of relative humidity at Sieshan station below 5000 m increases gradually with height, while the relative humidity at SETORS decreases and then increases and then decreases with height, and the relative humidity at Qumalai station increases and then decreases with height (Figure 4a–c). The relative humidity at 1400–3000 m is smaller than at other times, while the relative humidity at 08:00 is larger than at other times at every level of height, indicating that the ABL is richer in water vapor at 08:00 in Sieshan station (Figure 4a). The vertical variation of relative humidity at 14:00 in SETORS is different from the other moments, with the surface relative humidity smaller than the other moments then gradually increasing and then decreasing. The relative humidity first decreases slightly from the surface and then increases at the other three moments. The relative humidity at SETORS reaches a maximum of around 2000 m, close to 100%, and then gradually decreases (Figure 4b). The relative humidity at Qumalai station also reached a maximum around 2000 m, close to 90%. The relative humidity at the Qumalai station did not change significantly below 2000 m at 08:00, and the surface relative humidity at 14:00 and 20:00 was the smallest among the three stations (Figure 4c).

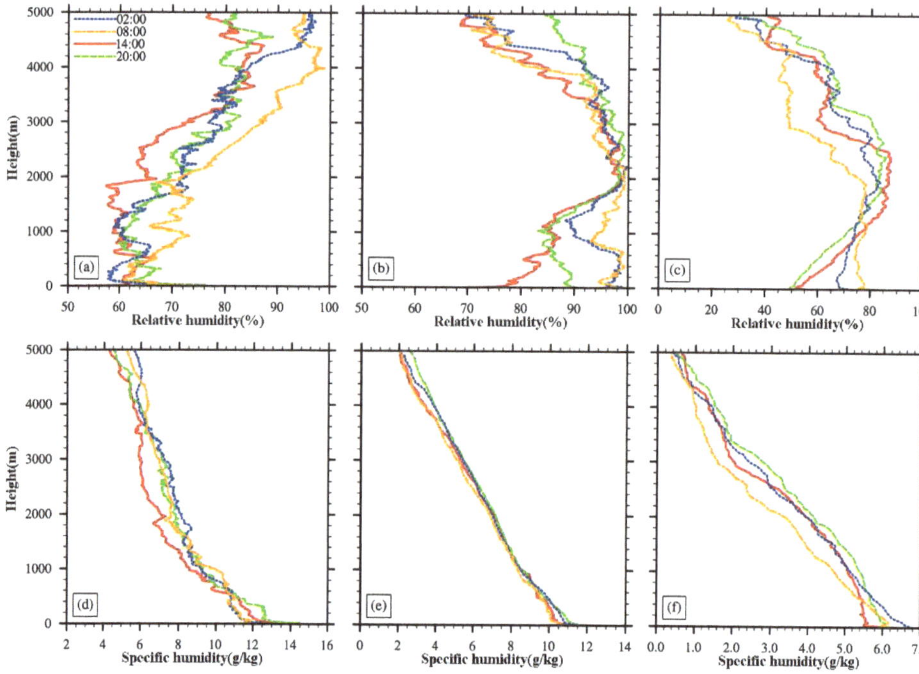

Figure 4. Vertical distribution characteristics of mean relative humidity and specific humidity at Sieshan stations (**a**,**d**), SETORS (**b**,**e**), and Qumalai station (**c**,**f**) (The 6 subfigures of **a**–**f** share a common legend).

Specific humidity is also an important parameter to characterize the distribution of water vapor in the vertical direction. As can be seen from the figure, the specific humidity of the near-surface layer is the smallest at Qumalai station, followed by SETORS, and the largest at Sieshan station (Figure 4d–f), which may be related to the geographical location, altitude, and ground heating field. The inverse humidity appears near the ground at 08:00 at Qumalai station (Figure 4f). The inverse temperature in the SBL and near the top of the CBL, low-level jet on water vapor transport, etc. These may form the inverse humidity within the near-ground layer [8].

Wind is one of the main meteorological elements of the energy and material transport exchange process in the ABL. Figure 5 shows the vertical distribution characteristics of wind direction and wind speed at the three stations. In addition to the 14:00 ground wind direction at Sieshan station, which is mainly easterly, the other wind directions are mainly southwesterly, and the wind direction at 1000–2000 m altitude is mainly southeasterly, and the wind direction above 2000 m gradually turns to southerly or southwesterly (Figure 5a). The wind direction below 1000 m varies greatly from day to day at SETORS, with easterly winds dominating at 14:00, southeasterly winds at 08:00 and 20:00, and southerly winds at 02:00. The wind direction above 1200 m changed less noticeably daily at SETORS, and the wind direction gradually changed to southerly or southwesterly (Figure 5b). Below 1800 m at Qumalai station, southeasterly winds predominate, while above 1800 m, they gradually change to southwesterly or westerly winds (Figure 5c).

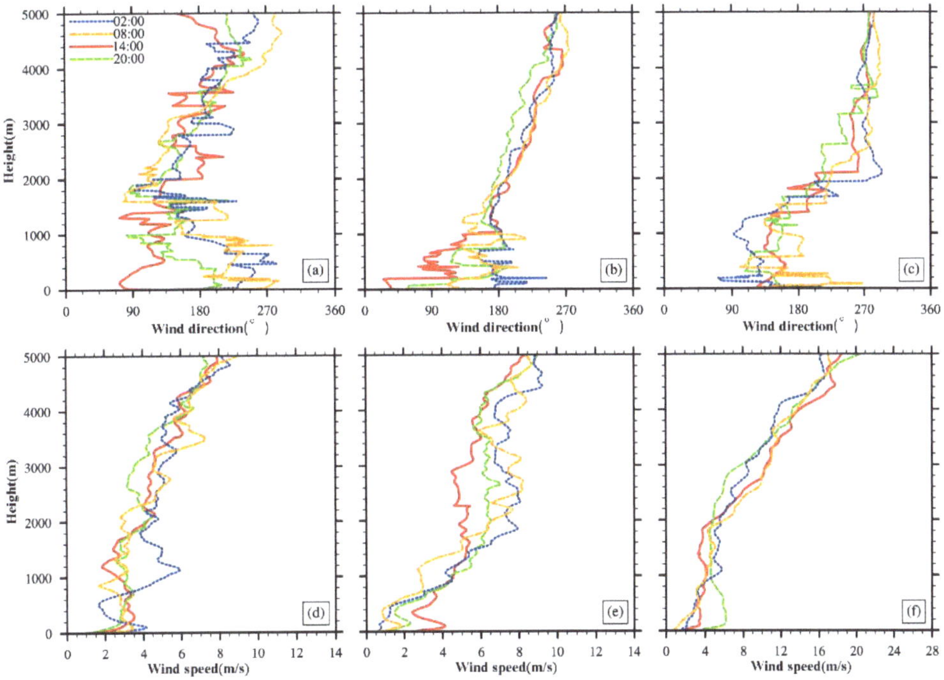

Figure 5. Vertical distribution characteristics of mean wind direction and wind speed at Sieshan stations (**a**,**d**), SETORS (**b**,**e**), and Qumalai station (**c**,**f**) (The 6 subfigures of **a**–**f** share a common legend).

Among the three stations, the wind speed at Qumalai station varied greatly with height, with a maximum wind speed of 20 m/s (Figure 5f). The wind speed varied less with height at Sieshan station, and the maximum wind speed could reach 9 m/s (Figure 5d). There was a sharp decrease in wind speed at 100 m at 02:00, and then, the wind speed started to increase sharply at 500 m and started to decrease at 1200 m (Figure 5d). At SETORS, except for 02:00, there was a sharp decrease in wind speed at 100 m, and then the wind speed increased sharply at 400 m, and the wind speed increased slowly with height above 1600 m at 14:00 (Figure 5e). At 20:00, the ground wind speed at Qumalai station reached a maximum of 4 m/s, and the wind speed began to increase sharply above 3000 m. At 02:00, 08:00, and 14:00, the wind speed increased sharply to around 2000 m, and the maximum wind speed was 20 m/s, which is the largest among the three stations (Figure 5f).

3.1.2. The ABLH Characteristics

Figure 6 shows the change process of the average ABLH at the three stations for four time periods, and it can be seen from the figure that the ABLH of each station reaches the maximum at 14:00. Combined with Figure 7a, the ABL type at night is mostly SBL at Sieshan station, and the ABLH is around 250 m (Figure 6). At 08:00, the ABLH type is SBL; it is mostly CBL at 14:00, and the ABLH can reach 1400 m during the day. The diurnal variation of the ABLH at SETORS and Sieshan station is similar. At 14:00, the ABLH at SETORS is slightly lower than that at Sieshan station, while the ABLH at SETORS is higher than that at Sieshan station at night (Figure 6). The ABL type at SETORS is mainly CBL at night (Figure 7b), and the frequency of both types of boundary layers is the same at 02:00, while the ABL is mainly CBL at 20:00 and also mainly CBL during the day (Figure 7b), which indicates that convection is developing vigorously and there is relatively strong convective activity even at night at SETORS, which may be related to the underlying surface and surface heating field. The ABLH is the highest among the three stations at 14:00 in Qumalai station, reaching about 1500 m. By 20:00, the ABLH is still high; compared with 14:00, it only decreases about 100 m (Figure 6), and combined with Figure 7c, we can see that the ABL type is dominated by CBL at 14:00 and 20:00, indicating that the convective activity in the afternoon continues into the evening for a longer time at Qumalai station. The ABLH is lower at 02:00 and 08:00, and the ABLH at 08:00 is significantly lower than that at 02:00 (Figure 6). Combined with Figure 7c, there are SBL at 02:00 and 08:00, indicating that the ABL is more stable during this period, and no convective activity is generated, or only weak convection is present during this period. The ABLH is the highest at Qumalai station among the three stations during the day, around 750 m and 800 m, respectively (Figure 6).

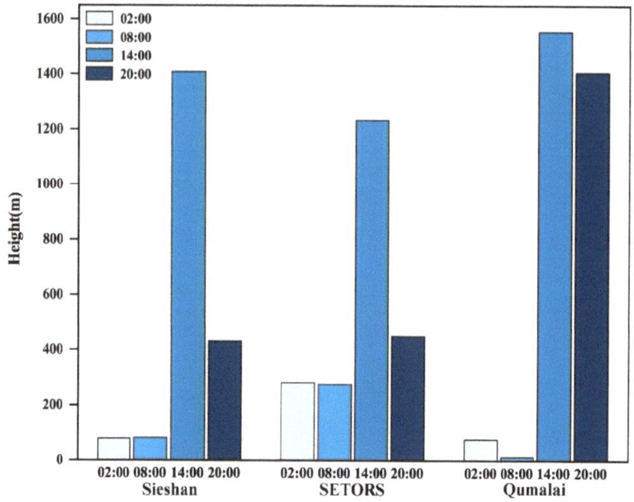

Figure 6. Diurnal variation of mean ABLH at three stations.

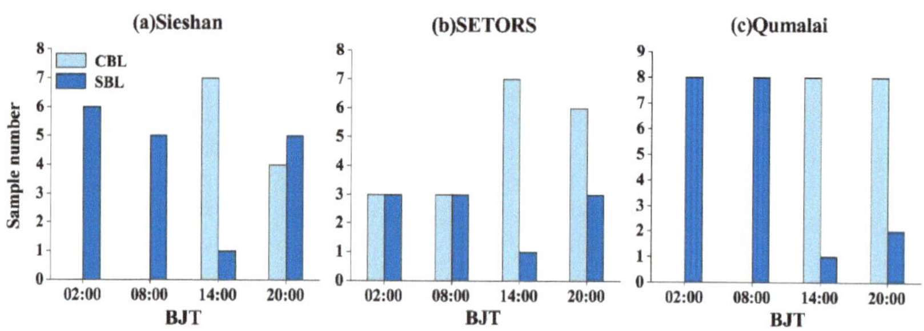

Figure 7. Diurnal variation of sample numbers of CBL and SBL at three stations (The 3 subfigures of **a–c** share a common legend).

Figure 8 shows the box plot of ABLH at each station for four time periods. It can be seen from the figure that the ABLH of the three stations varies greatly at different times. At 02:00 (Figure 8a), the ABLH is the highest compared with the other two stations at SETORS, mainly around 150–400 m. The ABLH is below 200 m at Sieshan station, while that is mainly below 100 m at Qumalai station. At 08:00 (Figure 8b), the ABLH at each station is similar to that at 02:00, with the ABLH at Qumalai station becoming significantly lower compared to that at 02:00, while the ABLH at SETORS is mostly slightly higher than that at 02:00. At 14:00 (Figure 8c), the ABLH is mainly below 2000 m at Sieshan station, and the ABLH is mainly around 1000–1500 m at SETORS, while ABLH is mainly below 2500 m at Qumalai station, and the ABLH varied within 500–2300 m, with a large range of variation. Except the ABLH slightly decreases at Qumalai station, the ABLH decreases significantly at the other two stations, and the ABLH mostly decreases to below 1000 m at these two stations. Compared with the ABLH at 20:00 and 02:00, the ABLH decreases sharply at Qumalai station, which indicates that the convective activity is strong during the day and weak at night at Qumalai station. The daily variation of ABLH is the smallest at SETORS.

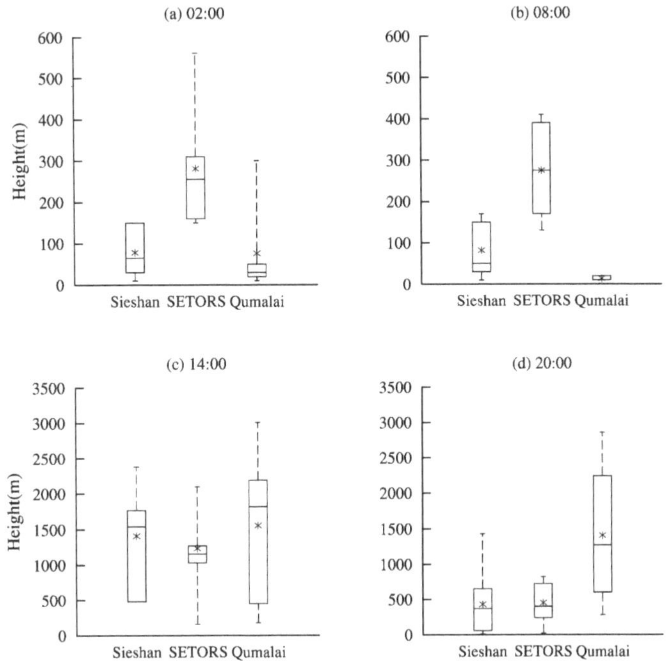

Figure 8. Box plot of ABLH at three stations, the * indicates outliers.

In general, the factors influencing the height of the atmospheric boundary layer include Topography: The height and complexity of the terrain affect the development and ABLH. For example, rugged mountainous areas often result in lower atmospheric boundary layer heights, while flat areas may have higher atmospheric boundary layer heights. Surface Temperature: Variations in surface temperature directly impact the height of the atmospheric boundary layer. Warmer surface temperatures during the day increase the atmospheric boundary layer height, while cooler surface temperatures at night decrease it. Solar Radiation: The intensity and distribution of solar radiation also influence changes in the height of the atmospheric boundary layer. Regions with intense radiation typically have higher atmospheric boundary layer heights. Vertical velocity and Horizontal Wind: Vertical velocity and horizontal wind in the atmosphere significantly affect the height of the atmospheric boundary layer. Upward velocity increases the boundary layer height, while downward velocity decreases it. Surface Features: Surface features, including vegetation cover and water distribution, also affect the height of the atmospheric boundary layer. Dense vegetation areas typically have lower ABLH.

The three representative sites selected in this study represent alpine grassland, forests, and mountainous cropland areas. Qumalai Station, located in the hinterland of the QTP, has flat terrain. Influenced by intense daytime surface heating, vigorous vertical convection develops, increasing the surface sensible heat flux and the temperature gradient in the atmospheric boundary layer. This gradient promotes the development of convection, resulting in the highest boundary layer height among the three stations, which can be sustained until evening. The cooling effect from nighttime surface radiation is evident, leading to the lowest boundary layer height. SETORS Station has a forested underlying surface, resulting in the lowest atmospheric boundary layer height among the three stations during the day. This is due to the dense vegetation reducing direct solar energy reaching the surface, which lowers the surface temperature and limits the upward extent of the atmospheric boundary layer. Additionally, the dense vegetation increases surface roughness, reducing wind speeds within the canopy layer and slowing vertical diffusion, contributing to the lower boundary layer height. Sieshan Station is situated in a lower-altitude mountainous cropland area with less intense solar radiation and less dense vegetation compared to SETORS station. This results in a moderate increase in surface temperature and an atmospheric boundary layer height between that of forests and grasslands.

3.2. Analysis of Factors Influencing the ABLH

3.2.1. Comparison of ABLH between Observed Data and ERA5 Reanalysis Data

Figure 9 shows the hourly variation of mean ABLH between the ERA5 reanalysis data and the observed data in June, from which it can be seen that the hourly variation of ABLH at the three stations is basically the same. The ABLH is lower in the morning and evening and reaches the highest at 15:00, among which the ABLH is the highest at Qumalai station, followed by the SETORS and finally the Sieshan station. At 02:00, the ABLH of ERA5 reanalysis data is smaller than that of observed data at SETORS and Sieshan station, while the opposite is true at Qumalai station. At 08:00, the ABLH of ERA5 reanalysis data is still smaller than that of the observed data at SETORS, while the ABLH of ERA5 reanalysis data is larger than that of the observed data at Qumalai station. At 14:00, the ERA5 reanalysis data of both SETORS and the Sieshan station show significant overestimation of the observed data, while at 20:00, the Sieshan and Qumalai stations show significant underestimation. In summary, the ABLH of ERA5 reanalysis data is smaller than that of the observed data in the morning and evening at Sieshan station and SETORS, and the opposite is true at noon. The ABLH of ERA5 reanalysis data is larger than that of the observed data in the morning and evening, and the difference between them is not obvious at noon in Qumalai station.

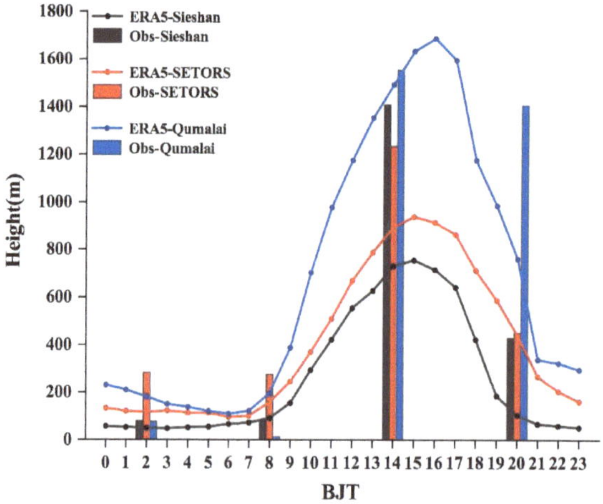

Figure 9. Hourly variation of the mean ABLH between the ERA5 reanalysis data and the observed data in June.

3.2.2. Spatial Distribution Characteristics of the Surface Heating Field (Latent Heat Flux (LE) and Sensible Heat Flux (H)), Surface Long-Wave Radiation (Rlu), and Vertical Velocity (w) in Four Time Periods

The H is the heat transport between the surface and atmosphere caused by turbulent motion near the ground, and LE is the heat transport between the surface and atmosphere caused by the phase change in water in the atmosphere. Therefore, H and LE are the key physical quantities that influence the formation and development of the atmospheric thermal boundary layer [10].

Figure 10 shows the spatial distribution characteristics of the surface heating field (LE and H), Rlu, and w in four time periods on the TP. In summer, the surface heating field of the plateau is dominated by LE. From 02:00 to 14:00, the heating of the surface by solar radiation gradually increases, and the LE and H increase. The LE and H are always larger in the eastern plateau than in the western plateau due to different underlying surfaces and altitudes (Figure 10a–c,e–g). At 20:00, the intensity of the surface heating field gradually decreases, and the LE and H in the western plateau are larger than those in the eastern plateau (Figure 10d,h), which indicates that during the period of gradual heating of the surface by solar radiation, the eastern plateau absorbs energy faster and exchanges energy faster, and the intensity of surface heating field rapidly decreases when there is no solar radiation heating. At 02:00 and 08:00, the Rlu value in the western plateau is smaller than that in the eastern plateau (Figure 10i,j), and the Rlu value reaches its maximum at 14:00, and there are large value areas of Rlu in the eastern and southwestern plateau, and Rlu value can reach more than 400 W/m^2 in the eastern part. At 20:00, the Rlu decreases and is mainly above 350 W/m^2 in the eastern plateau.

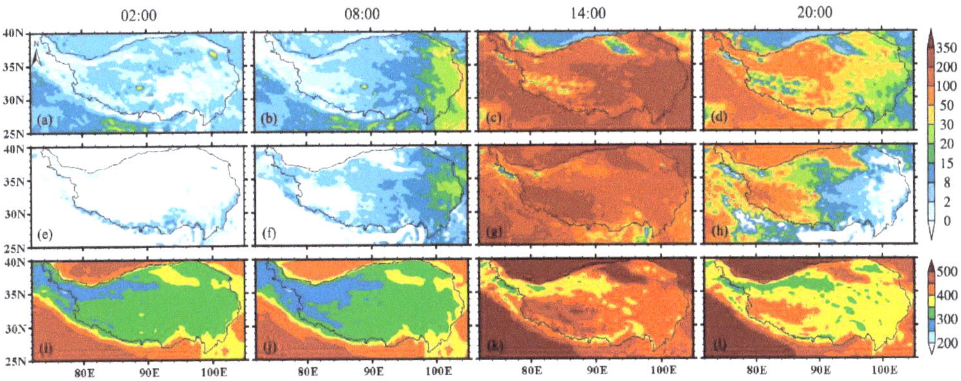

Figure 10. Spatial distribution of LE (**a–d**), H (**e–h**), and Rlu (**i–l**) for four time periods in June (unit: W/m^2).

Figure 11 shows the latitudinal profile of w, from which we can obtain that the w in the eastern plateau is larger than that in the western plateau. At 02:00 (Figure 11a), the w in the eastern plateau is larger than that in the western plateau, and it is mainly ascending motion, with the maximum value of w up to 0.12 Pa/s or more, and the area of large value of w is mainly between 600 and 300 hPa. At 08:00 (Figure 11b), the w in the western plateau becomes smaller, and a sinking motion occurs. The w in the eastern plateau continues to increase, but the vertical range becomes smaller, mainly concentrated between 600 and 400 hPa. At 14:00 (Figure 11c), the w reaches the maximum in the eastern plateau, which is about 0.18 Pa/s, and both the lateral and vertical ranges of w become larger. At 20:00 (Figure 11d), the w in the eastern plateau decreases compared to that in the afternoon, while the w in the central plateau increases, and the w is smaller and shows a sinking motion in the western plateau.

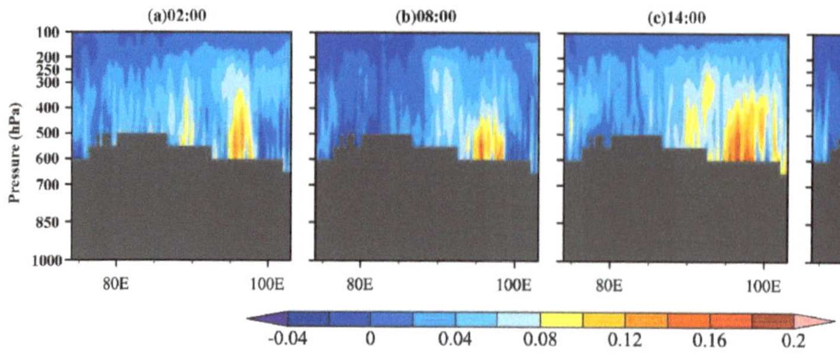

Figure 11. Latitudinal profiles of w at four times in June (here, positive w means rising, negative means sinking, unit: Pa/s).

3.2.3. Relationship between Surface Heating Field (LE and H), Rlu, w, and ABLH

Table 4 shows the R of the surface heating field, Rlu, and ABLH at the three stations. It can be seen that the R of LE and H at Sieshan station passed the significance test, indicating that the significant change of ABLH is mainly due to the influence of LE and H at Sieshan station, among which the correlation between LE and ABLH was higher. The most significant effect of H on the ABLH is found at SETORS. The correlation between Rlu and ABLH is higher at Qumalai station. The correlation between w and ABLH is not high in three stations, among which the correlation between w and ABLH is highest in Sieshan station, and the other two stations do not pass the significance test.

Table 4. R of surface heating field, Rlu, w, and ABLH (* indicates passing the significance test of 0.05).

Variables / ABLH	LE	H	Rlu	w
Sieshan	0.69 *	0.66 *	0.67 *	0.52 *
SETORS	0.83 *	0.86 *	0.82 *	0.07
Qumalai	0.63 *	0.53 *	0.75 *	0.16

Table 4 also shows that surface latent heat, sensible heat, and net radiation have the highest correlation with the ABLH at the SETORS station. This is because the dense vegetation releases a significant amount of latent heat during evapotranspiration, which raises the surrounding air temperature, aiding in convection development. This convection transports warm air upward, increasing the ABLH. At Sieshan Station in the cropland area, vertical velocity has the most significant impact on boundary layer height, possibly due to local circulations formed by complex terrain elevating the boundary layer height, with latent and sensible heat contributions being less dominant factors.

Figure 12 shows the daily variation of LE, H, and ABLH. It can be seen from the figure that LE and H increase at sunrise and decrease at sunset and reach the maximum value at 14:00, when the ABLH is also the highest. The LE is larger than H throughout the day and the peak difference is the largest at Sieshan station, and the peak LE is the largest among the three sites, reaching 300 W/m^2 (Figure 12a). The LE at night is comparable to H or even slightly greater than the LE at SETORS, and the peak LE is the smallest among the three stations, which is about 200 W/m^2 (Figure 12b). The peak H at Qumalai station is the largest among the three stations, reaching about 140 W/m^2. Under the effect of both LE and H, the heat exchange between the surface and the atmosphere is frequent, which promotes the development of ABLH (Figure 12c). The daily variation process of Rlu is basically consistent with LE and H, increasing at sunrise and decreasing at sunset, reaching the maximum at 14:00. Among the three stations, the daily variation of Rlu is obvious at Qumalai station; fluctuations are large (Figure 12c), followed by SETORS (Figure 12b), and there is least variation at Sieshan station (Figure 12a), it can be seen that

Rlu by solar radiation changes have a greater impact at Qumalai station. The Rlu heats up the atmosphere and contributes to the development of ABLH.

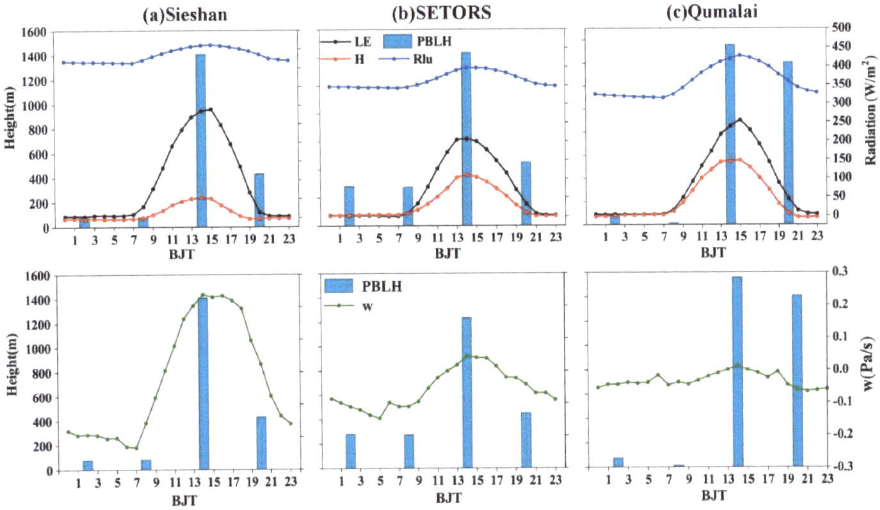

Figure 12. Average daily variation processes of LE, H, Rlu, w, and ABLH (top: LE, H, Rlu, and ABLH, bottom: w and ABLH) (The 3 subfigures of **a**–**c** share a common legend).

w is the component of velocity in the lead direction; in this paper, a positive w indicates upward motion, and a negative one indicates downward motion. Figure 12 shows the average daily variation of w and ABLH, and the w in this paper is taken as the value of the layer closest to the surface at the three stations. It can be seen from the figure that the w is negative in the morning and evening, when the air makes a downward motion and reaches the maximum value at each station around 14:00 when the air makes an upward motion, which is favorable to the development of ABLH. The daily variation of w at Sieshan station is basically consistent with the daily variation of ABLH (Figure 12a), which indicates that w has an effect on ABLH, while it has less effect at SETORS (Figure 12b) and almost no effect at Qumalai station (Figure 12c).

4. Conclusions

Based on sounding observations and ERA5 reanalysis data from the Qumalai station and SETORS in the eastern part of the TP and the Sieshan station located at the southeastern edge of the TP in June 2022 of the Second Tibetan Plateau Scientific Expedition and Research on the TP, this paper studied the characteristics of ABLH on different underlying surfaces in the eastern TP and the influencing factors of ABLH, and reached the following conclusions:

(1) Compared with the microwave radiometer data, the temperature difference is not obvious, and there are obvious temperature inversion phenomena in the lower layer of the observed data at Sieshan station. For the water vapor density, the difference between the four times microwave radiometer data is large, but the difference between the observed data is small at Sieshan station. The relative humidity of Sieshan station increases with height, while that increases with height and then decreases at SETORS and Qumalai station, reaching a maximum of 2200 m. The specific humidity near the ground at Qumalai station is the smallest, and the distribution of water vapor in the ABL is less. The wind direction above 2000 m is southerly or southwesterly, and the wind speed is mainly 4–8 m/s at Sieshan station. The wind direction above 1200 m is mainly southerly or southwesterly, and the wind speed changes slowly at this time at SETORS. The wind direction below 1800 m at Qumalai station is mainly southeast

wind, and the wind direction above 1800 m gradually changes to southwest wind or westerly wind, with the maximum wind speed up to 20 m/s.

(2) The ABL type is mainly CBL except for 14:00 at Sieshan station, and SBL at other times. The mean ABLH is 499.59 m at Sieshan station. The ABL type is mainly CBL, and the average ABLH is 560.11 m at SETORS. The ABL type is mainly CBL at 14:00 and 20:00, and SBL at other times at Qumalai station. The average height of boundary layer is 762.89 m at Qumalai station.

(3) The mean ABLH of the ERA5 reanalysis data in the morning and evening is smaller than observed data at Sieshan station and SETORS, whereas the opposite trend is observed at noon. At Qumalai station, the mean ABLH of the ERA5 reanalysis data in the morning and evening is larger than the observed data, with no significant differences at noon.

(4) In summer, the LE and H in the eastern plateau are less than those in the western plateau at 20:00, and the LE and H in the eastern plateau are greater than those in the western plateau at other three times. Rlu is greater in the eastern plateau than those in the western plateau at 02:00 and 08:00, and the opposite is true at 14:00 and 20:00. The w in the eastern plateau is greater than that in the western plateau. The diurnal variation of w in the eastern plateau is obvious, increasing first and then decreasing, mainly for upward motion.

(5) The influence of the surface heating field on the ABLH is most significant at Sieshan station and SETORS, while the ABLH is also influenced by the w at Sieshan station, and the ABLH is influenced by the Rlu at Qumalai station. All four influencing factors show a positive correlation with ABLH. The LE, H, and Rlu increase at sunrise and decrease at sunset and reach the maximum at around 14:00, when the w is positive and the air makes upward movement, prompting the development of ABLH.

The characteristics of the ABLH on different underlying surface types in the QTP can be summarized as follows:

Grassland: Grassland underlying surfaces are typically located in plateau areas with relatively flat terrain. During the day, warm surface temperatures increase the ABLH, while cold surface temperatures at night decrease it. Due to significant surface temperature variations, the boundary layer height experiences a large range of fluctuations, and it is relatively stable as it is less influenced by solar radiation and terrain.

Forest: The ABLH in forested areas is generally lower. This is because dense vegetation reduces direct solar radiation reaching the surface, resulting in lower surface temperatures. This, in turn, reduces the temperature gradient in the lower part of the atmospheric boundary layer, limiting its upward extent. Dense vegetation also increases surface roughness, slowing down vertical diffusion and contributing to a lower boundary layer height.

Cropland Areas: Underlying surfaces in mountainous cropland areas are relatively flat, with weaker solar radiation and lower vegetation cover. As a result, the boundary layer height falls between that of forests and grasslands. The increase in surface temperature is not as intense, leading to a relatively stable atmospheric boundary layer height.

The influence of different underlying surface types on the ABLH is primarily reflected in surface temperature variations, solar radiation intensity, vegetation cover, and terrain. Grasslands typically exhibit a larger range of boundary layer height variations, while forests and mountainous agricultural areas are relatively stable.

Author Contributions: Conceptualization, X.W., methodology and writing, J.M., formal analysis and validation, M.C. All authors have read and agreed to the published version of the manuscript.

Funding: The Second Tibetan Plateau Scientific Expedition and Research Program (STEP) (grant number 2019QZKK010304). The National Natural Science Foundation of China (grant number 41975096), the Open Grants of the State Key Laboratory of Severe Weather (2023LASW-B08) and the Technological Innovation Capacity Enhancement Program of the Chengdu University of Information Technology (grant number KYQN202203) provided financial support for this study.

Data Availability Statement: ERA5 reanalysis data (ERA5 hourly data on single levels from 1940 to present (copernicus.eu), ERA5 hourly data on pressure levels from 1940 to present (copernicus.eu), accessed on 1 November 2022).

Acknowledgments: The authors thank ERA5 reanalysis data from the ECMWF for providing support for this study.

Conflicts of Interest: The authors declare no conflicts of interest.

References

1. Su, Y.; Lü, S.; Fan, G. The characteristics analysis on the summer atmospheric boundary layer height and surface heat fluxes over the Qinghai-Tibetan Plateau. *Plateau Meteorol.* **2018**, *37*, 1470–1485.
2. Stull, R.B. *An Introduction to Boundary Layer Meteorology*; Springer Science & Business Media: Berlin/Heidelberg, Germany, 1988; Volume 13.
3. Li, Y.; Hu, Z.; Zhao, H. Overview on the Characteristic of Boundary Layer Structure in Tibetan Plateau. *Plateau Mt. Meteorol. Res.* **2012**, *32*, 91–96.
4. Wan, Y.; Zhang, Y.; Zhang, J.; Peng, Y. Influence of sensible heat on planetary boundary layer height in East Asia. *Plateau Meteorol.* **2017**, *36*, 173–182.
5. Baklanov, A.A.; Grisogono, B.; Bornstein, R.; Mahrt, L.; Zilitinkevich, S.S.; Taylor, P.; Larsen, S.E.; Rotach, M.W.; Fernando, H.J.S. The Nature, Theory, and Modeling of Atmospheric Planetary Boundary Layers. *Bull. Am. Meteorol. Soc.* **2011**, *92*, 123–128. [CrossRef]
6. Zhang, Q.; Zhang, J.; Qiao, J.; Wang, S. Relationship of atmospheric boundary layer depth with thermodynamic processes at the land surface in arid regions of China. *Sci. China Earth Sci.* **2011**, *54*, 1586–1594. [CrossRef]
7. Zhao, M. A Review of the Resear ch on the Effects of Boundary Layer and Land Surface Process on Heavy Rain in China. *Torrential Rain Disasters* **2008**, *2*, 186–190.
8. Ma, Y.; Hu, Z.; Wang, B.; Ma, W.; Chen, X.; Han, C.; Li, M.; Zhong, L.; Gu, L.; Sun, F.; et al. The Review of the Observation Experiments on Land-Atmosphere Interaction Progress on the Qinghai-Xizang (Tibetan) Plateau. *Plateau Meteorol.* **2021**, *40*, 1241–1262.
9. Xu, G.; Cui, C.; Zhou, Z.; Zhang, B.; Li, Y.; Zhao, X. Atmospheric boundary layer heights estimated from radiosonde observations over the Qing hai-Tibet Plateau and its downstream area. *Torrential Rain Disasters* **2014**, *33*, 217–227.
10. Fu, W.; Li, M.; Yin, S.; Lü, Z.; Wang, L.; Shu, L. Study on the Atmospheric Boundary Layer Structure of the Qinghai Xizang Plateau under the South Branch of the Westerly Wind and the Plateau Monsoon Circulation Field. *Plateau Meteorol.* **2022**, *41*, 190–203.
11. Song, X.; Zhang, H.; Liu, X.; Fan, S.; Liu, H.; Hu, F.; Li, S.; Zhou, M.; Bian, L.; Xu, X. Determination of atmospheric boundary layer height in unstable conditions over the middle Tibetan Plateau. *Acta Sci. Nat. Univ. Pekin.* **2006**, *42*, 328–333. [CrossRef]
12. Jiang, X.; Li, Y.; He, G. The climatic characteristics analysis of radiosonde data over east part of Tibetan Plateau. *Plateau Mt. Meteorol. Res.* **2008**, *28*, 1–9.
13. Li, M.; Ma, Y.; Hu, Z.; Ma, W.; Wang, J.; Ogino, S.-Y. Study on characteristics of atmospheric boundary layer over Naqu region of Northern Tibetan Plateau. *Plateau Meteorol.* **2004**, *23*, 728–733.
14. Li, M.; Dai, Y.; Ma, Y.; Zhong, L.; Lü, S. Analysis on structure of atmospheric boundary layer and energy exchange of surface layer over Mount Qomolangma region. *Plateau Meteorol.* **2006**, *25*, 807–813.
15. Li, M.; Ma, Y.; Ma, W.; Ishikawa, H.; Sun, F.; Ogino, S.-Y. Structural difference of atmospheric boundary layer between dry and rainy seasons over the central Tibetan Plateau. *J. Glaciol. Geocryol.* **2011**, *33*, 72–79.
16. Ma, W.; Dai, Y.; Ma, Y.; Hu, Z.; Li, M.; Wang, J. Analysis on the boundary layer and spatial profile of northern Tibetan Plateau area by radiosonde data. *J. Arid. Land Resour. Environ.* **2005**, 40–46.
17. Gu, L.; Hu, Z.; Lü, S.; Yao, J. Atmosphere Boundary Layers in Typical Days over the North Tibet Plateau during Summer. *J. Glaciol. Geocryol.* **2006**, 893–899.
18. Wang, S.; Ma, Y. The structure of atmospheric boundary Layer over mount Qomolangma in summer. *J. Glaciol. Geocryol.* **2008**, *30*, 681–687.
19. Chen, X.; Škerlak, B.; Rotach, M.W.; Añel, J.A.; Su, Z.; Ma, Y.; Li, M. Reasons for the extremely high-ranging planetary boundary layer over the western Tibetan Plateau in winter. *J. Atmos. Sci.* **2016**, *73*, 2021–2038. [CrossRef]
20. Li, W.; Xing, Y.; Ma, S. The analysis and comparison between GTS1 radiosonde made in China and RS92 Radiosonde of Vaisala company. *Meteorol. Mon.* **2009**, *35*, 97–102.
21. Ran, Y.; Li, X.; Lu, L. China Land Cover Classification at 1 km Spatial Resolution Based on a Multi-source Data Fusion Approach. *Adv. Earth Sci.* **2009**, *24*, 192–203.
22. Hersbach, H.; Bell, B.; Berrisford, P.; Horányi, A.; Sabater, J.M.; Nicolas, J.; Radu, R.; Schepers, D.; Simmons, A.; Soci, C. Global reanalysis: Goodbye ERA-Interim, hello ERA5. *ECMWF Newsl.* **2019**, *159*, 17–24.
23. Eresmaa, N.; Karppinen, A.; Joffre, S.; Räsänen, J.; Talvitie, H. Mixing height determination by ceilometer. *Atmos. Chem. Phys.* **2006**, *6*, 1485–1493. [CrossRef]
24. Seibert, P.; Beyrich, F.; Gryning, S.-E.; Joffre, S.; Rasmussen, A.; Tercier, P. Review and intercomparison of operational methods for the determination of the mixing height. *Atmos. Environ.* **2000**, *34*, 1001–1027. [CrossRef]

25. Yang, F. Comparison of Determination Methods and Characteristics of Boundary Layer Height in Semi-Arid Area. Master's Thesis, Lanzhou University, Lanzhou, China, 2018.
26. Yang, F. The Atmospheric Boundary Layer Height Study Based on Ground-Based Remote Sensing and Numerical Simulation. Master's Thesis, Nanjing University, Nanjing, China, 2015.
27. Holzworth, G.C. Estimates of mean maximum mixing depths in the contiguous United States. *Mon. Weather. Rev.* **1964**, *92*, 235–242. [CrossRef]
28. Holtslag, A.; De Bruijn, E.; Pan, H. A high resolution air mass transformation model for short-range weather forecasting. *Mon. Weather. Rev.* **1990**, *118*, 1561–1575. [CrossRef]
29. Sicard, M.; Pérez, C.; Rocadenbosch, F.; Baldasano, J.; García-Vizcaino, D. Mixed-layer depth determination in the Barcelona coastal area from regular lidar measurements: Methods, results and limitations. *Bound.-Layer Meteorol.* **2006**, *119*, 135–157. [CrossRef]
30. Zhang, Q.; Wang, S. A study on at mospheric boundary layer structure on a clear day in the arid region in northwest China. *Acta Meteorol. Sin.* **2008**, 599–608.

remote sensing

MDPI

Article

Spatiotemporal Distributions of the Thunderstorm and Lightning Structures over the Qinghai–Tibet Plateau

Yangxingyi Du [1], Dong Zheng [1,2,*], Yijun Zhang [3], Wen Yao [1], Liangtao Xu [1] and Xianggui Fang [4]

1 State Key Laboratory of Severe Weather, Chinese Academy of Meteorological Sciences, Beijing 100081, China; duyangxingyi21@163.com (Y.D.); yaowen@cma.gov.cn (W.Y.); xult@cma.gov.cn (L.X.)
2 Collaborative Innovation Center on Forecast and Evaluation of Meteorological Disaster (CIC-FEMD), Nanjing University of Information Science & Technology, Nanjing 210000, China
3 Institute of Atmospheric Sciences, Fudan University, Shanghai 200433, China; zhangyijun@fudan.edu.cn
4 Naqu Meteorological Service, Naqu 852000, China; fangxianggui2003@163.com
* Correspondence: zhengdong@cma.gov.cn

Abstract: Utilizing data from the Tropical Rainfall Measuring Mission (TRMM) satellite's precipitation radar (PR) and lightning imaging sensor (LIS), this study explores the spatiotemporal distributions of thunderstorm and lightning structures over the Qinghai–Tibet Plateau (QTP), an aspect that has not been explored previously. The structural aspects are crucial when considering the impact of thunderstorm and lightning activity in the atmospheric processes. Thunderstorms over the QTP show clear spatial variations in both vertical height and horizontal extension. In the southern region, the average heights of 20 dBZ and 30 dBZ echo tops typically exceed 11.2 and 9.3 km, respectively. Meanwhile, in the eastern part, the average coverage areas for reflectivity greater than 20 dBZ and 30 dBZ consistently surpass 1000 and 180 km^2, respectively. The spatial distribution of thunderstorm vertical development height relative to the surface aligns more closely with the horizontal extension, indicating stronger convection in the eastern QTP. The thunderstorm flash rate shows an eastward and northward prevalence, while the thunderstorm flash density peaks in the western and northeastern QTP, with a minimum in the southeast. Furthermore, in the eastern QTP, lightning duration, spatial expansion, and radiance are more pronounced, with the average values typically exceeding 0.22 s, 14.5 km, and 0.50 J m^{-2} sr^{-1} μm^{-1}, respectively. Monthly variations reveal heightened values during the summer season for thunderstorm vertical extension, areas with reflectivity greater than 30 dBZ, and lightning frequency. Diurnal variations highlight an afternoon increase in thunderstorm vertical and horizontal extension, lightning frequency, duration, and spatial scale. From a statistical perspective, under weak convective conditions, lightning length exhibits a positive correlation with thunderstorm convection intensity, contrasting with the opposite relationship suggested by previous studies. This article further analyzes and discusses the correlations between various thunderstorm and lightning structural parameters, enhancing our understanding of the distinctive features of thunderstorm and lightning activities in the QTP.

Keywords: Qinghai–Tibet Plateau; thunderstorm vertical development; thunderstorm horizontal extension; flash duration; flash spatial extension; flash radiance

Citation: Du, Y.; Zheng, D.; Zhang, Y.; Yao, W.; Xu, L.; Fang, X. Spatiotemporal Distributions of the Thunderstorm and Lightning Structures over the Qinghai–Tibet Plateau. *Remote Sens.* **2024**, *16*, 468. https://doi.org/10.3390/rs16030468

Academic Editor: Yuriy Kuleshov

Received: 20 December 2023
Revised: 19 January 2024
Accepted: 23 January 2024
Published: 25 January 2024

1. Introduction

The Qinghai–Tibet Plateau (QTP) exerts a significant influence on atmospheric circulation and climate change owing to its expansive topography and thermal effects [1–6]. Thunderstorms, as convective cloud systems that generate lightning activity, play a crucial role in various aspects, including QTP precipitation, the global radiation budget, and the exchange of materials between the troposphere and stratosphere [5,7,8]. Additionally, lightning activity serves as a major natural source of nitrogen oxides in the QTP, contributing to atmospheric chemical processes [9,10]. Beyond its meteorological significance, lightning over the QTP poses a notable natural disaster, with the per capita probability of lightning strike casualties ranking among the highest in China [11]. Consequently, meteorologists have taken their attention to QTP thunderstorms and lightning activities in recent years.

Previous studies have noted that the number of thunderstorm days and the frequency of thunderstorm events in the central and eastern QTP approaches the levels seen in South China [4,12]. However, the lightning activity in the QTP is visibly weak and is only significantly higher than that in the northwest arid regions of China [13,14]. Studies indicate that QTP thunderstorms exhibit weaker convection, smaller scale, shorter lifespans (around 1 h), and lower lightning frequency when compared to thunderstorms in the regions to the east and south of the QTP [15–18]. Additionally, lightning discharges in QTP thunderstorms are primarily associated with the lower positive and middle negative charge regions [15,19,20]. This is in contrast to normal thunderstorms on plains where the upper positive and middle negative charge layers contribute more lightning, especially intracloud (IC) lightning [21,22]. Furthermore, research indicates that lightning on the QTP exhibits shorter duration, smaller spatial extension scale, and weaker radiance than that observed in Central and Eastern China [17,18].

Some studies have explored the spatiotemporal distributions of thunderstorm and lightning activity across the QTP. Du et al. [23] examined the spatial and temporal patterns of thunderstorm activity on the QTP using the thunderstorm feature dataset established by Ma et al. [24]. This dataset was based on black body temperature (TBB) and cloud classification (CLC) products from the Fengyun-2E (FY-2E) geostationary satellite, along with lightning data from the World Wide Lightning Location Network (WWLLN). They identified three centers with frequent thunderstorms in the QTP: southeast (around 100°E, 29°N), south–central (around 92°E, 30°N), and southwest (around 88°E, 30°N), with the southeast exhibiting the highest frequency of thunderstorms. However, based on observations from the precipitation radar (PR) carried by the Tropical Rainfall Measuring Mission (TRMM) satellite, Qie et al. [25] suggested that the highest frequency of thunderstorms occurs in the central QTP (approximately 87–93°E, 31–35°N). It is evident that the spatial distribution of thunderstorm activity over the QTP, as provided by different datasets, varies.

Due to the limited availability of ground-based data on the QTP, much of the existing research on lightning activities in the QTP relies on data from the lightning imaging sensor (LIS) aboard the TRMM [26,27]. These studies indicate that approximately 95% of lightning in the QTP occurs from May to September, with peak activity in the months from June to August in different regions. The peak diurnal variation in lightning activity is around 17:00 (local time, LT). Moreover, the density in lightning activity is highest in the eastern and central QTP and lowest in the western and northern QTP. The spatial distribution of lightning activity exhibits the characteristic of advancing from east to west starting in May and then retreating from west to east starting in September. Ma et al. [28] found discrepancies in the spatiotemporal distributions of lightning between the observations of WWLLN and TRMM/LIS. For instance, WWLLN indicates relatively weak lightning activity in the northeastern QTP, while a region of relatively strong lightning activity is observed in the southwestern QTP, differing from the spatial distribution of LIS lightning. The primary and secondary monthly peaks of WWLLN lightning are in August and September, respectively, while those of LIS lightning occur in July and June. They proposed that the disparities in spatiotemporal distributions between LIS lightning and WWLLN lightning could be attributed to differences in their observation models and to the structural difference of the thunderstorms between different QPT regions.

In comparison to the extensive research on the spatial and temporal distributions of thunderstorm and lightning activities in the QTP, the spatiotemporal distributions of thunderstorm and lightning structures in the QTP remain largely unknown. The term "thunderstorm structure" here refers to the vertical and horizontal expansion as well as the parameters associated with the convective characteristics of thunderstorms, while "lightning structure" encompasses flash duration, spatial extension, flash radiance, and other aspects representing the scale and intensity of the lightning. Only Qie et al. [25] divided the QTP into three parts (east, central, and west) based on active regions of thunderstorm activity and compared the differences in the structure of thunderstorm clouds across these regions. Their findings revealed that the area of thunderstorm clouds is the largest in the east and the smallest in the west. Additionally, the echo depth of 40 dBZ exhibits the most significant difference among the three regions, with values of 3.0 km in the east, 2.5 km in the center, and 2.3 km in the west. Meanwhile, the structural aspects of thunderstorms and lightning play a crucial role in assessing the impact of thunderstorm and lightning activity in the QTP on weather, climate, and atmospheric chemistry. For instance, the larger scale and higher vertical development of thunderstorms undoubtedly exert a greater influence on the radiation budget and material transport. Additionally, the size and energy of lightning directly correlate with the quantity of nitrogen oxides produced [29–31]. Therefore, this study aims to unravel the spatiotemporal distribution characteristics of thunderstorm and lightning structures in the QTP and explore the relationship between different structural parameters and the potential explanations behind them. The ideal outcome of this study is to further refine our understanding of the distinctive features of thunderstorm and lightning activities in the QTP, deepening our comprehension of the unique relationship between these activities.

2. Data and Methodology

2.1. Data Description

The data utilized herein were originally derived from observations of the TRMM [32–34]. The TRMM satellite was launched in November 1997, and the mission was terminated in April 2015, contributing an approximately 17-year dataset of rainfall and lightning in tropical and subtropical areas worldwide. The utilized datasets cover from 1998 to 2014 in this study, and the data from August 2001, in which the orbit of the TRMM satellite increased from 350 km to 403 km, were excluded.

2.1.1. RPF Dataset

The TRMM-based radar precipitation feature (RPF) dataset was built by the University of Utah by collecting multiple observational datasets from several instruments aboard the TRMM satellite, including the PR, LIS, TRMM Microwave Imager (TMI), and Visible and Infrared Radiometers (VIRS) [35]. An RPF was defined as a contiguous area in which the rainfall rate observed by the PR was greater than zero [36]. The PR identified the cloud with a horizontal resolution of approximately 5 km and vertical resolution of 0.25 km. The RPFs over the QTP were divided into two categories, namely those with and without LIS lightning. The former corresponds to the thunderstorms. In this study, the position of the RPF, PR reflectivity-related vertical and horizontal extensions of the thunderstorm cloud, and information such as the flash rate and flash density of the thunderstorm obtained by combining with LIS observations are used; they are collectively referred to as the thunderstorm structure.

2.1.2. LIS Data

The lightning data utilized in this study were acquired from the LIS, one of the instruments aboard the TRMM satellite. The LIS captured light signals using a CCD with a resolution of 128 × 128 pixels. The field width of the LIS spanned approximately 500–550 km on the Earth's surface, with a spatial resolution ranging from 3 to 6 km [37,38], covering latitudes approximately between 38°N and 38°S. According to Boccippio et al. [32], the detection efficiency of the LIS was 73 ± 11% and 93 ± 4% during the day and night, respectively, and 88 ± 9% throughout the entire day.

The primary flash product of the LIS comprised events, groups, flashes, and so on. An event corresponded to a single detected pixel surpassing the background radiance threshold. Spatially continuous or single events within the same ~2 ms LIS frame were designated as a group, and groups occurring within 5.5 km and 330 ms of each other were clustered together as a flash [39]. LIS data not only provided information on lightning frequency and location but also included flash duration, footprint (area of the region illuminated by lightning), and radiance. Additionally, we calculated the flash length by extracting the largest horizontal distance between any two events involved in a flash. Flash duration, horizontal extension (footprint and length), and radiance are all characteristics associated with the lightning discharge process and channel development; collectively, they are referred to as lightning structure in this analysis. Following the methodology used in the work by You et al. [17], extreme flashes with durations greater than 3 s or equal to 0 s, as well as lengths larger than 1000 km or smaller than 4 km, were excluded.

2.1.3. Matching of RPF and LIS Data

When examining the relationship between thunderstorm structure and lightning structure, the alignment of RPF data and LIS data is essential. We employed the same methodology as You et al. [17]. Initially, parameters describing the shape (long axis, short axis, and rotation angle) and position of the thunderstorm were extracted from the RPF data to define an ellipse simulating the thunderstorm shape. Subsequently, the long and short axes of the original ellipse were extended by one-third to create a new elliptical region. Lightning falling within this extended elliptical region was then identified as associated with the respective RPF. You et al. [17] justified the one-third extension of the ellipse's short and long axes based on the following reasons: (1) the minimum detectable signal of TRMM/PR that constrained the RPF range was approximately above 17–18 dBZ [40], implying that the outer regions of storms potentially involved in lightning discharges were not encompassed in the RPF, and (2) the referenced position of flashes in the matching process was the centroid of pixels with radiance weighted, potentially located outside the RPF, with parts of flash channels even propagating through the RPF. Their tests indicated that the proportion of lightning ultimately matched with the RPF to the total lightning was comparable to considering PR and LIS swath widths (about 250 km and 600 km, respectively).

2.2. Analysis Region

The delineation of the analysis regions is indicated by the red lines in Figure 1a. While this region essentially covers the primary body of the QTP, it does not perfectly align with the entire QTP area. The determination of this region considered both terrain features and the comprehensive distribution characteristics of RPF data. The brown line in Figure 1b roughly follows the 3000 m isoheight, and the RPFs with lightning are denoted by gray dots on the map based on their locations. A distinct edge, approximately outlined by the red lines in Figure 1b, exhibits low-density RPFs with lightning and serves as a division between the samples over the QTP and those over the southern Himalayan front. Previous studies have highlighted significant differences in the properties and mechanisms of deep convection and lightning activity between the QTP and the southern Himalayan front [12,16–18,41,42]. Thus, to minimize contamination of our study area, it is essential to exclude thunderstorms over the latter region. The red line in Figure 1b is employed as the boundary for the south and west sides of the research region, while the 3000 m contour lines demarcate the boundaries for the north, east, and southeast sides. As an inherent limitation of the observation range of TRMM/PR, the effective northern analysis boundary is positioned at approximately 36°N (Figure 1b).

A total of 18,745 RPF thunderstorm samples were finally extracted in the analysis region, accounting for about 2.0% of all RPF samples there. There were 137,492 original LIS lightning samples, while 123,592 LIS flashes remained after quality control. The numbers of RPFs and flashes matched together are 14,650 and 40,762, respectively.

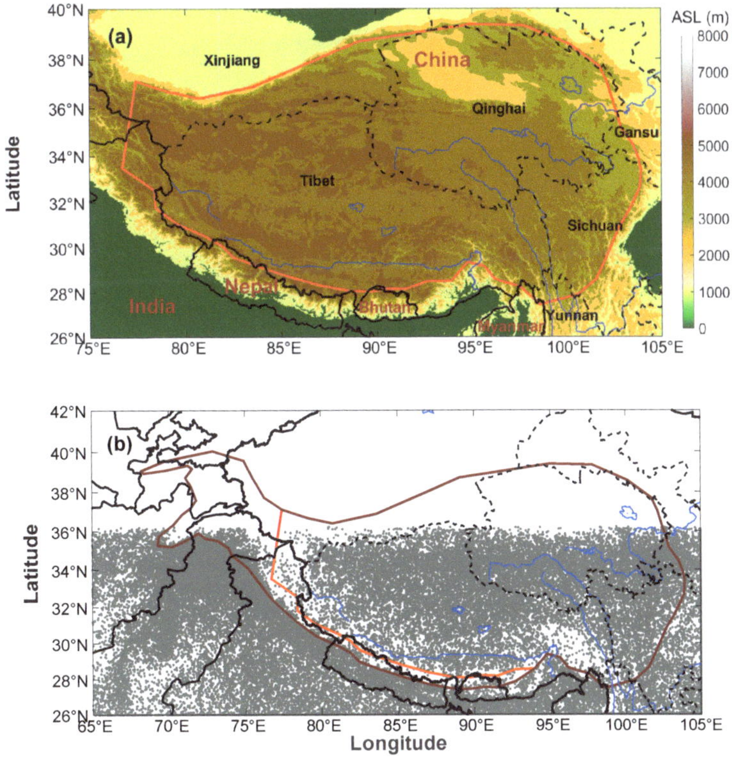

Figure 1. Study region in the Qinghai–Tibet Plateau (QTP). (**a**) Topographic map depicting the study area enclosed by the red line, with altitude represented by color shading. (**b**) Visualization of the study area selection. Gray dots represent samples of RPFs with lightning, brown solid lines indicate 3000 m contours, and red solid lines represent the adjusted analytical boundaries for the southern and western areas of the QTP.

2.3. Analysis Methods

The analysis of thunderstorm structure encompasses its vertical and horizontal extension, flash rate, and flash density. Simultaneously, the lightning structure is characterized by flash duration, length, footprint, and radiance. Our focus lies on their spatial and temporal distributions in the QTP. For spatial distribution analysis, we employed a $1° \times 1°$ grid, ensuring that the sample number of thunderstorms or flashes in a statistical grid was no less than 10. In examining monthly and diurnal changes in thunderstorm and lightning structure, a requirement was set that the number of thunderstorms and flashes in each statistical period should be no less than 20, with diurnal changes counted in 2 h intervals. Section 5 addresses the spatiotemporal correlation of different parameters, utilizing both Pearson correlation and Spearman correlation tests. The Pearson correlation gauges the linear correlation between two variables. The Spearman correlation evaluates the strength of the monotone relationship between two random variables, that is, the extent to which the two variables follow a trend of increasing or decreasing, even if no proportional relationship is maintained. Only results with the largest correlation coefficients are presented, and it is important to note that all shown correlation results have passed the significance level test of 5%.

3. Spatiotemporal Distributions of Thunderstorm Structure over the QTP

3.1. Vertical and Horizontal Extensions of Thunderstorms

We utilized the 20 and 30 dBZ PR echo top heights (abbreviated as H_{20dBZ} and H_{30dBZ}, respectively) and heights at the 20 and 30 dBZ PR echo tops relative to the ground (abbreviated as RH_{20dBZ} and RH_{30dBZ}, respectively) to characterize the vertical development characteristics of thunderstorms in the QTP. The areas of regions with PR reflectivity greater than 20 ($A_{>20dBZ}$) and 30 dBZ ($A_{>30dBZ}$) within thunderstorms were employed to characterize the horizontal extension of QTP thunderstorms.

It is worth noting that H_{20dBZ} and H_{30dBZ} were directly provided in the RPF data, while RH_{20dBZ} and RH_{30dBZ} were obtained by subtracting the topographic height of the corresponding location from H_{20dBZ} and H_{30dBZ}. Additionally, the RPF only provides the area of the region with a composite reflectivity greater than 20 dBZ. $A_{>30dBZ}$ was derived by selecting the maximum areas with reflectivity greater than 30 dBZ at all 16 altitude levels (1 km apart) provided by the RPF. Zheng and Zhang [18] reported, based on TRMM/RPF, that the maximum $A_{>30dBZ}$ of QTP thunderstorms tends to occur at an altitude of 5 km.

It is important to mention that we chose to analyze the echo top height and regional area defined by 30 rather than 40 dBZ, a threshold commonly used in the analysis of plain thunderstorms and typically associated with lightning occurrence. Among the RPFs with lightning in the analysis region, only about 42.2% had reflectivity greater than 40 dBZ, while approximately 96.2% had reflectivity greater than 30 dBZ. In comparison, Zheng and Zhang [18] reported that, among thunderstorms in Central and Eastern China and the southern Himalaya foothills, samples with reflectivity exceeding 40 and 30 dBZ both exceeded 90%. This highlights that the 30 dBZ threshold for QTP thunderstorms is more representative.

3.1.1. Spatial Distribution

Figure 2a,b depict the spatial distributions of the H_{20dBZ} and H_{30dBZ} of thunderstorms in the QTP, respectively. A noticeable north–south contrast in the vertical development of QTP thunderstorm clouds is evident, with southern thunderstorms generally reaching higher altitudes than their northern counterparts. In the regions to the south of the 32°N latitude, H_{20dBZ} generally exceeds 11.2 km, and H_{30dBZ} generally surpasses 9.3 km. H_{20dBZ} reveals two distinct regions with average values greater than 12 km (Figure 2a). One is situated in the southeastern QTP, centered around approximately 98°E, 29.5°N, coinciding with several prominent northwest–southeast-oriented mountain ranges. The other center is around 86.5°E, 29.5°N, with an average H_{20dBZ} exceeding 12.2 km at its core. These locations also represent centers with high values for H_{30dBZ}, averaging over 9.8 km (Figure 2b). Additionally, other significant centers with high H_{30dBZ} values are found at 95.5°E, 32°N; 100.5°E, 32°N; and 84.5°E, 30°N in the southern QTP. Regions with lower average H_{20dBZ} and H_{30dBZ} values are primarily located in the northeastern and northwestern QTP and on the west side of the great bend of the Brahmaputra River. In these areas, the average H_{20dBZ} is generally below 10.5 km, and the average H_{30dBZ} is generally below 9 km. Zheng and Zhang [18] reported that the average echo top heights for thunderstorms across the entire QTP was approximately 11.1 km for 20 dBZ and 9.3 km for 30 dBZ. Qie et al. [25], who divided the QTP into eastern, central, and western regions, found that the average 20 dBZ echo top heights of thunderstorms were about 11.1, 11.1, and 10.8 km, respectively, while the average 30 dBZ echo top heights were about 9.5, 9.4, and 9.2 km, respectively. Their results indicated no significant differences among these regions. This may be attributed to their division of regions in the east–west direction, which failed to capture the north–south contrast characteristic of the vertical development of thunderstorms in the QTP. Furthermore, the high-value region of thunderstorm vertical development in Figure 2a,b exhibit a spatial correlation with the high-value center of the probability distribution with a brightness temperature lower than −52 °C, as provided by Ma et al. [28] in their figure 4, based on geostationary satellite data.

Figure 2c,d illustrate the spatial distributions of RH_{20dBZ} and RH_{30dBZ} of QTP thunderstorms, respectively. As these values account for terrain height, they can offer insights into the depth of the updraft in QTP thunderstorms. In the region to the south of 32°N and east of 99°E on the QTP, the mean value of RH_{20dBZ} generally exceeds 6.5 km. Correspondingly, in the southeastern QTP where RH_{20dBZ} exceeds 7 km, RH_{30dBZ} is generally greater than 5 km. Lower values of RH_{20dBZ} and RH_{30dBZ} are primarily found in the northwestern QTP, specifically to the north of 30°N and west of 91°E, where values are generally below 6 km and 4.4 km, respectively. Overall, the distribution of RH_{20dBZ} and RH_{30dBZ} in the QTP is characterized by both north–south contrast (more noticeable in RH_{20dBZ}) and east–west contrast (more apparent in RH_{30dBZ}).

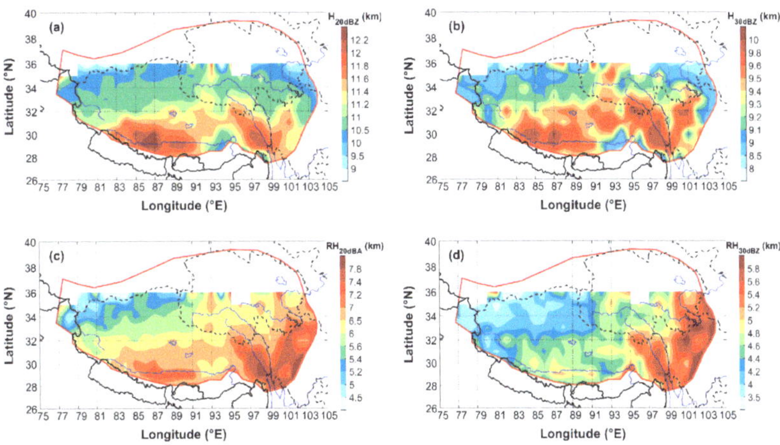

Figure 2. Geographic distributions of average (**a**) H_{20dBZ}, (**b**) H_{30dBZ}, (**c**) RH_{20dBZ}, and (**d**) RH_{30dBZ} of thunderstorms over the QTP.

Comparing the spatial distribution of H_{20dBZ} and H_{30dBZ} with RH_{20dBZ} and RH_{30dBZ} reveals significant differences in the northeast and east border regions of the QTP. While H_{20dBZ} *and* H_{30dBZ} are low in these regions, RH_{20dBZ} and RH_{30dBZ} are relatively high. In the southwestern QTP, specifically at 83–91°E and 29–31°N, local centers with high values of H_{20dBZ} and H_{30dBZ} are observed, but the values of RH_{20dBZ} and RH_{30dBZ} are moderate and not particularly prominent.

Figure 3 illustrates a significant east–west contrast in the horizontal extension of thunderstorms over the QTP. Specifically, the eastern QTP exhibits a larger horizontal extension of thunderstorms compared to the western QTP. In the eastern QTP, the average $A_{>20dBZ}$ is generally greater than 1000 km^2, and $A_{>30dBZ}$ is generally greater than 180 km^2. Conversely, the western QTP shows smaller values for both $A_{>20dBZ}$ and $A_{>30dBZ}$. The mean values of $A_{>20dBZ}$ and $A_{>30dBZ}$ exhibit highly similar spatial distribution patterns. The maximum value center is situated near the eastern boundary of the analysis region, around longitude 102°E and latitude 29–34°N. Here, $A_{>20dBZ}$ can exceed 2000 km^2, and $A_{>30dBZ}$ can surpass 600 km^2. Notably, there is a localized area centered around 97°E, 29.5°N where $A_{>20dBZ}$ is less than 1000 km^2, and $A_{>30dBZ}$ is less than 200 km^2, despite the high vertical development of thunderstorms (Figure 2). Thunderstorms in the western QTP exhibit minimal horizontal expansion and are generally smaller in the westward direction. For instance, in regions to the west of 85°E, $A_{>20dBZ}$ is generally less than 700 km^2, and $A_{>30dBZ}$ is generally less than 140 km^2.

Zheng and Zhang [18] reported an average $A_{>20dBZ}$ of 1182 km^2 for QTP thunderstorms. Qie et al. [25] highlighted significant variations in the horizontal expansion of thunderstorm clouds across the eastern, central, and western QTP, with the largest expansion observed in the east and the smallest in the west, aligning with the overall geographical distribution pattern presented in Figure 3. Additionally, Zheng et al. [43] discovered that the $A_{>20dBZ}$ regional statistical value for deep convective clouds, as observed by TRMM PR, was approximately one order of magnitude higher than that observed by the ground-based radar in their analysis of deep convection clouds in Naqu (a city located in the central QTP). They attributed this difference to the lower horizontal spatial resolution of TRMM PR compared to the ground-based radar, suggesting that smaller-scale convective clouds may not be captured by TRMM PR. Moreover, they found similarities between TRMM PR and ground-based radar observations at 20 and 30 dBZ echo top heights. Based on these findings, it is speculated that if smaller thunderstorm samples are considered, the actual horizontal expansion of QTP thunderstorms could be smaller than currently indicated in Figure 3a,b.

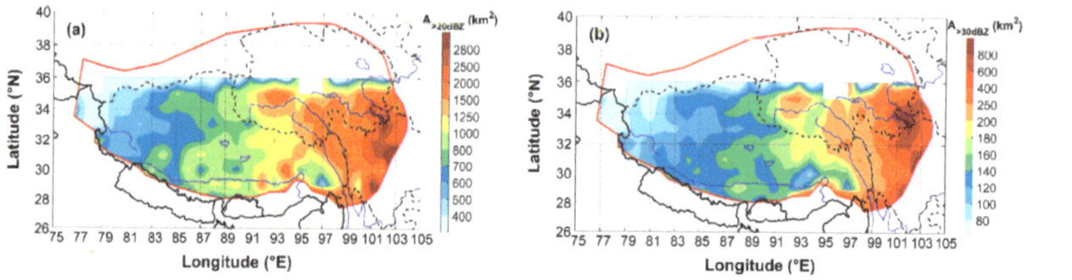

Figure 3. Geographic distributions of average (**a**) $A_{>20dBZ}$ and (**b**) $A_{>30dBZ}$ of thunderstorms over the QTP.

3.1.2. Temporal Variation

The monthly variations in the vertical and horizontal extensions of thunderstorms over the QTP are illustrated in Figure 4a. Both H_{20dBZ} and H_{30dBZ} exhibit a single-peak pattern in their monthly variations. Their average values increase from 9.5 and 7.8 km in March to a peak of 11.7 and 9.7 km in August, followed by a decline to 10.2 and 9.0 km in October, respectively. Similar to the monthly variation in H_{20dBZ} and H_{30dBZ}, RH_{20dBZ} and RH_{30dBZ} peak at 7.0 km and 5.0 km in August. However, the difference lies in the fact that RH_{20dBZ} and RH_{30dBZ} demonstrate an oscillating change trend from March to May. The valley value of RH_{20dBZ} is 5.5 km, occurring in May, while RH_{30dBZ} has a valley value of 4.0 km in March. In contrast, $A_{>20dBZ}$ and $A_{>30dBZ}$ display entirely different monthly variation patterns. $A_{>20dBZ}$ reaches its maximum value of about 2200 km^2 in March, subsequently dropping to around 1100 km^2 in May. It then fluctuates between approximately 1000 and 1400 km^2 from May to October, showing no distinct trend. The lowest value occurs in August, at about 1070 km^2. $A_{>30dBZ}$ experiences a rapid rise from June to approximately 240 km^2, fluctuating between about 230 km^2 and 250 km^2 until September. It then begins to decline, with the lowest value appearing in May (about 150 km^2). The variation observed, with $A_{>20dBZ}$ being smaller from May to October and larger in March and April, may be attributed to the predominance of thermally driven thunderstorms with smaller-scale samples during the former period. Conversely, thunderstorms in the latter period are more likely to occur under favorable circulation conditions, potentially exhibiting a larger horizontal extension but a weaker lightning yield. During their data control processes, Du et al. [23] and Ma and Zheng [24] both identified abnormal thunderstorm samples characterized by a relatively large area but a relatively low frequency of flashes, predominantly correlating with thunderstorms in cold seasons. On the other hand, the variation pattern of $A_{>30dBZ}$ suggests the sensitivity of 30 dBZ reflectivity to convective enhancement over the QTP.

The diurnal variations in thunderstorm vertical and horizontal extent parameters are depicted in Figure 4b. Both the mean vertical and horizontal extensions of thunderstorms exhibit similar diurnal patterns. For instance, they all reach their lowest values between 8:00 and 10:00 LT (H_{20dBZ}, H_{30dBZ}, RH_{20dBZ}, RH_{30dBZ}, and $A_{>30dBZ}$) or 10:00 and 12:00 LT ($A_{>20dBZ}$) and show a general increase during the afternoon. H_{20dBZ}, H_{30dBZ}, RH_{20dBZ}, and RH_{30dBZ} peak at 18:00–20:00 LT (11.7, 9.6, 7.0, and 5.0 km, respectively), while the peaks of $A_{>20dBZ}$ and $A_{>30dBZ}$ (3000 and 640 km^2, respectively) are reached at 2:00–4:00 LT and 0:00–2:00 LT, respectively. All parameters exhibit a decreasing trend between 4:00 LT and 10:00 LT. The diurnal variation in the vertical and horizontal extensions of thunderstorms underscores the significant role of solar heating in the development of QTP convection, which is a key factor contributing to the evident diurnal variation in thunderstorms, deep convection, and lightning activities [16,43,44].

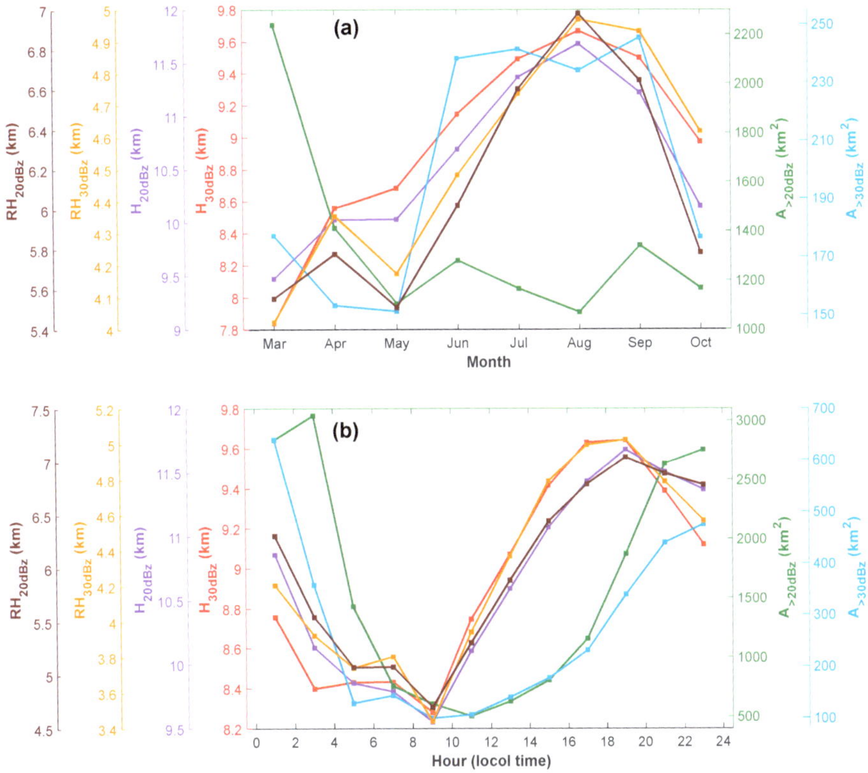

Figure 4. Temporal variations in average vertical and horizontal expansion characteristics of QTP thunderstorms: (**a**) monthly variation and (**b**) diurnal variation. The corresponding standard deviations are shown in Tables S1 and S2 in the Supplementary Materials.

3.2. Flash Rate and Density of Thunderstorm

We explored two parameters: flash rate and flash density of the thunderstorms. Flash rate of thunderstorm denotes the frequency of lightning flashes per minute within the thunderstorm, while flash density of thunderstorm is computed by dividing the flash rate by $A_{>20dBZ}$ and $A_{>30dBZ}$, respectively.

3.2.1. Spatial Distribution

The geographic distributions of parameters related to lightning activity within thunderstorms are presented in Figure 5. The flash rate of thunderstorms exhibits an initial west–east contrast followed by a north–south contrast (Figure 5a). Across most areas of central and western QTP, the average flash rates of thunderstorms are below 2.0 flashes per minute (fl min^{-1}), while near the southern and western borders of the analysis area, thunderstorms have average flash rates smaller than 1.4 fl min^{-1}. Over the eastern boundary area, average flash rates of thunderstorms can exceed 2.5 fl min^{-1}. Notably, over Naqu City (approximately located around 92°E, 32°N), the average flash rate of thunderstorm ranges between approximately 1.6 and 2.0 fl min^{-1}, aligning with the range of 1.0–2.0 fl min^{-1} suggested by Zhang et al. [45], who derived the range from investigations of thunderstorm cases over Naqu. Zheng and Zhang [18] reported an average flash rate of 1.82 fl min^{-1} for QTP thunderstorms, with a median of 0.95 fl min^{-1}.

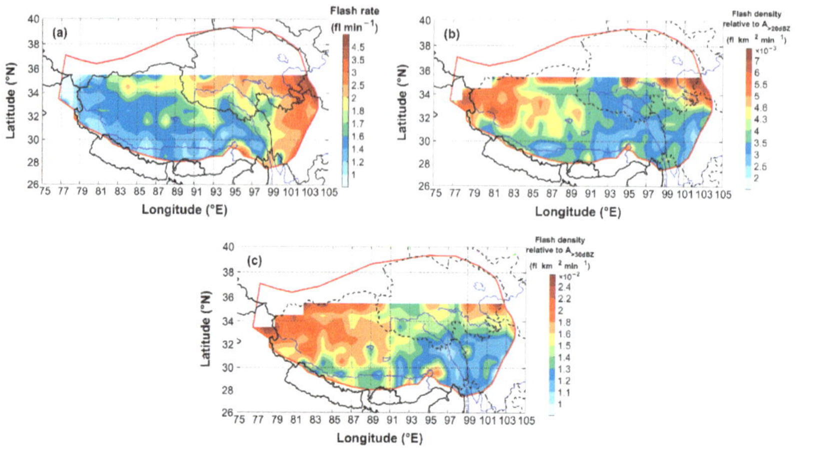

Figure 5. Geographic distributions of average (**a**) flash rate per minute, (**b**) flash density relative to $A_{>20dBZ}$ per minute, and (**c**) flash density relative to $A_{>30dBZ}$ per minute for thunderstorms over the QTP.

The flash densities of thunderstorms over the QTP (Figure 5b,c) show a distinct geographic distribution pattern compared to the flash rate of thunderstorm. Generally, thunderstorms over the western QTP and the northeastern corner of the analysis region exhibit higher flash density, with average flash densities relative to $A_{>20dBZ}$ and $A_{>30dBZ}$ generally exceeding 4×10^{-3} and 1.6×10^{-2} fl km^{-2} min^{-1}, respectively. Most areas over the central and eastern QTP have lower flash densities. The southeastern QTP features the lowest flash density of thunderstorms. Zheng and Zhang [18] reported a flash density relative to $A_{>20dBZ}$ of approximately 4.01×10^{-3} fl km^{-2} min^{-1} for thunderstorms in the QTP.

3.2.2. Temporal Variation

In the monthly variation (Figure 6a), the flash rate of thunderstorms strengthens from March to June, reaching its peak value of 2.1 fl min^{-1} in June. Subsequently, it gradually declines, with a brief increase in September before another decline. The peak flash density of thunderstorms relative to $A_{>20dBZ}$ occurs in July (4.3×10^{-3} fl km^{-2} min^{-1}). During the same period, $A_{>20dBZ}$ does not show substantial changes from May to September (the primary lightning season) (Figure 4a). The flash densities of thunderstorms relative to $A_{>30dBZ}$ peaks in May (1.8×10^{-2} fl km^{-2} min^{-1}) but decreases in the subsequent months when thunderstorms and lightning activity are relatively high.

The diurnal variation in the flash rate of thunderstorms (Figure 6b) exhibits a peak value at 20:00–22:00 LT (2.6 fl min^{-1}) and a valley value at 6:00–8:00 LT (1.3 fl min^{-1}), indicating a continuous enhancement trend after solar heating. The peak average flash densities relative to $A_{>20dBZ}$ and $A_{>30dBZ}$ (5.2×10^{-3} and 2.1×10^{-2} fl km^{-2} min^{-1}, respectively) occur at 10:00–12:00 LT and 8:00–10:00 LT, respectively, and show a decreasing trend in the afternoon. This pattern is likely influenced by the continuous increase in $A_{>20dBZ}$ and $A_{>30dBZ}$ in the afternoon (Figure 3b).

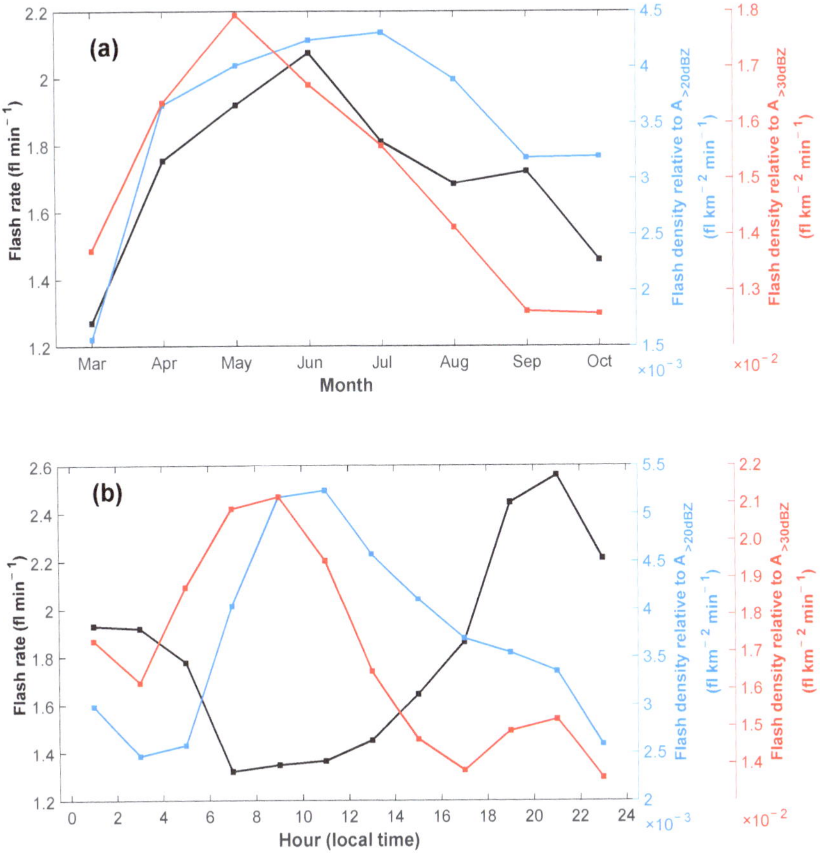

Figure 6. Temporal variations in average flash rate and flash density of thunderstorms: (**a**) monthly variation and (**b**) diurnal variation. The corresponding standard deviations are shown in Tables S1 and S2 in the Supplementary Materials.

4. Spatiotemporal Distributions of Lightning Structure over the QTP

4.1. Spatial Distribution

Figure 7 illustrates the geographical distribution pattern of lightning structure concerning flash duration, length, footprint, and radiance. While these parameters exhibit a general trend of being larger in the east compared to the west, notable differences persist in the distribution details.

Figure 7a reveals that flash duration is highest in the eastern part of the analysis area, particularly near the boundary, where the maximum can exceed 0.22 s. In the regions to the east of 87°N, the average flash duration generally surpasses 0.2 s. Moreover, in areas to the east of 92°N, there is a northwest–southeast band with an average flash duration greater than 0.21 s. Conversely, in the western QTP, roughly corresponding to the Ali region, flash duration is the smallest, with the average value generally below 0.19 s.

The spatial distributions of flash length and footprint, representing the horizontal extension of lightning, are depicted in Figure 7b,c. Their patterns exhibit similarities, showing larger values in the east and smaller values in the west. Furthermore, there is a decreasing trend from southeast to northwest. In the southeastern QTP (approximately 92–103°E and 28–33°N), the average spatial extension of lightning is the greatest, with flash lengths generally exceeding 14.5 km, and footprints reaching over 230 km^2. In the central region to the north of 30°N, the average flash length approximately ranges from about 13.5 to 14.5 km, with footprints between 200 and 230 km^2. The minimum lightning spatial extension is observed in the northeast and west of the analysis region, where the average flash length is generally less than 13.5 km and, in some areas, even less than 12.5 km. The average flash footprint is generally less than 200 km^2 in these regions. According to Zheng and Zhang [18], the median and mean flash lengths over the entire QTP are 12.6 and 13.9 km, respectively.

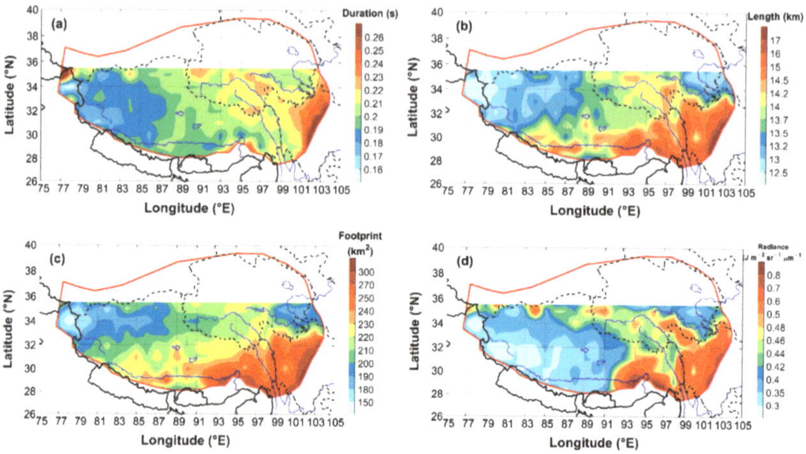

Figure 7. Geographic distributions of average (**a**) flash duration, (**b**) flash length, (**c**) flash footprint, and (**d**) flash radiance over the QTP.

Figure 7d illustrates the spatial distribution of flash radiance. Overall, this pattern is more analogous to the spatial distribution of lightning spatial extension. In the southeastern region near the boundary of the analysis area, the average flash radiance generally exceeds 0.48 J m^{-2} sr^{-1} µm^{-1}, with maximum values surpassing 0.8 J m^{-2} sr^{-1} µm^{-1}. Conversely, the northeastern area of the analysis region exhibits lower flash radiance, with average values generally below 0.46 J m^{-2} sr^{-1} µm^{-1}. Unlike the spatial distribution pattern of lightning spatial extension, a notable low flash radiance is observed in the southwestern area of the analysis region, approximately corresponding to the region to the south of 33°N and between 79 and 90°E. Here, the average flash radiance is generally lower than 0.35 J m^{-2} sr^{-1} µm^{-1}.

4.2. Temporal Variation

Figure 8a presents the monthly variations in flash structure, revealing distinctive patterns for lightning spatial extension compared to flash duration and radiance. Lightning spatial extension exhibit a large mean from November to March, corresponding to a period of reduced lightning activity [26,28]. Subsequently, it rapidly decreases from March to May, with the smallest recorded mean flash length (13.1 km) and footprint (201.6 km^2) occurring in May. As lightning activity increases from May to September [26,28], lightning spatial extension demonstrates an increasing trend, reaching a phased peak in September. During this peak, the average flash length is 14.6 km, and the average flash footprint is 235 km^2. The monthly variation trend of flash duration closely mirrors that of flash radiance (Figure 8a). Both parameters reach their valleys in July, coinciding with the peaks of thunderstorm and lightning activities [26]. During this period, the average flash duration and flash radiance are approximately 0.19 s and 0.39 J m^{-2} sr^{-1} µm^{-1}, respectively. In contrast, flash duration peaks in November, aligning with the highest values observed for flash spatial extension, with values of 0.29 s, and flash radiance peaks in March with value of 1 J m^{-2} sr^{-1} µm^{-1}.

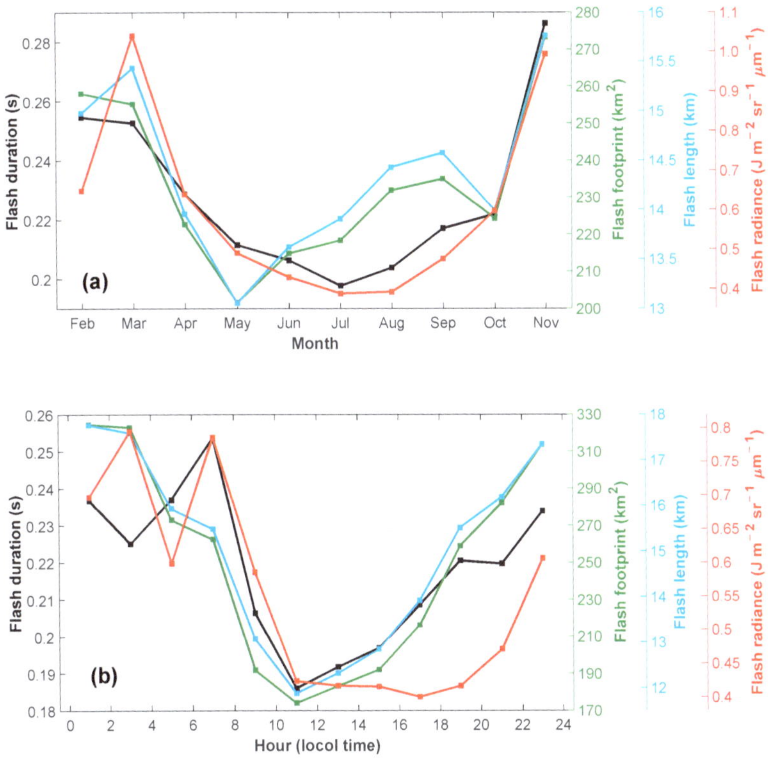

Figure 8. Monthly (**a**) and diurnal (**b**) variations in average lightning structural parameters. The corresponding standard deviations are shown in Tables S1 and S2 in the Supplementary Materials.

The diurnal variation in lightning structure is evident (Figure 8b). The spatial extension of lightning reaches its valley value at 10:00–12:00 LT, corresponding to mean flash length and footprint values of 11.9 km and 174.1 km^2, respectively. Prior to midnight, there is a noticeable growth trend in the spatial extension of lightning, with flash length and footprint peaking at 0:00 to 2:00 LT (17.8 km and 324.4 km^2, respectively), followed by a gradual decline. The diurnal variation in flash duration closely resembles that of the spatial extension of lightning, particularly the general increasing trend in the afternoon after reaching the valley (approximately 0.19 s) at 10:00–12:00 LT. In the early morning hours, flash duration exhibits fluctuations and ultimately peaks at 6:00–8:00 LT with a value of about 0.25 s. Flash radiance attains its peak value of 0.8 J m^{-2} sr^{-1} μm^{-1} at 2:00–4:00 LT, experiences fluctuations, and rapidly decreases at 6:00–12:00 LT. Unlike the spatial extension and duration of lightning, flash radiance exhibits a gradual decline after 12:00 LT until reaching the minimum value of 0.42 J m^{-2} sr^{-1} μm^{-1} at 16:00–18:00 LT, followed by a rapid increase to the peak value at 2:00–4:00 LT.

5. Discussion on the Correlations between Thunderstorm and Lightning Structures

5.1. Correlation between Vertical and Horizontal Extension of Thunderstorms

Upon comparing Figures 2 and 3, it becomes evident that RH_{20dBZ} and RH_{30dBZ} (Figure 2c,d) share similarities with $A_{>20dBZ}$ and $A_{>30dBZ}$ in their spatial distributions. All parameters exhibit larger values in the east and smaller values in the west, differing from H_{20dBZ} and H_{30dBZ}, which demonstrate larger values in the south and smaller values in the north. Computed based on parameter averages within 1° × 1° grid boxes, the spatial correlations of RH_{20dBZ} with $A_{>20dBZ}$ and $A_{>30dBZ}$ yield coefficients of 0.61 (Spearman correlation) and 0.68 (Spearman correlation), respectively. Similarly, the spatial correlation coefficients of RH_{30dBZ} with $A_{>20dBZ}$ and $A_{>30dBZ}$ are 0.75 (Spearman correlation) and 0.83 (Spearman correlation), respectively. Based on the analysis, it is evident that a fixed threshold reflectivity echo top height may not accurately capture the convective intensity differences in thunderstorms across various regions of the QTP, given the potential kilometer-level variations in topographic heights. RH_{20dBZ} and RH_{30dBZ} emerge as more reliable indicators of the convective intensity characteristics of QTP thunderstorms. This stems from their ability to account for the influence of ground height, providing a clearer reflection of the vertical extension depth of airflow. The thermodynamic driving forces play a pivotal role in QTP convective activity. In thermally driven individual thunderstorms, robust updrafts not only vertically transport hydrometeors but also induce horizontal diffusion of hydrometeors in the middle and upper layers, thereby expanding the cloud area. Even if the vertical development of the thunderstorm cloud remains relatively constant (potentially limited by the troposphere top), the horizontal expansion may continue to grow, especially during prolonged thunderstorm durations. Consequently, the positive correlation between the area of the QTP thunderstorm cloud and the strength and duration of the updraft becomes apparent. Therefore, the overall convection intensity in the eastern QTP should surpass that in the western QTP.

5.2. Correlations of Flash Rate and Density with Vertical and Horizontal Extension of Thunderstorms

The flash rate of a thunderstorm serves as a measure of its capacity to produce lightning. A comparison between Figure 2, Figure 3, and Figure 5a reveals a notable similarity in the spatial distribution patterns of the thunderstorm's flash rate and its horizontal extension. Broadly speaking, regions characterized by a large (small) horizontal extension of the thunderstorm tend to exhibit a high (low) flash rate of thunderstorm. Calculating the spatial correlation coefficients based on the mean flash rate of the thunderstorm and the means of $A_{>20dBZ}$ and $A_{>30dBZ}$ in $1^\circ \times 1^\circ$ grids, we determined these coefficients to be 0.73 (Pearson correlation) and 0.80 (Pearson correlation), respectively. Furthermore, their diurnal correlation coefficients were found to be 0.81 (Spearman correlation) and 0.84 (Spearman correlation), respectively. These findings align with the positive correlation between horizontal expansion and convective intensity. Intense convection induces the horizontal diffusion of hydrometeors, leading to the spread of high-density charges and the enlargement of high-density charge regions capable of supporting lightning discharges. Consequently, this results in a greater number of lightning flashes during periods of relatively strong convection.

The flash density of thunderstorms can be considered as a metric for assessing the efficiency of lightning generation by thunderstorms. It generally exhibits a larger magnitude in the west and a smaller magnitude in the east, inversely correlating with the relative height, horizontal extension, and flash rate of thunderstorms. Specifically, the spatial correlation coefficients of flash densities relative to $A_{>30dBZ}$ with RH_{20dBZ} and RH_{30dBZ} are -0.75 (Spearman correlation) and -0.74 (Spearman correlation), respectively. In diurnal variations, when convection intensifies in the afternoon (Figure 6b), the flash density of thunderstorms shows a decreasing trend, contrary to the variation trends observed in other thunderstorm structural parameters. As the charging region of thunderstorms is predominantly situated in the convective area, the horizontal expansion of thunderstorms, while extending the range of charged particles, results in a decrease in charge concentration with distance. This phenomenon contributes to a reduction in the average charge concentration across the entire area of thunderstorms and may be a significant factor underlying the overall inverse correlation between flash density and the flash rate of thunderstorms.

5.3. Relationships among Different Lightning Structural Parameters

The flash duration, horizontal extension scale, and radiance are roughly consistent in spatiotemporal distributions (Figures 7 and 8). In the spatial distribution in $1^\circ \times 1^\circ$ grid boxes, the correlation coefficient between flash length and flash radiance is 0.75 (Pearson correlation) and that between flash length and flash duration is 0.69 (Spearman correlation);

meanwhile, the correlation coefficient between flash radiance and flash duration is 0.71 (Spearman correlation). Analysis based on flash samples shows the corresponding correlations of 0.76 (Spearman correlation), 0.41 (Pearson correlation), and 0.54 (Spearman correlation), respectively. It can be seen that flash spatial extension is more correlated with flash radiance, while they have relatively weak relationships with flash duration, which agrees with previous studies employed in other regions [17,46] or the thunderstorm case [47].

The lack of a significant correlation between flash duration and flash spatial extension can be attributed to various factors. Certain discharge processes may occur within pre-existing channels, resulting in limited channel extension but prolonged flash duration. Additionally, variations in the propagation speed of the lightning channel can occur in different environments [47]. Hence, even if the lightning channel expands to the same scale, the time required for this expansion may differ. The relatively strong association between flash spatial expansion and radiance may be explained by a larger lightning scale indicating a broader range and involvement of more charges in the discharge, thereby contributing to an intensified cumulative flash radiance.

5.4. Association of Lightning Structure with Thunderstorm Structure

In recent years, several studies have investigated the relationship between lightning structure and thunderstorm structure. Bruning and MacGorman [48] reported that, in supercell thunderstorms, the horizontal extension of lightning is small in the region around and near the updraft, while in the region far away from the updraft, the horizontal extension of lightning is large. Subsequent studies have supported similar findings [47,49–53]. Some studies [47,50,54,55] proposed that, in areas with intense convection, strong updraft, wind shear, and entrainment can result in fragmented charge regions. They use the term "charge pockets" to describe this charge distribution pattern, where lightning is frequently initiated due to the proximity of different-polarity charge regions. However, lightning tends to have a small horizontal extension because its channels are constrained by the small scale of the charge regions. Conversely, in regions with relatively weak convection, a vertically layered charge structure predominates. In this case, charge regions mainly expand in the horizontal direction, leading to a relatively large horizontal extension. In such situations, the flash rate of the thunderstorm is relatively low, but the lightning channels can propagate over a considerable range, forming a large horizontal extension. Studies have noted that this inverse correspondence between convection intensity and flash extension also holds when comparing some different types of thunderstorms, such as winter and summer thunderstorms [56], marine and land thunderstorms [17,46,57–59], supercells and normal thunderstorms [47,60], and morning and afternoon thunderstorms [61]. This forementioned concept is utilized to explain differences in convection intensity, flash rate of thunderstorm, and flash spatial extension between various types of thunderstorms; specifically, thunderstorms characterized by strong and weak convections are more likely to be dominated by the patterns of charge pockets and horizontally broad charge regions, respectively [18,61].

Interestingly, when examining the spatial and temporal distributions of thunderstorm structure and lightning structure over the QTP, the relationship between thunderstorm convection intensity and flash extension is not strictly inverse. For the subsequent analysis, we utilize flash length as a representation of the spatial extension of lightning.

In the spatial distributions, the flash length (Figure 7b) exhibits a rough similarity to the RH_{20dBZ} and RH_{30dBZ} of thunderstorms (Figure 2c,d). For instance, they all indicate large values in the southeastern QTP, secondary large values in the southern QTP, and small values in the western and northwestern QTP. Considering their mean values on the $1^\circ \times 1^\circ$ grid, the Spearman correlation coefficients between flash length and RH_{20dBZ}, as well as between flash length and RH_{30dBZ}, are 0.50 and 0.44, respectively, indicating a somewhat positive correlation.

In terms of temporal variation, the extension of lightning and the convection intensity of the thunderstorm also exhibit somewhat synchronous changes. For instance, in their monthly variations, the vertical extension of thunderstorms (Figure 4a) and lightning extension (Figure 8a) both show an overall increasing trend from May to September. Concerning diurnal variation, the correlation appears to be even more pronounced; flash length (Figure 8b) and the vertical and horizontal extension of thunderstorms (Figure 4b) all demonstrate an increasing trend in the afternoon and a decreasing trend in the early morning. We observed that the thunderstorm structural parameters slightly precede the lightning extension in diurnal variation (comparing peak and valley). If the relevant thunderstorm structural parameters were shifted backward by 2 h, the correlation coefficients of flash length with RH_{20dBZ}, RH_{30dBZ}, $A_{>20dBZ}$, and $A_{>30dBZ}$ were 0.69 (Pearson correlation), 0.69 (Pearson correlation), 0.91 (Spearman correlation), and 0.94 (Spearman correlation), respectively.

In terms of climatic statistics, the convection intensity and lightning extension seem to be positively correlated in QTP, contrary to the inverse correspondence suggested by previous studies. Both You et al. [17] and Zheng and Zhang [18] expressed a view in their discussions: within the context of generally weak convective intensity, convective enhancement primarily causes intensifying microphysical and electrification processes. This leads to an increase in the electrification rate within the cloud and the expansion of the high charge concentration region, resulting in a simultaneous increase in lightning rate and the spatial extension of lightning. Only when the convective intensity surpasses a certain threshold does the charge region begin to fragment, leading to an increase in lightning rate and a decrease in lightning scale.

Generally, the convection intensity of QTP thunderstorms or convective systems is weaker than that in other regions [16,18,42,43]. Additionally, Zheng and Zhang [18] compared the scales of lightning extension and the regions with radar echo greater than 30 dBZ of QTP thunderstorms in the mixed-phase region, found them to be comparable, and suggested that the charge region within QTP thunderstorms should be relatively intact in distribution rather than fragmented. Therefore, we speculate that the positive correspondence between thunderstorm convection intensity and lightning extension in spatiotemporal distributions on QTP may be related to the overall weak convection of QTP thunderstorms, supporting the explanation by You et al. [17] and Zheng and Zhang [18].

Furthermore, from the perspective of thunderstorm samples, we matched each thunderstorm structural parameter to its corresponding average lightning extension scale to understand the correlation between them. However, we found that the correlation was very weak. Possible reasons for this weak correlation include significant differences in the structures of thunderstorms and flashes among different regions of QTP and the limited representativeness of lightning structure due to the short TRMM/LIS observation time (80–90 s) with the typically low lightning frequency of QTP thunderstorms (usually 1–2 fl min^{-1}), leading to a small lightning sample number for an RPF with lightning. To pursue statistical regularity, we sorted the samples based on the values of the thunderstorm structural parameters, divided the samples into 10 categories using the 10, 20, . . . , 90 deciles of the values as thresholds, and then calculated the average values of the thunderstorm structure parameters and flash length in each category, analyzing their linear correlations. The result is shown in Figure 9.

Figure 9 presents compelling information. Firstly, it is evident that, as thunderstorms H_{20dBZ} and RH_{20dBZ} increase, there is a corresponding upward trend in flash length (Figure 9a,b). Conversely, as thunderstorms H_{30dBZ} and RH_{30dBZ} increase, there is an overall decreasing trend in flash length (Figure 9c,d).

These analytical findings may suggest the intricate relationship between the convective intensity of thunderstorms and lightning extension on the QTP when considering individual samples. H_{20dBZ} and RH_{20dBZ}, as well as H_{30dBZ} and RH_{30dBZ}, could be associated with varying levels of convection intensity. For example, in the study of deep convective clouds in Naqu using ground-based radar, Zheng et al. [43] discovered that, although H_{20dBZ} and H_{30dBZ} exhibit similar diurnal variation trends, H_{30dBZ} experiences slower growth during its intensification stage and declines more rapidly during its decay stage compared to H_{20dBZ}. This suggests that weak updrafts may contribute to H_{20dBZ} having higher values, but for QTP thunderstorms, only relatively strong updrafts may support H_{30dBZ} in reaching significant heights. Consequently, the positive correlation between H_{20dBZ}, RH_{20dBZ}, and lightning extension may imply that, for QTP thunderstorms with weak convection, enhanced convection positively corresponds to the lightning extension. Conversely, thunderstorms with relatively strong convection in the QTP may follow the trend of increasing convection and decreasing lightning extension scale, aligning with the suggestion from previous studies.

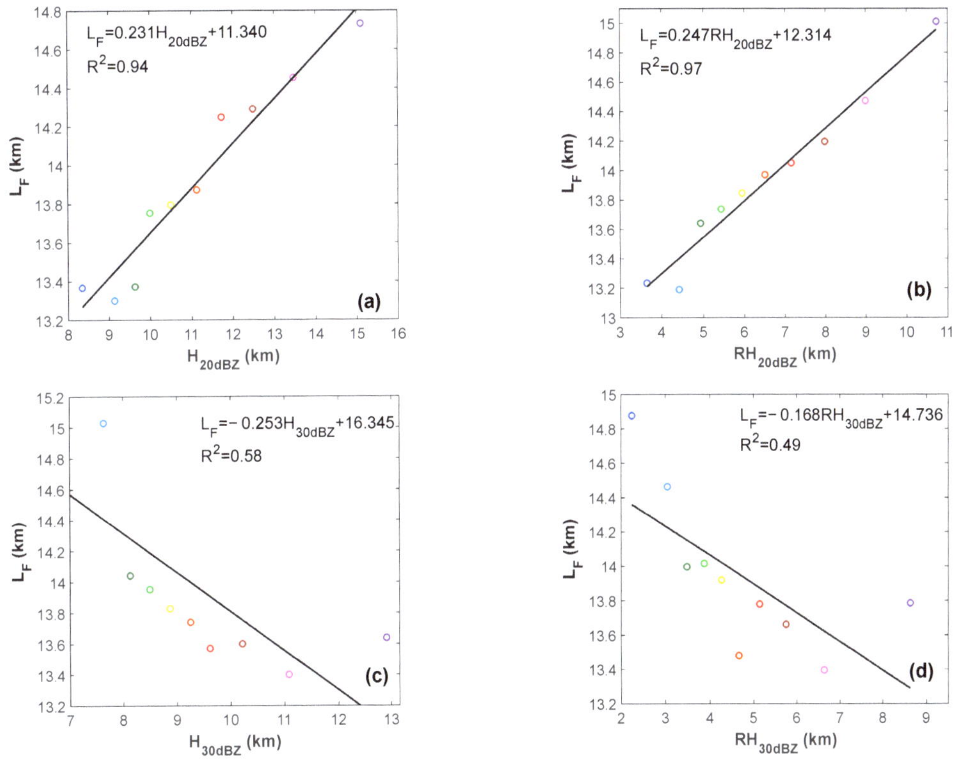

Figure 9. The correlation between the mean value of flash length (L_F) and (**a**) H_{20dBZ}, (**b**) RH_{20dBZ}, (**c**) H_{30dBZ}, and (**d**) RH_{30dBZ} based on sample classification.

6. Conclusions

Utilizing the TRMM/RPF dataset and TRMM/LIS lightning observation data, we investigated the spatiotemporal distributions of thunderstorm and lightning structures over the QTP and examined the interrelationships among various structural parameters. This aspect, crucial when assessing the impact of thunderstorm and lightning activity on atmospheric processes, had not been explored previously. The following conclusions were drawn.

The vertical extension of southern QTP thunderstorms surpasses that in the northern region, with average H_{20dBZ} and H_{30dBZ} generally exceeding 11.2 and 9.3 km, respectively, south of 32°N latitude. The thunderstorm vertical extension relative to the surface shows a discernible east–high and west–low pattern, with average RH_{20dBZ} above 7 km and RH_{30dBZ} above 5 km in the eastern and southeastern QTP. Monthly variations indicate an increase in the thunderstorm vertical extension from May to August. Diurnal variations show an upward trend in the afternoon and a decline in the early morning for the thunderstorm vertical extension.

The horizontal extension of thunderstorms is larger in the eastern QTP than the western QTP. East of 91°E, the average $A_{>20dBZ}$ and $A_{>30dBZ}$ consistently exceed 1000 and 180 km^2, respectively. During the summer season, the average $A_{>20dBZ}$ and $A_{>30dBZ}$ fluctuate between approximately 1000 and 1400 km^2 and between about 230 and 250 km^2, respectively. In diurnal variation, the horizontal extension of thunderstorms increases in the afternoon and diminishes in the early morning, with the peak and valley showing a lag compared to the vertical extension.

Thunderstorm flash rates are higher in the eastern QTP than the western QTP, and in the northern QTP compared to the southern QTP. Central–western QTP thunderstorms mostly have flash rates below 2.0 fl min^{-1}, while the thunderstorm in specific areas in the eastern QTP exceed 2.5 fl min^{-1}. Conversely, flash density of thunderstorms is greater in the western and northeastern QTP and smaller in the southeast. Both flash rate and density of thunderstorms experience an initial increase followed by a decline from March to October. In diurnal variation, the thunderstorm flash rate rises from 8:00 to 22:00 LT, while the thunderstorm flash density peaks in the morning and declines in the afternoon.

Flash duration, spatial extension, and radiance show larger averages in the eastern QTP than the western QTP. The average flash duration can exceed 0.22 s near the eastern border, and in the southeastern QTP, the average flash length and radiance may exceed 14.5 km and 0.50 J m^{-2} sr^{-1} μm^{-1}, respectively. Flash duration in the western QTP, flash length in the northeastern and western QTP, and flash radiance in the southwestern QTP are relatively smaller, generally below 0.19 s, 13.5 km, and 0.35 J m^{-2} sr^{-1} μm^{-1}, respectively. Flash duration and radiance are smaller in the season with active lightning, while the horizontal extension of lightning increases from May to September. In the afternoon, lightning structural parameters exhibit an overall increasing trend, except for flash radiance.

The 20 and 30 dBZ echo top heights relative to the surface are effective indicators of thunderstorm convective intensity. In spatial distribution, RH_{20dBZ} and RH_{30dBZ} show similarities to $A_{>20dBZ}$ and $A_{>30dBZ}$, indicating stronger convective intensity in the eastern QTP than in the western QTP. The thunderstorm flash rate (density) generally correlates positively (negatively) with the relative vertical and horizontal extensions of thunderstorms in their spatiotemporal distributions. Among lightning structure parameters, horizontal extension and radiance exhibit a stronger correlation. Flash length shows a positive spatial and temporal correlation with the RH_{20dBZ} and RH_{30dBZ} of QTP thunderstorms. In classified statistical analysis, flash length correlates positively with H_{20dBZ} and RH_{20dBZ} and negatively with $H3_{0dBZ}$ and RH_{30dBZ}. This outcome suggests that the lightning extension and convection intensity may change in the same direction when the convection intensity is relatively weak, and in the opposite direction when the convection intensity is relatively strong.

Supplementary Materials: The following supporting information can be downloaded at: https://www.mdpi.com/article/10.3390/rs16030468/s1.

Author Contributions: Conceptualization, D.Z.; methodology, D.Z. and Y.D.; software and validation, Y.D. and X.F.; formal analysis, Y.D. and D.Z.; investigation, D.Z., Y.D., Y.Z., W.Y. and L.X.; resources, D.Z., Y.D., Y.Z., W.Y., L.X. and X.F.; data curation, Y.D. and D.Z.; writing—original draft preparation, Y.D., D.Z. and Y.Z.; writing—review and editing, D.Z. and Y.Z.; visualization, Y.D.; supervision, D.Z., Y.Z., W.Y. and L.X.; project administration, D.Z., Y.Z. and W.Y.; funding acquisition, D.Z., Y.Z. and X.F. All authors have read and agreed to the published version of the manuscript.

Funding: This study is funded by the National Natural Science Foundation of China (U2342215), the Second Tibetan Plateau Scientific Expedition and Research Program (2019QZKK0104) and Regional Collaborative Innovation Project of Science and Technology of Tibet Autonomous Region (QYXTZX-NQ2023-01).

Data Availability Statement: The data associated with this paper can be accessed from https://doi.org/10.5281/zenodo.10408363 (accessed on 12 December 2023) or from the corresponding author.

Conflicts of Interest: The authors declare no conflicts of interest.

References

1. Bian, J.; Li, D.; Bai, Z.; Li, Q.; Lyu, D.; Zhou, X. Transport of Asian surface pollutants to the global stratosphere from the Tibetan Plateau region during the Asian summer monsoon. *Natl. Sci. Rev.* **2020**, *7*, 516–533. [CrossRef]
2. Liu, Y.; Lu, M.; Yang, H.; Duan, A.; He, B.; Yang, S.; Wu, G. Land–atmosphere–ocean coupling associated with the Tibetan Plateau and its climate impacts. *Natl. Sci. Rev.* **2020**, *7*, 534–552. [CrossRef] [PubMed]
3. Wu, G. Land–air coupling over the Tibetan Plateau and its climate impacts. *Natl. Sci. Rev.* **2020**, *7*, 485. [CrossRef] [PubMed]
4. Xu, X.D.; Zhou, M.Y.; Chen, J.Y.; Bian, L.G.; Zhang, G.Z.; Liu, H.Z.; Li, S.M.; Zhang, H.S.; Zhao, Y.J.; Suolongduoji; et al. A comprehensive physical pattern of land-air dynamic and thermal structure on the Qinghai-Xizang Plateau. *Sci. China Ser. D-Earth Sci.* **2002**, *45*, 577–594. [CrossRef]
5. Zhao, P.; Li, Y.; Guo, X.; Xu, X.; Liu, Y.; Tang, S.; Xiao, W.; Shi, C.; Ma, Y.; Yu, X.; et al. The Tibetan Plateau Surface-Atmosphere Coupling System and Its Weather and Climate Effects: The Third Tibetan Plateau Atmospheric Science Experiment. *J. Meteorol. Res.* **2019**, *33*, 375–399. [CrossRef]
6. Zhao, P.; Xu, X.; Chen, F.; Guo, X.; Zheng, X.; Liu, L.; Hong, Y.; Li, Y.; La, Z.; Peng, H.; et al. The Third Atmospheric Scientific Experiment for Understanding the Earth–Atmosphere Coupled System over the Tibetan Plateau and Its Effects. *Bull. Am. Meteorol. Soc.* **2018**, *99*, 757–776. [CrossRef]
7. Fu, Y.; Ma, Y.; Zhong, L.; Yang, Y.; Guo, X.; Wang, C.; Xu, X.; Yang, K.; Xu, X.; Liu, L.; et al. Land-surface processes and summer-cloud-precipitation characteristics in the Tibetan Plateau and their effects on downstream weather: A review and perspective. *Natl. Sci. Rev.* **2020**, *7*, 500–515. [CrossRef]
8. Fu, Y.F.; Liu, G.S.; Wu, G.X.; Yu, R.C.; Xu, Y.P.; Wang, Y.; Li, R.; Liu, Q. Tower mast of precipitation over the central Tibetan Plateau summer. *Geophys. Res. Lett.* **2006**, *33*. [CrossRef]
9. Gordillo-Vázquez, F.J.; Pérez-Invernón, F.J.; Huntrieser, H.; Smith, A.K. Comparison of Six Lightning Parameterizations in CAM5 and the Impact on Global Atmospheric Chemistry. *Earth Space Sci.* **2019**, *6*, 2317–2346. [CrossRef]
10. Guo, F.X.; Ju, X.Y.; Bao, M.; Lu, G.Y.; Liu, Z.P.; Li, Y.W.; Mu, Y.J. Relationship between lightning activity and tropospheric nitrogen dioxide and the estimation of lightning-produced nitrogen oxides over China. *Adv. Atmos. Sci.* **2017**, *34*, 235–245. [CrossRef]
11. Zhang, W.J.; Meng, Q.; Ma, M.; Zhang, Y.J. Lightning casualties and damages in China from 1997 to 2009. *Nat. Hazards* **2011**, *57*, 465–476. [CrossRef]
12. Li, J.L.; Wu, X.K.; Yuan, T.; Qie, X.S.; Yang, J. The temporal and spatial distribution of thunderstorms in Asia Monsoon Region based on the TRMM multi-sensor database. *Chin. J. Geophys. Chin. Ed.* **2019**, *62*, 4098–4109. [CrossRef]
13. Xu, M.Y.; Qie, X.S.; Pang, W.J.; Shi, G.Y.; Liang, L.; Sun, Z.L.; Yuan, S.F.; Zhu, K.X.; Zhao, P.T. Lightning climatology across the Chinese continent from 2010 to 2020. *Atmos. Res.* **2022**, *275*, 106251. [CrossRef]
14. Yuan, T.; Qie, X. Spatial and temporal distributions of lightning activities in China from satellite observation. *Plateau Meteorol.* **2024**, *23*, 488–494.
15. Qie, X.S.; Zhang, T.L.; Chen, C.P.; Zhang, G.S.; Zhang, T.; Wei, W.Z. The lower positive charge center and its effect on lightning discharges on the Tibetan Plateau. *Geophys. Res. Lett.* **2005**, *32*. [CrossRef]
16. Wu, X.; Qie, X.; Yuan, T. Regional distribution and diurnal variation of deep convective systems over the Asian monsoon region. *Sci. China-Earth Sci.* **2013**, *56*, 843–854. [CrossRef]
17. You, J.; Zheng, D.; Zhang, Y.; Yao, W.; Meng, Q. Duration, spatial size and radiance of lightning flashes over the Asia-Pacific region based on TRMM/LIS observations. *Atmos. Res.* **2019**, *223*, 98–113. [CrossRef]
18. Zheng, D.; Zhang, Y. New Insights Into the Correlation Between Lightning Flash Rate and Size in Thunderstorms. *Geophys. Res. Lett.* **2021**, *48*, e2021GL096085. [CrossRef]
19. Guo, F.; Wang, M.; Huang, Z.; Li, Y.; Mu, Y.; Lian, C.; Zeng, F. The model analysis of lightning charge structure characters and attributions over Qinghai-Tibet Plateau. *Plateau Meteorol.* **2018**, *37*, 911–922.
20. Li, Y.J.; Zhang, G.S.; Zhang, Y.J. Evolution of the Charge Structure and Lightning Discharge Characteristics of a Qinghai-Tibet Plateau Thunderstorm Dominated by Negative Cloud-to-Ground Flashes. *J. Geophys. Res. Atmos.* **2020**, *125*, e2019JD031129. [CrossRef]
21. Proctor, D.E. Regions where lightning flashes began. *J. Geophys. Res. Atmos.* **1991**, *96*, 5099–5112. [CrossRef]
22. Williams, E.R. The tripole structure of thunderstorms. *J. Geophys. Res. Atmos.* **1989**, *94*, 13151–13167. [CrossRef]
23. Du, Y.; Zheng, D.; Ma, R.; Zhang, Y.; Lyu, W.; Yao, W.; Zhang, W.; Ciren, L.; Cuomu, D. Thunderstorm Activity over the Qinghai–Tibet Plateau Indicated by the Combined Data of the FY-2E Geostationary Satellite and WWLLN. *Remote Sens.* **2022**, *14*, 2855. [CrossRef]
24. Ma, R.; Zheng, D. A dataset of thunderstorm features of the FY-2E full-disc observation area from 2010 to 2018. *China Sci. Data* **2023**, *8*. [CrossRef]
25. Qie, X.; Wei, L.; Zhu, K.; Qie, K.; Xu, C.; Sun, Z.; Jiang, R.; Zhang, H.; Yuan, S. Regional differences of convection structure of thunderclouds over the Tibetan Plateau. *Atmos. Res.* **2022**, *278*, 106338. [CrossRef]
26. Qie, X.S.; Toumi, R.; Yuan, T. Lightning activities on the Tibetan Plateau as observed by the lightning imaging sensor. *J. Geophys. Res. Atmos.* **2003**, *108*. [CrossRef]
27. Qie, X.S.; Toumi, R.; Zhou, Y.J. Lightning activity on the central Tibetan Plateau and its response to convective available potential energy. *Chin. Sci. Bull.* **2003**, *48*, 296–299. [CrossRef]

28. Ma, R.; Zheng, D.; Zhang, Y.; Yao, W.; Zhang, W.; Cuomu, D. Spatiotemporal Lightning Activity Detected by WWLLN over the Tibetan Plateau and Its Comparison with LIS Lightning. *J. Atmos. Ocean. Technol.* **2021**, *38*, 511–523. [CrossRef]
29. Bruning, E.C.; Thomas, R.J. Lightning channel length and flash energy determined from moments of the flash area distribution. *J. Geophys. Res. Atmos.* **2015**, *120*, 8925–8940. [CrossRef]
30. Mecikalski, R.M.; Carey, L.D. Radar Reflectivity and Altitude Distributions of Lightning as a Function of IC, CG, and HY Flashes: Implications for LNOx Production. *J. Geophys. Res. Atmos.* **2018**, *123*, 12796–12813. [CrossRef]
31. Zhang, R.; Zhang, G.S.; Li, Y.J.; Wang, Y.H.; Wu, B.; Yu, H.; Liu, Y.X. Estimate of NOx production in the lightning channel based on three-dimensional lightning locating system. *Sci. China-Earth Sci.* **2014**, *57*, 1613–1625. [CrossRef]
32. Boccippio, D.J.; Koshak, W.J.; Blakeslee, R.J. Performance assessment of the optical transient detector and lightning Imaging sensor. Part I: Predicted diurnal variability. *J. Atmos. Ocean. Technol.* **2002**, *19*, 1318–1332. [CrossRef]
33. Kummerow, C.; Barnes, W.; Kozu, T.; Shiue, J.; Simpson, J. The tropical rainfall measuring mission (TRMM) sensor package. *J. Atmos. Ocean. Technol.* **1998**, *15*, 809–817. [CrossRef]
34. Kummerow, C.; Simpson, J.; Thiele, O.; Barnes, W.; Chang, A.T.C.; Stocker, E.; Adler, R.F.; Hou, A.; Kakar, R.; Wentz, F.; et al. The status of the Tropical Rainfall Measuring Mission (TRMM) after two years in orbit. *J. Appl. Meteorol.* **2000**, *39*, 1965–1982. [CrossRef]
35. Liu, C.; Zipser, E.J.; Cecil, D.J.; Nesbitt, S.W.; Sherwood, S. A Cloud and Precipitation Feature Database from Nine Years of TRMM Observations. *J. Appl. Meteorol. Climatol.* **2008**, *47*, 2712–2728. [CrossRef]
36. Iguchi, T.; Meneghini, R.; Awaka, J.; Kozu, T.; Okamoto, K. Rain profiling algorithm for TRMM precipitation radar data. In *Remote Sensing and Applications: Earth, Atmosphere and Oceans*; Igarashi, T., Nakamura, K., Shimoda, H., Tanaka, T., Burrows, J.P., Nakajima, T., Talagrand, O., Naeije, M.C., Eds.; Advances in Space Research-Series; Elsevier: Amsterdam, The Netherlands, 2000; Volume 25, pp. 973–976.
37. Cecil, D.J.; Buechler, D.E.; Blakeslee, R.J. Gridded lightning climatology from TRMM-LIS and OTD: Dataset description. *Atmos. Res.* **2014**, *135*, 404–414. [CrossRef]
38. Christian, H.J.; Blakeslee, R.J.; Boccippio, D.J.; Boeck, W.L.; Buechler, D.E.; Driscoll, K.T.; Goodman, S.J.; Hall, J.M.; Koshak, W.J.; Mach, D.M.; et al. Global frequency and distribution of lightning as observed from space by the Optical Transient Detector. *J. Geophys. Res. Atmos.* **2003**, *108*, ACL4-1–ACL4-15. [CrossRef]
39. Mach, D.M.; Christian, H.J.; Blakeslee, R.J.; Boccippio, D.J.; Goodman, S.J.; Boeck, W.L. Performance assessment of the Optical Transient Detector and Lightning Imaging Sensor. *J. Geophys. Res. Atmos.* **2007**, *112*. [CrossRef]
40. Park, S.; Jung, S.-H.; Lee, G. Cross validation of TRMM PR reflectivity profiles using 3D reflectivity composite from the ground-based radar network over the Korean Peninsula. *J. Hydrometeorol.* **2015**, *16*, 668–687. [CrossRef]
41. Houze, R.A.; Wilton, D.C.; Smull, B.F. Monsoon convection in the Himalayan region as seen by the TRMM Precipitation Radar. *Q. J. R. Meteorol. Soc.* **2007**, *133*, 1389–1411. [CrossRef]
42. Qie, X.; Wu, X.; Yuan, T.; Bian, J.; Lu, D. Comprehensive pattern of deep convective systems over the Tibetan Plateau–South Asian monsoon region based on TRMM data. *J. Clim.* **2014**, *27*, 6612–6626. [CrossRef]
43. Zheng, D.; Fan, P.; Zhang, Y.; Yao, W.; Fang, X.; Ran, R. Deep convective clouds observed by ground-based radar over Naqu, Qinghai–Tibet Plateau. *Atmos. Res.* **2023**, *293*, 106930. [CrossRef]
44. Chen, G.; Zheng, Y.; Xiao, T. Distribution and spatiotemporal variations of deep convective clouds over China during the warm season. *Meteorolog. Mon.* **2011**, *37*, 75–84.
45. Zhang, T.; Qie, X.; Yan, M. The lightning characteristics of thunderstorm over Tibetan Plateau and its origin discussion. *Plateau Meteorol.* **2007**, *26*, 774–782.
46. Peterson, M.; Liu, C.T. Characteristics of lightning flashes with exceptional illuminated areas, durations, and optical powers and surrounding storm properties in the tropics and inner subtropics. *J. Geophys. Res. Atmos.* **2013**, *118*, 11727–11740. [CrossRef]
47. Zheng, D.; Zhang, Y.; Meng, Q. Properties of Negative Initial Leaders and Lightning Flash Size in a Cluster of Supercells. *J. Geophys. Res. Atmos.* **2018**, *123*, 12857–12876. [CrossRef]
48. Bruning, E.C.; MacGorman, D.R. Theory and Observations of Controls on Lightning Flash Size Spectra. *J. Atmos. Sci.* **2013**, *70*, 4012–4029. [CrossRef]
49. Bruning, E.; Salinas, V.; Nixon, C.; Berksth, S.; Brothers, M.; Mansell, E.; Chmielewski, V.C. The coupling of thunderstorm kinetic and electrical energetics. In Proceedings of the XVI International Conference on Atmospheric Electricity, Nara, Japan, 17–22 June 2018.
50. Calhoun, K.M.; MacGorman, D.R.; Ziegler, C.L.; Biggerstaff, M.I. Evolution of Lightning Activity and Storm Charge Relative to Dual-Doppler Analysis of a High-Precipitation Supercell Storm. *Mon. Weather Rev.* **2013**, *141*, 2199–2223. [CrossRef]
51. MacGorman, D. Meteorological aspects of lightning and thunderstorm electrification. In Proceedings of the XVI International Conference on Atmospheric Electricity, Nara, Japan, 17–22 June 2018.
52. Schultz, C.J.; Carey, L.D.; Schultz, E.V.; Blakeslee, R.J. Insight into the Kinematic and Microphysical Processes that Control Lightning Jumps. *Weather Forecast.* **2015**, *30*, 1591–1621. [CrossRef]
53. Souza, J.C.S.; Bruning, E.C. Assessment of Turbulence Intensity in Different Spots of Lightning Flash Propagation. *Geophys. Res. Lett.* **2021**, *48*, e2021GL095923. [CrossRef]

54. Calhoun, K.M.; Mansell, E.R.; MacGorman, D.R.; Dowell, D.C. Numerical Simulations of Lightning and Storm Charge of the 29–30 May 2004 Geary, Oklahoma, Supercell Thunderstorm Using EnKF Mobile Radar Data Assimilation. *Mon. Weather Rev.* **2014**, *142*, 3977–3997. [CrossRef]
55. Thiel, K.C.; Calhoun, K.M.; Reinhart, A.E.; MacGorman, D.R. GLM and ABI Characteristics of Severe and Convective Storms. *J. Geophys. Res. Atmos.* **2020**, *125*, e2020JD032858. [CrossRef]
56. López, J.A.; Pineda, N.; Montanyà, J.; van der Velde, O.; Fabró, F.; Romero, D. Spatio-temporal dimension of lightning flashes based on three-dimensional Lightning Mapping Array. *Atmos. Res.* **2017**, *197*, 255–264. [CrossRef]
57. Beirle, S.; Koshak, W.; Blakeslee, R.; Wagner, T. Global patterns of lightning properties derived by OTD and LIS. *Nat. Hazards Earth Syst. Sci.* **2014**, *14*, 2715–2726. [CrossRef]
58. Peterson, M.; Deierling, W.; Liu, C.T.; Mach, D.; Kalb, C. The properties of optical lightning flashes and the clouds they illuminate. *J. Geophys. Res. Atmos.* **2017**, *122*, 423–442. [CrossRef]
59. Said, R.K.; Cohen, M.B.; Inan, U.S. Highly intense lightning over the oceans: Estimated peak currents from global GLD360 observations. *J. Geophys. Res. Atmos.* **2013**, *118*, 6905–6915. [CrossRef]
60. Mecikalski, R.M.; Carey, L.D. Radar Reflectivity and Altitude Distributions of Lightning Flashes as a Function of Three Main Storm Types. *J. Geophys. Res. Atmos.* **2018**, *123*, 12814–12828. [CrossRef]
61. Chronis, T.; Cummins, K.; Said, R.; Koshak, W.; McCaul, E.; Williams, E.R.; Stano, G.T.; Grant, M. Climatological diurnal variation of negative CG lightning peak current over the continental United States. *J. Geophys. Res. Atmos.* **2015**, *120*, 582–589. [CrossRef]

remote sensing

Article

Improving Predictions of Tibetan Plateau Summer Precipitation Using a Sea Surface Temperature Analog-Based Correction Method

Lin Wang [1], Hong-Li Ren [1,2,*], Xiangde Xu [1], Li Gao [3], Bin Chen [1], Jian Li [1], Huizheng Che [4], Yaqiang Wang [5] and Xiaoye Zhang [4]

1 State Key Laboratory of Severe Weather, Institute of Tibetan Plateau Meteorology, Chinese Academy of Meteorological Sciences, Beijing 100081, China
2 Collaborative Innovation Center on Forecast and Evaluation of Meteorological Disasters (CIC-FEMD), Nanjing University of Information Science and Technology, Nanjing 210044, China
3 Ensemble Forecasting Division, CMA Earth System Modeling and Prediction Center (CEMC), Beijing 100081, China
4 State Key Laboratory of Severe Weather, Key Laboratory of Atmospheric Chemistry of CMA, Chinese Academy of Meteorological Sciences, Beijing 100081, China
5 Institute of Artificial Intelligence for Meteorological, Chinese Academy of Meteorological Sciences, Beijing 100081, China
* Correspondence: renhl@cma.gov.cn

Abstract: Boreal summer precipitation over the Tibetan Plateau (TP) is difficult to predict in current climate models and has become a challenging issue. To address this issue, a new analog-based correction method has been developed. Our analysis reveals a substantial correlation between the prediction errors of TP summer precipitation (TPSP) and previous February anomalies of sea surface temperature (SST) in the key regions of tropical oceans. Consequently, these SST anomalies can be selected as effective predictors for correcting prediction errors. With remote-sensing-based and observational datasets employed as benchmarks, the new method was validated using the rolling-independent validation method for the period 1992–2018. The results clearly demonstrate that the new SST analog-based correction method of dynamical models can evidently improve prediction skills of summer precipitation in most TP regions. In comparison to the original model predictions, the method exhibits higher skills in terms of temporal and spatial skill scores. This study offers a valuable tool for effectively improving the TPSP prediction in dynamical models.

Keywords: Tibetan Plateau; summer precipitation prediction; analog-based correction; prediction errors; multi-model ensemble

Citation: Wang, L.; Ren, H.-L.; Xu, X.; Gao, L.; Chen, B.; Li, J.; Che, H.; Wang, Y.; Zhang, X. Improving Predictions of Tibetan Plateau Summer Precipitation Using a Sea Surface Temperature Analog-Based Correction Method. *Remote Sens.* **2023**, *15*, 5669. https://doi.org/10.3390/rs15245669

Academic Editor: Joan Bech

Received: 16 October 2023
Revised: 27 November 2023
Accepted: 6 December 2023
Published: 8 December 2023

1. Introduction

The Tibetan Plateau (TP), often referred to as the "Third Pole" and the "Roof of the World" [1,2], is characterized by its high altitude, diverse topography, and unique climatic conditions. As one of the least affected areas by human activities, the TP possesses fragile vegetation displaying a high sensitivity to global warming [3], where precipitation is of greater significance than temperature throughout its diverse ecosystems [4]. The TP is also known as the "Asian Water Tower", as it serves as the source of major river systems, including the Yellow River, Yangtze, Indus, and Ganges. It plays an indispensable role in providing water resources and ensuring ecosystem security for the surrounding regions [5]. As a substantial heat source, it also has an effect on the formation of summer circulation [6–8]. The variations in TP precipitation, coupled with the release of substantial latent heat, have significant effects not only on river discharge, glacier mass balance, and human water supply [9–11] but also exert further influence on our climate regionally and even globally [12–14]. Therefore, enhancing the predictive capability of climate models in predicting TPSP is crucial for research and operation.

Seasonal precipitation prediction in the TP is influenced by various factors. Substantial advancement has been achieved in understanding the variability of TPSP and the potential impact factors underlying its influence. Via exciting atmospheric teleconnection, tropical sea surface temperature anomalies (SSTAs) have the potential to exert an effect on the TPSP [15–19]. El Niño–Southern Oscillation (ENSO), as the main driver of inter-annual climate variability, typically has indirect and modulated impacts on TPSP under different phases and types of ENSO [16,18–20]. In the positive phase of the Indian Ocean basin mode (IOBM), characterized by uniform warming of SST over the basin, the land–sea thermal contrast between the Eurasian continent and the Indian Ocean weakens. As a result of this weakening impact, precipitation over the TP decreases in early summer [21,22]. Some other climate modes, like the North Atlantic Oscillation (NAO) and Arctic Oscillation (AO), also provide certain predictability sources for the TPSP [23–25].

Although climate models have emerged as powerful tools for operational prediction, their useful predictions with accepted skills are mainly located in the tropics. In contrast, the climate prediction skill for extratropics is relatively limited, except for places with predictability originating from the tropics [26,27]. Among meteorological variables, precipitation usually has limited applicability of prediction. The TP region, with its unique geographical characteristics, presents quite large challenges for climate prediction. Encouragingly, significant advancements in model resolution, physics parameterizations, and dynamic core studies [28–32] have greatly reduced model biases associated with TP precipitation. Nevertheless, the current operational prediction skill for the TP precipitation still falls far from the expected level. For example, Wang et al. (2022) demonstrated that climate models still exhibit limited skill in seasonal to inter-annual prediction for the TPSP, especially across the central–eastern TP [33]. Thus, improvements are urgently needed.

A generally effective approach to enhancing the low prediction skill is a Multi-Model Ensemble (MME). It has been extensively employed in some international projects such as the "DEMETER" [34] and "ENSEMBLES" [35] for seasonal to inter-annual prediction. Notably, some famous institutions like the National Center for Environmental Prediction (NCEP), APEC Climate Center (APCC), International Research Institute for Climate and Society (IRI), and the European Centre for Medium-Range Weather Forecasts (ECMWF) currently provide seasonal prediction products on the basis of the MME approach. For instance, the North American MME (NMME) was developed to provide abundant information to improve intra-seasonal to inter-annual prediction around the world [36–39]. The China MME (CMME) project has been built to supply multi-element predictions since 2018 [40].

On the other hand, various statistical or empirical correction methods have been developed in the past decades, aiming to improve model predictions [41–51]. Model prediction errors are flow-dependent, which can vary with changing climate states, and they have been found to be correlated to physical predictors [52]. In the case of linear systems, the development of statistical models that utilize historical data to correct model errors has proven effective in improving predictions. Commonly used correction methods include mean bias correction [47,53], regression analysis [54,55], coupled field techniques [46,56], and the inter-annual increment method [57–59]. For nonlinear systems, however, simply having a large amount of data is not sufficient, as more data relevance is required. As an alternative, analog states from historical data with properties identical to the current state can be considered. These analog states are available for estimation and subsequent correction of state-dependent errors. Ren and Chou hypothesized that there is a certain degree of similarity between flow-dependent errors arising from similar initial climate states [60]. Based on this hypothesis, the analog-based correction (ABC) of errors dynamical-statistical prediction method was developed and applied using dynamical models and historical data [59–62]. Ren et al. [63] and Liu and Ren [64] successfully employed the ABC method for significantly improving the ENSO prediction. These have provided a valuable perspective to deal with the issues of predicting TPSP by employing the framework of ABC.

In this study, to utilize the advantages of the ABC method, we aim to develop a novel scheme focused on improving the prediction of the TPSP. Section 2 describes the methods and data employed in this investigation. Section 3 gives the results. A brief discussion is provided in Section 4. Section 5 summarizes the major findings.

2. Data and Method

2.1. Model Datasets

The NMME comprises climate models from multiple institutions. Table 1 provides a concise description of each model selected for this research. The first column of the table indicates the institution responsible for each model and its name. Since its establishment in 2011, the NMME has been offering monthly real-time forecasts. It also offers retrospective forecasts dating back to 1982 [36,65,66]. The comprehensive information and long-term coverage offered by the NMME dataset are well suited for our research. This research encompasses a time span from 1982 to 2018. The models included in the NMME have numerous ensemble members. All members of each model were utilized to calculate the ensemble mean. The NMME mean was obtained by performing ensemble averaging on the selected models. The spatial resolutions of the hindcasts for the NMME models are $1.0^{\circ} \times 1.0^{\circ}$ latitude/longitude, with the distance range within the TP estimated to be approximately 78–99 km (latitude) and 111 km (longitude). Please refer to the references mentioned in Table 1 for more details about each individual model. In this study, we define summer as the average of the months from June to August, indicated as (JJA). Thus, predictions initialized in May are defined as having a 1-month lead. The other lead times follow in a similar manner.

Table 1. Description of models.

Model	Time	Members	Lead	Reference
CMC1-CanCM3	1982–2018	10	0–11	[67]
CMC2-CanCM4	1982–2018	10	0–11	[67]
COLA-RSMAS-CCSM3	1982–2018	6	0–11	[68]
COLA-RSMAS-CCSM4	1982–2018	10	0–11	[69]
GFDL_CM2p1_aer04	1982–2018	10	0–11	[70]
GFDL_CM2p5_FLOR_A06	1982–2018	12	0–11	[70]
GFDL_CM2p5_FLOR_B01	1982–2018	12	0–11	[70]
NCEP-CFSv2	1982–2018	24	0–9	[71]

2.2. Observational Datasets

We utilized three sets of observed precipitation data: CN05.1 [72], CRU v4.05 [73], and GPCP v2.3 [74]. These datasets were selected due to their widespread use and reliability in capturing precipitation patterns. CN05.1 is a dataset produced by the National Meteorological Information Center of China, which combines rain gauge observations with spatial interpolation techniques using an "anomaly approach" based on over 2400 stations [72]. The dataset provides monthly precipitation at a resolution of $0.25^{\circ} \times 0.25^{\circ}$ latitude/longitude, with an estimated distance range within the TP of approximately 19.5–24.75 km (latitude) and 27.75 km (longitude). The gridded data obtained via interpolation from meteorological station observations may inevitably contain errors during the interpolation process. Despite the potential errors, the CN05.1 dataset is widely recognized and extensively employed in meteorological and climate research. It provides high-quality ground observation data that are valuable for analyzing climate change, studying extreme weather events, and validating climate models. CRU v4.05, developed by the Climatic Research Unit, is another widely used dataset. It incorporates various sources of observed data, including rain gauges, weather stations, and satellite observations. CRU v4.05 provides monthly precipitation at a resolution of $0.5^{\circ} \times 0.5^{\circ}$ latitude/longitude, with an estimated distance range within the TP of approximately 39–45.5 km (latitude) and 55.5 km (longitude). The GPCP v2.3 dataset is a collaborative effort among several organizations, including NASA,

NOAA, and the University of Maryland. It combines rain gauge measurements, satellite observations, and other data sources to generate global precipitation estimates. GPCP v2.3 provides monthly precipitation data at a resolution of 2.5° × 2.5° latitude/longitude, with an estimated distance range within the TP of approximately 195–247.5 km (latitude) and 277.5 km (longitude). To address the comparability between the observational datasets and the model datasets, we employed an interpolation technique to interpolate all observed datasets to a consistent spatial resolution of 1.0° × 1.0°, which matches the resolution of the model data. Specifically, we used bilinear interpolation to estimate the values at the grid points in the interpolated dataset. To ensure robustness and representativeness, we performed ensemble averaging by combining the three observed datasets. Ensemble averaging involved calculating the mean value across the three datasets at each grid point, resulting in a composite precipitation dataset. In this research, the average of observations is used to evaluate the model results and calculate the historical prediction errors. The SST data were sourced from the National Oceanic and Atmospheric Administration (NOAA) Optimum Interpolation (OI) SST version 2. Climatology is subtracted across the entire period (1982–2018) to calculate anomalies. At different calendar months and lead times, the predicted anomalies are derived by subtracting their own corresponding climatology.

2.3. Analog-Based Correction Method

A brief overview of the ABC method [60,63] is given here:

$$\widehat{Pred}(\psi_0) = Pred(\psi_0) + \sum_{i=1}^{n} w_i \left[Obs\left(\tilde{\psi}_i\right) - Pred\left(\tilde{\psi}_i\right) \right] \quad (1)$$

where *Pred* and *Obs* refer to model predictions and observations, respectively. In the equation, w_i is the weight function. "~" represents the historical analog time. ψ_0 indicates the current initial state and $\tilde{\psi}_i$ is the analogy initial state (i = 1, 2 … n, where n denotes the overall number of analogs chosen). For different ψ_0, one can target the historical analog $\tilde{\psi}_i$ that is well suited for an estimation of current errors. $\widehat{Pred}(\psi_0)$ represents the corrected model prediction. To the right, the first term corresponds to the original model prediction, while the second term represents the prediction correction. The prediction correction term is determined by utilizing historical analog prediction errors from the same model. Furthermore, it should be noted that prediction objectives are expressed in the equation as functions $Pred(\psi)$ rather than prediction variables ψ, indicating that the method can be used to correct either prediction variables (such as geopotential height) or their functions (such as precipitation). Particularly, in this study, the variable ψ denotes SST and $Obs\left(\tilde{\psi}_i\right)$ denotes the observed precipitation corresponding to the predicted precipitation $Pred\left(\tilde{\psi}_i\right)$ for the initial value $\tilde{\psi}_i$.

2.4. Evaluation Methods

The Temporal Correlation Coefficient (TCC) and the Pattern Correlation Coefficient (PCC) are used to quantify model predictability. We define $x_{l,\,k}$ and $f_{l,\,k}$ as the anomalies for observation and prediction in space (l) and time (k); P is the number of space samples; and Q is the number of time samples. They are calculated as follows:

$$\mathrm{TCC}_l = \frac{\sum_{k=1}^{Q} \left(x_{l,\,k} - \overline{x_l}\right)\left(f_{l,\,k} - \overline{f_l}\right)}{\sqrt{\sum_{k=1}^{Q} \left(x_{l,\,k} - \overline{x_l}\right)^2} \times \sqrt{\sum_{k=1}^{Q} \left(f_{l,\,k} - \overline{f_l}\right)^2}}, \quad (2)$$

$$\mathrm{PCC}_k = \frac{\sum_{l=1}^{P} \left(x_{l,\,k} - \overline{x_k}\right)\left(f_{l,\,k} - \overline{f_k}\right)}{\sqrt{\sum_{l=1}^{P} \left(x_{l,\,k} - \overline{x_k}\right)^2} \times \sqrt{\sum_{l=1}^{P} \left(f_{l,\,k} - \overline{f_k}\right)^2}} \quad (3)$$

The range of the TCC and PCC are from −1.0 to 1.0, and a large positive (negative) value indicates a highly similar (opposite) correlation between prediction and observation.

3. Results

3.1. Evaluation of Model Direct Predictions

Figure 1 shows the TCC skills directly from the individual NMME models, as well as their mean for predicting the TPSP. The individual models exhibit low prediction skills, characterized by significant correlation coefficients limited to small areas in the southwestern TP. The NMME-mean shows higher positive TCC skills, although this improvement is mainly observed in the southwestern TP. The central–eastern TP is characterized by high population density and extensive economic and agricultural activities, indicating the urgent need for improvements in predicting precipitation over this region. The subsequent analysis focuses on identifying effective predictors of TPSP. Next, a new analog-based correction method of dynamical models is developed, utilizing potential predictors.

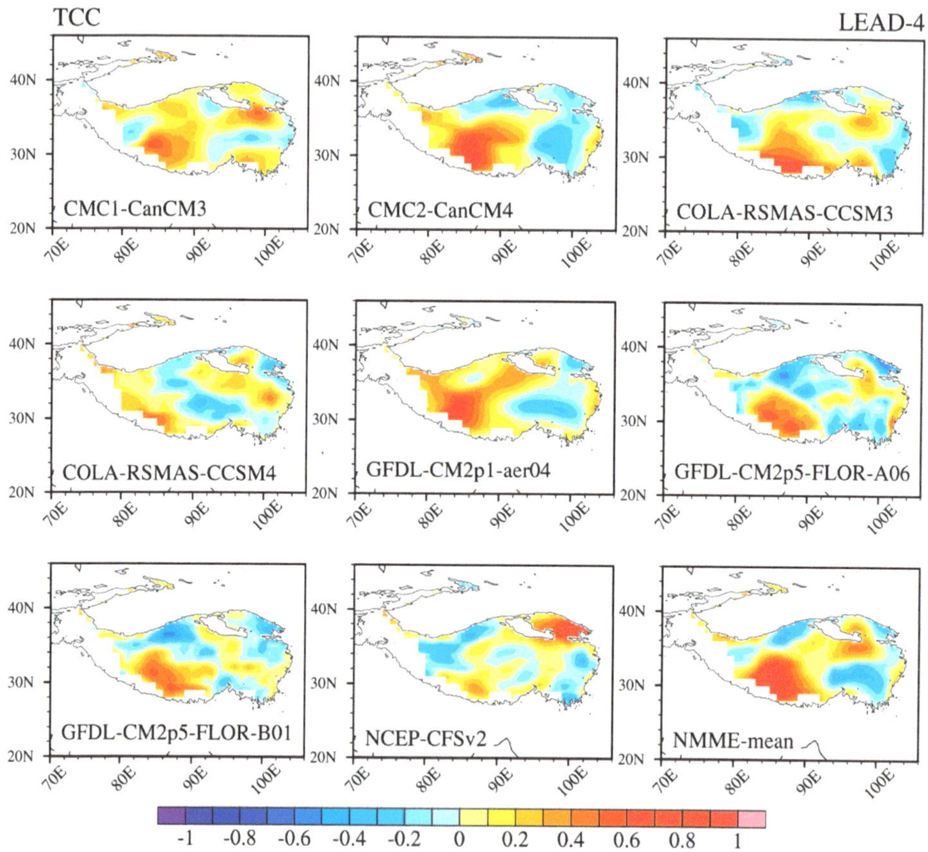

Figure 1. TCC skills between the observed and predicted TPSP, derived from the original NMME models and NMME-mean at four-month lead during 1992–2018.

3.2. The Establishment of the SST Analog-Based Correction Method

Oceanic basin-wide tropical SSTAs have been recognized as critical climate-forcing factors that significantly contribute to climate prediction due to their long-term sustainability with respect to atmospheric processes. Existing studies have extensively examined the oceanic factors that contribute to the inter-annual variability of TPSP, such as ENSO,

NAO/AO, and IOBM (e.g., [15–19,21,23]). Specifically, positive anomalies in SST lead to positive anomalies in the upper tropospheric height field and result in anomalous anticyclonic circulation. The SSTAs are modulated by the influence of winter ENSO and Indian Ocean dipole events. Collectively, these oceanic processes regulate the transport of water vapor, subsequently influencing the TPSP [24].

Figure 2 provides the 5-year-out running correlation between the previous February global SSTAs and the TP area-averaged summer precipitation. The delayed influence of the SST was investigated first, with an analysis of the correlation from the TP-averaged precipitation to global SSTAs in February from 1982 to 2018 (shown in Figure 2a). The previous February SSTAs in the Indian Ocean (IO) and North Atlantic Ocean (ATL) were found to be significantly related to the TPSP. The maximum correlation coefficients with the TP-averaged precipitation index are 0.53 and 0.55, respectively. It is essential to recognize that the relationship between two climate systems can exhibit temporal instability, limiting climate predictability (e.g., [75–79]). Taking these considerations into account, further analysis of running correlation coefficients was conducted to determine the optimal predictor domains. These coefficients were evaluated by consecutively excluding 5 years at a time from the time series and calculating correlations between the predictor and precipitation for each omission; a series of correlation values was obtained. The minimal running correlations refer to the reliability of the predictor–precipitation relationship, as illustrated in Figure 2b. Accordingly, the SSTAs identified by the red boxes in Figure 2b were chosen as predictors and labeled as the ATL-SST and IO-SST.

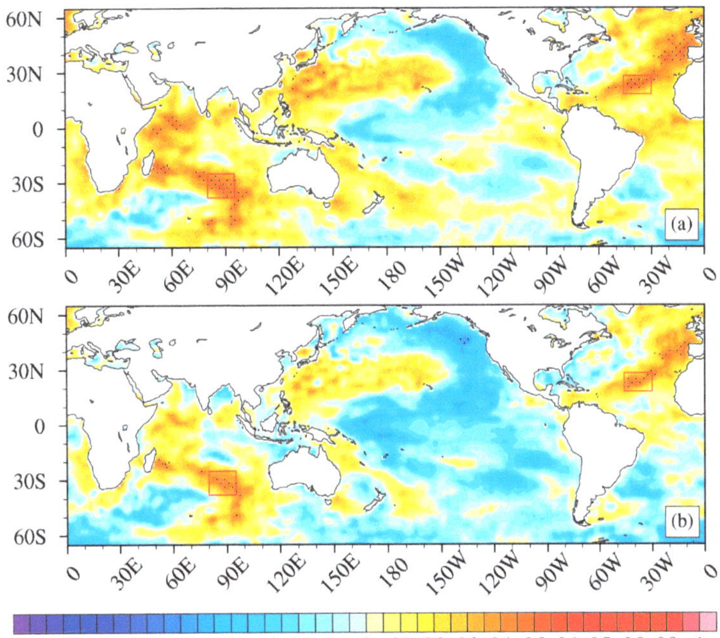

Figure 2. (**a**) The correlation map and (**b**) the 5-year-out running correlation map between the previous February global SSTAs and the TP area-averaged summer precipitation during the whole period of 1982–2018. The dotted areas denote the statistical significance of Student's *t*-test at the 99% confidence level.

Figure 3 further provides the correlation between the observational region-averaged SST indices in the two key regions (hereafter referred to as IO-index and ATL-index) and the precipitation prediction errors in each individual model [55]. Positive correlations can be found throughout the TP, with the ATL-index demonstrating relatively higher correlation coefficients compared to the IO-index. Moreover, the spatial distributions of correlation coefficients generally exhibit a high degree of similarity among these model results, with the majority of PCCs between SST-TP precipitation correlations for the NMME-mean and individual models exceeding 0.8 (Figure 4). These findings imply that SSTAs within the key regions influence the formation and development of prediction errors, providing insights into prediction error spatial distribution and variability.

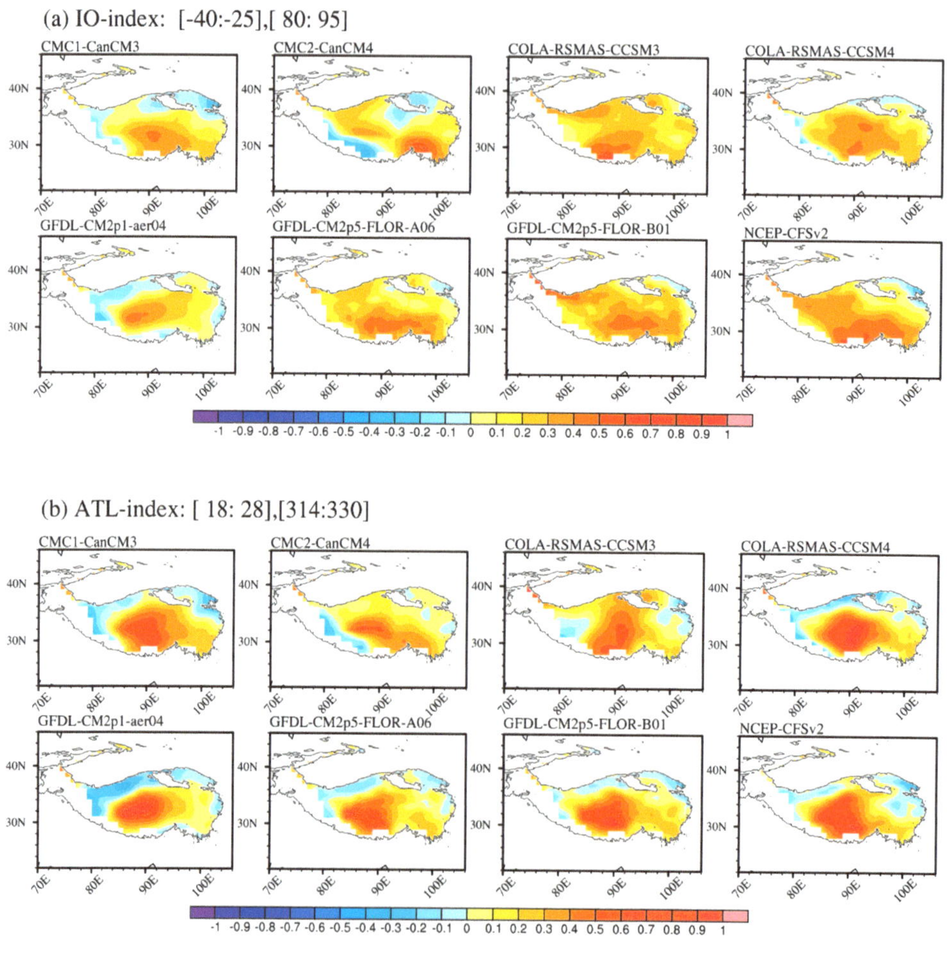

Figure 3. The correlation maps between prediction errors of JJA precipitation at a four-month lead and the observed (**a**) IO-index and (**b**) ATL-index during the whole period of 1982–2018.

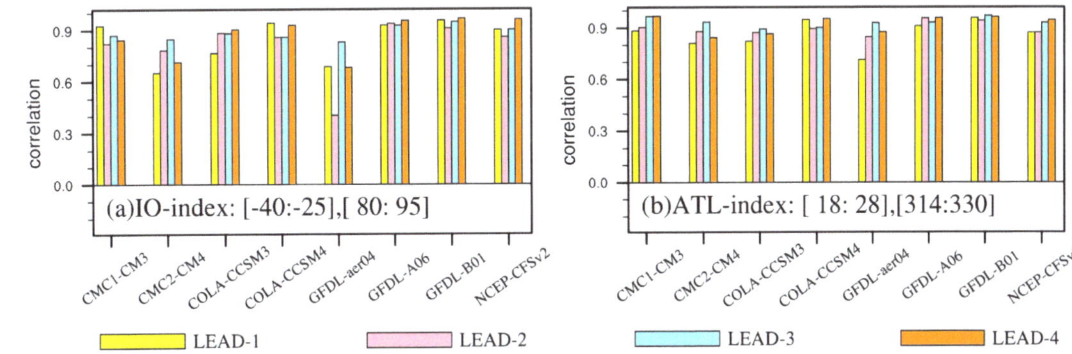

Figure 4. The PCCs between the NMME-mean and individual models for the correlation maps (obtained from Figure 3) of the JJA precipitation prediction errors and the observed (**a**) IO-index and (**b**) ATL-index.

In some cases, there may not be available historical years that are directly comparable or considered the "most analogous" to the current year. To address this issue, a flexible approach has been adopted for the final predictions, which involves selecting analog states by evaluating the PCCs that capture the similarity between present and historical SSTAs within the predictor domain, with a threshold set at 0.7. When the correlation between the SSTAs within the predictor domain exceeds 0.7, the corrected precipitation is utilized as the ultimate prediction result. Conversely, when the correlation is less than 0.7, the original model-predicted precipitation is used as the ultimate prediction for that year. This allows for adaptability in capturing the influence of both high and low correlations between the SSTAs in the predictor domain and prediction errors, resulting in more reliable predictions. Based on the aforementioned analysis, two predictors were selected. Consequently, the final predictions are determined by averaging the results of these two selected predictors, referred to as ensemble–mean (EM) predictions.

3.3. Validation of the New Method

To validate the new method, rolling-independent reforecast experiments were adopted in this study. This approach involves using only historical observational data from previous years to be utilized for training. For example, for the year 1992 (1993), only observational data from 1982 to 1991 (1982–1992) are utilized to train the model, and so on. Subsequently, a combined dataset spanning from 1992 to 2018, incorporating both hindcasts and operational predictions, is employed for evaluating how well the correction method performs.

Figures 5 and 6 depict the TCC skills of the individual models for JJA mean precipitation after applying the new method, using the predictors of the IO-SST and the ATL-SST, during the rolling-independent period from 1992 to 2018. In comparison to the original predictions, which demonstrate low prediction skill over the central–eastern TP (Figure 1), the new method predictions exhibit significantly improved skill in predicting JJA precipitation. Figures 5b and 6b show the TCC differences between the corrected and the original predictions, highlighting the substantial improvements achieved by the new method in predicting the TPSP. For the corrected predictions of the IO-SST as a predictor, positive values are found throughout the central–eastern TP regions in most models with the highest TCC improvement (over 0.5). When the ATL-SST is used as a predictor, the COLA-RSMAS-CCSM4 model demonstrates the most significant improvement in TCC with a maximum value of 0.81, while the majority of other models also exhibit improvements exceeding 0.5. Using ATL-SST as a predictor leads to a skill improvement, notably in the southwestern TP region in Figure 6, even if the original models already show remarkable skill in this area. This finding is further supported by the skills of area-averaged TCC for both the original and corrected NMME-mean, as shown in Figure S1. Additionally, it can be observed from

Figure S1 that the EM predictions combining the two predictors are generally superior to those of each individual predictor alone.

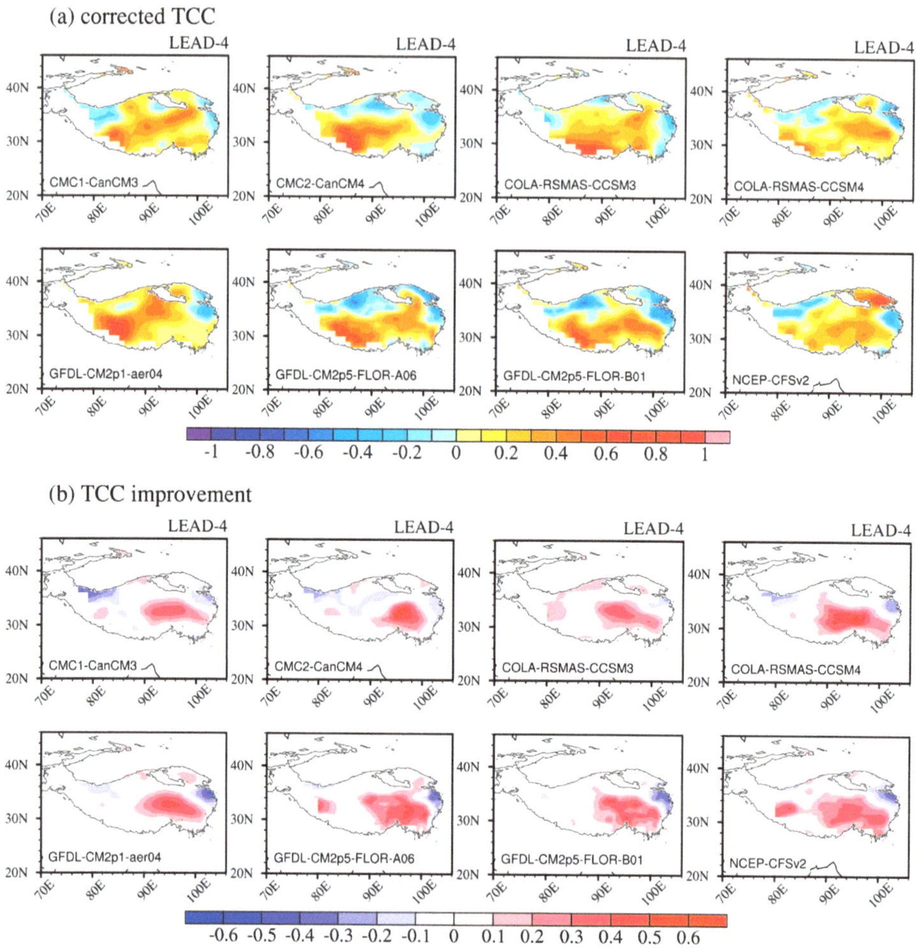

Figure 5. (**a**) Patterns of TCC skills for the new method of the IO-SST as predictor (top 2 rows) and (**b**) the skill differences between the corrected and original models (bottom 2 rows) at four-month lead during the period of 1992–2018.

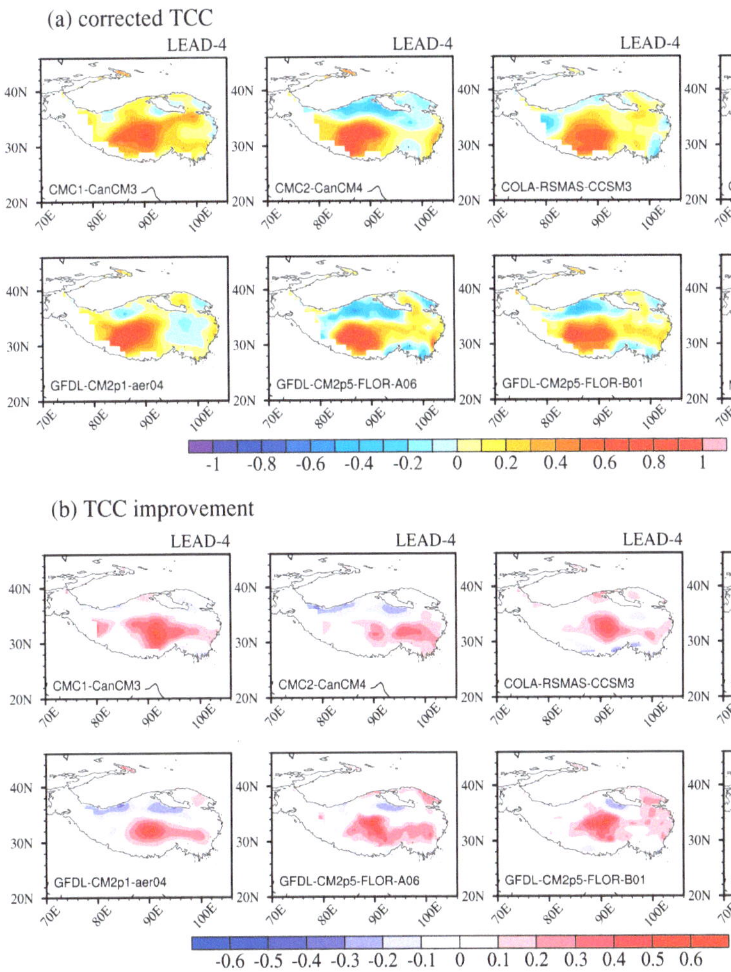

Figure 6. (**a**) Patterns of TCC skills for the new method of the ATL-SST as predictor (top 2 rows) and (**b**) the skill differences between the corrected and original models (bottom 2 rows) at four-month lead, during the period of 1992–2018.

Figure 7 illustrates the spatial patterns of the TCCs of the corrected NMME-mean, as well as the differences between the corrected NMME-mean and the original NMME-mean, for each lead month ranging from 1 to 4. In contrast to the limited improvement observed in the southwestern TP of the original NMME-mean, the corrected NMME-mean exhibits widespread positive TCCs across almost the whole TP region, except for some northern peripheral areas. As seen in Figure 7b, TCCs have significantly improved, with values exceeding 0.6.

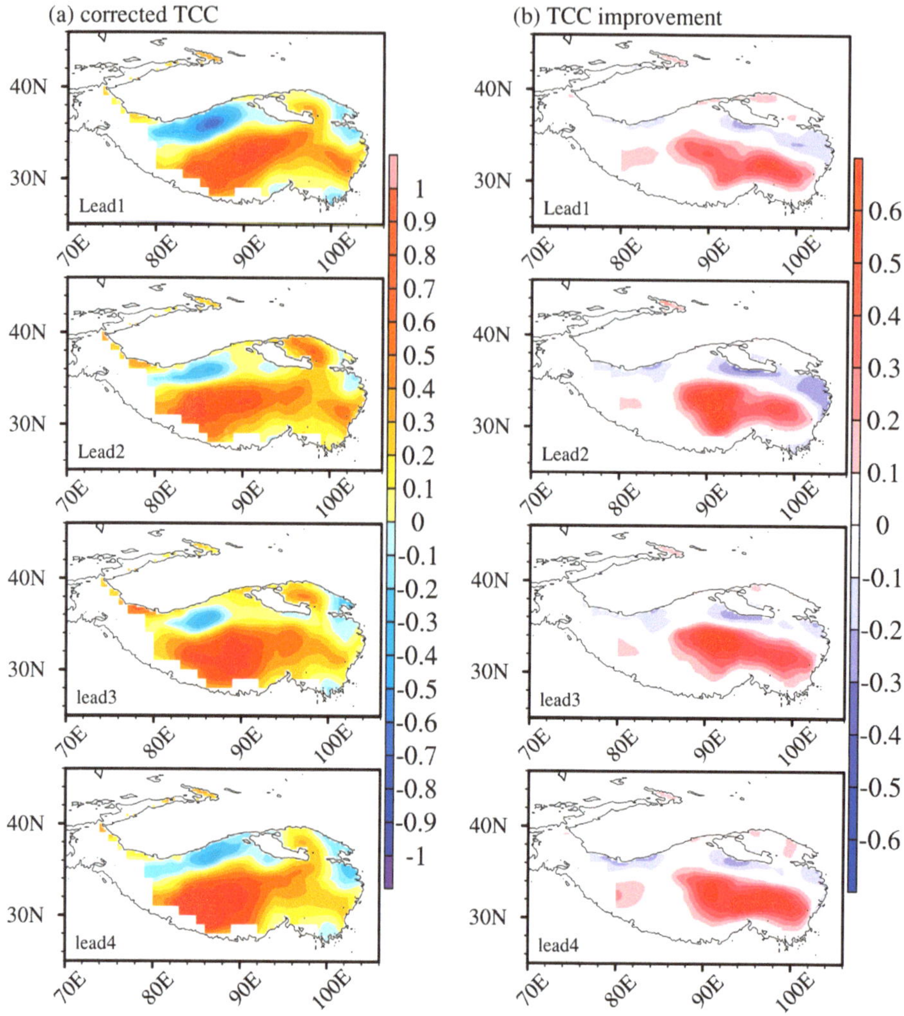

Figure 7. Patterns of TCC skills (**a**) for the corrected NMME-mean from the new method of combining two selected predictors and the skill differences (**b**) between the corrected and original NMME-mean predictions during the period of 1992–2018. The first to fourth rows reflect one- to four-month leads, respectively.

Figure 8 presents the PCCs from the observed anomalous precipitation with respect to the predictions generated by both the original NMME-mean and the new method. In most years, the NMME-mean based on the new method exhibits larger or very similar PCCs compared to the original NMME-mean. The long-term averaged PCCs over 1992–2018 increased from approximately 0.16 (0.17, 0.20, and 0.16) of the uncorrected NMME-mean to approximately 0.25 (0.23, 0.27, and 0.28) of the original NMME-mean at lead time of one month (two months, three months, and four months). Figure S2 shows substantial improvements in the long-term averaged PCCs for the individual models derived from the new method. The results obtained from this study indicate that the new method has the ability to significantly improve the prediction of the TPSP.

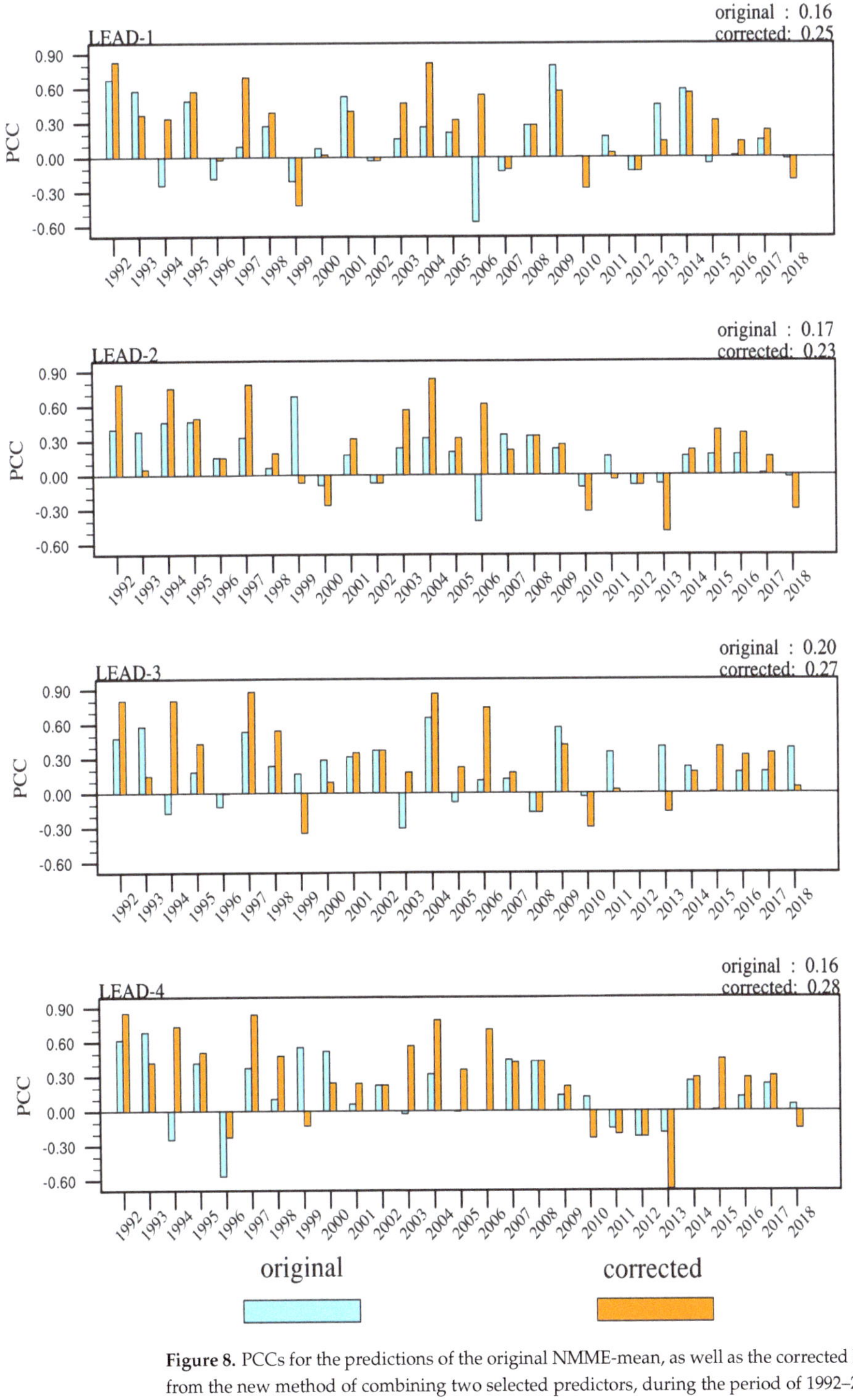

Figure 8. PCCs for the predictions of the original NMME-mean, as well as the corrected NMME-mean from the new method of combining two selected predictors, during the period of 1992–2018. The first to fourth rows reflect one- to four-month leads, respectively. The long-term averaged PCC skills are given in the top-right for each panel.

4. Discussion

The TP poses considerable challenges for accurately predicting summer precipitation due to its unique geographical features, complex topography, and atmospheric circulation patterns. Current dynamical models have limitations in accurately capturing TPSP. This study introduces a new analog-based correction method that has been developed to address this challenge and improve the prediction of the TPSP. The method is motivated by the hypothesis that the flow-dependent model prediction errors exhibit some degree of similarity when the prior climate states of the models are analogous to one another. When the new method is applied to the NMME model hindcasts dataset, the verification against independent observations shows that the new SST analog-based correction method of dynamical models can evidently improve prediction skills of summer precipitation in most TP regions.

Over the past few years, dynamic models have become the most powerful tool in climate predictions. However, it is inevitable that these models will have inherent errors in their predictions. Specifically, prediction errors refer to the disparities between the predicted values and the corresponding observed values. The errors are known to vary with the climate state, which can be estimated statistically using related physical variables as predictors. In other words, the variability of the climate system directly or indirectly influences the configuration and evolution of prediction errors, and such influences will further be reflected in the distribution and change in prediction errors. This serves as the fundamental premise of the present study. Our study examined the correlation between preceding February SSTAs and prediction errors of TPSP. The analysis reveals a significant relationship between prediction errors and preceding February SSTAs in key regions of the Indian Ocean and North Atlantic Ocean. Additionally, a high degree of similarity in the prediction errors is observed among different dynamical models utilized in this study. In simpler terms, when the SSTAs in February exhibit similarities, it is possible for the model prediction errors to exhibit similar patterns as well. This formed the basis of our method, and we utilized this relationship to estimate error weights by combining them. Building on these findings, the analog-based correction method has been applied to predict summer precipitation using the identified SST predictors. The rolling-independent validation technique has been used from 1992 to 2018 to verify the new method. The findings of this study demonstrate promising results for the proposed correction method in predicting summer precipitation over the TP. Both the IO-SST and ATL-SST exhibit superior performance with positive TCCs over most areas of the TP, albeit with regional differences. The EM predictions using both the IO-SST and ATL-SST predictors demonstrate superior performance compared to each predictor individually. This approach provided a novel perspective and practical application for error correction in climate prediction, ultimately contributing to the advancement of the field.

Our findings provided insights that, despite the overall positive performance exhibited by the IO-SST and ATL-SST, there are regional differences indicating that individual predictors may exert specific influences on particular areas of the TP. The regional variations could be attributed to the complex interactions between atmospheric and oceanic processes, as well as the specific geographical features and atmospheric circulation patterns in different sub-regions of the TP. Therefore, future prediction research will consider dividing the entire TP into several sub-regions. In addition to our focus on SST predictors, previous studies, as mentioned in our introduction, have extensively investigated the influence of ENSO on TP precipitation. We also explored the application of an ENSO-based ABC method to predict TP summer precipitation. However, despite our efforts, this approach has not yielded a significant improvement in prediction skills. Notably, Liu and Ren (2023) found that EP El Niño and CP La Niña have contrasting effects on summer precipitation in the southwestern TP, resulting in a trans-type inversion characterized by significant decreases and increases, respectively [19]. These findings highlight the complexity of the relationship between ENSO and TP precipitation, which is may also reflected in model errors. Considering the intricate nature of this relationship, our future investigations aim to delve deeper into the characteristics of TP precipitation under different ENSO event types and apply the findings to improve the TPSP prediction in dynamical models. Beyond that, it is possible to consider that nonlinear prediction errors can be objectively isolated and addressed via alternative nonlinear methodologies, including deep learning techniques [80].

Although the new method for TP precipitation shows significant improvement, it is important to acknowledge that its performance may vary across different time periods. The variation can arise from uncertainties between the predictor and predictand at interdecadal or decadal scales [81]. Therefore, achieving stable and highly skilled TP precipitation prediction remains a major challenge. In this study, the analog-based correction mechanism is exclusively established using the preceding SST signals. Nevertheless, it is imperative to acknowledge the intricate and extensive drivers of TPSP variability. Other boundary signals, including snow cover and soil moisture [82–85], may also influence TPSP. In addition, we used bilinear interpolation to estimate the values at the grid points in the interpolated dataset. It is also important to acknowledge the limitations and potential biases associated with interpolation techniques. Bilinear interpolation assumes a uniform distribution of data in space and equal distances between grid points. In reality, there may be discrepancies in spatial resolution or grid layout between the observed datasets and the model data, which could introduce interpolation bias. Although it is not possible to eliminate this type of bias entirely, in order to improve the reliability of the observational data, we employed multiple datasets. This ensemble averaging approach involved calculating the mean value across the three datasets at each grid point, resulting in a composite precipitation dataset. This helped to reduce the impact of individual dataset biases and improve the representativeness of the interpolated dataset. In addition, in our study, we also compared the skills verified by directly using each set of observational data, respectively. It is evident that the overall distribution of TCC skills remains relatively consistent. These findings align with the results obtained from multiple observational ensembles. Despite the inherent variability in the observation data, we made diligent efforts to account for its influence, and via comprehensive analysis, we demonstrated the robustness of our error correction method from multiple perspectives.

5. Conclusions

This study introduced an analog-based correction method aimed at improving summer precipitation prediction over the TP. The new method has been successfully applied to predict precipitation for the models included in the NMME dataset. The performance of the new method is validated by utilizing remote-sensing-based and observational datasets as benchmarks. Three main conclusions were obtained as follows:

(1) The prediction skill for summer precipitation over the TP in current climate models is constrained, with notable limitations observed in the central–eastern TP region. Additionally, the prediction errors demonstrate a pronounced level of consistency across various dynamical models.
(2) The prediction errors of TPSP exhibit a significant correlation with the previous February anomalies of SST in the key regions of the tropical Ocean. Both the ATL-SST and IO-SST can be considered effective predictors for correcting the prediction errors while noting the presence of regional differences.
(3) The prediction skill for summer precipitation over the TP exhibits notable improvements via the application of the new SST analog-based correction method, as demonstrated by higher temporal and spatial skill scores obtained from the rolling-independent verification. This study provides a valuable tool for enhancing the prediction of TPSP within dynamical models.

Supplementary Materials: The following supporting information can be downloaded at https://www.mdpi.com/article/10.3390/rs15245669/s1. Figure S1: The area-averaged TCCs for the original and corrected NMME-mean during the period of 1992–2018; Figure S2: Long-term averaged

PCCs between the observation and the original predictions, as well as the corrected predictions from the new method of two selected predictors during the period of 1992–2018.

Author Contributions: Conceptualization, H.-L.R.; methodology, H.-L.R. and L.W.; software, L.W.; validation, B.C., L.G., J.L., H.C., Y.W. and X.Z.; formal analysis, H.-L.R. and L.W.; investigation, L.W. and H.-L.R.; writing—original draft preparation, L.W. and. H.-L.R.; writing—review and editing, L.W., H.-L.R., X.X., B.C., L.G., J.L., H.C., Y.W. and X.Z.; supervision, H.-L.R.; funding acquisition, H.-L.R. All authors have read and agreed to the published version of the manuscript.

Funding: This research was jointly funded by the China National Natural Science Foundation (grant numbers U2242206 and 42305045), the Special Project for Innovation and Development of CMA (Grant number CXFZ2023J050) and the Basic Research and Operational Special Project of CAMS (grant numbers 2023Z003, 2023Z024, and 2021Z007).

Data Availability Statement: The data used in this study can be downloaded from https://psl.noaa.gov/data/gridded/data.noaa.oisst.v2.highres.html (accessed on 13 September 2023) for monthly SST dataset; https://psl.noaa.gov/data/gridded/data.gpcp.html (accessed on 13 September 2023) for GPCP monthly precipitation dataset; and https://www.uea.ac.uk/web/groups-and-centres/climatic-research-unit/data/ (accessed on 13 September 2023) for CRU monthly precipitation dataset. All NMME data for this study can be freely obtained from the database hosted at IRI, http://iridl.ldeo.columbia.edu/SOURCES/.Models/.NMME/#expert (accessed on 13 September 2023). The CN05.1 precipitation dataset was obtained by the authors via an agreement with the China Meteorological Administration for the current study and so is not publicly available.

Acknowledgments: We thank the National Meteorological Information Center of China Meteorological Administration for providing precipitation data.

Conflicts of Interest: The authors declare no conflict of interest.

References

1. Qiu, J. China: The third pole. *Nature* **2008**, *454*, 393–396. [CrossRef]
2. Yao, T.; Xue, Y.; Chen, D.; Chen, F.; Thompson, L.; Cui, P.; Koike, T.; Lau, W.K.-M.; Lettenmaier, D.; Mosbrugger, V.; et al. Recent Third Pole's Rapid Warming Accompanies Cryospheric Melt and Water Cycle Intensification and Interactions between Monsoon and Environment: Multidisciplinary Approach with Observations, Modeling, and Analysis. *Bull. Am. Meteorol. Soc.* **2019**, *100*, 423–444. [CrossRef]
3. Nowak, A.S.; Nobis, M. Distribution patterns, floristic structure and habitat requirements of the alpine river plant community Stuckenietum amblyphyllae ass. nova (Potametea) in the Pamir Alai Mountains (Tajikistan). *Acta Soc. Bot. Pol.* **2012**, *81*, 101–108. [CrossRef]
4. Sun, J.; Cheng, G.; Li, W.; Sha, Y.; Yang, Y. On the Variation of NDVI with the Principal Climatic Elements in the Tibetan Plateau. *Remote Sens.* **2013**, *5*, 1894–1911. [CrossRef]
5. Immerzeel, W.W.; van Beek, L.P.; Bierkens, M.F. Climate change will affect the Asian water towers. *Science* **2010**, *328*, 382–1385. [CrossRef]
6. Flohn, H. Large-scale aspects of the 'summer monsoon' in South and East Asia. *J. Meteorol. Soc. Jpn.* **1957**, *75*, 180–186. [CrossRef]
7. Yeh, T.C.; Lo, S.W.; Chu, P.C. On the heat balance and circulation structure in troposphere over Tibetan Plateau. *Acta Meteorol. Sin.* **1957**, *28*, 108–121. (In Chinese)
8. Ueda, H.; Kamahori, H.; Yamazaki, N. Seasonal contrasting features of heat and moisture budgets between the eastern and western Tibetan plateau during the GAME IOP. *J. Clim.* **2003**, *16*, 2309–2324. [CrossRef]
9. Bibi, S.; Wang, L.; Li, X.; Zhou, J.; Chen, D.; Yao, T. Climatic and associated cryospheric, biospheric, and hydrological changes on the Tibetan Plateau: A review. *Int. J. Climatol.* **2018**, *38*, e1–e17. [CrossRef]
10. Dong, W.; Lin, Y.; Wright, J.S.; Ming, Y.; Xie, Y.; Wang, B.; Luo, Y.; Huang, W.; Huang, J.; Wang, L.; et al. Summer rainfall over the southwestern Tibetan Plateau controlled by deep convection over the Indian subcontinent. *Nat. Commun.* **2016**, *7*, 10925. [CrossRef] [PubMed]
11. Yao, T.; Thompson, L.; Yang, W.; Yu, W.-S.; Gao, Y.; Guo, X.-J.; Yang, X.; Duan, K.; Zhao, H.; Xu, B.; et al. Different glacier status with atmospheric circulations in Tibetan Plateau and surroundings. *Nat. Clim. Chang.* **2012**, *2*, 663–667. [CrossRef]
12. Duan, A.; Wu, G. Weakening trend in the atmospheric heat source over the Tibetan Plateau during recent decades. Part I: Observations. *J. Clim.* **2008**, *21*, 3149–3164. [CrossRef]
13. Wu, Z.; Li, J.; Jiang, Z.; Ma, T. Modulation of the Tibetan Plateau Snow Cover on the ENSO Teleconnections: From the East Asian Summer Monsoon Perspective. *J. Clim.* **2012**, *25*, 2481–2489. [CrossRef]
14. Jiang, X.; Li, Y.; Yang, S.; Yang, K.; Chen, J. Interannual Variation of Summer Atmospheric Heat Source over the Tibetan Plateau and the Role of Convection around the Western Maritime Continent. *J. Clim.* **2016**, *29*, 121–138. [CrossRef]

15. Gao, X.; Wang, M.; Filippo, G. Climate Change over China in the 21st Century as Simulated by BCC_CSM1.1-RegCM4.0. *Atmos. Ocean. Sci. Lett.* **2013**, *6*, 381–386.
16. Wang, C.; Ma, Z. Quasi-3-yr cycle of rainy season precipitation in Tibet related to different types of ENSO during 1981–2015. *J. Meteorol. Res.* **2018**, *32*, 181–190. [CrossRef]
17. Hu, S.; Zhou, T.; Wu, B. Impact of Developing ENSO on Tibetan Plateau Summer Rainfall. *J. Clim.* **2021**, *34*, 3385–3400. [CrossRef]
18. Hu, S.; Wu, B.; Zhou, T.; Yu, Y. Dominant Anomalous Circulation Patterns of Tibetan Plateau Summer Climate Generated by ENSO-Forced and ENSO-Independent Teleconnections. *J. Clim.* **2022**, *35*, 1679–1694. [CrossRef]
19. Liu, M.; Ren, H.-L.; Wang, R.; Ma, J.; Mao, X. Distinct Impacts of Two Types of Developing El Niño–Southern Oscillations on Tibetan Plateau Summer Precipitation. *Remote Sens.* **2023**, *15*, 4030. [CrossRef]
20. Yang, X.; Yao, T.; Deji; Zhao, H.; Xu, B. Possible ENSO Influences on the Northwestern Tibetan Plateau Revealed by Annually Resolved Ice Core Records. *J. Geophys. Res. Atmos.* **2018**, *123*, 3857–3870. [CrossRef]
21. Chen, X.; You, Q. Effect of Indian Ocean SST on Tibetan Plateau Precipitation in the Early Rainy Season. *J. Clim.* **2017**, *30*, 8973–8985. [CrossRef]
22. Zhao, Y.; Duan, A.; Wu, G. Interannual variability of late-spring circulation and diabatic heating over the Tibetan Plateau associated with Indian Ocean forcing. *Adv. Atmos. Sci.* **2018**, *35*, 927–941. [CrossRef]
23. Liu, X.; Yin, Z. Spatial and Temporal Variation of Summer Precipitation over the Eastern Tibetan Plateau and the North Atlantic Oscillation. *J. Clim.* **2001**, *14*, 2896–2909. [CrossRef]
24. He, K.; Liu, G.; Wu, R.; Nan, S.; Wang, S.; Zhou, C.; Qi, D.; Mao, X.; Wang, H.; Wei, X. Oceanic and land relay effects in the link between spring sea surface temperatures in the Indian Ocean and summer precipitation over the Tibetan Plateau. *Atmos. Res.* **2022**, *266*, 105953. [CrossRef]
25. Wang, Z.; Duan, A.; Yang, S.; Ullah, K. Atmospheric moisture budget and its regulation on the variability of summer precipitation over the Tibetan Plateau. *J. Geophys. Res. Atmos.* **2017**, *122*, 614–630. [CrossRef]
26. Wang, B.; Lee, J.-Y.; Kang, I.-S.; Shukla, J.; Park, C.K.; Kumar, A.; Schemm, J.; Cocke, S.; Kug, J.-S.; Luo, J.-J.; et al. Advance and prospectus of seasonal prediction: Assessment of the APCC/ CliPAS 14-model ensemble retrospective seasonal prediction (1980–2004). *Clim. Dyn.* **2009**, *33*, 93–117. [CrossRef]
27. Wang, B.; Lee, J.-Y.; Xiang, B. Asian summer monsoon rainfall predictability: A predictable mode analysis. *Clim. Dyn.* **2015**, *44*, 61–74. [CrossRef]
28. Yu, R.; Li, J.; Zhang, Y.; Chen, H. Improvement of rainfall simulation on the steep edge of the Tibetan Plateau by using a finite-difference transport scheme in CAM5. *Clim. Dyn.* **2015**, *45*, 2937–2948. [CrossRef]
29. Bao, Q.; Wu, X.; Li, J.; Wang, L.; He, B.; Wang, X.; Liu, Y.; Wu, G. Outlook for El Niño and the Indian Ocean Dipole in autumn-winter 2018–2019. *Chin. Sci. Bull.* **2019**, *64*, 73–78. [CrossRef]
30. Bao, Q.; Li, J. Progress in climate modeling of precipitation over the Tibetan Plateau. *Natl. Sci. Rev.* **2020**, *7*, 486–487. [CrossRef]
31. He, B.; Bao, Q.; Wang, X.; Zhou, L.; Wu, X.; Liu, Y.; Wu, G.; Chen, K.; He, S.; Hu, W.; et al. CAS FGOALS-f3-L Model Datasets for CMIP6 Historical Atmospheric Model Intercomparison Project Simulation. *Adv. Atmos. Sci.* **2019**, *36*, 771–778. [CrossRef]
32. Lee, W.L.; Liou, K.N.; Wang, C. Impact of 3-D topography on surface radiation budget over the Tibetan Plateau. *Theor. Appl. Climatol.* **2013**, *113*, 95–103. [CrossRef]
33. Wang, L.; Ren, H.-L.; Xu, X.; Huang, B.; Wu, J.; Liu, J. Seasonal-interannual predictions of summer precipitation over the Tibetan Plateau in North American Multimodel Ensemble. *Geophys. Res. Lett.* **2022**, *49*, e2022GL100294. [CrossRef]
34. Palmer, T.; Alessandri, A.; Andersen, U.; Cantelaube, P.; Davey, M.; Délécluse, P.; Déqué, M.; Díez, E.; Doblas-Reyes, F.J.; Feddersen, H.; et al. Development of a European multimodel ensemble system for seasonal to interannual prediction (DEMETER). *Bull. Am. Meteorol. Soc.* **2004**, *85*, 853–872. [CrossRef]
35. Hewitt, C.D.; Griggs, D.J. Ensembles-based predictions of climate change and their impacts (ENSEMBLES). *EOS Trans. AGU* **2004**, *85*, 566.
36. Kirtman, B.P.; Min, D.; Infanti, J.M.; Kinter, J.L.; Paolino, D.A.; Zhang, Q.; van den Dool, H.; Saha, S.; Mendez, M.; Becker, E.; et al. Phase-1 seasonal to interannual prediction, phase-2 toward developing intra-seasonal prediction. *Bull. Am. Meteorol. Soc.* **2014**, *95*, 585–601. [CrossRef]
37. Shukla, S.; Jason, R.; Hoell, A.; Funk, C.; Franklin, R.C.; Kirtman, B. Assessing North American multimodel ensemble (NMME) seasonal forecast skill to assist in the early warning of anomalous hydrometeorological events over East Africa. *Clim. Dyn.* **2019**, *53*, 7411–7427. [CrossRef]
38. Slater, L.J.; Villarini, G.; Bradley, A.A. Weighting of NMME temperature and precipitation forecasts across Europe. *J. Hydrol.* **2017**, *552*, 646–659. [CrossRef]
39. Giannini, A.; Ali, A.; Kelley, C.P.; Lamptey, B.L.; Minoungou, B.; Ndiaye, O. Advances in the lead time of Sahel rainfall prediction with the North American Multimodel Ensemble. *Geophys. Res. Lett.* **2020**, *47*, e2020GL087341. [CrossRef]
40. Ren, H.L.; Wu, Y.J.; Bao, Q.; Ma, J.H.; Li, Q.P.; Wan, J.H.; Li, Q.; Wu, X.; Liu, Y.; Tian, B.; et al. China multi-model ensemble prediction system version 1.0 (CMMEv1.0) and its application to flood-season prediction in 2018. *J. Meteorol. Res.* **2019**, *31*, 542–554.
41. Wigley, T.M.L.; Jones, P.D.; Briffa, K.R.; Smith, G. Obtaining sub-grid-scale information from coarse-resolution general circulation model output. *J. Geophys. Res. Atmos.* **1990**, *95*, 1943–1953. [CrossRef]

42. Zeng, Q.C.; Zhang, B.L.; Yuan, C.G.; Lu, P.S.; Yang, F.L.; Li, X.; Wang, H.J. A note on some methods suitable for verifying and correcting the prediction of climate anomaly. *Adv. Atmos. Sci.* **1994**, *11*, 121–127.
43. Zeng, Q.C.; Yuan, C.G.; Xu, L.; Zhang, R.H.; Yang, F.L.; Zhang, B.L.; Lu, P.S.; Bi, X.Q.; Wang, H.J. Seasonal and Extraseasonal predictions of summer monsoon precipitation by GCMs. *Adv. Atmos. Sci.* **1997**, *14*, 163–176.
44. Busuioc, A.; Von Storch, H.; Schnur, R. Verification of GCM-generated regional seasonal precipitation for current climate and of statistical downscaling estimates under changing climate conditions. *J. Clim.* **1999**, *12*, 258–272. [CrossRef]
45. Wang, H.J.; Fan, K. A New Scheme for Improving the Seasonal Prediction of Summer Precipitation Anomalies. *Wea. Forecast.* **2009**, *24*, 548–554. [CrossRef]
46. Liu, Y.; Fan, K. An application of hybrid downscaling model to forecast summer precipitation at stations in China. *Atmos. Res.* **2014**, *143*, 17–30. [CrossRef]
47. Yu, H.P.; Zhang, Q.; Wei, Y.; Liu, C.X.; Ren, Y.; Yue, P.; Zhou, J. Bias-corrections on aridity index simulations of climate models by observational constraints. *Int. J. Climatol.* **2022**, *42*, 889–907. [CrossRef]
48. Wu, J.; Ren, H.L.; Zhang, P.; Wang, Y.; Liu, Y.; Zhao, C.; Li, Q. The dynamical-statistical subseasonal prediction of precipitation over China based on the BCC new-generation coupled model. *Clim. Dyn.* **2022**, *59*, 1213–1232. [CrossRef]
49. Ma, J.H.; Sun, J.Q. New statistical prediction scheme for monthly precipitation variability in the rainy season over northeastern China. *Int. J. Climatol.* **2021**, *41*, 5805–5819. [CrossRef]
50. Ma, J.H.; Sun, J.Q.; Liu, C.Z. A hybrid statistical-dynamical prediction scheme for summer monthly precipitation over Northeast China. *Meteorol. Appl.* **2022**, *29*, e2057. [CrossRef]
51. Zhu, Y.; Sun, J.; Ma, J. A hybrid statistical-dynamical prediction model for summer precipitation in northwestern China based on NCEP CFSv2. *Atmos. Res.* **2023**, *283*, 106567. [CrossRef]
52. Ren, H.-L. Predictor-based error correction method in short-term climate prediction. *Prog. Nat. Sci.* **2008**, *18*, 129–135. [CrossRef]
53. Yang, X.Q.; Anderson, J.L. Correction of systematic errors in coupled GCM forecasts. *J. Clim.* **2000**, *13*, 2072–2085. [CrossRef]
54. Gu, W.; Chen, L.; Li, W.; Chen, D. Development of a downscaling method in China regional summer precipitation prediction. *Acta Meteorol. Sin.* **2011**, *25*, 303–315. [CrossRef]
55. Samadi, S.; Wilson, C.A.; Moradkhani, H. Uncertainty analysis of statistical downscaling models using Hadley centre coupled model. *Theor. Appl. Climatol.* **2013**, *4*, 673–690. [CrossRef]
56. Paul, S.; Liu, C.M.; Chen, J.M.; Lin, S.H. Development of a statistical downscaling model for projecting monthly rainfall over East Asia from a general circulation model output. *J. Geophys. Res. Atmos.* **2008**, *113*, D15117. [CrossRef]
57. Fan, K.; Wang, H.J.; Choi, Y.-J. A physically-based statistical forecast model for the middle-lower reaches of the Yangtze River Valley summer rainfall. *Chin. Sci. Bull.* **2008**, *53*, 602–609. [CrossRef]
58. Fan, K.; Liu, Y.; Chen, H.P. Improving the prediction of the East Asian summer monsoon: New approaches. *Wea. Forecast.* **2012**, *27*, 1017–1030. (In Chinese) [CrossRef]
59. Ren, H.-L.; Chou, J.F. Analog correction method of errors by combining statistical and dynamical methods. *Acta Meteorol. Sin.* **2006**, *20*, 367–373.
60. Ren, H.-L.; Chou, J.F. Strategy and methodology of dynamical analogue prediction. *Sci. China Ser. D Earth Sci.* **2007**, *50*, 1589–1599. [CrossRef]
61. Ren, H.-L.; Chou, J.F.; Huang, J.P.; Zhang, P. Theoretical basis and application of analogue-dynamical model in the Lorenz system. *Adv. Atmos. Sci.* **2009**, *26*, 67–77. [CrossRef]
62. Yu, H.P.; Huang, J.P.; Chou, J.F. Improvement of Medium-Range Forecasts Using the Analog-Dynamical Method. *Mon. Wea. Rev.* **2014**, *142*, 1570–1587. [CrossRef]
63. Ren, H.-L.; Liu, Y.; Jin, F.-F.; Yan, Y.P.; Liu, X.W. Application of the analogue-based correction of errors method in ENSO prediction. *Atmos. Ocean. Sci. Lett.* **2014**, *7*, 157–161.
64. Liu, Y.; Ren, H.-L. Improving ENSO prediction in CFSv2 with an analogue-based correction method. *Int. J. Climatol.* **2017**, *37*, 5035–5046. [CrossRef]
65. Becker, E.; den Dool, H.V.; Zhang, Q. Predictability and Forecast Skill in NMME. *J. Clim.* **2014**, *27*, 5891–5906. [CrossRef]
66. Becker, E.; Kirtman, B.P.; Pegion, K. Evolution of the North American multi-model ensemble. *Geophys. Res. Lett.* **2020**, *47*, e2020GL087408. [CrossRef]
67. Merryfield, W.J.; Lee, W.S.; Boer, G.J.; Kharin, V.V.; Scinocca, J.F.; Flato, G.M.; Ajayamohan, R.S.; Fyfe, J.C.; Tang, Y.; Polavarapu, S. The Canadian Seasonal to Interannual Prediction System. Part I: Models and initialization. *Mon. Weather Rev.* **2013**, *141*, 2910–2945. [CrossRef]
68. Kirtman, B.P.; Min, D. Multimodel Ensemble ENSO Prediction with CCSM and CFS. *Mon. Weather Rev.* **2009**, *137*, 2908–2930. [CrossRef]
69. Infanti, J.; Kirtman, B. Southeastern US Rainfall Prediction in the North American Multi-Model Ensemble. *J. Hydrol.* **2014**, *15*, 529–550.
70. Vecchi, G.A.; Delworth, T.; Gudgel, R.; Kapnick, S.; Rosati, A.; Wittenberg, A.T.; Zeng, F.; Anderson, W.; Balaji, V.; Dixon, K.; et al. On the Seasonal Forecasting of Regional Tropical Cyclone Activity. *J. Clim.* **2014**, *27*, 7994–8016. [CrossRef]
71. Saha, S.; Moorthi, S.; Wu, X.R.; Wang, J.D.; Nadiga, S.; Tripp, P.; Behringer, D.; Hou, Y.-T.; Chuang, H.-Y.; Iredell, M.; et al. The NCEP Climate Forecast System Version 2. *J. Clim.* **2014**, *27*, 2185–2208. [CrossRef]

72. Wu, J.; Gao, X. A gridded daily observation dataset over China region and comparison with the other datasets. *Chin. J. Geophys.* **2013**, *56*, 1102–1111. (In Chinese)
73. Harris, I.C.; Jones, P.D.; Osborn, T. *CRU TS4.05: Climatic Research Unit (CRU) Time-Series (TS) Version 4.05 of High-Resolution Gridded Data of Month-by-Month Variation in Climate (Jan. 1901–Dec. 2020)*; NERC EDS Centre for Environmental Data Analysis: Chicago, IL, USA, 2021.
74. Adler, R.F.; Sapiano, M.R.P.; Huffman, G.J.; Wang, J.-J.; Gu, G.; Bolvin, D.; Chiu, L.; Schneider, U.; Becker, A.; Nelkin, E.; et al. The Global Precipitation Climatology Project (GPCP) Monthly Analysis (New Version 2.3) and a Review of 2017 Global Precipitation. *Atmosphere* **2018**, *9*, 138. [CrossRef]
75. Kumar, K.K.; Rajagopalan, B.; Cane, M.A. On the weakening relationship between the Indian monsoon and ENSO. *Science* **1999**, *284*, 2156–2159. [CrossRef]
76. Wang, H.J. The instability of the East Asian summer monsoon ENSO relations. *Adv. Atmos. Sci.* **2002**, *19*, 1–11.
77. Sun, J.Q.; Wang, H.J.; Yuan, W. Decadal variations of the relationship between the summer North Atlantic Oscillation and middle East Asian air temperature. *J. Geophys. Res.* **2008**, *113*, D15107. [CrossRef]
78. Sun, J.Q.; Wang, H.J.; Yuan, W. Linkage of the boreal spring Antarctic Oscillation to the West African Summer Monsoon. *J. Meteorol. Soc. Jpn.* **2010**, *88*, 15–28. [CrossRef]
79. Sun, J.Q.; Ahn, B.J. A GCM-based forecasting model for the landfall of tropical cyclones in China. *Adv. Atmos. Sci.* **2011**, *28*, 1049–1055. [CrossRef]
80. Ham, Y.G.; Kim, J.H.; Luo, J.J. Deep learning for multi-year ENSO forecasts. *Nature* **2019**, *573*, 568–572. [CrossRef]
81. Wang, B.; Xiang, B.; Li, J.; Webster, P.J.; Rajeevan, M.N.; Liu, J.; Ha, K.-J. Rethinking Indian monsoon rainfall prediction in the context of recent global warming. *Nat. Commun.* **2015**, *6*, 7154. [CrossRef]
82. Xu, J.J.; Lu, J. Precipitation over the Qinghai-Xizang Plateau in summer and its association with the Eurasian snow cover. *J. Nanjing Inst. Meteorol.* **1992**, *15*, 517–524. (In Chinese)
83. Wu, B.Y.; Zhang, R.H.; Wang, B.; Rosanne, D.A. On the association between spring Arctic sea ice concentration and Chinese summer rainfall. *Geophys. Res. Lett.* **2009**, *36*, L09501. [CrossRef]
84. Wu, R.G.; Kirtman, B.P. Observed relationship of spring and summer east Asian rainfall with winter and spring Eurasian snow. *J. Clim.* **2007**, *20*, 1285–1304. [CrossRef]
85. Yang, S.; Xu, L.Z. Linkage between Eurasian winter snow cover and regional Chinese summer rainfall. *Int. J. Climatol.* **1994**, *14*, 739–750. [CrossRef]

remote sensing

MDPI

Article

Long-Term Characteristics of Surface Soil Moisture over the Tibetan Plateau and Its Response to Climate Change

Chenxia Zhu [1], Shijie Li [1,*], Daniel Fiifi Tawia Hagan [1,2], Xikun Wei [1], Donghan Feng [3], Jiao Lu [4], Waheed Ullah [5] and Guojie Wang [1]

[1] Collaborative Innovation Center on Forecast and Evaluation of Meteorological Disasters, Nanjing University of Information Science & Technology, Nanjing 210044, China; zhuchenxia@nuist.edu.cn (C.Z.); dans7messiah@nuist.edu.cn (D.F.T.H.); xikunw@nuist.edu.cn (X.W.); gwang@nuist.edu.cn (G.W.)
[2] Hydro-Climate Extremes Lab, Ghent University, 9000 Ghent, Belgium
[3] Shandong Climate Center, Jinan 250031, China; fengdonghan974570@cma.cn
[4] School of Atmospheric Science & Remote Sensing, Wuxi University, Wuxi 214105, China; jiao_lu@cwxu.edu.cn
[5] Faculty of Defense and Security, Rabdan Academy, Abu Dhabi 114646, United Arab Emirates; wullah@ra.ac.ae
* Correspondence: lishijie@nuist.edu.cn

Abstract: Soil moisture over the Tibetan Plateau (TP) can affect hydrological cycles on local and remote scales through land–atmosphere interactions. However, TP long-term surface soil moisture characteristics and their response to climate change are still unclear. In this study, we firstly evaluate two satellite-based products—SSM/I (the Special Sensor Microwave Imagers) and ECV COMBINED (the Essential Climate Variable combined)—and three reanalysis products—ERA5-Land (the fifth generation of the land component of the European Centre for Medium-Range Weather Forecasts atmospheric reanalysis), MERRA2 (the second version of Modern-Era Retrospective Analysis for Research and Applications), and GLDAS Noah (the Noah land surface model driven by Global Land Data Assimilation System)—against two in situ observation networks. SSM/I and GLDAS Noah outperform the other soil moisture products, followed by MERRA2 and ECV COMBINED, and ERA5-Land has a certain degree of uncertainty in evaluating TP surface soil moisture. Analysis of long-term soil moisture characteristics during 1988–2008 shows that annual and seasonal mean soil moisture have similar spatial distributions of soil moisture decreasing from southeast to northwest. Additionally, a significant increasing trend of soil moisture is found in most of the TP region. With a non-linear machine learning method, we quantify the contribution of each climatic variable to warm-season soil moisture. It indicates that precipitation dominates soil moisture changes rather than air temperature. Pixel-wise partial correlation coefficients further show that there are significant positive correlations between precipitation and soil moisture over most of the TP region. The results of this study will help to understand the role of TP soil moisture in land–atmosphere coupling and hydrological cycles under climate change.

Keywords: soil moisture; Tibetan Plateau; climate change; machine learning

Citation: Zhu, C.; Li, S.; Hagan, D.F.T.; Wei, X.; Feng, D.; Lu, J.; Ullah, W.; Wang, G. Long-Term Characteristics of Surface Soil Moisture over the Tibetan Plateau and Its Response to Climate Change. *Remote Sens.* **2023**, *15*, 4414. https://doi.org/10.3390/rs15184414

Academic Editor: Raffaele Albano

Received: 12 July 2023
Revised: 31 August 2023
Accepted: 5 September 2023
Published: 7 September 2023

1. Introduction

Soil moisture (SM) stored in the unsaturated zone, second only to sea surface temperature (SST), can affect the global water and energy cycle, as well as land–atmosphere coupling, although it only accounts for 0.005% of Earth's water by volume [1–3]. The basic mechanism is that changes in soil moisture directly control surface energy partitioning [4,5]. Decreasing soil moisture tends to reduce evapotranspiration and increase sensible heat according to surface energy balance, making the near-surface environment drier and warmer [6]. If soil moisture increases, more evapotranspiration and less sensible heat can make cloud formation and precipitation [7]. Soil moisture can also non-locally affect precipitation through modulating mesoscale circulations [8]. Decreasing soil moisture can strengthen land–atmosphere coupling [9], improve the possibility of extreme events (e.g., droughts and heatwaves) [10], and limit the vegetation growth and crop field [11,12].

The Tibetan Plateau (TP), known as "The third pole of the world" and "Asia's water towers", meets the water demand of about 2 billion people over its surrounding downstream regions [13]. However, the sustainability of TP terrestrial water storage has been threatened by climate change [14,15]. It is urgent to fully understand the responses of the water cycle to climate change over the TP and its underlying mechanism. As one of the main hydrological components, TP soil moisture can not only affect local land–atmosphere interactions through surface energy partitioning [16], but also remotely affect weather and climate through large-scale atmospheric circulations [17,18].

Soil moisture consists of surface and root-zone soil moisture [16]. The former represents the water content of the upper soil layer, and the latter means the water which is available to plant roots. So far, site-based observations are the most effective method for obtaining true soil moisture at different layers [2]. Plenty of techniques are used to measure soil moisture with ground instruments, such as the gravimetric method [19], time domain reflectometry [20], neutron probes [21], and heat pulse sensors [22]. However, the distribution of observation sites is quite sparse, especially for the TP region with harsh environmental conditions and complex topography [23], despite the fact that numerous scholars tried to establish global long-term observation networks with ground measurements [24].

Satellite observations provide an alternative way to estimate global-scale soil moisture. It has the advantages of low-cost, quick revisiting period and all-weather [25]. There are several widely used soil moisture products retrieved by multiple satellite sensors: the Special Sensor Microwave Imager (SSM/I), Soil Moisture Active Passive (SMAP), Soil Moisture and Ocean Salinity (SMOS), Advanced Scatterometer (ASCAT), and Advanced Microwave Scanning Radiometer-Earth Observing System/2 (AMSR-E/2) [26–31]. Additionally, the Essential Climate Variable combined (ECV COMBINED) soil moisture product with good performance is produced by European Space Agency Climate Change Initiative (ESA CCI) by merging the above several satellite observations [32].

Satellite-based soil moisture products have been widely used to study the soil moisture-climate relationships [33]. However, many scholars point out soil moisture retrieved with remote sensing only reflects surface moisture at a very shallow upper layer (0~5 cm) [33,34]. Considering this, they try to estimate root-zone soil moisture as an alternative based on satellite-based soil moisture [35], or directly use root-zone soil moisture simulated by climate models or land surface models [36]. For example, global long-term root-zone reanalysis data for ERA5-Land, MERRA2, and GLDAS Noah are widely used [37–39]. However, surface soil moisture and root-zone soil moisture are strongly connected. Feldman et al. [40] point out that satellite-based soil moisture has a good potential to capture the dynamics of vegetation water uptake, especially for grass and crop vegetation types. It is due to the existing hydraulic connectivity and spatiotemporal memory between surface and root-zone soil moisture [41,42]. Similar results were found in the interpretation of soil moisture–evapotranspiration coupling for surface and root-zone soil moisture [43]. Therefore, efforts are made to study surface soil moisture here.

There are some limitations and uncertainties existing in soil moisture products. For satellite-based surface soil moisture, its performance may be affected by the reference parameters used in the retrieval algorithm [26,44], and the accuracy of original microwave observations (subject to dense vegetation, surface water, anthropogenic radio-frequency interference, etc.) [32,45]. For reanalysis soil moisture, its performance can be affected by the assimilated reference observations, model parameterizations (e.g., deep convection and soil properties) as well as the representation of topography [46,47]. Therefore, it is necessary to evaluate the performance of soil moisture products before practical applications. Numerous research teams have made great efforts to try to establish long-term soil moisture observation networks over the TP with ground techniques [23,48–51], which provides the opportunity for achieving TP soil moisture validation. Related evaluation studies of soil moisture products have been conducted over the TP against observation networks, such as the Naqu network, Maqu network, Pali network, Ngari network, and the Shiquanhe network [47,52–54].

To our knowledge, until now, long-term annual and seasonal trends in soil moisture over the TP region have not been fully estimated, due to remote sensing products mostly starting after the 2000s [55]. At the background of climate warming, long-term soil moisture characteristics over the southeastern TP have been abruptly changed since the 1880s, showing enhanced variability and a decreasing trend [56]. Air temperature and precipitation are thought to be two main factors of controlling soil moisture changes over the TP [56,57]. Air temperature and precipitation have significantly changed during recent decades. After the 2000s, the rising rate of surface temperature over the TP region is more than 0.29 °C decade^{-1}, especially for glacier regions [58]. Increasing precipitation during the most recent three decades is found in most of the TP region on annual and seasonal scales [59]. Recent studies show that precipitation, not air temperature, controls soil moisture changes over the TP [60,61]. These conclusions were obtained with linear correlations, but the relationships between soil moisture and climate are nonlinear [62]. Recently, machine learning-based attribution methods have been used to study the nonlinear relationships [63].

In this study, we firstly evaluate the performance of five common long-term surface soil moisture products against two sets of observations over the TP. These products include two satellite-based products (SSM/I and ECV COMBINED) and three reanalysis products (ERA5-Land, MERRA2 and GLDAS Noah). Soil moisture ground observations come from the coordinated enhanced observing period (CEOP) Asia-Australia monsoon project (CAMP) on the Tibetan Plateau (CAMP/Tibet) and a synthesis dataset released by Zhao Lin and Hu Guojie et al. (Zhao2021). Then, the mean and trend patterns of TP soil moisture are shown during 1988–2008 for these products on the annual and seasonal scales. Finally, we establish the nonlinear relationship between soil moisture and climatic variables during the warm season (May to October) with machine learning (e.g., random forest), and quantify the importance of each variable on soil moisture over the TP.

2. Study Area

TP is known as "The third pole of the world", which has the largest frozen water storage after the polar regions. It is also termed "Asia's water towers", because it is the source of 13 major river systems that feed almost 2 billion people [13]. It is located in the southwestern part of China at 80–105°E and 28–37°N, and it is the highest plateau in the world, with an average elevation of above 4000 m (Figure 1c).

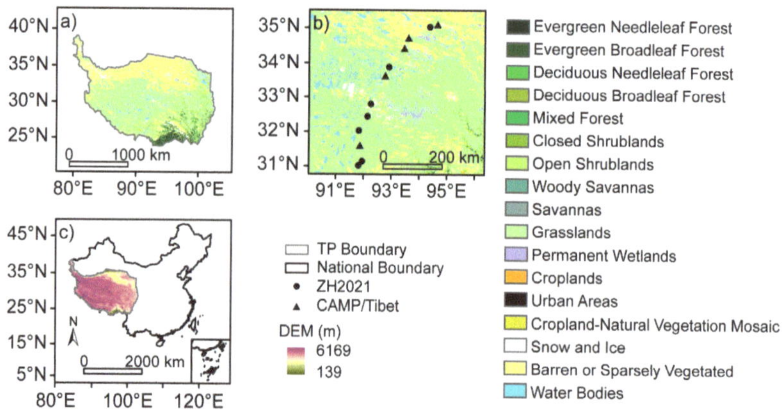

Figure 1. Overview of the study area. (**a**) Land cover classifications over the Tibetan Plateau (TP). (**b**) Locations of the seven CAMP/Tibet sites (black triangles) and five ZH2021 sites (black circles) over the TP. (**c**) TP elevation in meters and its geographical location in China.

The elevated topography of the TP exerts great dynamic and thermal effects on the weather and climate of local and surrounding regions [64]. According to long-term climate data records during 1961–2010, the annual average air temperature over the TP is between −2.2~0 °C [65]. The highest air temperature occurs in July, which can reach 15 °C, while the lowest temperature in January can be as low as −10.3 °C. The precipitation over the TP is strongly affected by the westerlies and Asian summer monsoon [66], with approximately 80% of the precipitation occurring during the summer monsoon period. The annual mean precipitation over the TP is between 415 and 512 mm [67]. Due to the continuous increase in elevation, the total precipitation gradually decreases from southeast (with up to 900 mm) to northwest (only up to 60 mm) [26].

According to a Multi-source Integrated Chinese Land Cover (MICLCover) map generated by Ran et al. [68], there are 17 land cover classifications over the TP. Grasslands dominate the TP region, mainly distributed south of 35°N. Secondly, there are barren or sparsely vegetated areas, mainly distributed over the northern TP. There are also evergreen forests at the southeastern edge, which may be related to abundant summer monsoon precipitation. There are some permanent wetlands, water bodies and snow and ice spreading over the TP as well (Figure 1a).

3. Data and Methods

3.1. Data

In this study, two sets of in situ measurements (Table 1) are used to evaluate multi-source (i.e., satellite and reanalysis) soil moisture products (Table 2) over the TP. The following are detailed descriptions.

Table 1. Information on in situ measurements.

	Sites Name	Longitude (°E)	Latitude (°N)	Time Range	Depth (m)
CAMP/Tibet	Amdo	91.62	32.24	01/10/2002–04/09/2003	−2.79, −2.58, −2.00, −1.60, −1.30, −1.00, −0.80, −0.60, −0.40, −0.20, −0.04
	BJ	91.90	31.37	01/10/2002–31/03/2004	−2.50, −2.10, −2.00, −1.60, −1.30, −1.00, −0.80, −0.60, −0.40, −0.20, −0.04
	D66	93.78	35.52	01/10/2002–31/03/2004	−2.50, −2.25, −2.00, −1.60, −1.30, −1.00, −0.80, −0.60, −0.40, −0.20, −0.04
	D105	91.94	33.06	01/10/2002–31/03/2004	−1.90, −1.85, −1.80, −1.60, −1.30, −1.00, −0.80, −0.60, −0.40, −0.20, −0.04
	D110	91.87	32.69	01/10/2002–31/03/2004	−1.80, −1.60, −1.30, −1.00, −0.80, −0.60, −0.40, −0.20,−0.04, 0.00
	MS3608	91.78	31.23	01/10/2002–31/03/2004	−2.42, −2.00, −1.96, −1.60, −1.30, −1.00, −0.80, −0.60, −0.40, −0.20, −0.04
	Tuotuohe	92.44	34.22	01/10/2002–31/03/2004	−2.71, −2.30, −2.00, −1.60, −1.30, −1.00, −0.80, −0.60, −0.40, −0.20, −0.04
ZH2021	Ch04	91.74	31.82	07/06/2001–31/12/2006	−1.20, −0.95, −0.50, −0.15
	Ch06	94.06	35.62	01/01/2005–31/12/2016	−1.80, −1.00, −0.40, −0.10
	QT01	93.04	35.14	01/01/2004–01/10/2014	−1.80, −1.00, −0.50, −0.10
	QT03	92.92	34.82	01/01/2004–31/12/2014	−2.40, −1.80, −0.90, −0.50, −0.05
	QT05	92.34	33.96	01/01/2004–31/12/2008	−2.70, −2.10, −0.90, −0.60, −0.10

3.1.1. In Situ Measurements

CAMP/Tibet

The coordinated enhanced observing period (CEOP) Asia-Australia monsoon project (CAMP) on the Tibetan Plateau (CAMP/Tibet) starts from 1 October 2002 to 30 September 2004. A meso-scale observational network of CAMP/Tibet is set up in the central plateau, which produces a large amount of data for the study of energy and water cycle over the TP [69]. Surface soil moisture observations with the depth of −0.04 m at seven stations (black triangles in Figure 1b) in soil moisture and soil temperature (SMTMS) networks belonging to this meso-scale network are used in this study. At each station, hourly soil moisture data are provided with the unit of percent (%), and such data are converted to volumetric soil moisture (m^3 m^{-3}) by dividing by 100. Readers can refer to CAMP/Tibet in Table 1 for detailed information about the location, time range and depths for each CAMP/Tibet station. CAMP/Tibet observations have already been widely used in many scientific studies [70,71].

ZH2021

Zhao Lin and Hu Guojie et al. have released a synthesis dataset of permafrost thermal state for the Qinghai–Tibet (Xizang) Plateau in China, which is abbreviated as ZH2021 hereafter [48]. This dataset consists of meteorological variables (air temperature, precipitation, specific humidity, et al.), deep ground temperature, soil moisture, and soil temperature data with quality control. In this study, we only used daily surface (top-layer) soil moisture observations at five stations for evaluation (black circles in Figure 1b). Detailed information can be seen from ZH2021 in Table 1.

3.1.2. Satellite and Reanalysis Datasets

SSM/I

Based on a radiative transfer model, van der Velde et al. [26] have retrieved long-term soil moisture data at a depth of about the top 3 cm of the soil, specifically over the TP using brightness temperatures observed by the Special Sensor Microwave Imagers (SSM/I) sensors (F08, F11 and F13), facilitated through the Defense Meteorological Satellite Programme. The brightness temperature observations after calibration and bias corrections were provided by the Precipitation Research Group of Colorado State University. When validating against in situ measurements at the Naqu network and comparing with GLDAS Noah simulations, SSM/I soil moisture performs well. The data gaps in SSM/I are filled with a three-dimensional distinct transform method [72]. The temporal resolution is daily (F08 satellite operates from July 1987 to December 1991 at UTC 06:12, F11 satellite operates from December 1991 to May 1995 at UTC 18:11, and F13 operates from May 1995 to December 2008 at UTC 17:42), and the spatial resolution is $0.25° \times 0.25°$. Due to the limitation of temporal range for SSM/I soil moisture, our study period is 1988–2008. Readers can refer to van der Velde et al. [26] for more detailed information.

ECV COMBINED

The Essential Climate Variable combined (ECV COMBINED) soil moisture developed by the European Space Agency Climate Change Initiative (ESA CCI) project combines various single-sensor active and passive microwave soil moisture products [32]. Since its first release in 2012, ECV COMBINED soil moisture has been continuously updated 13 times, and it has now been updated to ECV COMBINED v07.1 [73]. Here, we use the latest version ECV COMBINED v07.1, which merges the retrieved soil moisture products from five active and twelve passive microwave sensors. Due to the inclusion of three new sensors of FengYun 3C, FengYun 3D and ASCAT-C, this dataset has improved in temporal and spatial coverage, making it the most accurate ESA CCI soil moisture to date. This global daily soil moisture data start from 01/11/1978 to 31/12/2021, with a spatial resolution of $0.25° \times 0.25°$.

ERA5-Land

The European Center for Medium-Range Weather Forecasts (ECMWF) has produced a global dataset for the land component of the fifth generation of European ReAnalysis (ERA5), hereafter referred to as ERA5-Land [74]. ERA5-Land provides many high spatiotemporal resolution surface variables from 1950 to present. As one of the variables, soil moisture in ERA5-Land is divided into four layers (0–7, 7–28, 28–100, and 100–289 cm). In this study, we use ERA5-Land soil moisture simulations at the depth of 0–7 cm to represent surface soil moisture. Its spatial resolution is $0.1^{\circ} \times 0.1^{\circ}$, and the temporal resolution includes hourly and monthly.

MERRA2

The second version of Modern-Era Retrospective Analysis for Research and Applications (MERRA2) is the latest reanalysis data produced by NASA's Global Modeling and Assimilation Office (GMAO) [46]. MERRA2 optimizes the Goddard Earth Observing System (GEOS) model and analysis scheme and assimilates more observation types, resulting in improved data quality compared with its predecessor—MERRA. Validating against 119 observations over China, Hagan et al. [75] also proves that MERRA2 has better skill than the older generations with higher correlations and lower root mean square errors. MERRA2 global data start from 1980, with a spatial resolution of $0.625^{\circ} \times 0.5^{\circ}$. The soil column of MERRA2 is divided into three soil layers, namely, the 0–5 cm surface layer, the 0–100 cm root zone, and the prof layer from the surface down to the bedrock. In this study, MERRA2 surface soil moisture content (SFMC) at the depth of 0–5 cm is used. The hourly (M2T1NXLND) and monthly (M2TMNXLND) MERRA2 soil moisture can be accessible through the NASA Goddard Earth Sciences Data Information Services Center (GES DISC).

GLDAS Noah

The reanalysis data, Global Land Data Assimilation System version 2 (GLDAS-2) with 3 components, GLDAS-2.0, GLDAS-2.1 and GLDAS-2.2, have been released by NASA [38]. The GLDAS-2.0 0.25 degree 3-hourly (GLADS_NOAH025_3H) and monthly (GLDAS_NOAH025_M) product with open-loop simulation (no data assimilation) produced using Noah Model 3.6 is used in this study. The model is forced by the Princeton meteorological forcing data, which start from 1948 to 2014. The spatial resolution is 0.25 degree. It produces a series of land surface parameters, including soil moisture at the depths of 0–10, 10–40, 40–100 and 100–200 cm. Only the surface 0–10 cm soil moisture over the TP region is used for evaluation. It should be noted that the unit of GLDAS soil moisture is kg m^{-2}. In comparison with the other products, the unit has been transferred to volumetric soil moisture (m^3 m^{-3}) according to Zhu and Shi [76]. Previous studies have documented the good performance of GLDAS Noah soil moisture over the TP region [77,78].

Table 2. Overview of the soil moisture datasets used in this study.

Datasets	Temporal Coverage	Temporal Resolution	Spatial Resolution	Unit	References
Satellite datasets					
SSM/I	07/1987–12/2008	daily	$0.25^{\circ} \times 0.25^{\circ}$	m^3 m^{-3}	van der Velde et al. [26]
ECV COMBINED	11/1978–12/2021	daily	$0.25^{\circ} \times 0.25^{\circ}$	m^3 m^{-3}	Dorigo et al. [32]
Reanalysis datasets					
ERA5-Land	01/1950–present	hourly; monthly	$0.1^{\circ} \times 0.1^{\circ}$	m^3 m^{-3}	Muñoz-Sabater et al. [74]
MERRA2	01/1980–present	hourly; monthly	$0.625^{\circ} \times 0.5^{\circ}$	m^3 m^{-3}	Gelaro et al. [46]
GLDAS Noah	01/1948–01/2015	3 hourly; monthly	$0.25^{\circ} \times 0.25^{\circ}$	kg m^{-2}	Rodell et al. [38]

3.1.3. Meteorological Data and Ancillary Data

The China Meteorological Forcing Dataset (CMFD) produced by He et al. [79] is of a high spatiotemporal resolution (3-hourly, $0.1^{\circ} \times 0.1^{\circ}$) with a duration from 1979 to 2018. The CMFD merges multi-source data, including remote sensing products, in situ station data and reanalysis data. Through validation, the CMFD is of superior quality than the GLDAS because of the participation of more stations in China. The dataset includes seven near-surface variables: precipitation rate, 2-m air temperature, surface pressure, specific humidity, 10-m wind speed, downward shortwave and longwave radiation. All variables except surface pressure are used to study the response of soil moisture to climatic variables over the TP.

Some ancillary data are also used here. The elevation data are the Global 30 Arc-Second Elevation (GTOPO30) with a horizontal grid spacing of approximately 1 km supported by the United States Geological Survey (USGS). The GTOPO30 elevation data are completed in late 1996. As for land cover classifications over the TP, we use a multi-source Integrated Chinese Land Cover (MICLCover) map of 17 land cover types with a spatial resolution of 1 km developed by Ran et al. [68]. This MICLCover map is generated using the decision-fuse method, which combines land cover/land use classification maps including a 1:1,000,000 vegetation map, a 1:100,000 land use map, a 1:1,000,000 swamp-wetland map, a glacier map and a Moderate-Resolution Imaging Spectroradiometer land cover map in 2001. Validation against ground sample sites across China shows that the overall accuracy of the MICLCover map is 71%, higher than the accuracy of other land cover maps.

3.2. Methods

3.2.1. Data Pre-Processing

To accomplish the data evaluation, the spatio-temporal characteristics of soil moisture products and in situ observations need to be matched. Due to the limited availability of SSM/I data, the time range for all soil moisture products is 1988–2008. For CAMP/Tibet observations with the duration of 01/10/2002–31/03/2004, we use the only full year, 2003, for comparison. Since the temporal range of ZH2021 observations at each station differs, with the earliest observation starting on 07/06/2001, the corresponding time ranges from products are selected based on in situ stations. The temporal resolutions for all data are first uniformly converted to daily for inter-comparison. According to the transit time 17:42 for the SSM/I satellite F13 from May 1995 to December 2008, hourly CAMP/Tibet data in 2003 are converted to daily data by selecting the value at 18:00. Similarly, the hourly ERA5-Land (18:00) and MERRA2 (17:30) products and the 3-hourly GLDAS Noah (18:00) product are converted to daily data by selecting the nearest time to 17:42. Daily data with missing values can be aggregated to monthly averages when at least five valid values are available within a month [26]. Regarding spatial matching, the product data are obtained for corresponding pixels/grids based on site locations (longitude, latitude) [45].

When using machine learning to attribute soil moisture changes, the spatial resolutions of all these soil moisture products remain unchanged, while climatic variables from the CMFD are resampled with bilinear interpolation to obtain the same spatial resolution as each soil moisture product.

3.2.2. Evaluation Metrics

Five evaluation metrics are used to evaluate the performance of soil moisture products, as follows:

$$R = \frac{\sum_{i=1}^{n}\left((\theta_i - \overline{\theta})(O_i - \overline{O})\right)}{\sqrt{\sum_{i=1}^{n}(\theta_i - \overline{\theta})^2}\sqrt{\sum_{i=1}^{n}(O_i - \overline{O})^2}} \tag{1}$$

$$BIAS = \frac{1}{n}\sum_{i=1}^{n}(\theta_i - O_i) \tag{2}$$

$$MAE = \frac{1}{n}\sum_{i=1}^{n}|\theta_i - O_i| \tag{3}$$

$$RMSE = \sqrt{\frac{\sum_{i=1}^{n}(\theta_i - O_i)^2}{n}} \tag{4}$$

$$ubRMSE = \sqrt{RMSE^2 - BIAS^2} \tag{5}$$

where R, BIAS, MAE, RMSE, and ubRMSE represent the Pearson correlation, bias, mean absolute error, root mean square error, and unbiased root mean square error, respectively. θ and O represent soil moisture values of the products and site observations. n is the count of available samples, while i indicates the given time point. $\overline{\theta}$ and $\overline{O}$ indicate the average values of all samples for the products and site observations.

3.2.3. Trend Analysis

The annual trend and the seasonal trend (spring, summer, autumn and winter) of soil moisture over the TP region during 1988–2008 for each product are calculated with Theil–Sen's slope method, which removes the effect of skewed data compared to the common linear regression method [80]. Numerous scientists have used it to calculate the trend of hydrological variables [81]. The widely used Mann–Kendall method is used to identify the significance level of soil moisture trends [82,83].

3.2.4. Random Forest

The Random Forest regression method was used to identify the relative importance of each climatic variable on soil moisture change over the TP. This Random Forest model is a robust machine learning method with multi-decision trees' ensemble for classification or regression using a bagging function [84]. It can effectively lower the uncertainties of overfitting problems. In this study, we use the "TreeBagger" function of MATLAB R2021b software to realize the Random Forest regression. Firstly, we randomly divide warming season (May–October) soil moisture values in each product and climatic variables from the CMFD during 1988–2008 into two parts: two-thirds data for training and one-third for validation. All gridded soil moisture and climatic variables over the whole TP region are trained in the Random Forest model, together. Leaf node count is set to 5, regression trees' size is set to 300.

3.2.5. Partial Correlation

In the Results section, we compare the relative importance of each climatic variable for all soil moisture products and find that precipitation is the most important variable in controlling soil moisture changes over the TP region. Therefore, we calculate the partial correlation between warming season soil moisture in each product and precipitation on the grid scale, excluding the effects of other climatic variables. The "partialcorr" function of MATLAB R2021b software is used to calculate the partial correlation.

4. Results

4.1. The Evaluation of Multi-Source Soil Moisture Products

Before studying soil moisture characteristics over the TP, we firstly evaluate the performance of these products used in this study against two in situ observation networks: CAMP/Tibet and ZH2021. In Figure 2, SSM/I and GLDAS Noah have the best fitting results, with R^2 being more than 0.3 for two observation networks; more specifically, these two products' R^2 are more than 0.4 using ZH2021 as the reference, followed by MERRA2 with R^2 of 0.18 and 0.25 and ECV COMBINED with R^2 of 0.23 and 0.11. ERA5-Land has the lowest R^2, 0.11 in CAMP/Tibet and 0.01 in ZH2021. The overall fitting results show that all products tend to overestimate soil moisture with different magnitudes, especially for ERA5-Land, with almost all samples being above the 1:1 line.

Next, in detail, we use five different metrics to evaluate the performance of used soil moisture products (Table 3). SSM/I and GLDAS Noah have the best performance with $R > 0.56$, BIAS close to zero, $MAE < 0.09$, $RMSE < 0.11$ and $ubRMSE < 0.10$. SSM/I marginally performs better than GLDAS Noah using ZH2021 as the reference, while GLDAS Noah is better than SSM/I with CAMP/Tibet as the benchmark. The performance of ECV COMBINED is very close to SSM/I and GLDAS Noah in terms of all metrics, especially for being better than SSM/I for RMSE and ubRMSE using CAMP/Tibet as the reference. Although MERRA2's R of 0.50 is higher than ECV COMBINED's R of

0.34 in ZH2021, the remaining metrics (BIAS, MAE, RMSE, ubRMSE) of MERRA2 are worse than ECV COMBINED. ERA5-Land has the lowest R, the same as in Figure 2, and the highest BIAS, MAE, RMSE and ubRMSE, compared to the other products. It indicates that ERA5-Land has great uncertainties in representing surface soil moisture over the TP, which may be related to the characterization of snow fields at high altitudes (> 3300 m a.s.l.) in the land model [74].

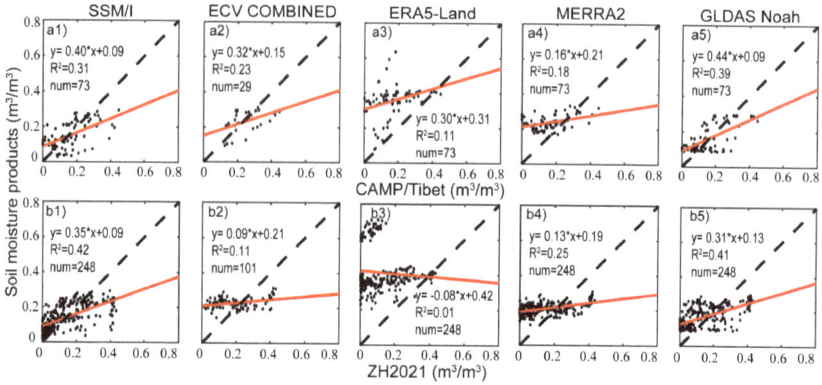

Figure 2. The scatterplots of monthly soil moisture values of multi-source products (SSM/I, ECV COMBINED, ERA5-Land, MERRA2, and GLDAS Noah) against two sources of in situ observation networks: (**a1–a5**) CAMP/Tibet and (**b1–b5**) ZH2021. The num means the count of available monthly soil moisture values between products and ground observations. The black dashed lines indicate the 1:1 line.

Table 3. Different metrics of evaluating multi-source soil moisture products with two sources of in situ observation networks: CAMP/Tibet and ZH2021.

	R	BIAS	MAE	RMSE	ubRMSE
		CAMP/Tibet			
SSM/I	0.56	−0.01	0.07	0.09	0.09
ECV COMBINED	0.48	−0.02	0.07	0.08	0.08
ERA5-Land	0.33	0.19	0.19	0.23	0.12
MERRA2	0.42	0.08	0.10	0.12	0.10
GLDAS Noah	0.63	0.00	0.07	0.08	0.08
		ZH2021			
SSM/I	0.65	0.00	0.08	0.10	0.10
ECV COMBINED	0.34	0.02	0.10	0.12	0.11
ERA5-Land	−0.09	0.28	0.28	0.33	0.18
MERRA2	0.50	0.07	0.13	0.14	0.12
GLDAS Noah	0.64	0.03	0.09	0.11	0.10

4.2. Long-Term Soil Moisture Characteristics over the TP

4.2.1. The Annual and Seasonal Mean Soil Moisture during 1988–2008

The multi-year mean soil moisture patterns during 1988–2008 over the TP using different products are shown in Figure 3. We find that all products except ECV COMBINED show similar spatial distributions, with higher values in the southeast of the TP region and lower values in the northwest. The multi-year mean soil moisture gradually increases from southeast to northwest. Only a few pixels with soil moisture values exist in ECV COMBINED. In terms of magnitudes, ERA5-Land has the highest values (more than 0.2 m^3/m^3 over most of the TP region), MERRA2 and GLDAS Noah have values ranging from 0.1 to 0.3 m^3/m^3, while SSM/I has the lowest values (less than 0.15 m^3/m^3 over most of the TP region).

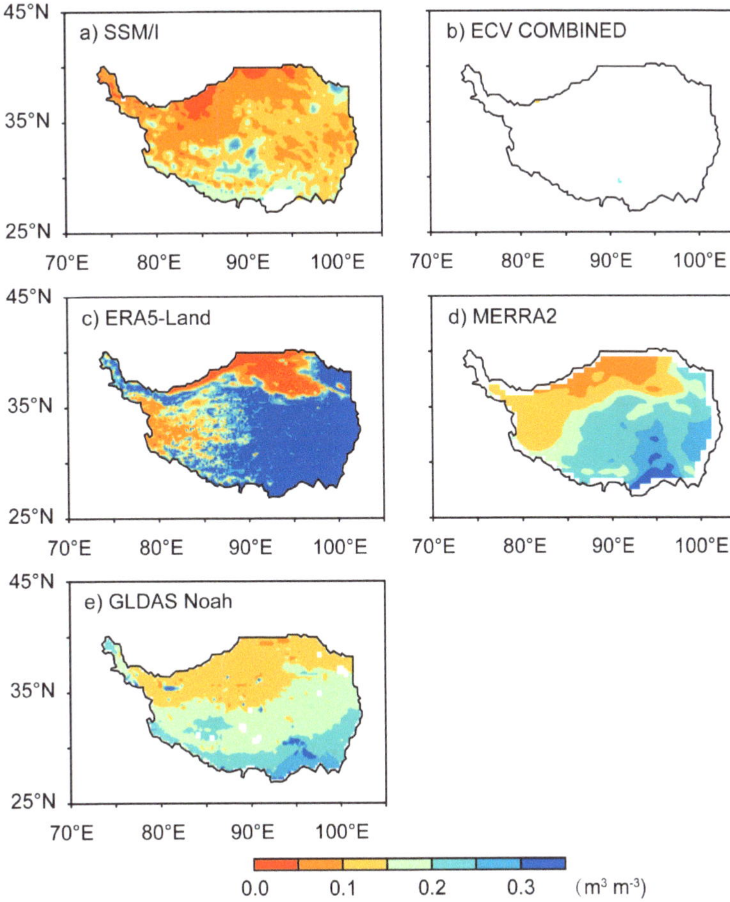

Figure 3. The annual mean soil moisture during 1988–2008 over the TP for (**a**) SSM/I, (**b**) ECV COMBINED, (**c**) ERA5-Land, (**d**) MERRA2, and (**e**) GLDAS Noah.

We also explore the consistencies of seasonal mean soil moisture over the TP region for all products (Figure 4). ECV COMBINED with lots of missing multi-year mean values is mostly due to missing values in winter. There are plenty of values in the other three seasons, especially for summer and autumn seasons (Figure 4e–h). By comparing these seasonal mean values among the products, the following are shown to be consistent results: like multi-year mean values, the similar spatial distributions are found in all four seasons for all the other products, with the southeast region having higher values and the northeast region having lower values; mean values of summer and autumn have the highest values for all products, compared to spring and winter. The differences among these products are as follows: there are little seasonal differences of mean values for ERA5-Land and MERRA2; however, the obvious transition features are found in the remaining products of SSM/I, ECV COMBINED and GLDAS Noah, with the mean values firstly increasing from spring to summer and then decreasing from summer to winter.

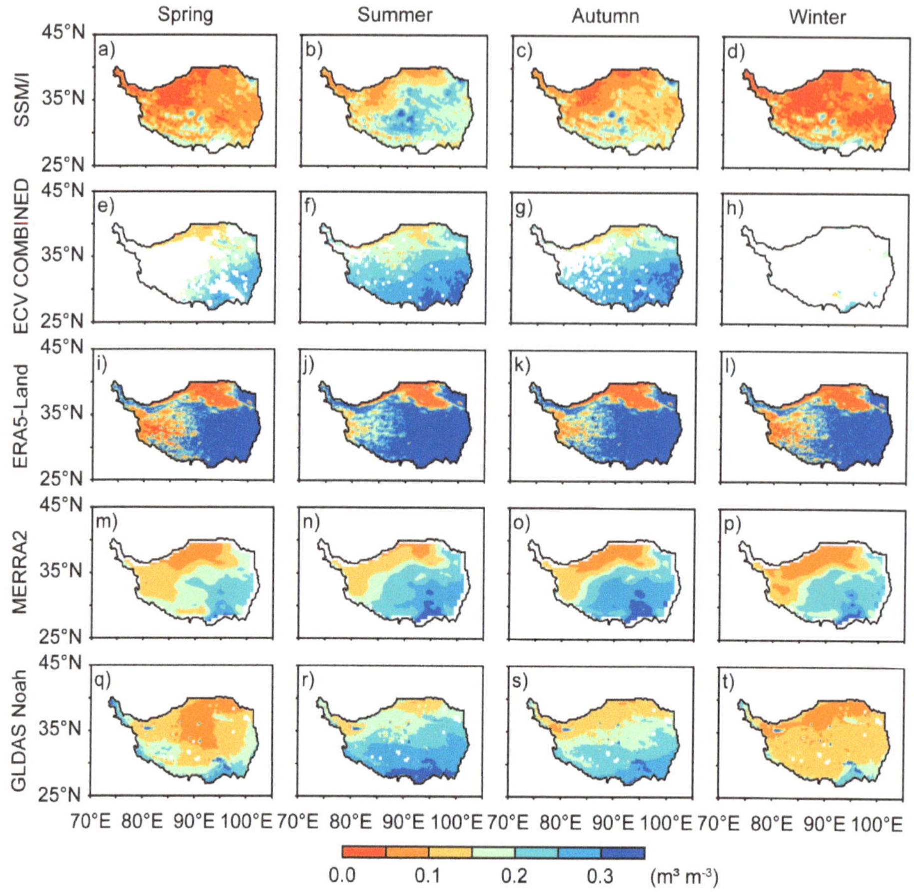

Figure 4. The seasonal mean soil moisture during 1988–2008 over the TP in spring, summer, autumn, and winter for (**a–d**) SSM/I, (**e–h**) ECV COMBINED, (**i–l**) ERA5-Land, (**m–p**) MERRA2, and (**q–t**) GLDAS Noah.

4.2.2. The Annual and Seasonal Trend of Soil Moisture during 1988–2008

Figure 5 shows the spatial distribution of soil moisture annual trend over the TP during 1988–2008 for all products. Except for ECV COMBINED, all products generally show significant increasing trends over the TP region. SSM/I has the increasing trend of about 2×10^{-3} m^3/m^3 over most of the TP region, especially in the central and western region with more than 2×10^{-3} m^3/m^3. ECV COMBINED does not have definitive trends since the distribution of its trend values is quite sparse. Significant increasing trends are found in the western TP region for ERA5-Land, while a few decreasing values are also found in the northern edge of the TP. For MERRA2, overall increasing trends exist in most of the TP region, especially for southwest, while decreasing trends are found in the southeast. GLDAS Noah has increasing trends in east and west of the TP region. It should be noted that most negative values in the figure have not passed the significant test at the significance level of 0.05. In terms of magnitudes, ERA5-Land has the biggest increasing trend, followed by MERRA2 and SSM/I, while GLDAS Noah shows the smallest increasing trend.

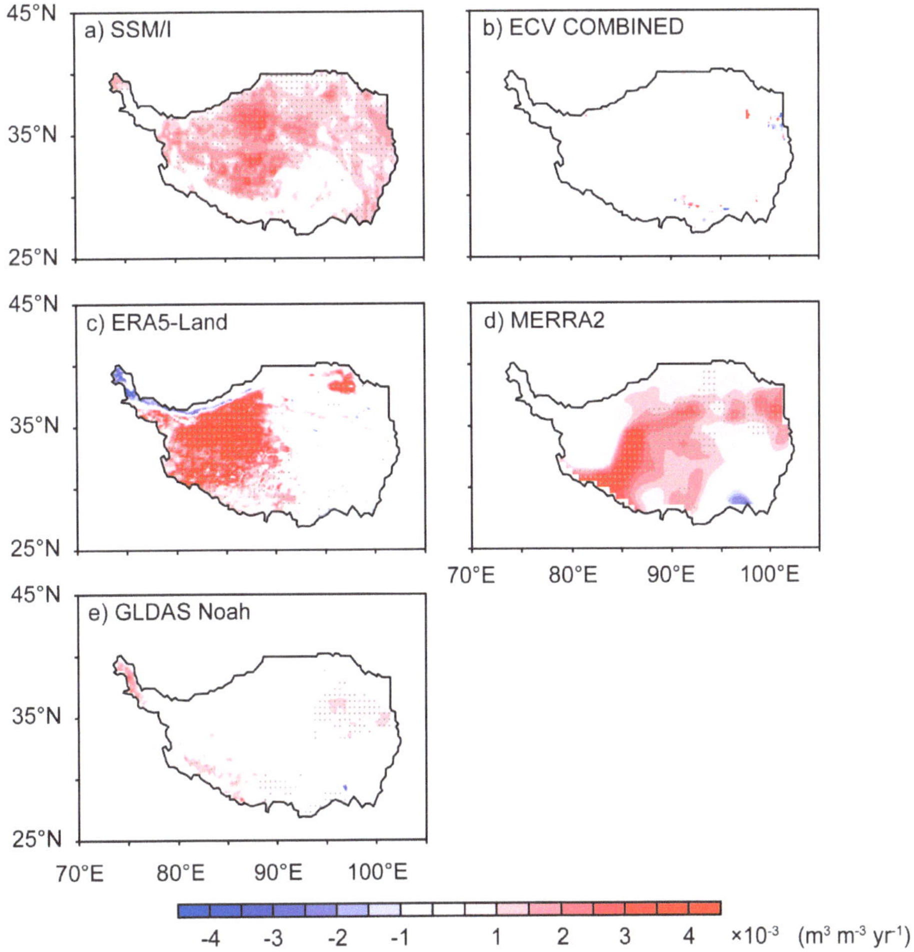

Figure 5. The annual trend of soil moisture during 1988–2008 over the TP region for (**a**) SSM/I, (**b**) ECV COMBINED, (**c**) ERA5-Land, (**d**) MERRA2, and (**e**) GLDAS Noah. The dotted region passed the significant test level (alpha = 0.05).

Soil moisture seasonal trends during 1988–2008 for all products are shown in Figure 6. Similar to the spatial distribution of soil moisture annual trend, each seasonal trend for all products generally shows an increasing pattern. The biggest increasing trends can be found in each season of ERA5-Land with more than 3×10^{-3} m^3/m^3, while GLDAS Noah has the smallest increasing trend of about 1×10^{-3} m^3/m^3. SSM/I has an increasing trend over most of the TP region, especially in the summer season, with more than 3×10^{-3} m^3/m^3. For ECV COMBINED, increasing trends are found in the east for spring, west for summer, and middle for autumn. For ERA5-Land, the biggest increasing trends with 3×10^{-3} m^3/m^3 are found in the west of the TP region for all seasons. Increasing trends are found in most of the TP region except for some west and southeast regions for MERRA2. Increasing trends of GLDAS Noah are found in the southwest TP in spring, eastern TP in summer, northeast and southwest TP in autumn, and some scattered regions in winter. Few negative values showing decreasing trends have passed the significance test at the significance level of 0.05.

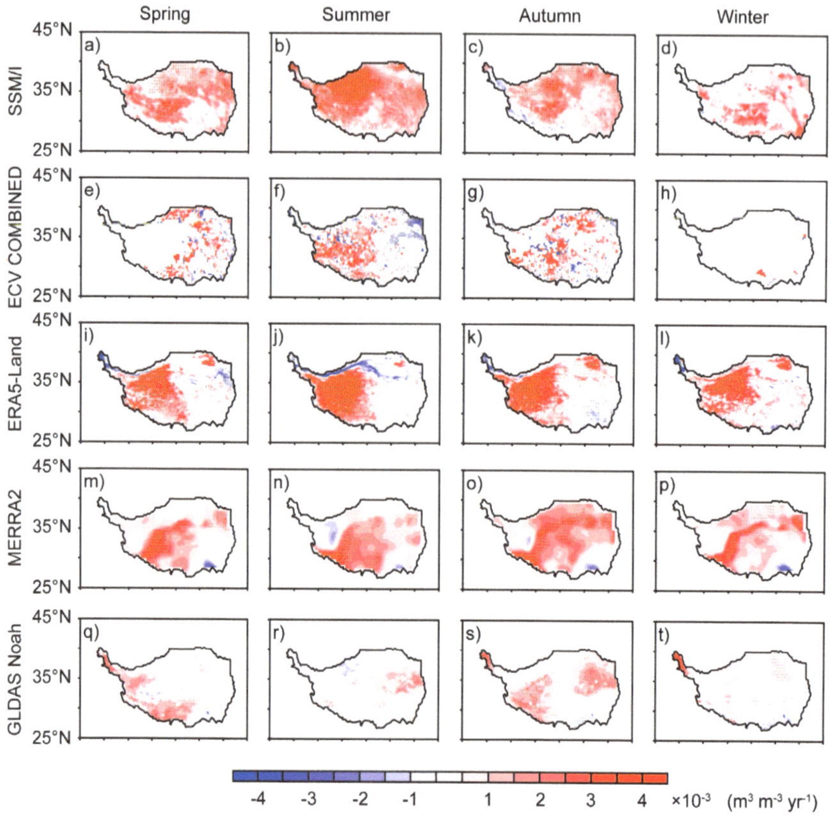

Figure 6. The seasonal trend of soil moisture during 1988–2008 over TP region in spring, summer, autumn, and winter for (**a–d**) SSM/I, (**e–h**) ECV COMBINED, (**i–l**) ERA5-Land, (**m–p**) MERRA2, and (**q–t**) GLDAS Noah. The dotted region passed the significance level of 0.05.

4.3. Quantifying the Relative Importance of Each Climatic Variable on TP Soil Moisture

The above-mentioned results indicate that soil moisture over the TP region is significantly increasing during 1988–2008 for all products on the annual and seasonal scales. Until now, it is unclear that which variable dominates TP soil moisture changes. This subsection estimates each climatic variable's relative importance on TP soil moisture during warm seasons (Figure 7). The relative importance (%) of each variable shown in the figure is calculated with its absolute importance divided by the sum of absolute contributions of all climatic variables (precipitation, longwave radiation, shortwave radiation, air temperature, specific humidity and wind speed), while the absolute importance is obtained with the Random Forest regression (RF) method (see Section 3.2.4 for details). The sum percentages of all variables are equal to 100% for each product. The attribution results indicate that precipitation is the dominant factor for controlling the soil moisture of SSM/I, ECV COMBINED and MERRA2, especially for MERRA2, which has the highest contribution of about 39%, while precipitation and wind speed have considerable contributions to soil moisture changes, with more than 20% for ERA5-Land and GLDAS Noah. Additionally, the other variables' (i.e., longwave radiation, shortwave radiation, air temperature, specific humidity) contributions are less than 20%, especially for longwave radiation, with about 10% contributions for all products.

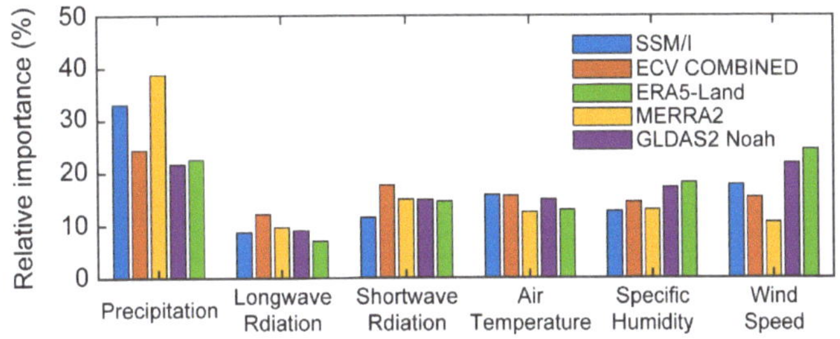

Figure 7. The relative importance of each climatic variable on soil moisture over the TP region during warm seasons of 1988–2008 for SSM/I, ECV COMBINED, ERA5-Land, MERRA2, and GLDAS Noah. The climatic variables are precipitation, longwave radiation, shortwave radiation, air temperature, specific humidity and wind speed.

The importance in each variable is calculated by taking all pixels as a whole. The result indicates that precipitation contributes more than air temperature. Then, we calculate the partial correlation between TP soil moisture and precipitation on the grid scale (Figure 8). The partial correlation results show that precipitation has significant positive correlations with soil moisture over most of the TP region during 1988–2008 for all products. In terms of the magnitude, MERRA2 has the highest values, with more than 0.8, followed by ERA5-Land and ECV COMBINED, with correlations of 0.4–0.8; SSM/I and GLDAS Noah have the lowest correlations, with 0.2–0.4.

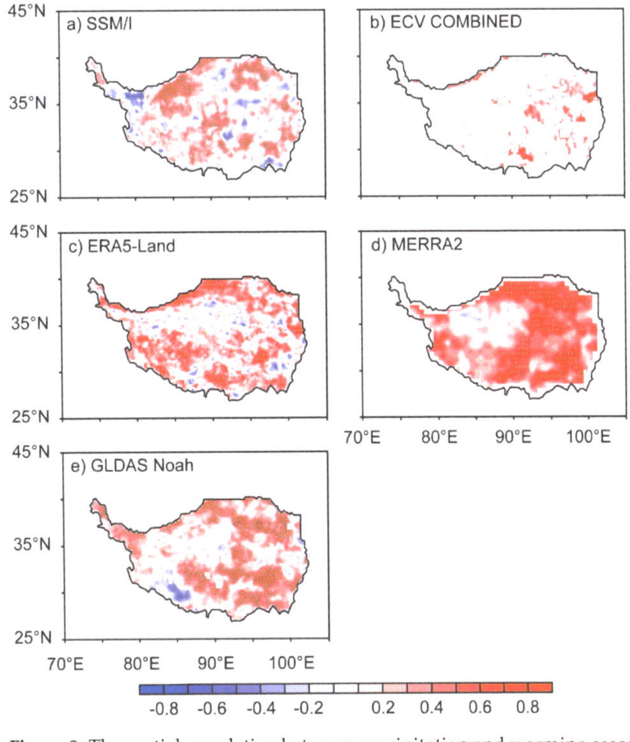

Figure 8. The partial correlation between precipitation and warming season soil moisture over TP region during 1988–2008 for SSM/I, ECV COMBINED, ERA5-Land, MERRA2, and GLDAS Noah. The dotted region passed the significant test level (alpha = 0.05).

Lastly, we evaluate the performance of the trained random forest regression model in each product (Figure 9). All products except for ECV COMBINED have the R^2 with about or over 0.5, while ECV COMBINED's R^2 is 0.30. The difference may be due to there not being enough pixels to train the model, compared to the other products. For RMSE, ECV COMBINED has the second lowest value, with 0.39, compared to SSM/I, MERRA2, and GLDAS Noah, all with more than 0.40. Additionally, ERA5-Land has the lowest RMSE (0.38) and the highest R^2 (0.74), indicating that it has the best trained model. The best performance of the ERA5-Land trained model here shows the greatest model-regressed nonlinear relationship between meteorological forcing and ERA5-Land soil moisture. It should be noted that the performance of the trained model has nothing to do with the performance/quality of the product itself. These two performance evaluations are with different purposes, and better performance here does not necessarily mean better performance there. The overall results show that all products have relatively good performance in the trained model.

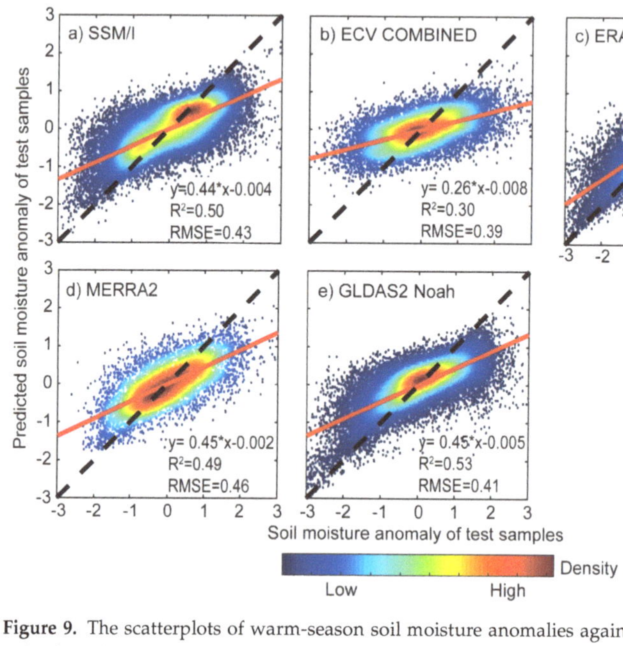

Figure 9. The scatterplots of warm-season soil moisture anomalies against those predicted with trained random forest regression method over TP region during 1988–2008 for SSM/I, ECV COMBINED, ERA5-Land, MERRA2, and GLDAS Noah.

5. Discussions

We evaluate the performance of five multi-source long-term soil moisture products with two sets of in situ observations networks. The evaluation results indicate that satellite-based SSM/I soil moisture observations perform well over the TP with higher R and lower BIAS, MAE, RMSE, and ubRMSE, while ECV COMBINED is no better than SSM/I, which may be due to too many missing values impacting the final calculation of the metrics. The reanalysis-based products tend to overestimate soil moisture values (Figure 2 and Table 3), especially for ERA5-Land. Similar overestimation results based on reanalysis products are also found in Xing et al. [55] and Wang et al. [61]. Wang et al. [61] point out that ERA5 and GLDAS obviously overestimate actual soil moisture values compared to satellite-based products. The overestimation of reanalysis products mainly depends on uncertainties of the model structure, parameterization, assimilation scheme and forcing data [85]. In addition, the mismatching between the coarse pixel of these products and the effective footprint of in situ observation sites, as well as inconsistent depths between the products and in situ observation sites, may also introduce some uncertainties into data evaluation [45]. Since soil moisture over the TP can be variable in space, there may also be uncertainty in evaluating the performance of soil moisture products in the entire TP using in situ observations of CAMP/Tibet and ZH2021, which are mainly distributed in the central plateau.

Through analyzing the characteristics, we discover the significant increasing trend of soil moisture over the TP. The wetting phenomenon has also been detected by van der Velde et al. [26]. Further, we attribute the increasing soil moisture over the TP region during recent decades to climatic variables with the nonlinear regression method—Random Forest. Here, we only focus on the warm season rather than the annual scale to avoid the effect of snow ablation on the attributed results. The attributed results indicate that precipitation dominates soil moisture changes over the TP rather than air temperature. Precipitation is the main source of moisture to the TP region, despite the fact that rising air temperature can reduce soil moisture with enhanced evapotranspiration. Precipitation's dominant role in soil moisture over the TP has also been confirmed by Wang et al. [60] and Dai et al. [86].

Longwave radiation contributes the least to warm-season soil moisture changes over the TP region. A study conducted by Nayak et al. [87] performed sensitive experiments in India using the Noah Land surface model and found that the annual surface layer soil moisture change is the highest when introducing perturbations in longwave radiation. Our study is inconsistent with the study of Nayak et al. regarding the lowest/highest contribution of downward longwave radiation to soil moisture changes. One possible reason for the inconsistency is that the two studies focus on different timescales [88]; the spatial heterogeneity of downward longwave radiation may be another possible explanation [89]. In the near future, sensitive experiments with regional models should be carried out over the specific TP region to investigate the soil moisture change with perturbations in different climatic factors.

Although the trained Random Forest (RF) model has successfully predicted soil moisture for all products, the estimating performance of R^2 is not high enough. This may be related to uncertainties from the selected input meteorological forcing data or the used machine learning method. Another state-of-the-art Extreme Gradient Boosting (XGBoost), which has also been widely used in related ecohydrological studies [90,91], may be an alternative method to improve the performance.

6. Conclusions

In this study, we firstly evaluated the performance of multi-source soil moisture products over the TP region with the two sets of in situ observations; then, we studied the long-term characteristics of soil moisture; lastly, the responses of soil moisture over the TP region to climatic variables were estimated with a machine learning method. The conclusions are as follows:

(a) SSM/I and GLDAS Noah soil moisture data outperform ECV COMBINED, ERA5-Land and MERRA2, which were followed by ECV COMBINED and MERRA2. ERA5-Land might not be proper for evaluating surface soil moisture over the TP, as it has the greatest difference from in situ observations, which may be due to uncertainties from the representation of snow fields at high altitudes (> 3300 m a.s.l.). The reanalysis products tend to overestimate soil moisture compared to satellite-based products.

(b) All products show a similar spatial distribution of annual mean soil moisture—decreasing from the southeast to the northwest—over the TP. For seasonal mean soil moisture, the soil moisture of all products except for ERA5-Land firstly increases from spring to summer and then decreases from autumn to winter. ERA5-Land almost has similar spatial distributions among four different seasons. A significant increase in soil moisture is found over most of the TP region during 1988–2008, especially in middle-west regions, on annual and seasonal scales for all products.

(c) With the Random Forest nonlinear regression method, we quantify the contribution of each climatic variable to warm-season soil moisture over the TP region. The attribution results indicate that precipitation controls soil moisture changes over the TP region rather than air temperature. Further, pixel-wise partial correlations between precipitation and soil moisture show that increasing precipitation would increase soil moisture over most of the TP region.

Author Contributions: This paper is a collaborative work by all of the authors. Conceptualization, C.Z., S.L. and D.F.T.H.; data curation, X.W. and D.F.; methodology, C.Z. and S.L.; formal analysis, C.Z. and S.L.; visualization, C.Z. and S.L.; writing—original draft preparation, C.Z. and S.L.; writing—review and editing, D.F.T.H., J.L. and W.U.; supervision, G.W.; funding acquisition, G.W. All authors have read and agreed to the published version of the manuscript.

Funding: This research is financially supported by the "Sino-German Cooperation Group" project in the Sino-German Science Center of the National Natural Science Foundation of China, grant number GZ1447.

Data Availability Statement: The CAMP/Tibet data can be available from https://data.eol.ucar.edu/dataset/76.129 (UCAR/NCAR–Earth Observing Laboratory, et al., 2011) (accessed on 2 November 2021). The ZH2021 data can be downloaded from the National Tibetan Plateau/Third Pole Environment Data Center (https://doi.org/10.11888/Geocry.tpdc.271107) (accessed on 23 March 2023). SSM/I soil moisture can be available from the National Tibetan Plateau Data Center (https://doi.org/10.11888/Soil.tpdc.271611) (accessed on 15 August 2022). ECV COMBINED soil moisture is available at https://www.esa-soilmoisture-cci.org/data (accessed on 23 February 2023). ERA5-Land hourly and monthly data can be available at https://doi.org/10.24381/cds.e2161bac and https://doi.org/10.24381/cds.68d2bb30 (accessed on 3 February 2023). The hourly (M2T1NXLND) and monthly (M2TMNXLND) MERRA2 soil moisture can be accessible through the NASA Goddard Earth Sciences Data Information Services Center (GES DISC) at https://doi.org/10.5067/RKPHT8KC1Y1T and https://doi.org/10.5067/8S35XF81C28F (accessed on 5 February 2023). GLDAS Noah data are accessible from https://doi.org/10.5067/342OHQM9AK6Q and https://doi.org/10.5067/9SQ1B3ZXP2C5 (accessed on 5 February 2023). The China Meteorological Forcing Dataset (CMFD) is obtained from the National Tibetan Plateau Data Center (TPDC) (https://doi.org/10.11888/AtmosphericPhysics.tpe.249369.file) (accessed on 22 February 2023). The GTOPO30 elevation data can be downloaded through EarthExplorer (https://earthexplorer.usgs.gov/) (accessed on 3 January 2023). The MICLCover data are accessible from https://cstr.cn/18406.11.Socioeco.tpdc.270467 (accessed on 4 January 2023).

Acknowledgments: The authors are grateful to the anonymous reviewers and editors for their constructive comments and suggestions.

Conflicts of Interest: The authors declare no conflict of interest.

References

1. Bras, R.L. *Hydrology: An Introduction to Hydrologic Science;* Addison-Wesley: Boston, MA, USA, 1990.
2. Peng, J.; Loew, A.; Merlin, O.; Verhoest, N.E.C. A review of spatial downscaling of satellite remotely sensed soil moisture. *Rev. Geophys.* **2017**, *55*, 341–366. [CrossRef]
3. Peng, J.; Albergel, C.; Balenzano, A.; Brocca, L.; Cartus, O.; Cosh, M.H.; Crow, W.T.; Dabrowska-Zielinska, K.; Dadson, S.; Davidson, M.W. A roadmap for high-resolution satellite soil moisture applications–confronting product characteristics with user requirements. *Remote Sens. Environ.* **2021**, *252*, 112162. [CrossRef]
4. Koster, R.D.; Dirmeyer, P.A.; Guo, Z.; Bonan, G.; Chan, E.; Cox, P.; Gordon, C.; Kanae, S.; Kowalczyk, E.; Lawrence, D. Regions of strong coupling between soil moisture and precipitation. *Science* **2004**, *305*, 1138–1140. [CrossRef] [PubMed]
5. Miralles, D.G.; Teuling, A.J.; Van Heerwaarden, C.C.; Vilà-Guerau de Arellano, J. Mega-heatwave temperatures due to combined soil desiccation and atmospheric heat accumulation. *Nat. Geosci.* **2014**, *7*, 345–349. [CrossRef]
6. Dirmeyer, P.A.; Cash, B.A.; Kinter, J.L., III; Stan, C.; Jung, T.; Marx, L.; Towers, P.; Wedi, N.; Adams, J.M.; Altshuler, E.L. Evidence for enhanced land–atmosphere feedback in a warming climate. *J. Hydrometeorol.* **2012**, *13*, 981–995. [CrossRef]
7. Wei, J.; Dirmeyer, P.A. Dissecting soil moisture-precipitation coupling. *Geophys. Res. Lett.* **2012**, *39*, L19711. [CrossRef]
8. Taylor, C.M.; de Jeu, R.A.; Guichard, F.; Harris, P.P.; Dorigo, W.A. Afternoon rain more likely over drier soils. *Nature* **2012**, *489*, 423–426. [CrossRef]
9. Hsu, H.; Dirmeyer, P.A. Soil moisture-evaporation coupling shifts into new gears under increasing CO_2. *Nat. Commun.* **2023**, *14*, 1162. [CrossRef]
10. Miralles, D.G.; Gentine, P.; Seneviratne, S.I.; Teuling, A.J. Land–atmospheric feedbacks during droughts and heatwaves: State of the science and current challenges. *Ann. NY Acad. Sci.* **2019**, *1436*, 19–35. [CrossRef]
11. Yao, Y.; Liu, Y.; Zhou, S.; Song, J.; Fu, B. Soil moisture determines the recovery time of ecosystems from drought. *Glob. Chang. Biol.* **2023**, *29*, 3562–3574. [CrossRef]

12. Li, S.; Wang, G.; Zhu, C.; Lu, J.; Ullah, W.; Hagan, D.F.T.; Kattel, G.; Liu, Y.; Zhang, Z.; Song, Y. Vegetation growth due to CO_2 fertilization is threatened by increasing vapor pressure deficit. *J. Hydrol.* **2023**, *619*, 129292. [CrossRef]
13. Yao, T.; Bolch, T.; Chen, D.; Gao, J.; Immerzeel, W.; Piao, S.; Su, F.; Thompson, L.; Wada, Y.; Wang, L. The imbalance of the Asian water tower. *Nat. Rev. Earth Environ.* **2022**, *3*, 618–632. [CrossRef]
14. Zhang, Q.; Shen, Z.; Pokhrel, Y.; Farinotti, D.; Singh, V.P.; Xu, C.-Y.; Wu, W.; Wang, G. Oceanic climate changes threaten the sustainability of Asia's water tower. *Nature* **2023**, *615*, 87–93. [CrossRef] [PubMed]
15. Li, X.; Long, D.; Scanlon, B.R.; Mann, M.E.; Li, X.; Tian, F.; Sun, Z.; Wang, G. Climate change threatens terrestrial water storage over the Tibetan Plateau. *Nat. Clim. Chang.* **2022**, *12*, 801–807. [CrossRef]
16. Seneviratne, S.I.; Corti, T.; Davin, E.L.; Hirschi, M.; Jaeger, E.B.; Lehner, I.; Orlowsky, B.; Teuling, A.J. Investigating soil moisture–climate interactions in a changing climate: A review. *Earth-Sci. Rev.* **2010**, *99*, 125–161. [CrossRef]
17. Ullah, W.; Guojie, W.; Gao, Z.; Tawia Hagan, D.F.; Bhatti, A.S.; Zhua, C. Observed linkage between Tibetan Plateau soil moisture and South Asian summer precipitation and the possible mechanism. *J. Clim.* **2021**, *34*, 361–377. [CrossRef]
18. Zhu, C.; Ullah, W.; Wang, G.; Lu, J.; Li, S.; Feng, A.; Hagan, D.F.T.; Jiang, T.; Su, B. Diagnosing potential impacts of Tibetan Plateau spring soil moisture anomalies on summer precipitation and floods in the Yangtze River basin. *J. Geophys. Res. Atmos.* **2023**, *128*, e2022JD037671. [CrossRef]
19. Wang, A.; Shi, X. A multilayer soil moisture dataset based on the gravimetric method in China and its characteristics. *J. Hydrometeorol.* **2019**, *20*, 1721–1736. [CrossRef]
20. Robinson, D.A.; Jones, S.B.; Wraith, J.M.; Or, D.; Friedman, S.P. A review of advances in dielectric and electrical conductivity measurement in soils using time domain reflectometry. *Vadose Zone J.* **2003**, *2*, 444–475. [CrossRef]
21. Hollinger, S.E.; Isard, S.A. A soil moisture climatology of Illinois. *J. Clim.* **1994**, *7*, 822–833. [CrossRef]
22. Valente, A.; Morais, R.; Tuli, A.; Hopmans, J.; Kluitenberg, G. Multi-functional probe for small-scale simultaneous measurements of soil thermal properties, water content, and electrical conductivity. *Sens. Actuators A Phys.* **2006**, *132*, 70–77. [CrossRef]
23. Ma, Y.; Hu, Z.; Xie, Z.; Ma, W.; Wang, B.; Chen, X.; Li, M.; Zhong, L.; Sun, F.; Gu, L. A long-term (2005–2016) dataset of hourly integrated land–atmosphere interaction observations on the Tibetan Plateau. *Earth Syst. Sci. Data* **2020**, *12*, 2937–2957. [CrossRef]
24. Dorigo, W.; Wagner, W.; Hohensinn, R.; Hahn, S.; Paulik, C.; Xaver, A.; Gruber, A.; Drusch, M.; Mecklenburg, S.; van Oevelen, P. The International Soil Moisture Network: A data hosting facility for global in situ soil moisture measurements. *Hydrol. Earth Syst. Sci.* **2011**, *15*, 1675–1698. [CrossRef]
25. Leng, P.; Li, Z.-L.; Duan, S.-B.; Gao, M.-F.; Huo, H.-Y. A practical approach for deriving all-weather soil moisture content using combined satellite and meteorological data. *ISPRS J. Photogramm. Remote Sens.* **2017**, *131*, 40–51. [CrossRef]
26. Van Der Velde, R.; Salama, M.; Pellarin, T.; Ofwono, M.; Ma, Y.; Su, Z. Long term soil moisture mapping over the Tibetan plateau using Special Sensor Microwave/Imager. *Hydrol. Earth Syst. Sci.* **2014**, *18*, 1323–1337. [CrossRef]
27. Yao, P.; Lu, H.; Shi, J.; Zhao, T.; Yang, K.; Cosh, M.H.; Gianotti, D.J.S.; Entekhabi, D. A long term global daily soil moisture dataset derived from AMSR-E and AMSR2 (2002–2019). *Sci. Data* **2021**, *8*, 143. [CrossRef] [PubMed]
28. Entekhabi, D.; Njoku, E.G.; O'neill, P.E.; Kellogg, K.H.; Crow, W.T.; Edelstein, W.N.; Entin, J.K.; Goodman, S.D.; Jackson, T.J.; Johnson, J. The soil moisture active passive (SMAP) mission. *Proc. IEEE* **2010**, *98*, 704–716. [CrossRef]
29. Kerr, Y.H.; Waldteufel, P.; Wigneron, J.-P.; Delwart, S.; Cabot, F.; Boutin, J.; Escorihuela, M.-J.; Font, J.; Reul, N.; Gruhier, C. The SMOS mission: New tool for monitoring key elements ofthe global water cycle. *Proc. IEEE* **2010**, *98*, 666–687. [CrossRef]
30. Wagner, W.; Hahn, S.; Kidd, R.; Melzer, T.; Bartalis, Z.; Hasenauer, S.; Figa, J.; De Rosnay, P.; Jann, A.; Schneider, S. The ASCAT soil moisture product: A review of its. *Meteorol. Z.* **2013**, *22*, 1–29. [CrossRef]
31. Koike, T.; Nakamura, Y.; Kaihotsu, I.; Davaa, G.; Matsuura, N.; Tamagawa, K.; Fujii, H. Development of an advanced microwave scanning radiometer (AMSR-E) algorithm for soil moisture and vegetation water content. *Proc. Hydraul. Eng.* **2004**, *48*, 217–222. [CrossRef]
32. Dorigo, W.; Wagner, W.; Albergel, C.; Albrecht, F.; Balsamo, G.; Brocca, L.; Chung, D.; Ertl, M.; Forkel, M.; Gruber, A. ESA CCI Soil Moisture for improved Earth system understanding: State-of-the art and future directions. *Remote Sens. Environ.* **2017**, *203*, 185–215. [CrossRef]
33. Denissen, J.M.; Teuling, A.J.; Reichstein, M.; Orth, R. Critical soil moisture derived from satellite observations over Europe. *J. Geophys. Res. Atmos.* **2020**, *125*, e2019JD031672. [CrossRef]
34. Bassiouni, M.; Good, S.P.; Still, C.J.; Higgins, C.W. Plant water uptake thresholds inferred from satellite soil moisture. *Geophys. Res. Lett.* **2020**, *47*, e2020GL087077. [CrossRef]
35. Ford, T.; Harris, E.; Quiring, S. Estimating root zone soil moisture using near-surface observations from SMOS. *Hydrol. Earth Syst. Sci.* **2014**, *18*, 139–154. [CrossRef]
36. Farahmand, A.; Reager, J.; Madani, N. Drought cascade in the terrestrial water cycle: Evidence from remote sensing. *Geophys. Res. Lett.* **2021**, *48*, e2021GL093482. [CrossRef]
37. Muñoz-Sabater, J.; Dutra, E.; Balsamo, G.; Boussetta, S.; Zsoter, E.; Albergel, C.; Agusti-Panareda, A. ERA5-Land: An improved version of the ERA5 reanalysis land component. In Proceedings of the Joint ISWG and LSA-SAF Workshop IPMA, Lisbon, Portugal, 26–28 June 2018; pp. 26–28.

38. Rodell, M.; Houser, P.; Jambor, U.; Gottschalck, J.; Mitchell, K.; Meng, C.-J.; Arsenault, K.; Cosgrove, B.; Radakovich, J.; Bosilovich, M. The global land data assimilation system. *Bull. Am. Meteorol. Soc.* **2004**, *85*, 381–394. [CrossRef]
39. Reichle, R.H.; Liu, Q.; Koster, R.D.; Draper, C.S.; Mahanama, S.P.; Partyka, G.S. Land surface precipitation in MERRA-2. *J. Clim.* **2017**, *30*, 1643–1664. [CrossRef]
40. Feldman, A.F.; Short Gianotti, D.J.; Dong, J.; Akbar, R.; Crow, W.T.; McColl, K.A.; Konings, A.G.; Nippert, J.B.; Tumber-Dávila, S.J.; Holbrook, N.M. Remotely sensed soil moisture can capture dynamics relevant to plant water uptake. *Water Resour. Res.* **2023**, *59*, e2022WR033814. [CrossRef]
41. Akbar, R.; Short Gianotti, D.; McColl, K.A.; Haghighi, E.; Salvucci, G.D.; Entekhabi, D. Hydrological storage length scales represented by remote sensing estimates of soil moisture and precipitation. *Water Resour. Res.* **2018**, *54*, 1476–1492. [CrossRef]
42. Wang, J.; Qi, L.; He, J.-H.; Wu, Z.-W. Relationship between spring soil moisture in the Tibetan Plateau and summer precipitation in the Yangtze river basin and its possible mechanism. *Chin. J. Geophys.* **2016**, *59*, 3985–3995.
43. Berg, A.; Sheffield, J. Historic and projected changes in coupling between soil moisture and evapotranspiration (ET) in CMIP5 models confounded by the role of different ET components. *J. Geophys. Res. Atmos.* **2019**, *124*, 5791–5806. [CrossRef]
44. Bauer-Marschallinger, B.; Freeman, V.; Cao, S.; Paulik, C.; Schaufler, S.; Stachl, T.; Modanesi, S.; Massari, C.; Ciabatta, L.; Brocca, L. Toward global soil moisture monitoring with Sentinel-1: Harnessing assets and overcoming obstacles. *IEEE Trans. Geosci. Remote Sens.* **2018**, *57*, 520–539. [CrossRef]
45. Fan, L.; Xing, Z.; De Lannoy, G.; Frappart, F.; Peng, J.; Zeng, J.; Li, X.; Yang, K.; Zhao, T.; Shi, J. Evaluation of satellite and reanalysis estimates of surface and root-zone soil moisture in croplands of Jiangsu Province, China. *Remote Sens. Environ.* **2022**, *282*, 113283. [CrossRef]
46. Gelaro, R.; McCarty, W.; Suárez, M.J.; Todling, R.; Molod, A.; Takacs, L.; Randles, C.A.; Darmenov, A.; Bosilovich, M.G.; Reichle, R. The modern-era retrospective analysis for research and applications, version 2 (MERRA-2). *J. Clim.* **2017**, *30*, 5419–5454. [CrossRef] [PubMed]
47. Bi, H.; Ma, J.; Zheng, W.; Zeng, J. Comparison of soil moisture in GLDAS model simulations and in situ observations over the Tibetan Plateau. *J. Geophys. Res. Atmos.* **2016**, *121*, 2658–2678. [CrossRef]
48. Zhao, L.; Zou, D.; Hu, G.; Wu, T.; Du, E.; Liu, G.; Xiao, Y.; Li, R.; Pang, Q.; Qiao, Y. A synthesis dataset of permafrost thermal state for the Qinghai–Tibet (Xizang) Plateau, China. *Earth Syst. Sci. Data* **2021**, *13*, 4207–4218. [CrossRef]
49. Zhang, P.; Zheng, D.; van der Velde, R.; Wen, J.; Zeng, Y.; Wang, X.; Wang, Z.; Chen, J.; Su, Z. Status of the Tibetan Plateau observatory (Tibet-Obs) and a 10-year (2009–2019) surface soil moisture dataset. *Earth Syst. Sci. Data* **2021**, *13*, 3075–3102. [CrossRef]
50. Yang, K.; Qin, J.; Zhao, L.; Chen, Y.; Tang, W.; Han, M.; Lazhu; Chen, Z.; Lv, N.; Ding, B. A multiscale soil moisture and freeze–thaw monitoring network on the third pole. *Bull. Am. Meteorol. Soc.* **2013**, *94*, 1907–1916.
51. Su, Z.; Wen, J.; Dente, L.; Van Der Velde, R.; Wang, L.; Ma, Y.; Yang, K.; Hu, Z. The Tibetan Plateau observatory of plateau scale soil moisture and soil temperature (Tibet-Obs) for quantifying uncertainties in coarse resolution satellite and model products. *Hydrol. Earth Syst. Sci.* **2011**, *15*, 2303–2316. [CrossRef]
52. Liu, J.; Chai, L.; Lu, Z.; Qu, Y.; Wang, J.; Yang, S. Validation of Five Passive Microwave Remotely Sensed Soil Moisture Products over the Qinghai-Tibet Plateau, China. In Proceedings of the IGARSS 2019—2019 IEEE International Geoscience and Remote Sensing Symposium, Yokohama, Japan, 28 July–2 August 2019; pp. 6182–6185.
53. Wang, G.; Hagan, D.F.T.; Lou, D.; Chen, T. Evaluation of soil moisture derived from FY3B microwave brightness temperature over the Tibetan Plateau. *Remote Sens. Lett.* **2016**, *7*, 817–826. [CrossRef]
54. Dente, L.; Su, Z.; Wen, J. Validation of SMOS soil moisture products over the Maqu and Twente regions. *Sensors* **2012**, *12*, 9965–9986. [CrossRef]
55. Xing, Z.; Fan, L.; Zhao, L.; De Lannoy, G.; Frappart, F.; Peng, J.; Li, X.; Zeng, J.; Al-Yaari, A.; Yang, K. A first assessment of satellite and reanalysis estimates of surface and root-zone soil moisture over the permafrost region of Qinghai-Tibet Plateau. *Remote Sens. Environ.* **2021**, *265*, 112666. [CrossRef]
56. Wang, L.; Liu, H.; Grießinger, J.; Chen, D.; Sun, C.; Fang, C. Enhanced Variability and Declining Trend of Soil Moisture Since the 1880s on the Southeastern Tibetan Plateau. *Water Resour. Res.* **2023**, *59*, e2022WR033953. [CrossRef]
57. Meng, X.; Deng, M.; Talib, J.; Taylor, C.; Wu, P.; Lyu, S.; Chen, H.; Li, Z.; Zhao, L. Diagnosing product variability in the soil moisture response to precipitation on the Tibetan Plateau. *J. Hydrometeorol.* **2023**, *24*, 625–639. [CrossRef]
58. Chen, W.; Yao, T.; Zhang, G.; Woolway, R.I.; Yang, W.; Xu, F.; Zhou, T. Glacier surface heatwaves over the Tibetan Plateau. *Geophys. Res. Lett.* **2023**, *50*, e2022GL101115. [CrossRef]
59. Wu, P.; Liu, Y.-J.; Wang, J.; Ding, Y.-H. Revisiting the variations of precipitation and water vapour budget over the Tibetan Plateau. *Adv. Clim. Chang. Res.* **2023**, *14*, 77–84. [CrossRef]
60. Wang, L.; Lu, J.; Zhou, R.; Duan, G.; Wen, Z. Analysis of Soil Moisture Change Characteristics and Influencing Factors of Grassland on the Tibetan Plateau. *Remote Sens.* **2023**, *15*, 298. [CrossRef]
61. Wang, H.; Zan, B.; Wei, J.; Song, Y.; Mao, Q. Spatiotemporal characteristics of soil Moisture and land–atmosphere coupling over the Tibetan Plateau derived from three gridded datasets. *Remote Sens.* **2022**, *14*, 5819. [CrossRef]

62. Tawia Hagan, D.F.; Wang, G.; San Liang, X.; Dolman, H.A. A time-varying causality formalism based on the Liang–Kleeman information flow for analyzing directed interactions in nonstationary climate systems. *J. Clim.* **2019**, *32*, 7521–7537. [CrossRef]
63. Li, W.; Migliavacca, M.; Forkel, M.; Denissen, J.M.; Reichstein, M.; Yang, H.; Duveiller, G.; Weber, U.; Orth, R. Widespread increasing vegetation sensitivity to soil moisture. *Nat. Commun.* **2022**, *13*, 3959. [CrossRef]
64. Sun, H.; Liu, X. Impacts of dynamic and thermal forcing by the Tibetan Plateau on the precipitation distribution in the Asian arid and monsoon regions. *Clim. Dyn.* **2021**, *56*, 2339–2358. [CrossRef]
65. Qu, Y.; Zhu, Z.; Chai, L.; Liu, S.; Montzka, C.; Liu, J.; Yang, X.; Lu, Z.; Jin, R.; Li, X. Rebuilding a microwave soil moisture product using random forest adopting AMSR-E/AMSR2 brightness temperature and SMAP over the Qinghai–Tibet Plateau, China. *Remote Sens.* **2019**, *11*, 683. [CrossRef]
66. Cui, A.; Lu, H.; Liu, X.; Shen, C.; Xu, D.; Xu, B.; Wu, N. Tibetan plateau precipitation modulated by the periodically coupled westerlies and Asian monsoon. *Geophys. Res. Lett.* **2021**, *48*, e2020GL091543. [CrossRef]
67. Yang, S.; Zeng, J.; Fan, W.; Cui, Y. Evaluating Root-Zone Soil Moisture Products from GLEAM, GLDAS, and ERA5 Based on In Situ Observations and Triple Collocation Method over the Tibetan Plateau. *J. Hydrometeorol.* **2022**, *23*, 1861–1878. [CrossRef]
68. Ran, Y.; Li, X.; Lu, L.; Li, Z. Large-scale land cover mapping with the integration of multi-source information based on the Dempster–Shafer theory. *Int. J. Geogr. Inf. Sci.* **2012**, *26*, 169–191. [CrossRef]
69. Ma, Y.; Su, Z.; Koike, T.; Yao, T.; Ishikawa, H.; Ueno, K.i.; Menenti, M. On measuring and remote sensing surface energy partitioning over the Tibetan Plateau—from GAME/Tibet to CAMP/Tibet. *Phys. Chem. Earth Parts A/B/C* **2003**, *28*, 63–74. [CrossRef]
70. Jiang, H.; Yi, Y.; Zhang, W.; Yang, K.; Chen, D. Sensitivity of soil freeze/thaw dynamics to environmental conditions at different spatial scales in the central Tibetan Plateau. *Sci. Total Environ.* **2020**, *734*, 139261. [CrossRef] [PubMed]
71. Ge, N.; Zhong, L.; Ma, Y.; Cheng, M.; Wang, X.; Zou, M.; Huang, Z. Estimation of land surface heat fluxes based on Landsat 7 ETM+ data and field measurements over the northern Tibetan Plateau. *Remote Sens.* **2019**, *11*, 2899. [CrossRef]
72. Wang, G.; Garcia, D.; Liu, Y.; De Jeu, R.; Dolman, A.J. A three-dimensional gap filling method for large geophysical datasets: Application to global satellite soil moisture observations. *Environ. Model. Softw.* **2012**, *30*, 139–142. [CrossRef]
73. Gruber, A.; Scanlon, T.; van der Schalie, R.; Wagner, W.; Dorigo, W. Evolution of the ESA CCI Soil Moisture climate data records and their underlying merging methodology. *Earth Syst. Sci. Data* **2019**, *11*, 717–739. [CrossRef]
74. Muñoz-Sabater, J.; Dutra, E.; Agustí-Panareda, A.; Albergel, C.; Arduini, G.; Balsamo, G.; Boussetta, S.; Choulga, M.; Harrigan, S.; Hersbach, H. ERA5-Land: A state-of-the-art global reanalysis dataset for land applications. *Earth Syst. Sci. Data* **2021**, *13*, 4349–4383. [CrossRef]
75. Hagan, D.F.T.; Parinussa, R.M.; Wang, G.; Draper, C.S. An evaluation of soil moisture anomalies from global model-based datasets over the people's republic of China. *Water* **2019**, *12*, 117. [CrossRef]
76. Zhu, Z.; Shi, C. Simulation and evaluation of CLDAS and GLDAS soil moisture data in China. *Sci. Technol. Eng.* **2014**, *14*, 138–144.
77. Deng, M.; Meng, X.; Li, Z.; Lyv, Y.; Lei, H.; Zhao, L.; Zhao, S.; Ge, J.; Jing, H. Responses of soil moisture to regional climate change over the three rivers source region on the tibetan plateau. *Int. J. Climatol.* **2020**, *40*, 2403–2417. [CrossRef]
78. Mingshan, D.; Xianhong, M.; Yingsai, M.; Yingying, A. Analysis on soil moisture characteristics of Tibetan Plateau based on GLDAS. *J. Arid Meteorol.* **2018**, *36*, 595.
79. He, J.; Yang, K.; Tang, W.; Lu, H.; Qin, J.; Chen, Y.; Li, X. The first high-resolution meteorological forcing dataset for land process studies over China. *Sci. Data* **2020**, *7*, 25. [CrossRef] [PubMed]
80. Wilcox, R.R. *Fundamentals of Modern Statistical Methods: Substantially Improving Power and Accuracy*; Springer: Berlin/Heidelberg, Germany, 2010; Volume 249.
81. Lu, J.; Wang, G.; Li, S.; Feng, A.; Zhan, M.; Jiang, T.; Su, B.; Wang, Y. Projected land evaporation and its response to vegetation greening over China under multiple scenarios in the CMIP6 models. *J. Geophys. Res. Biogeosci.* **2021**, *126*, e2021JG006327. [CrossRef]
82. Mann, H.B. Nonparametric tests against trend. *Econom. J. Econom. Soc.* **1945**, *13*, 245–259. [CrossRef]
83. Kendall, M.G. *Rank Correlation Methods*; Griffin: London, UK, 1975.
84. Rial, M.; Cortizas, A.M.; Rodríguez-Lado, L. Understanding the spatial distribution of factors controlling topsoil organic carbon content in European soils. *Sci. Total Environ.* **2017**, *609*, 1411–1422. [CrossRef]
85. Chen, Y.; Yang, K.; Qin, J.; Zhao, L.; Tang, W.; Han, M. Evaluation of AMSR-E retrievals and GLDAS simulations against observations of a soil moisture network on the central Tibetan Plateau. *J. Geophys. Res. Atmos.* **2013**, *118*, 4466–4475. [CrossRef]
86. Dai, L.; Fu, R.; Guo, X.; Du, Y.; Zhang, F.; Cao, G. Soil moisture variations in response to precipitation across different vegetation types on the northeastern Qinghai-Tibet plateau. *Front. Plant Sci.* **2022**, *13*, 854152. [CrossRef]
87. Nayak, H.P.; Nayak, S.; Maity, S.; Patra, N.; Singh, K.S.; Dutta, S. Sensitivity of Land Surface Processes and Its Variation during Contrasting Seasons over India. *Atmosphere* **2022**, *13*, 1382. [CrossRef]
88. Zhang, T.; Shen, S.; Cheng, C. Impact of radiations on the long-range correlation of soil moisture: A case study of the A'rou superstation in the Heihe River Basin. *J. Geogr. Sci.* **2019**, *29*, 1491–1506. [CrossRef]
89. Yang, K.; He, J.; Tang, W.; Qin, J.; Cheng, C.C. On downward shortwave and longwave radiations over high altitude regions: Observation and modeling in the Tibetan Plateau. *Agric. For. Meteorol.* **2010**, *150*, 38–46. [CrossRef]

90. Bartold, M.; Kluczek, M. A Machine Learning Approach for Mapping Chlorophyll Fluorescence at Inland Wetlands. *Remote Sens.* **2023**, *15*, 2392. [CrossRef]
91. Karthikeyan, L.; Mishra, A.K. Multi-layer high-resolution soil moisture estimation using machine learning over the United States. *Remote Sens. Environ.* **2021**, *266*, 112706. [CrossRef]

remote sensing

Article

Implications for Validation of IMERG Satellite Precipitation in a Complex Mountainous Region

Luhan Li [1,2], Xuelong Chen [2,*], Yaoming Ma [2,1,3,4,5,6], Wenqing Zhao [2], Hongchao Zuo [1], Yajing Liu [2], Dianbin Cao [7] and Xin Xu [2]

1 College of Atmospheric Science, Lanzhou University, Lanzhou 730000, China; lilh18@lzu.edu.cn (L.L.); zuohch@lzu.edu.cn (H.Z.)
2 Land-Atmosphere Interaction and Its Climatic Effects Group, State Key Laboratory of Tibetan Plateau Earth System, Environment and Resources (TPESER), Institute of Tibetan Plateau Research, Chinese Academy of Sciences, Beijing 100101, China; ymma@itpcas.ac.cn (Y.M.); zhaowq@itpcas.ac.cn (W.Z.); liuyajing@itpcas.ac.cn (Y.L.); xuxin@itpcas.ac.cn (X.X.)
3 College of Earth and Planetary Sciences, University of Chinese Academy of Sciences, Beijing 100049, China
4 National Observation and Research Station for Qomolongma Special Atmospheric Processes and Environmental Changes, Dingri 858200, China
5 Kathmandu Center of Research and Education, Chinese Academy of Sciences, Beijing 100101, China
6 China-Pakistan Joint Research Center on Earth Sciences, Chinese Academy of Sciences, Islamabad 45320, Pakistan
7 Alpine Paleoecology and Human Adaptation Group, State Key Laboratory of Tibetan Plateau Earth System, Environment and Resources (TPESER), Institute of Tibetan Plateau Research, Chinese Academy of Sciences, Beijing 100101, China; dbcao@itpcas.ac.cn
* Correspondence: x.chen@itpcas.ac.cn

Abstract: Satellite-based precipitation retrievals such as the Integrated Multi-satellite Retrievals for Global Precipitation Measurement (IMERG), provide alternative data in mountainous regions. In this study, we evaluated IMERG in the Yarlung Tsangbo Grand Canyon (YGC) using ground observations. It was found that IMERG underestimated the total rainfall primarily due to under-detection of rainfall events, with misses being more prevalent than false alarms. We analyzed the relationships between the probability of detection (POD), false alarm ratio (FAR), bias in detection (BID), and Heidke skill score (HSS) and terrain factors. It was found that the POD decreased with elevation, leading to increased underestimation of rainfall events at higher elevations, and the FAR was higher in valley sites. In terms of the hit events, IMERG overestimated the light rainfall events and underestimated the heavy rainfall events and the negative bias in the hit events decreased with elevation. IMERG could capture the early morning peak precipitation in the YGC region but underestimated the amplitude of the diurnal variation. This bias was inherent at the sensor level, and the Global Precipitation Climatology Center (GPCC) calibration partially improved the underestimation. However, this improvement was not sufficient for the YGC region. This study fills the gap in IMERG validation in a complex mountainous region and has implications for users and developers.

Keywords: GPM; Yarlung Tsangbo Grand Canyon; underestimation; validation; diurnal variation; rainfall events under-detection

Citation: Li, L.; Chen, X.; Ma, Y.; Zhao, W.; Zuo, H.; Liu, Y.; Cao, D.; Xu, X. Implications for Validation of IMERG Satellite Precipitation in a Complex Mountainous Region. *Remote Sens.* **2023**, *15*, 4380. https://doi.org/10.3390/rs15184380

Academic Editor: Seon Ki Park

Received: 10 July 2023
Revised: 25 August 2023
Accepted: 4 September 2023
Published: 6 September 2023

1. Introduction

The Tibetan Plateau (TP), known as the Third Pole [1], is the world's highest plateau. The Asian monsoon passes over the southeastern Tibetan Plateau (SETP) and has an important impact on the climate and environment of the plateau. The Yarlung Tsangbo Grand Canyon (YGC) in the SETP is crucial to the TP region's water cycle and is recognized as one of the largest water vapor canyons [2]. The YGC plays a critical role in transporting water vapor from the South Asian monsoon to the plateau [2]. Thus, examining the water and energy circulation processes in the YGC region is of great significance due to its strong influence on the TP climate system.

One of the key factors affecting the water and energy cycles is precipitation, for which conventional observations are primarily dependent on rain gauges and ground-based radars. However, rain gauges face challenges in accurately estimating precipitation over large areas with complex terrain. Moreover, in the mountainous regions, the available radars must overcome issues such as beam blockage, anomalous propagation errors, and imprecise backscatter to rain rate relationships [3]. The scarcity of long-term, well-distributed precipitation observations over the complex terrain of the TP highlights the necessity for alternative data sources. Satellite-based precipitation retrievals can fill these data gaps [4].

Currently, global satellite-based rainfall products utilize a combination of microwave, infrared (IR), and integrated microwave and IR observations from multiple satellite missions using a diverse range of merging methods [3]. Early space-based precipitation retrieval efforts concentrated on estimating rainfall through infrared measurements of cloud tops from geosynchronous satellites. Nonetheless, the accuracy was limited due to the indirect association between surface rain rates and cloud-top temperatures [5]. Although infrared sensors on geostationary satellites can provide high temporal resolution precipitation estimates, microwave sensors are preferred for measuring precipitation because their radiative signatures are more closely connected to the precipitating particles. Currently, the most widely used satellite precipitation products (SPPs) include the Tropical Rainfall Measurement Mission (TRMM), Precipitation Estimation from Remotely Sensed Information using Artificial Neural Networks (PERSIANN) [6], Climate Prediction Center morphing technology (CMORPH) [7], Global Satellite Mapping of Precipitation (GSMaP) [8], Climate Hazards Group InfraRed Precipitation with Station data (CHIRPS) [9], Multi-Source Weighted-Ensemble Precipitation (MSWEP) [10], and Global Precipitation Measurement (GPM) Mission [11].

The GPM Mission, as the successor to the TRMM, aims to deliver uniform global precipitation products using a diverse array of microwave sensors within a consistent framework. The GPM Level 3 product, the Integrated Multi-satellite Retrievals for GPM (IMERG), combines intermittent precipitation estimates from all constellation microwave sensors with the more frequent but less accurate IR-based observations from geosynchronous satellites and monthly surface precipitation gauge data. This process creates a consistently calibrated, uniformly gridded global precipitation dataset, as well as relevant error and metadata information [3]. Compared to other datasets, IMERG has superior temporal (0.5 h) and spatial ($0.1^{\circ} \times 0.1^{\circ}$) resolutions, providing a more refined data source for studying climate change and enhancing weather forecasts on the TP. Consequently, the performance of IMERG in the plateau region has attracted considerable attention. Previous evaluations of IMERG over the TP have focused on whether IMERG serves as a suitable successor to the TRMM Multi-satellite Precipitation Analysis (TMPA) and comparing its performance to other precipitation products for the plateau region.

Throughout the TP, compared to the TMPA, IMERG inherits the spatial precipitation pattern and the ability to capture the northward progression of the Indian Monsoon lifecycle over the TP from its predecessor [12]. IMERG has a higher similarity to ground-based observations, enhanced detection capabilities, and reduced errors, and it mitigates the TMPA's overestimation over the TP. Simultaneously, IMERG improves the detection of light rainfall events and the potential for solid precipitation [12,13]. This is closely related to the increased radar frequency and the inclusion of the imager's high-frequency channel in the GPM platform [14]. Notably, IMERG can adequately depict diurnal precipitation characteristics [2]; however, discrepancies arise when describing the timing of the maximum intensity, and overestimation of the maximum precipitation rate occurs [13]. Compared to other satellite products, IMERG's performance is more prominent, which is mainly reflected in its ability to detect precipitation and non-precipitation events [15], precipitation data correction ability [16], and hydrological modeling potential [17,18].

Nevertheless, IMERG still has some shortcomings. First, IMERG overestimates the precipitation frequency in multiple regions of the TP, such as the eastern edge of the TP and the Sichuan Basin [17], the southeastern area during the rainy season [18], and the Yarlung

Tsangbo River region [19]. Second, IMERG's detection capabilities are insufficient at high altitudes [12,20,21]. Third, IMERG overestimates light rainfall events and underestimates moderate to heavy rainfall events [19]. This phenomenon is not only present in IMERG but is also a common issue with most satellite products [15,20,22].

Existing research has acknowledged the advantages of IMERG in the TP region while also pointing out several remaining issues. However, the current evaluation of IMERG in the TP region still has the following shortcomings.

- Spatial scale: Previous studies have mainly focused on the entire plateau or river basin scales, i.e., research on the Yarlung Tsangbo River [18,22]. Due to the limited number of ground observation stations in the downstream region of the Yarlung Tsangbo River, satellite precipitation studies have mainly concentrated on the upstream region [20]. Our study area, the YGC, is the downstream region of the Yarlung Tsangbo River and is an important region for water vapor transport on the SETP. However, due to its remote location and sparse observation sites, evaluation in this area has not yet received sufficient attention. Moreover, the microclimate characteristics and ecosystems in the YGC differ significantly from those in other basins along the Yarlung Tsangbo River, warranting further investigation. Therefore, we conducted a detailed assessment of IMERG at the grid scale specifically for the YGC region.
- Temporal scale: Existing IMERG validation studies in the TP region have mainly focused on the daily, monthly, seasonal, and annual scales, and only a few studies explored the hourly scale and diurnal variation characteristics [12,13]. However, the half-hourly temporal scale of IMERG is crucial for studying mountain areas such as the YGC, which experience high variations in rainfall frequency [2]. The rain gauge data used in this study were obtained from the Second Tibetan Plateau Scientific Expedition and Research Program (STEP) titled Investigation of the Water Vapor Channel of the Yarlung Tsangbo Grand Canyon (INVC) [23]. The ground-based precipitation observations are available at half-hourly scales, allowing our assessment to be refined to sub-hourly scales.
- Elevation and terrain factors: In the TP region, the accuracy of IMERG precipitation products varies with elevation [17]. Ma et al. [24] reported that IMERG performs better at elevations of 3000–4000 m, while Zhang et al. [13] found differences in the correlation coefficients and relative biases between IMERG and observations at elevations of around 3500 m. Some studies have argued that IMERG performs poorly at high elevations [12,20,21], while others have reported the opposite view [17]. This phenomenon is due to the varying accuracy of the same product in the southern, southeastern, central-eastern, and northeastern TP regions [15]. Moreover, the TP region has complex and heterogeneous terrain with strong local influencing factors. The relationship between IMERG and the terrain in existing studies remains unclear. However, elevation and terrain are the key factors affecting IMERG detection. Therefore, in this study, we also focused on exploring how the elevation and terrain in the YGC region affect the consistency between IMERG and ground-based observations.
- Instrument-level assessment: Tan et al. [5] pointed out that identifying the sources of satellite uncertainties during satellite evaluation can promote satellite development and improvement. Consequently, some studies have investigated the integration process of IMERG and conducted evaluations from the instrument and algorithm perspectives [5]. Currently, few assessments have been conducted from this perspective. In this study, we evaluated the diurnal and seasonal variations derived from different IMERG sensors, which will help IMERG users and developers to better understand the sources of the uncertainties in the IMERG product for the YGC region.

In this study, based on the aforementioned aspects, we utilized ground observation data for the YGC region of the SETP to evaluate IMERG's half-hourly and daily scale data. The evaluation encompasses precipitation event detection, rainfall amount estimation, and diurnal and seasonal variation reproduction capabilities. Additionally, we investigated the influences of altitude and topography factors on IMERG's detection performance and quantified the product's uncertainties from the instrument perspective. This effectively fills the gap in IMERG validation for the Yarlung Tsangbo River Basin, improves our understanding of the impact of the South Asian Monsoon on the TP's water cycle, and has significant implications for satellite precipitation users and algorithm developers.

The rest of this article is organized as follows. Section 2 describes the study area and the datasets and methods used in this study. Section 3 presents the evaluation results for IMERG against the rain gauge data at half-hourly and daily time scales, including rainfall events, rainfall amounts, different rainfall intensities, daily cycles, seasonal variation, and sensor level performance, as well as an analysis of the influence of the topographic factors on IMERG's performance. The discussion and the conclusions are presented in Sections 4 and 5.

2. Datasets and Methods

2.1. Study Area and Datasets

The Yarlung Tsangbo River flows from west to east through the southern TP, tracing a dramatic bend around Mount Namcha Barwa, the highest peak in the eastern Himalayas. After passing through Motog County and finally crossing into India, it forms one of the world's deepest canyons [2]. The YGC is approximately 496.3 km long and has a maximum depth of 5382 m [25]. Positioned between the eastern Himalayan Mountains and the Hengduan Mountains, it forms a concave ridge with two windward ridge arms shaping a funnel-like valley (Figure 1a). Its north-south orientation amplifies a channel effect. The warm and humid air from the Indian Ocean goes up the Brahmaputra River, passes through the lower valley of the Yarlung Tsangbo River, and enters into the TP [25,26]. The canyon's influence on the TP's climate predominantly manifests in precipitation patterns. A moisture-rich "wet tongue" extends into the plateau following the canyon's path, resulting in precipitation rates nearly double those of similar latitude plateau peripheries. For the beginning of the rainy season, the plateau is 1–2 months earlier than other areas at the same latitude [25]. Moreover, the channel's favorable hydrothermal conditions create the northern hemisphere's northernmost tropical mountainous environment in the valley. Here, numerous tropical species exceed their typical latitudinal or altitudinal ranges, reaching their distribution's furthest northern or highest limits. Thus, the YGC mediates and impacts the Himalayas' geography, landscape, glaciers, and geological hazards, serving as a pivotal zone for studying climate, topography, and ecosystem interactions.

The study area is situated at 94–96°E, 29–31°N (Figure 1b). Eighteen observation sites (blue dots in Figure 1b, hereafter referred to as INVC-RG) were installed in the YGC, supported by the INVC program [23]. The INVC established these observation stations at various altitudes ranging from 511 to 3328 m [27]. These stations not only cover a significant portion of the YGC but also extend the coverage of the observation area, such as Pailong and Danka, in Yigong Zangbo and Palong Zangbo Valley. This expanded data coverage provides more possibilities for analyzing and modeling dynamic precipitation variations. Furthermore, we incorporated data from four observation sites (black triangles in Figure 1b, hereafter referred to as CMA-RG) managed by the China Meteorological Administration (CMA). Detailed information about these twenty-two observation sites is presented in Table 1.

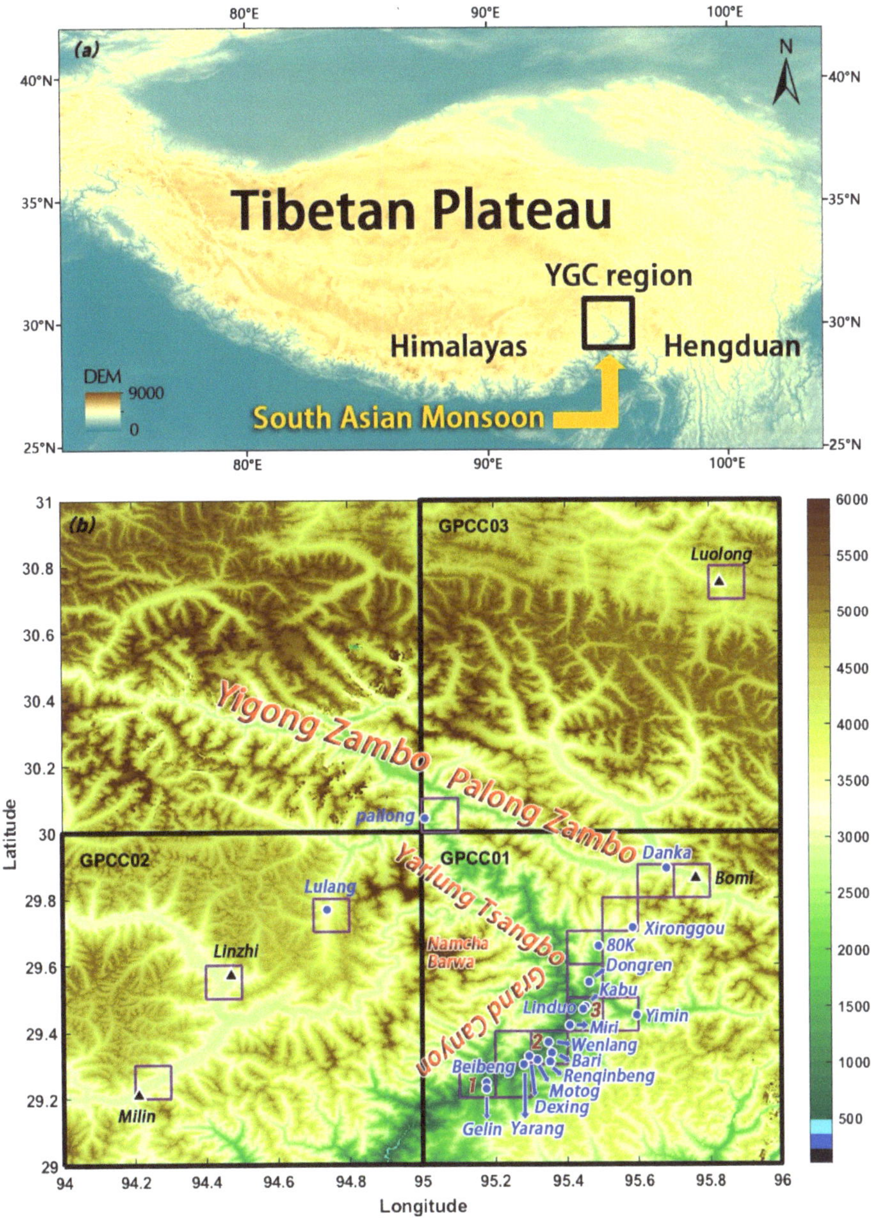

Figure 1. (**a**) Locations of the Yarlung Tsangbo Grand Canyon (YGC), the Himalayas, and the Hengduan Mountains on the Tibetan Plateau (TP) and DEM (shaded; unit: m) of the TP. (**b**) DEM of the YGC region (shaded; unit: m) and the locations of 22 rain gauge sites. The blue dots denote the rain gauge sites (from Chen et al. [2]), and black triangles represent the national meteorological stations. The small purple boxes are the IMERG $0.1^\circ \times 0.1^\circ$ grids. The IMERG grids with multiple RGs (IMERG01, IMERG02, and IMERG03) are marked with red numbers (1, 2, 3). The big black boxes denote the GPCC $1^\circ \times 1^\circ$ grids, which are distinguished by GPCC01, GPCC02, and GPCC03. Locations of Yigong Zambo, Palong Zambo, YGC, and the Mount Namcha Barwa are marked in red font.

Table 1. Rain gauge sites in the YGC region.

No.	Site	IMERG GRID	GPCC GRID	Lat.	Lon.	Elevation (m)	Relative Elevation (m)	Analysis Period
1	Beibeng	IMERG01	GPCC01	29.245	95.175	853	243	Jan. 2019–Sep. 2021
2	Gelin	IMERG01	GPCC01	29.226	95.175	1789	1179	Nov. 2020–Sep. 2021
3	Yarang		GPCC01	29.299	95.280	757	141	Jan. 2019–Sep. 2021
4	Dexing		GPCC01	29.324	95.294	737	85	Jul. 2020–Sep. 2021
5	Wenlang	IMERG02	GPCC01	29.366	95.347	851	168	Jul. 2020–Sep. 2021
6	Motog	IMERG02	GPCC01	29.313	95.317	1300	617	Jan. 2019–Oct. 2020
7	Bari	IMERG02	GPCC01	29.334	95.357	1700	1017	Oct. 2020–Sep. 2021
8	Renqinbeng	IMERG02	GPCC01	29.308	95.352	2058	1375	Aug. 2020–Sep. 2021
9	Miri	IMERG03	GPCC01	29.418	95.406	832	129	Jan. 2019–Sep. 2021
10	Linduo	IMERG03	GPCC01	29.465	95.443	840	137	Oct. 2020–Sep. 2021
11	Kabu	IMERG03	GPCC01	29.473	95.450	1425	722	Jan. 2019–Sep. 2021
12	Yimin		GPCC01	29.447	95.595	1751	424	Jul. 2020–Sep. 2021
13	Dongren		GPCC01	29.547	95.461	1185	416	Jan. 2019–Sep. 2021
14	80K		GPCC01	29.655	95.488	2109	895	Jan. 2019–Sep. 2021
15	Xironggou		GPCC01	29.710	95.585	2786	265	Jan. 2019–Aug. 2021
16	Danka		GPCC01	29.890	95.680	2700	65	Jan. 2019–Sep. 2021
17	Bomi		GPCC01	29.860	95.760	2749	110	Jan. 2018–Sep. 2021
18	Milin		GPCC02	29.210	94.210	2952	92	Jan. 2018–Sep. 2021
19	Linzhi		GPCC02	29.570	94.470	3001	102	Jan. 2018–Sep. 2021
20	Lulang		GPCC02	29.766	94.738	3328	146	Jan. 2019–Sep. 2021
21	Pailong		GPCC03	30.041	95.010	2081	197	Jan. 2019–Sep. 2021
22	Luolong		GPCC03	30.750	95.830	3640	81	Jan. 2018–Sep. 2021

Notation: The sites are numbered from south to north.

The ground-based precipitation data utilized in this study were primarily derived from rain gauges (hereafter referred to as RGs) with the following two main sources. The INVC employs tipping-bucket RGs with a minimum detection unit of 0.2 mm/tip. The data from INVC-RGs were used to analyze the half-hourly precipitation, daily precipitation, diurnal variation of the precipitation (DVP), and seasonal variation. It can be downloaded from the website: http://data.tpdc.ac.cn/zh-hans/disallow/e68f1de1-3a13-4ae1-90e0-9e3a3f57f912/, accessed on 26 February 2023. The data from CMA-RGs were used to analyze the daily precipitation and seasonal variation, which can be downloaded from http://www.cma.gov.cn, accessed on 27 January 2023.

The 22 stations utilized in this study are independent of the Global Precipitation Climatology Center (GPCC) observation network [28]. In terms of quality control, the physically incorrect values were removed, and apparent erroneous station data were eliminated through comparison with neighboring RGs. Due to the inherent deformation and calibration algorithms, IMERG inevitably generates a substantial amount of very small precipitation values, which are commonly considered to be noise [29]. The precipitation/no-precipitation thresholds for half-hourly and daily data were set to 0.2 mm/0.5 h and 0.2 mm/d. Events below these thresholds were classified as no-precipitation events.

The satellite data used in this study were obtained from the IMERG-Final product (V06 version) with a spatial resolution of $0.1^\circ \times 0.1^\circ$. The GPM core observatory serves as a calibration and evaluation tool, integrating all passive microwave (PMW) and IR-based precipitation products into IMERG. Consequently, the IMERG data originates from two sources: (1) the integration of inter-calibrated microwave precipitation estimates from the GPM microwave imager (GMI), TRMM microwave imager (TMI), and all constellation partner sensors; and (2) when low Earth orbit (LEO)-PMW data are sparse, Kalman filter-generated geosynchronous Earth orbit (GEO)-IR precipitation estimates. Additionally, gauge-based analysis was used to provide crucial regionalization and bias correction of the satellite estimates [14].

IMERG has three versions: the early multi-satellite product (IMERG-E, available approximately 4 h after observation), the late multi-satellite product (IMERG-L, available approximately 14 h after observation), and the final satellite measuring instrument product (IMERG-F, available approximately 3.5 months after observation). Lei et al. [17] compared the three versions in the eastern TP region and found that IMERG-F is in better agreement with gauge observations. For all operational versions, the complete IMERG products encompass calibration against monthly precipitation analyses. However, the distinction lies in the calibration methodology: climatological calibration is applied to the early and late products, while monthly calibration is employed for the final product. Consequently, the climatological coefficients of the final product vary according to the month and location and are recommended for general applications [14]. Due to the intricate terrain of the YGC region, acquiring foundational data poses significant challenges. In the analysis and verification of the YGC area, there is a shortage of high spatiotemporal resolution observational data. IMERG-F's monthly gauge calibration, in comparison to the IMERG-E and IMERG-L versions, is notably more localized. This aspect provides us with increased potential for the validation of subsequent meteorological simulations. Therefore, the objective of this study is to investigate the role of calibration within IMERG-F and to assess the usability of the final product. This exploration aims to furnish more sound data sources for future research. Therefore, the IMERG-F product (hereinafter referred to as IMERG) was applied in this study.

In the half-hourly analysis, the primary data used were IMERG's half-hourly precipitationCal, the multi-satellite precipitation estimate adjusted using the GPCC meteorological gauge data every month. The precipitationCal data is also the officially recommended data for general use. Moreover, in the DVP analysis, the following data were also used: (1) precipitationUncal (hereinafter referred to as Uncal), which is the multi-satellite precipitation estimate without GPCC gauge calibration; (2) HQprecipitation (hereinafter referred to as HQ), which is the merged microwave-only precipitation estimate; and (3) IRprecipitation (hereinafter referred to as IR), which is the IR precipitation estimate [14]. IMERG's one-day and monthly precipitation products were utilized for the daily and seasonal analysis. The IMERG product can be downloaded from https://disc.gsfc.nasa.gov/, accessed on 13 December 2022, and the GPCC product can be downloaded from https://opendata.dwd.de/climate_environment/GPCC/html/download_gate.html, accessed on 8 November 2022.

2.2. Performance Analysis Methods

In order to analyze the performance of the IMERG, different evaluation metrics were used, two terrain indices were identified, and the effect of terrain on the detection capability of the IMERG was studied.

The complex topography of the TP significantly influences the precipitation variations at observation stations due to elevation and terrain undulation [30–32]. In our investigation, we explored the impact of topographical factors on IMERG's detection in the YGC region by employing site elevation and relative elevation above the valley bottom. The elevations for the INVC-RGs were sourced from global positioning system measurements, while those for the CMA-RGs were defined by the CMA. The relative elevation was defined as the site elevation minus the minimum digital elevation model (DEM) value of the IMERG grid corresponding to the site's location, which indicates the relative position of the site in the IMERG grid where it is located. Lower relative elevation means the site is closer to the valley. The DEM data were extracted from the ASTER GDEM dataset, featuring a spatial resolution of 30 m (1 arc-second). The elevations and the relative elevations with sites are shown in Figure 2.

For the comparison of the satellite and ground-based precipitation observations, we employed the method by selecting the IMERG grid cell closest to the RG station and directly comparing the ground-based precipitation measurements with the corresponding IMERG grid values [33]. Notably, when an IMERG grid cell contained multiple RG stations, we refrained from spatially averaging the RG data [29].

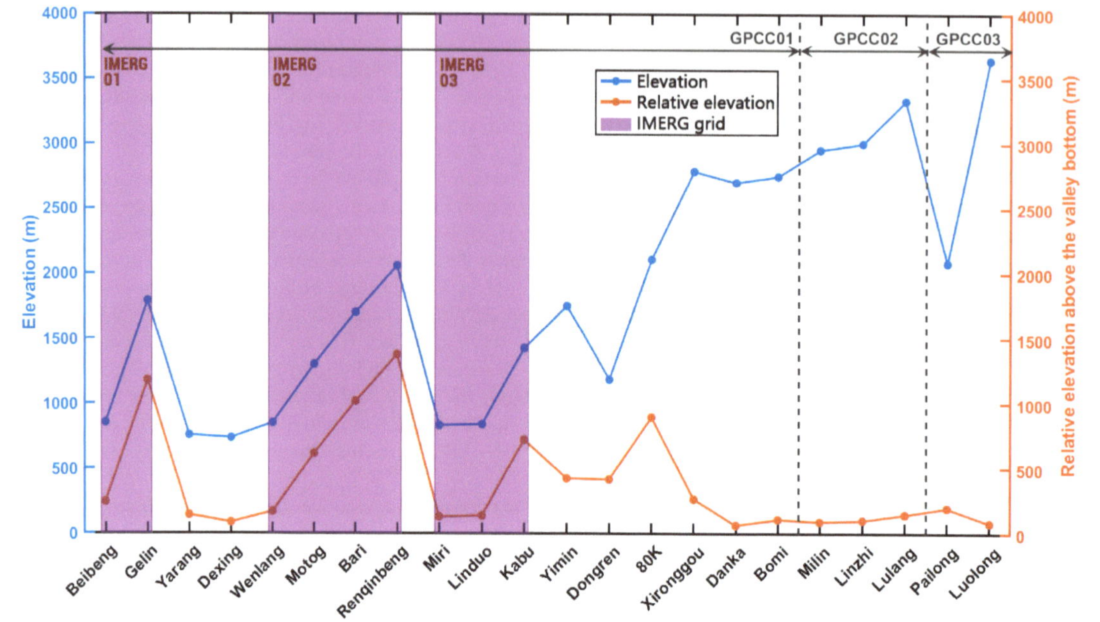

Figure 2. The elevations (unit: m) and the relative elevations (unit: m) of 22 rain gauge sites. The purple shading denotes the IMERG01, IMERG02, and IMERG03 grids, and the double arrows demonstrate the range covered by GPCC01, GPCC02, and GPCC03 grids.

In this study, we employed the metrics proposed by Tang et al. [34] and Tan et al. [5] to evaluate the IMERG products. Our approach involved calculating contingency tables that included hits (H), misses (M), false alarms (F), and correct negatives (C) and comparing RG measurements with IMERG rainfall estimates. A hit was recorded when both the RG and IMERG indicated rainfall; a miss was recorded when the RG registered rainfall but the IMERG did not; a false alarm was recorded when IMERG identified rainfall but the RG did not; and a correct negative was recorded when neither the RG nor IMERG reported rainfall. We classified an instance as "raining" when the rain rate reached at least 0.2 mm for half-hourly and daily rainfall. The formulas of the statistical metrics are listed in Table 2.

Table 2. List of the statistical metrics used for evaluating IMERG.

Statistic Metrics	Formula	Perfect Score
Probability of detection (*POD*)	$POD = \frac{H}{H+M}$	1
False alarm ratio (*FAR*)	$FAR = \frac{F}{H+F}$	0
Bias in detection (*BID*)	$BID = \frac{H+F}{H+M}$	1
Heidke skill score (*HSS*)	$HSS = \frac{H+C-E}{N-E}$	1
Expected number of instances (*E*)	$E = \frac{1}{N}[(H+M)(H+F)+(C+M)(C+F)]$	——
Normalized mean error (*NME*)	$NME = \frac{\frac{1}{n}\sum_i (y_i - x_i)}{\frac{1}{n}\sum_i x_i}$	0
Normalized mean absolute error (*NMAE*)	$NMAE = \frac{\frac{1}{n}\sum_i \lvert y_i - x_i \rvert}{\frac{1}{n}\sum_i x_i}$	0
Correlation coefficient (*CC*)	$CC = \frac{\sum_{i=1}^{n}(x_i-\overline{x})(y_i-\overline{y})}{\sqrt{\sum_{i=1}^{n}(x_i-\overline{x})^2}\sqrt{\sum_{i=1}^{n}(y_i-\overline{y})^2}}$	1

Notation: *H*, hits; *M*, misses; *F*, false alarms; *C*, correct negatives; *N*, the sample size; *E*, the expected number of instances that can be correctly identified based solely on random chance; x_i and y_i represent the rain rates for the RG and IMERG, respectively, and n signifies the number of hits. "——" indicates no perfect score.

The probability of detection (POD) measures the proportion of actual rainfall events correctly detected by the estimate, the false alarm ratio (FAR) quantifies the ratio of incorrectly estimated rainfall events, the bias in detection (BID) assesses the tendency of the IMERG estimation to overestimate (BID > 1) or underestimate (BID < 1) the number of rainfall events, and the Heidke skill score (HSS) provides a comprehensive assessment that whether the IMERG estimate performs worse (HSS < 0) or better (HSS > 0) than random chance relative to the RG. After evaluating the rainfall events, we concentrated on the hits and analyzed the rain rates within three conventional measures: the Normalized Mean Error (NME), which evaluates the degree of systematic error in the estimation method; the Normalized Mean Absolute Error (NMAE), which assesses the level of random error; and the Pearson Correlation Coefficient (CC), which quantifies the linear agreement between the estimated rainfall and the reference data.

Regarding the precipitation intensity, we adopted the percentile method based on the approaches of Sharifi et al. [35] and Yu et al. [36]. The half-hourly and daily precipitation samples from RGs were arranged in ascending order. The 50th, 70th, 90th, and 98th percentiles of the ground-based observation data were set as the thresholds for light, moderate, heavy, and extreme precipitation, respectively. The classification of the rainfall intensities was as follows: (1) half-hourly precipitation: light rainfall 0.2–0.4 mm/0.5 h, moderate rainfall 0.4–0.8 mm/0.5 h, heavy rainfall 0.8–1.6 mm/0.5 h, rainstorm 1.6–3.2 mm/0.5 h, extreme rainfall > 3.2 mm/0.5 h; (2) daily precipitation: light rainfall 0.2–3.8 mm/d, moderate rainfall 3.8–8.4 mm/d, heavy rainfall 8.4–20.8 mm/d, rainstorm 20.8–40.2 mm/d, extreme rainfall > 40.2 mm/d.

3. Results

3.1. Assessment of Precipitation Events

First, we assessed the capability of IMERG to accurately determine the occurrence of precipitation at each RG. Figure 3a presents an evaluation of the half-hourly precipitation events based on 18 INVC-RGs. The statistical results indicate that the half-hourly precipitation events at each site in the YGC region accounted for 6.5% to 22.2% of all of the available data. The hit rate of IMERG ranged from 1.4% to 5.5% for precipitation events and from 75.4% to 90.3% for non-precipitation events. The errors in IMERG's rainfall detection included misses and false alarms, accounting for 4.6–18.9% and 2.1–4.1% of the total events, respectively. For every site within the YGC region, the proportion of misses by IMERG exceeded that of the false alarms. The miss-to-false ratio was the lowest at Lulang (approximately 1.4) and highest at Xironggou (up to 9). This suggests that despite a certain degree of compensation between misses and false alarms, significant under-detection errors persist in IMERG.

Figure 3b assesses the ability of IMERG to capture daily precipitation events observed by 18 INVC-RGs and four CMA-RGs. The daily precipitation is accumulated from half-hourly data, which increases the probability of the precipitation per unit of time, thus increasing the proportion of precipitation events, decreasing the proportion of non-precipitation events, and enhancing the likelihood of IMERG detecting rainfall. By comparing Figure 3a,b, it can be seen that as the timescale increases, the proportions of misses, hits, and false alarms all increase significantly, but a predominance of misses and the proportion of hits is slightly larger than that of false alarms.

Based on the aforementioned evaluation indicators, the four skill scores of each RG were further calculated (Table 3). IMERG detects 10–35% of the half-hourly rainfall events, while on a daily scale, the rate increases to 21–34%. This improvement mirrors the overall trend of the POD, but not all of the stations exhibit this behavior. Half-hourly POD is higher than daily POD to the south of Dongren (Site No. 1–13); however to the north of 80K, daily POD is higher than half-hourly POD (Site No. 14–16, 20, 21), thereby revealing a noticeable north-south differentiation, which is consistent with the spatial correlation analysis of rainfall reported by Chen et al. [2].

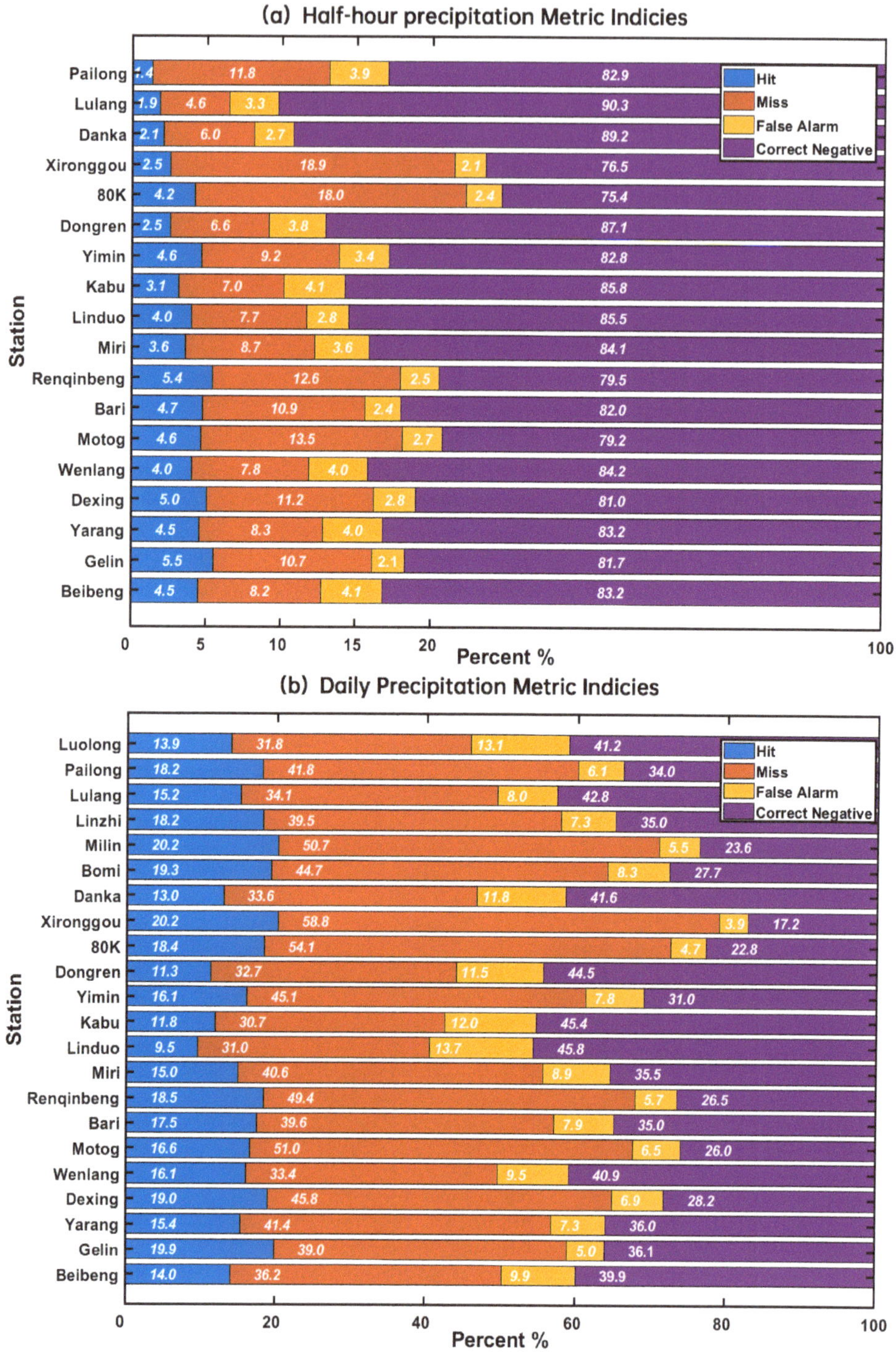

Figure 3. Hits, misses, false alarms, and correct negatives for each site on the (**a**) half-hourly and (**b**) daily temporal scales. The white numbers indicate the percentages.

Table 3. Probability of detection (POD), false alarm ratio (FAR), bias in detection (BID), and Heidke skill score (HSS) on the half-hourly and daily scales for IMERG.

No.	Site	IMERG GRID	GPCC GRID	Half-Hourly				Daily			
				POD	FAR	BID	HSS	POD	FAR	BID	HSS
1	Beibeng	IMERG01	GPCC01	0.35	0.48	0.67	0.35	0.28	0.41	0.48	0.08
2	Gelin	IMERG01	GPCC01	0.34	0.28	0.47	0.4	0.34	0.2	0.42	0.19
3	Yarang		GPCC01	0.35	0.47	0.66	0.36	0.27	0.32	0.4	0.09
4	Dexing		GPCC01	0.31	0.35	0.48	0.35	0.29	0.27	0.4	0.08
5	Wenlang	IMERG02	GPCC01	0.34	0.5	0.68	0.34	0.33	0.37	0.52	0.14
6	Motog	IMERG02	GPCC01	0.26	0.37	0.4	0.29	0.25	0.28	0.34	0.03
7	Bari	IMERG02	GPCC01	0.3	0.33	0.46	0.35	0.31	0.31	0.45	0.11
8	Renqinbeng	IMERG02	GPCC01	0.3	0.32	0.44	0.34	0.27	0.23	0.36	0.07
9	Miri	IMERG03	GPCC01	0.29	0.51	0.59	0.3	0.27	0.37	0.43	0.06
10	Linduo	IMERG03	GPCC01	0.34	0.42	0.58	0.38	0.23	0.59	0.57	0
11	Kabu	IMERG03	GPCC01	0.31	0.57	0.71	0.3	0.28	0.5	0.56	0.07
12	Yimin		GPCC01	0.34	0.42	0.58	0.36	0.26	0.33	0.39	0.05
13	Dongren		GPCC01	0.28	0.6	0.69	0.27	0.26	0.51	0.52	0.05
14	80K		GPCC01	0.19	0.36	0.29	0.21	0.25	0.2	0.32	0.05
15	Xironggou		GPCC01	0.12	0.45	0.21	0.13	0.26	0.16	0.3	0.04
16	Danka		GPCC01	0.26	0.56	0.58	0.28	0.28	0.48	0.53	0.06
17	Bomi		GPCC01	——	——	——	——	0.3	0.3	0.43	0.06
18	Milin		GPCC02	——	——	——	——	0.29	0.21	0.36	0.07
19	Linzhi		GPCC02	——	——	——	——	0.32	0.29	0.44	0.13
20	Lulang		GPCC02	0.29	0.64	0.8	0.28	0.31	0.34	0.47	0.15
21	Pailong		GPCC03	0.1	0.74	0.4	0.08	0.3	0.25	0.4	0.13
22	Luolong		GPCC03	——	——	——	——	0.3	0.49	0.59	0.06

Notation: The sites are numbered from south to north. "——" indicates no data.

As the denominator of the POD is determined by both hits and misses, a high proportion of misses within a day can still cause a decrease in the POD score of the daily rainfall, even if IMERG hits a certain number of events within the half-hourly rainfall.

For IMERG's precipitation estimates, the FAR is 28–74% for half-hourly rainfall and approximately 21–59% for daily rainfall (Table 3). Nearly all of the sites exhibit a decrease in the FAR as the temporal scale increases. It is noteworthy that when multiple RGs are located within the same IMERG grid, the satellite estimates have a relatively higher FAR in lower altitude areas. In the YGC area, the maximum yearly rainfall difference within a single grid can reach up to 536 mm [2], indicating significant spatial variability at sub-grid scales. This variability underscores the challenges posed by the low site density and representativeness in the validation of IMERG. IMERG consistently underestimates the occurrence of rainfall across the entire YGC region (BID < 1), corresponding with a high miss ratio. This is a common phenomenon in orographic precipitation [37,38]. Overall, IMERG's estimates have a lower total hit ratio for daily rainfall events, and its comprehensive ability, measured by the HSS, is less than ideal. Therefore, when using IMERG to determine the occurrence of daily rainfall events, a cautious approach should be adopted for the YGC region.

3.2. *Assessment of Rainfall Quantification*

To delve deeper into the system's performance in quantifying rainfall, we will focus on rainfall events correctly hit by IMERG and assess the rainfall amounts (Table 4). The NME and NMAE are used to represent the systematic and random errors of IMERG relative to the RG data, respectively, while the CC is used to measure their concurrence.

Table 4. Evaluation of half-hourly and accumulated daily precipitation for the hit events between the IMERG and RGs.

No.	Site	IMERG GRID	GPCC GRID	Half-Hourly			Daily		
				NME	NMAE	CC	NME	NMAE	CC
1	Beibeng	IMERG01	GPCC01	−0.31	0.71	0.2	−0.72	1	−0.23
2	Gelin	IMERG01	GPCC01	−0.28	0.79	0.1	−0.42	1.31	***−0.2***
3	Yarang		GPCC01	−0.22	0.72	0.16	−0.45	1.16	−0.18
4	Dexing		GPCC01	−0.28	0.75	0.12	−0.44	1.21	***−0.18***
5	Wenlang	IMERG02	GPCC01	−0.25	0.73	0.18	−0.23	1.36	***−0.18***
6	Motog	IMERG02	GPCC01	−0.16	0.73	0.17	−0.4	1.11	***−0.08***
7	Bari	IMERG02	GPCC01	−0.35	0.72	0.24	−0.12	1.49	***−0.24***
8	Renqinbeng	IMERG02	GPCC01	−0.34	0.73	0.15	−0.35	1.3	***−0.23***
9	Miri	IMERG03	GPCC01	−0.31	0.69	0.11	−0.27	1.23	***−0.12***
10	Linduo	IMERG03	GPCC01	−0.47	0.71	0.16	−0.8	0.9	***−0.12***
11	Kabu	IMERG03	GPCC01	−0.33	0.68	0.14	−0.4	1.19	***−0.13***
12	Yimin		GPCC01	−0.09	0.75	0.14	0.05	1.48	***−0.18***
13	Dongren		GPCC01	−0.16	0.75	0.15	−0.44	1.15	***−0.06***
14	80K		GPCC01	−0.45	0.7	0.1	−0.63	1.02	***−0.08***
15	Xironggou		GPCC01	−0.16	0.87	***0.06***	−0.55	1.05	***−0.07***
16	Danka		GPCC01	0.26	0.97	0.12	−0.23	1.1	***−0.03***
17	Bomi		GPCC01	——	——	——	−0.34	1.16	***−0.13***
18	Milin		GPCC02	——	——	——	0.11	1.36	***−0.08***
19	Linzhi		GPCC02	——	——	——	−0.44	1.02	***−0.04***
20	Lulang		GPCC02	−0.06	0.82	0.11	0.64	2	***−0.09***
21	Pailong		GPCC03	0.34	0.89	***0.03***	−0.12	1.1	***−0.09***
22	Luolong		GPCC03	——	——	——	0.28	1.47	***0.01***

Notation: The sites are numbered from south to north. NME, normalized mean error; NMAE, normalized mean absolute error; CC, correlation coefficient. The bold and italicized numbers denote the sites that do not pass the significance test. "——" indicates no data.

IMERG exhibits a systematic negative bias (NME < 0) of the half-hourly rainfall (ranging from −0.06 to −0.45), thereby underestimating the rainfall in the YGC area (Table 4). This negative bias is larger at lower altitudes and decreases with increasing altitude, which is due to the fact that sites located at lower elevations are usually at the southern end of the range, where there is more water vapor.

When the rainfall is accumulated on a daily scale (Table 4), the systematic negative bias is compounded, leading to an increase in IMERG's negative bias at most stations. In addition, the random error on the daily scale is also amplified during the accumulation, leading to larger random errors (NAME 0.9 to 2) for all of the RGs on the daily rainfall scale. This makes it difficult for the correlation between IMERG and the RGs at the daily rainfall scale to pass the significance test. Hence, the negative correlation itself is not reliable. In conclusion, despite the successful prediction of rainfall events, IMERG's rainfall estimation still presents a significant systematic negative bias and substantial random errors. Compared to the daily rainfall scale, IMERG performs better on the half-hourly scale, providing more possibilities for short-term weather studies in the YGC area.

3.3. Influence of Terrain on IMERG Detection

As a typical orographically induced precipitation region, the YGC area exhibits a spatial precipitation gradient that changes with altitude (Figure 2). We selected the elevation and the relative elevation as the terrain factors to investigate the correlations between IMERG's evaluation indicators and the terrain (Figures 4–6). Considering the similarities of IMERG products subjected to monthly-scale correction under the same GPCC grid and the potential influence of RG data from different sources, we divided the data into three categories for the correlation computation: (1) r-all, representing all INVC-RGs and

CMA-RGs points; (2) r-st18, only INVC-RGs points; and (3) r-st16, INVC-RGs points within the GPCC01 grid.

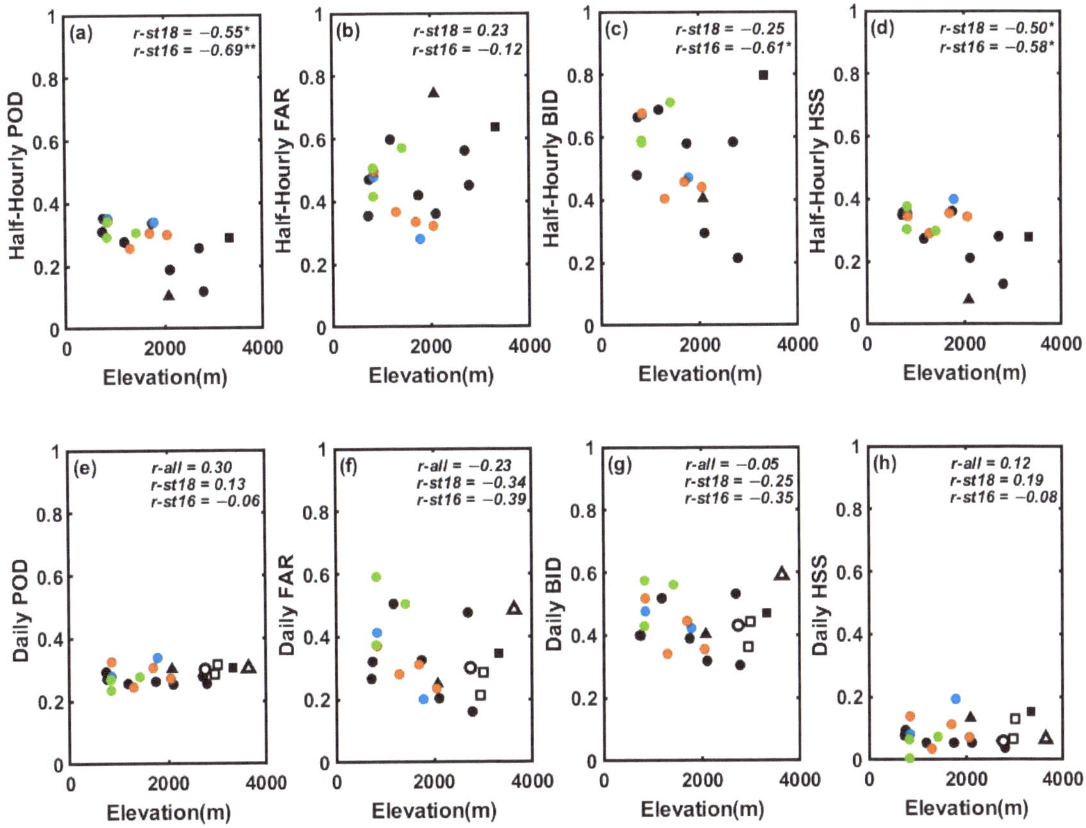

Figure 4. Scatter plots of the elevations (unit: m) of the sites versus the four scores (POD, FAR, BID, and HSS) at the (**a**–**d**) half-hourly and (**e**–**h**) daily scales. The circles represent the sites in the GPCC01 grid, the squares represent sites in the GPCC02 grid, and the triangles represent the sites in the GPCC03 grid. The hollow marks indicate the stations operated by the China Meteorological Agency (CMA), and the solid marks indicate non-CMA sites. The blue circles indicate the IMERG01 grid, the red circles indicate the IMERG02 grid, and the green circles indicate the IMERG03 grid. The correlations (r) between the four scores and the DEMs are noted in the upper right corner of each panel: r-all represents the correlations derived from all 22 sites at the daily scale, r-st18 represents the correlations derived from the 18 non-CMA sites, and r-st16 represents the correlations of the 16 non-CMA sites in the GPCC01 grid. * indicates a p-value of 0.05, and ** indicates a p-value of 0.01.

Figure 4 illustrates the relationship between IMERG's ability to capture precipitation events and the elevation. As the elevation increases, the ability of IMERG to detect precipitation events (POD) decreases, the underestimation of the precipitation events (BID) increases, and hence, the accuracy and reliability (HSS) of the satellite precipitation estimation decrease. The impact of elevation primarily occurs on the half-hourly scale, and as the temporal step accumulates, this correlation becomes less noticeable. The relationship between the FAR and elevation is somewhat unclear, which may be associated with the limited number of stations. Future work may investigate this issue by increasing the number of high-altitude sites.

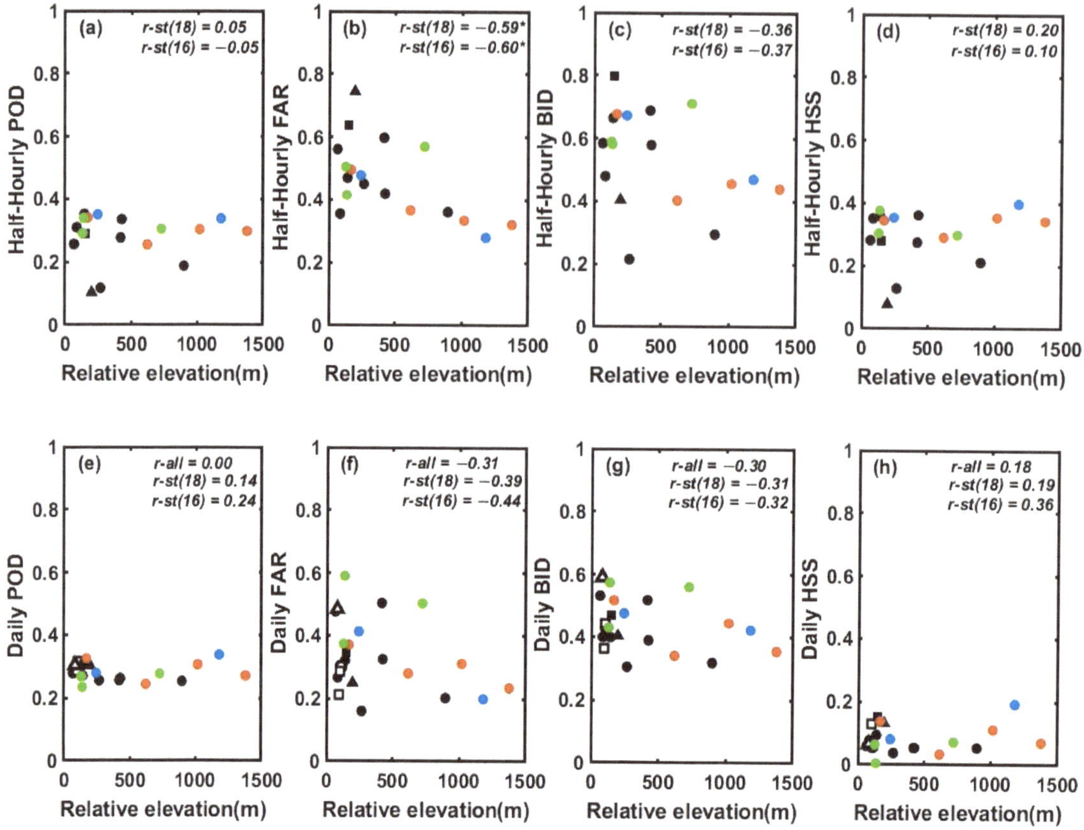

Figure 5. Scatter plots of the relative elevations (unit: m) of the sites versus the four scores (POD, FAR, BID, and HSS) at the (**a–d**) half-hourly and (**e–h**) daily scales. The circles represent the sites in the GPCC01 grid, the squares represent sites in the GPCC02 grid, and the triangles represent the sites in the GPCC03 grid. The hollow marks indicate the stations operated by the China Meteorological Agency (CMA), and the solid marks indicate non-CMA sites. The blue circles indicate the IMERG01 grid, the red circles indicate the IMERG02 grid, and the green circles indicate the IMERG03 grid. The correlations (r) between the four scores and the DEMs are noted in the upper right corner of each panel: r-all represents the correlations derived from all 22 sites at the daily scale, r-st18 represents the correlations derived from the 18 non-CMA sites, and r-st16 represents the correlations of the 16 non-CMA sites in the GPCC01 grid. * indicates a p-value of 0.05.

As shown in Figure 5b, we found a significant negative correlation (−60%) between the FAR and relative elevation in the case of half-hourly precipitation events, implying that IMERG has a higher FAR in the valley sites. By dividing the half-hourly data into day and night, we could see the difference in the correlation between day and night (Figure A1). The negative correlation between FAR and relative elevation is more pronounced during the daytime. It might be due to greater mountain–valley wind speeds amplifying the effect of topography on FAR during the day.

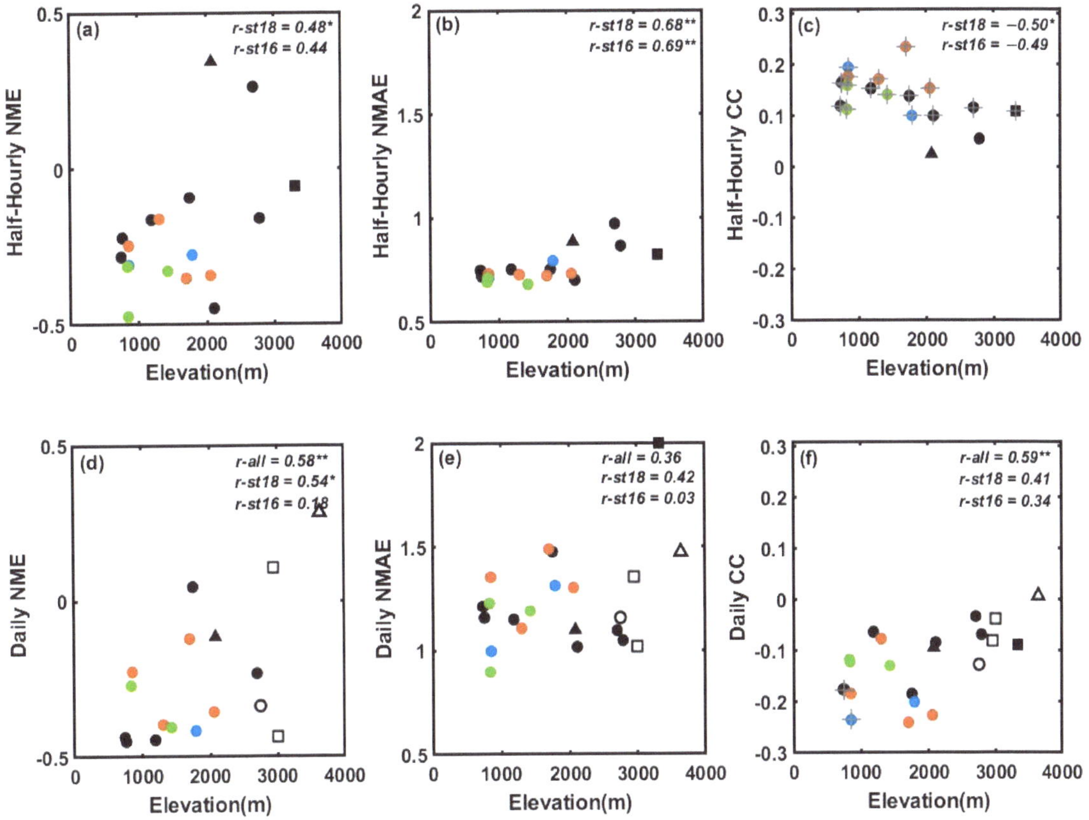

Figure 6. Scatter plots of the elevations (unit: m) of the sites versus the normalized mean error (NME), normalized mean absolute error (NMAE), and correlation coefficient (CC) at the (**a–c**) half-hourly and (**d–f**) daily scales. The circles represent the sites in the GPCC01 grid, the squares represent sites in the GPCC02 grid, and the triangles represent the sites in the GPCC03 grid. The hollow marks indicate the stations operated by the China Meteorological Agency (CMA), and the solid marks indicate non-CMA sites. The blue circles indicate the IMERG01 grid, the red circles indicate the IMERG02 grid, and the green circles indicate the IMERG03 grid. The correlations (r) between the four scores and the DEMs are noted in the upper right corner of each panel: r-all represents the correlations derived from all 22 sites at the daily scale, r-st18 rep-resents the correlations derived from the 18 non-CMA sites, and r-st16 represents the correlations of the 16 non-CMA sites in the GPCC01 grid. * indicates a p-value of 0.05, and ** indicates a p-value of 0.01. The grey + symbol on the circles in (**c**,**f**) indicates that the site reaches a p-value of 0.05.

Next, we focused on hit events and analyzed the NME, NMAE, and CC with respect to the site's elevation and relative elevation. While elevation shows some correlations with these metrics, the relative elevation demonstrates no more significant correlations, so we only show the relationship between the metrics and elevation in Figure 6. These results indicate that the influence of the site's position on whether IMERG identifies precipitation events as occurring outweighs its influence on the estimated rainfall volume. As can be seen from Figure 6a,d, even if IMERG correctly captures the occurrence of precipitation events, there is still a systematic negative bias at lower altitude sites in terms of rainfall estimation. As the altitude increases, this negative bias gradually decreases. Within the 2000–4000 m altitude range, IMERG's systematic bias is close to 0 at some stations while exhibiting a significant positive bias for others. The half-hourly NMAE exhibits a strong positive correlation with the elevation (up to 68%) (Figure 6b), implying that IMERG has a larger random error in high-altitude areas. A larger random bias occurred in daily precipitation than that on half-hourly scales (Figure 6e). Correspondingly, IMERG's NMAE is correlated with the CC for the half-hourly precipitation, but there is not a reliable correlation between the CC and DEM for the daily rainfall. Thus, the reliability of IMERG's half-hourly precipitation is higher.

Overall, we found that the ability of IMERG to detect precipitation events decreases with increasing elevation, and IMERG exhibits a higher FAR in valley areas. In hit events, increasing elevation improves the negative bias of IMERG rainfall.

3.4. IMERG Performance under Different Rainfall Intensities

We examined IMERG's ability to capture various rainfall intensities and the distribution of the accumulated precipitation when precipitation events are hit by IMERG (Figure 7). Figure 7a demonstrates that IMERG tends to overestimate the occurrence of light and moderate rainfall events while underestimating the occurrence of events with rainfall intensities exceeding 0.8 mm/0.5 h. This pattern is also reflected in the accumulated rainfall estimates (Figure 7b). IMERG performs commendably for extreme rainfall ($p > 3.2$ mm/0.5 h) in terms of both the number of precipitation events and the comparison of the accumulated rainfall. Overall, IMERG tends to overestimate light to moderate rainfall events and underestimate heavy rainfall to rainstorm events, causing a shift in the precipitation distribution towards lower intensities and an overall underestimation bias. This phenomenon has also been observed in IMERG evaluations along the eastern coast of the U.S. and has been identified as a systematic bias in IMERG [5].

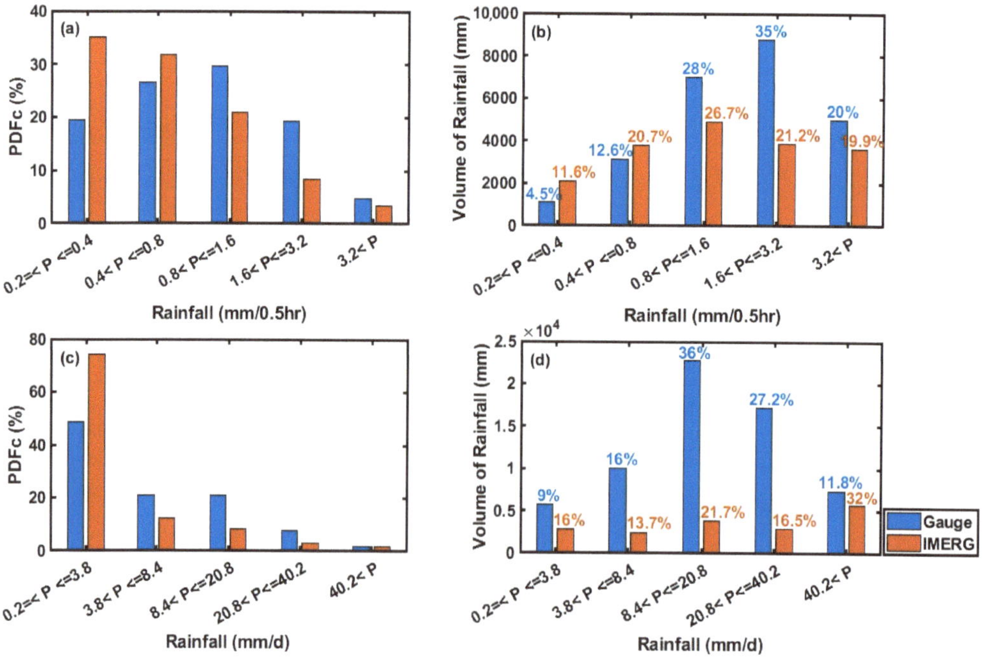

Figure 7. (**a**) Probability density function by occurrence (PDFc) and (**b**) Volume of rainfall and probability density function by volume (PDFv, the percentage number is on the histogram) of the half-hourly rainfall with different intensities for hit events. (**c**) PDFc and (**d**) Volume of rainfall and PDFv of daily rainfall. The blue bars represent the rain gauges, and the red bars represent IMERG.

After the precipitation is accumulated into the daily precipitation, the hit ratio increases, and the distribution of IMERG's daily rainfall shifts towards lighter precipitation (Figure 7c). IMERG overestimates the occurrence of light rainfall ($p \leq 3.8$ mm/d), and it underestimates moderate to heavy and extreme rainfall. It was found that IMERG underestimates the total rainfall amounts for all precipitation intensities (Figure 7d), and the distribution patterns of the accumulated precipitation of IMERG under different rainfall intensities exhibit a certain discrepancy compared to ground observations. For ground-based daily precipitation, heavy rainfall ($8.4 < p \leq 20.8$ mm/d) contributes the most (about 36%), followed by rainstorms, moderate rainfall, extreme rainfall, and light rainfall. However, the highest contributions in IMERG come from extreme rainfall, heavy rainfall, rainstorms, light rainfall, and moderate rainfall. Therefore, caution is advised when utilizing IMERG's daily precipitation data to study local rainfall intensity distributions.

3.5. Diurnal Variation Assessment

Prior studies, including those conducted by Zhou, Kim, and Guan [38], have acknowledged IMERG's effectiveness in representing large-scale diurnal rainfall patterns. They noticed that IMERG's diurnal and semidiurnal data mostly aligned with the reference values, but they also highlighted IMERG's tendency to underestimate the peak precipitation in mountainous regions and its propensity to detect precipitation prematurely in mesoscale convective systems. Chen et al. [2] revealed a distinct diurnal variation pattern within the YGC region, which differed from that on the central plateau. This pattern shows that the DVP reaches its lowest point at around 15:00 local standard time, then slowly increases and peaks in the early morning. Therefore, it is essential to examine whether IMERG can accurately capture this unique DVP in the YGC region.

The evaluation conducted in this paper is concentrated on two areas (Table 5): (1) the DVP phase, including the ability to capture the peak rainfall times and the level of synchronicity between the two datasets (indicated by the CC), and (2) the systematic bias of the DVP (indicated by the NME). The diurnal variation in Figure 8 was obtained by averaging the daily variation of all available data, and the CC and the NME were calculated based on the mean diurnal variation. IMERG is a comprehensive dataset that was produced by combining microwave sensor data, infrared data, and gauge calibration data. These individual datasets contribute positively to the final IMERG product, but they may also introduce their own biases. Hence, when analyzing the DVP, it is vital to assess not only the calibrated IMERG data (hereinafter referred to as Cal) but also the HQ, IR, and Uncal data.

By studying the NMEs of the HQ, IR, Uncal, and Cal datasets, we found that all of these datasets demonstrated negative biases, with only a few exceptions (such as Cal NME at Lulang station in Table 5). This implies a systemic underestimation of the rainfall at the sensor level. By comparing the HQ and IR data, we found that the IR generally exhibited a better correlation with the DVP (Table 5). IR data are constantly gathered, providing regular updates on the presence or absence of rainfall at specific time intervals. In contrast, HQ data are collected via sporadic microwave detection. As a result, HQ data often exhibit poorer correlation coefficients and, in some cases, even uncorrelated results (bold and italicized in Table 5), particularly at stations experiencing pronounced diurnal variations and larger Cal negative biases. When the HQ and IR data were combined to generate the Uncal data, we noticed an improvement in the negative bias and an increase in the CC. This suggests that the integration of microwave and infrared data improves the chances of detecting rainfall events, thereby enhancing the depiction of the diurnal phase. By comparing the Cal and Uncal data, it was found that there was no significant change in the CC, but the negative bias was substantially improved. For instance, at Lulang station, it increased from −32% to 5%. This confirms that the GPCC's primary aim is to fine-tune the rainfall volume rather than altering its occurrence.

Table 5. Evaluation of IMERG diurnal variation of the precipitation (DVP).

No.	Site	IMERG GRID	GPCC GRID	HQ		IR		Uncal		Cal	
				NME	CC	NME	CC	NME	CC	NME	CC
1	Beibeng	IMERG01	GPCC01	−0.7	0.64	−0.69	0.43	−0.66	0.74	−0.41	0.7
2	Gelin	IMERG01	GPCC01	−0.75	0.59	−0.75	0.41	−0.72	0.73	−0.51	0.69
3	Yarang		GPCC01	−0.66	0.56	−0.66	0.59	−0.6	0.79	−0.33	0.75
4	Dexing		GPCC01	−0.71	0.51	−0.71	***0.39***	−0.71	0.71	−0.52	0.73
5	Wenlang	IMERG02	GPCC01	−0.61	0.58	−0.64	0.53	−0.65	0.85	−0.42	0.87
6	Motog	IMERG02	GPCC01	−0.65	0.51	−0.68	0.66	−0.69	0.77	−0.48	0.76
7	Bari	IMERG02	GPCC01	−0.68	0.67	−0.71	0.47	−0.72	0.9	−0.53	0.92
8	Renqinbeng	IMERG02	GPCC01	−0.73	0.68	−0.75	0.69	−0.76	0.91	−0.6	0.9
9	Miri	IMERG03	GPCC01	−0.56	0.55	−0.64	0.62	−0.58	0.78	−0.36	0.74
10	Linduo	IMERG03	GPCC01	−0.66	0.63	−0.72	0.6	−0.68	0.83	−0.51	0.81
11	Kabu	IMERG03	GPCC01	−0.48	0.58	−0.58	0.64	−0.51	0.78	−0.25	0.74
12	Yimin		GPCC01	−0.54	***0.32***	−0.65	0.74	−0.55	0.82	−0.34	0.79
13	Dongren		GPCC01	−0.43	***0.24***	−0.5	0.43	−0.43	0.5	−0.16	0.49
14	80K		GPCC01	−0.85	0.5	−0.85	0.55	−0.83	0.77	−0.75	0.78
15	Xironggou		GPCC01	−0.86	***0.4***	−0.85	0.55	−0.84	0.84	−0.76	0.83
16	Danka		GPCC01	−0.51	***0.39***	−0.49	***0.26***	−0.48	0.6	−0.17	0.56
20	Lulang		GPCC02	−0.4	***0.11***	−0.35	0.6	−0.32	0.56	0.05	0.53
21	Pailong		GPCC03	−0.56	***0.09***	−0.57	***0.19***	−0.56	***0.37***	−0.3	0.44

Notation: HQ, GPM microwave; IR, GPM infrared; Uncal, Un-calibrated IMERG; Cal, IMERG calibrated; NME, normalized mean error; CC, correlation coefficient. The bold and italicized numbers denote that the sites do not pass the significance test. The sites are numbered from south to north.

Figure 8 illustrates that both IMERG and the RGs commonly display an early morning surge in rainfall. This can be attributed to a mix of high-frequency and high-intensity rainfall [2]. The DVP at the RGs is primarily manifested in two forms. (1) Concentrated precipitation peak: The minimum rainfall tends to occur around 16:00, and then, the rainfall escalates and reaches a peak between 3:00 and 7:00. This pattern is observable at Site No.1−3, 9−14, 16, and 21 (Figure 8a,b,e–j,l). (2) Multiple precipitation peaks: The primary peak still occurs in the early morning, but it is accompanied by various secondary peaks (as observed in Site No.4−8, 15, and 20 in Figure 8c,d,i,k). IMERG tends to represent multiple peaks. As the stations are positioned along the trajectory of the water vapor transmission, the time of the rainfall peak is delayed from south to north, with IMERG predominantly centering the rainfall peak at around 4:00. Thus, the satellite data often struggle to capture the variations in the precipitation peaks over time.

We observe that the stations with higher correlation coefficients (i.e., CC > 0.6) typically experienced larger rainfall volumes. This implies that increased rainfall at a station simplifies the task for IMERG in identifying DVP patterns. In contrast, the stations with less daily rainfall and a flatter DVP tended to have lower CCs. In the IMERG02 grid, four stations share the same basin and water vapor channel and are devoid of disturbances from other water vapor sources. As the altitude increases, the rainfall measured by the RG increases, transitioning from the valley to the slope. This increases the negative bias and enhances IMERG's correlation with the DVP. Thus, within the same IMERG grid, satellites can more readily capture the DVP volume at valley stations. However, their proficiency in reflecting the phase of the diurnal variations is not as good for stations on slopes or peaks. This is consistent with our prior discussion of the satellite's capability to pinpoint the timing of a rainfall event.

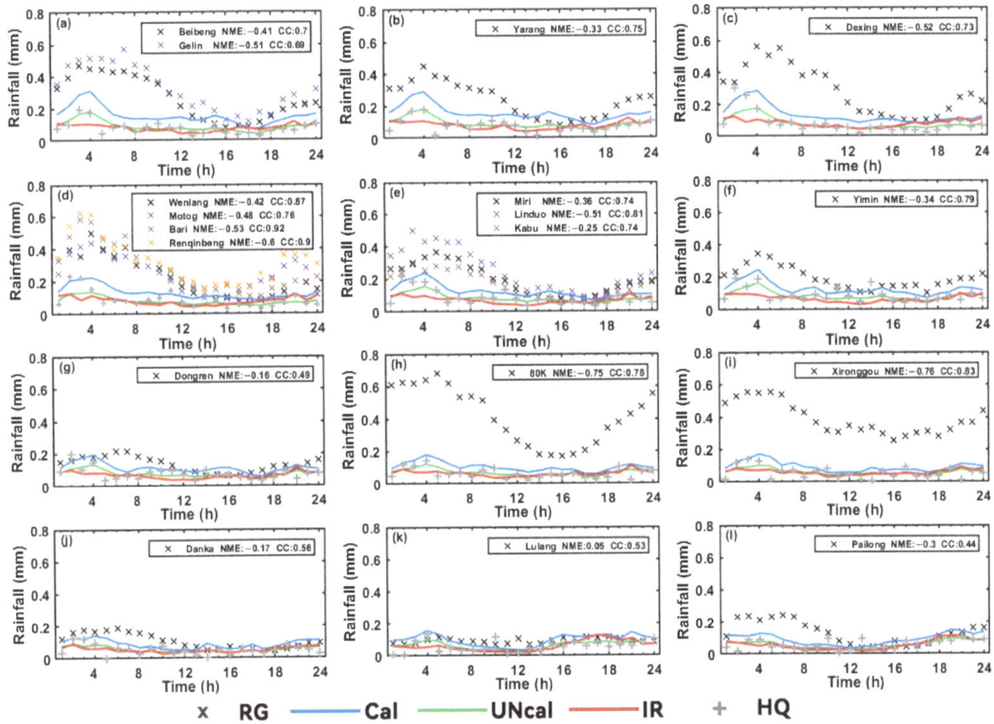

Figure 8. The diurnal variation in the rainfall derived from the rain gauges and IMERG. Each panel shows the result of one IMERG grid. The x symbols represent the rain gauge data, the blue lines represent the Cal data (IMERG calibrated), the green lines represent the Uncal data (Un-calibrated IMERG), the red lines represent the IR data (GPM infrared rainfall), and the grey + symbols represent the HQ data (GPM microwave rainfall). The Correlation coefficient (CC) and normalized mean error (NME) of the Cal data against the rain gauge data are noted in the upper right corner of each panel. All of the times are Beijing time. (**a**,**d**,**e**) correspond to IMERG01, IMERG02, IMERG03, respectively. (**a**–**g**) correspond to stations south of Dongren, and (**h**–**l**) correspond to stations north of 80K.

3.6. Assessment of Seasonal Variation

We have confirmed that the calibration process using GPCC data enhanced the accuracy of IMERG outputs relative to station observations. Regarding the calibration process, it is evident that the GPCC's monthly data significantly influences the quality of the IMERG data across different time scales. This relationship is validated in Table 6; that is, the correlation between the GPCC and IMERG exceeds 92%, and the relative bias ranges from −29% to +32%. As a result, any bias between the GPCC and station measurements would directly affect the quality of the IMERG data.

Table 6 shows that there is considerable variability in the NME and CC between the GPCC and the various stations, which suggests a certain level of representational issue with the GPCC's 1° resolution at the station level. Therefore, it is crucial to offer a higher-resolution GPCC product for IMERG within the YGC region. The seasonal traits depicted by the GPCC (Figure 9) show that during the monsoon season (June–September), the main rainfall peak is in July, and there is a secondary peak in September. In the pre-monsoon period (March–May), a weak peak occurs in April. These seasonal features are also reflected in IMERG's seasonal variations (Figure 10). The seasonal variation at some RG stations corresponds well with IMERG's representation, with the CC reaching 0.99 at Linzhi station (Table 6). However, discrepancies between the seasonal variation at some stations and satellite estimates do exist, such as during the monsoon season with the primary peak in June (Renqinbeng, Yimin in Figure 10d,f), the primary peak in September (Pailong in Figure 10o), and pre-monsoon peaks in April (Dexing, Wenlang, Renqinbeng, Yimin, and Danka in Figure 10c,d,f,j) and March (Dongren in Figure 10g).

Table 6. Evaluation of seasonal variations of the GPCC against IMERG, the GPCC against gauge stations, and IMERG against gauge stations.

No.	Site	IMERG GRID	GPCC GRID	GPCC vs. IMERG		GPCC vs. Stations		IMERG vs. Stations	
				NME	CC	NME	CC	NME	CC
1	Beibeng	IMERG01	GPCC01	−0.28	0.96	−0.67	0.95	−0.54	0.95
2	Gelin	IMERG01	GPCC01	−0.28	0.96	−0.63	0.75	−0.49	0.84
3	Yarang		GPCC01	−0.29	0.95	−0.62	0.93	−0.47	0.96
4	Dexing		GPCC01	−0.16	0.94	−0.59	0.88	−0.52	0.87
5	Wenlang	IMERG02	GPCC01	−0.15	0.95	−0.52	0.88	−0.44	0.97
6	Motog	IMERG02	GPCC01	−0.15	0.95	−0.72	0.96	−0.67	0.93
7	Bari	IMERG02	GPCC01	−0.15	0.95	−0.59	0.8	−0.52	0.92
8	Renqinbeng	IMERG02	GPCC01	−0.15	0.95	−0.66	0.87	−0.6	0.89
9	Miri	IMERG03	GPCC01	−0.08	0.95	−0.57	0.97	−0.53	0.96
10	Linduo	IMERG03	GPCC01	−0.08	0.95	−0.6	0.78	−0.57	0.77
11	Kabu	IMERG03	GPCC01	−0.08	0.95	−0.56	0.94	−0.52	0.97
12	Yimin		GPCC01	−0.11	0.94	−0.39	0.82	−0.32	0.88
13	Dongren		GPCC01	0.02	0.96	−0.42	0.76	−0.43	0.79
14	80K		GPCC01	0.1	0.95	−0.76	0.95	−0.78	0.94
15	Xironggou		GPCC01	0.18	0.92	−0.74	0.97	−0.78	0.94
16	Danka		GPCC01	0.29	0.94	0.09	0.89	−0.16	0.89
17	Bomi		GPCC01	0.32	0.95	0.01	0.89	−0.23	0.86
18	Milin		GPCC02	−0.08	0.99	0.12	0.96	0.22	0.96
19	Linzhi		GPCC02	0.15	0.99	0.1	0.99	−0.04	0.99
20	Lulang		GPCC02	0.08	0.98	−0.33	0.96	−0.38	0.96
21	Pailong		GPCC03	−0.07	0.97	−0.5	0.95	−0.47	0.9
22	Luolong		GPCC03	0.25	0.95	0.31	0.94	0.05	0.98

Notation: NME, normalized mean error; CC, correlation coefficient. The sites are numbered from south to north.

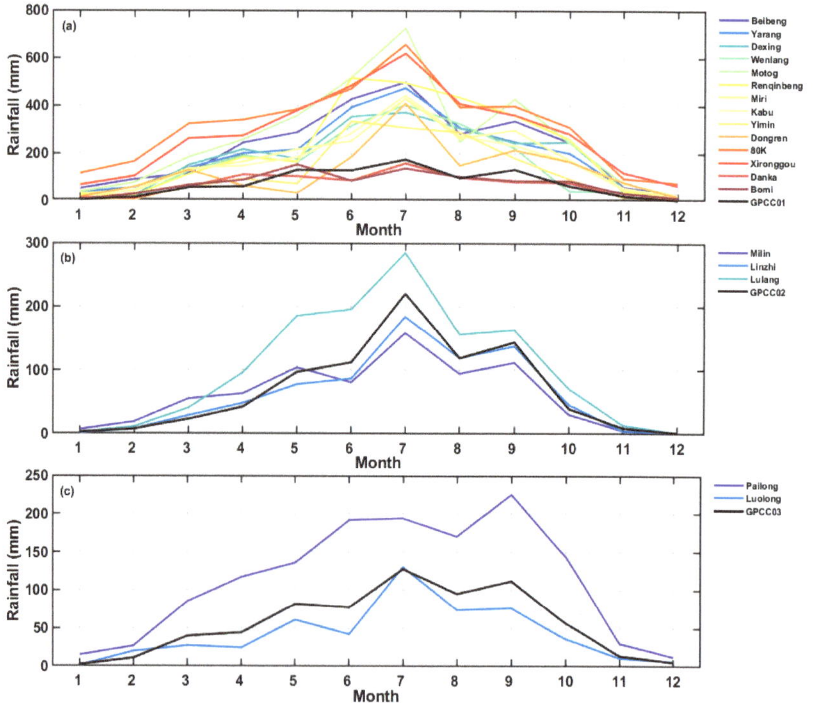

Figure 9. Seasonal variation comparison of rain gauges and GPCC grid value for (**a**) grid GPCC01, (**b**) grid GPCC02, and (**c**) grid GPCC03.

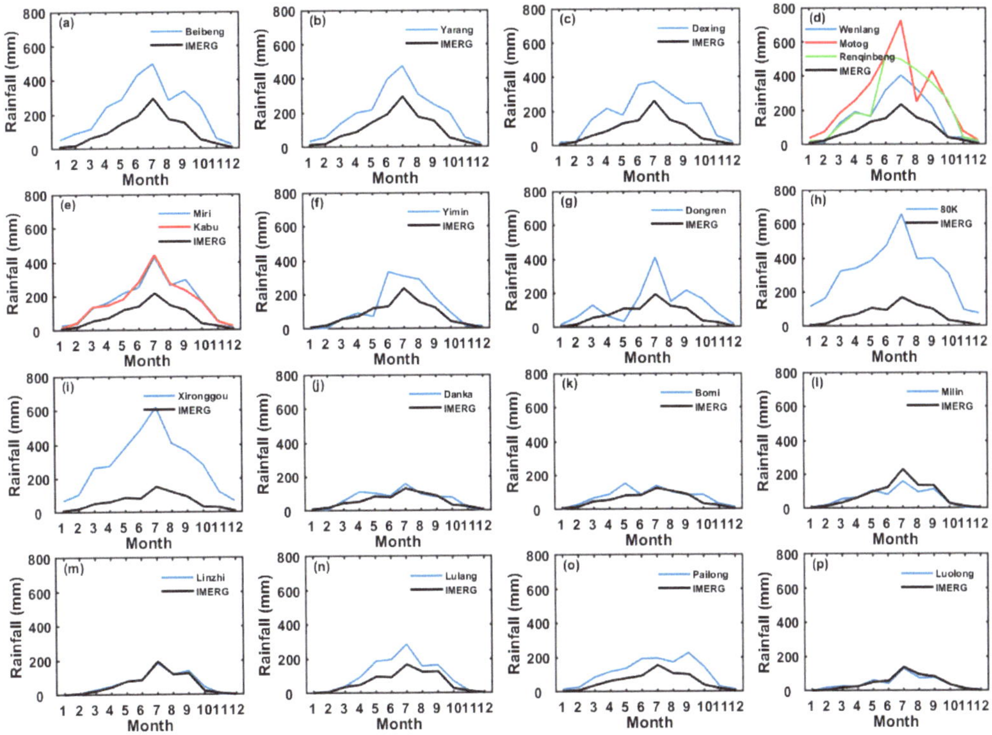

Figure 10. Seasonal variation comparison of rain gauges and IMERG grid value. Each panel shows the result of one IMERG grid. The sites are numbered from south to north. (**a**–**p**) correspond to the different IMERG grids.

4. Discussion

An evaluation of IMERG's performance in relation to rainfall events, quantity of rainfall, rainfall intensity, diurnal variations, and seasonal variations reveals that its applicability in the YGC region is influenced by a combination of the region's hydrometeorological characteristics and topography. The results are discussed in terms.

(1) The underestimation in IMERG exists at the GPM sensor level.

In the IMERG observational system, the quality of IMERG data is influenced by the PMW sensor, the IR data, and the gauge data used for calibration. Our evaluation demonstrated that the ratio of the miss rate is higher than the false alarm, which is also detected on the U.S. Eastern Shore [5]. Tan et al. [5] pointed out that this is associated with the PMW instruments. The performance of the onboard Ku-PR is essential for ensuring data quality. Yet, significant under-detection errors were noted in Ku-PR evaluations [37]. Arulraj and Barros [37] pinpointed two main factors causing this underestimation: firstly, ground clutter impacts on the 1–2 km reflectivity profile, lead to non-uniform beam filling, causing the underestimation of low-level and bottom-level precipitation systems. Secondly, seeder-feeder interactions are prevalent in mountainous regions when two layers of cloud happen. Zhou et al. [39] reported a bimodal distribution of cloud base heights in the YGC region. The seeder-feeder interactions might have a high chance of happening in this region, which generally leads to enhanced precipitation recorded by rain RGs near minor mountain areas [40,41]. Ku-PR struggles to capture these enhanced lower-level echoes from the seeder cloud [37]. This explains why the GPM has an underestimation for this region. Wang et al. [42] reported that the raindrop size distribution in Motog was maritime-like. However, the raindrop size distribution in the GPM-IMERG algorithm for the YGC region might be a different one. Uncertainties in raindrop size distribution can lead to satellite precipitation underestimations [37]. As such, future research should intensify observations on cloud microphysics characteristics within the YGC region to formulate a specific raindrop spectral parameter tailored to its conditions.

IMERG's tendency to overestimate light rainfall events and underestimate heavy rainfall events can be traced back to issues at the instrument level, observed in PMW sensors such as TMI, the Special Sensor Microwave Imager/Sounder (SSMIS), and the Microwave Humidity Sounder (MHM) [5]. Additionally, the integration of PMW with geo-IR estimates happens in two ways [14]: First, morphing the LEO-PMW data. Second, in cases where LEO-PMW data are sparse, geo-IR precipitation estimates are incorporated using a Kalman filter. Tan et al. [5] observed that part of the IR's underestimation can be attributed to the morphing process, which tends to propagate the PMW's underestimation to the IR data. Both underestimation of the PMW and IR data was observed by us in the diurnal variations (as seen in Figure 8); hence, the Uncal, which is the combination of PMW (HQ) and IR, also revealed an underestimation.

(2) Higher FAR in the valley stations.

The negative correlation between FAR and the relative elevation was observed in Figure 5, which represents the FAR in the valley is higher than that at the mountain peaks. The water vapor in the valley largely stems from warm, moist air masses transported from the Indian Ocean [42]. When these moist air masses from the Indian Ocean enter into the valley, they might not always reach the lifting condensation level to generate precipitation immediately and form stratiform clouds or fog in the valley. Wang et al. [42] demonstrated that stratiform rainfall constitutes over 95% of total events at Motog station. Arulraj and Barros [37] reported that the GPM satellite has high false alarms for stratiform rainfall. Due to orographic lifting—the process where terrain forces the air to rise—these moist air masses are channeled up the slopes, inducing more precipitation at higher elevations. This orographic effect augments the probability of IMERG misidentifying non-precipitating clouds as rain clouds in the valley.

In scenarios of light precipitation, the spatiotemporal intermittency is prominent. If fewer PMWs traverse the valley region, IMERG might lean towards using the instantaneous precipitation rate, deduced linearly from PMW signals bracketing the 30 min interval. This method might occasionally lead to satellite estimates indicating rain, even when ground observations at the valley affirm a lack of precipitation. Further corroborating this, Arulraj and Barros [37] pinpointed the morphing of PMW data as a chief contributor to false alarms, culminating in misguided rainfall indications during clear periods. These factors can be invoked to elucidate the pronounced negative correlation between FAR and relative elevation.

(3) In the evaluation of rainfall events, why do topographic correlations stand out on half-hourly scales but fade on daily scales?

The POD, FAR, BID, and HSS have a significant correlation with terrains on half-hourly scales, but that is not significant on daily scales. Statistically, the POD represents IMERG's capability to detect precipitation events within a set timeframe. The odds of the satellite capturing rainfall events rise with the increasing time scales. For instance, if the half-hourly POD of site A and site B is 50% and 30%, the daily POD of the two sites will be 100%. This indicates that the variability of POD, influenced by elevation, becomes negligible on a daily scale. The YGC region, serving as a crucial conduit for water vapor transport, is molded by both expansive weather systems such as the South Asian monsoon and westerlies, as well as terrain-induced circulations of various scales. On the half-hour scale, terrain-induced circulations result in noticeable spatial heterogeneity. However, the spatial pattern of precipitation on the daily scale is more homogeneous. This means that the variations of POD, FAR, BID, and HSS among different sites could be diminished from the half-hourly to daily scale.

(4) The benefit of GPCC gauge calibration in IMERG algorithm.

The GPCC exhibits a substantial negative bias south of Galongla Mountain (sites No. 1–15). However, in the northern region of the mountain, the bias is relatively minor. This is mainly due to the fact that the southern part of Galongla Mountain has long suffered from a dearth of rain gauge observation data before the establishment of the INVC-RGs. Conversely, the presence of CMA-RGs in the northern part of Galongla Mountain has enhanced the quality of GPCC and further improved IMERG-Final products in these areas. Thus, satellite validation work in the downstream region of the YGC is imperative and urgently needed.

By comparing the INVC-RG and CMA-RG in GPCC02 and GPCC03 grids, we identified the importance of choosing precipitation observation stations (Figure 9b,c). Lulang and Linzhi are located in the same grid of GPCC02 and influenced by the same water vapor branch originating from the YGC valley, but the value of the GPCC02 grid is very close to Linzhi but differs from that at Lulang (Table 6). This is primarily because Lulang station, situated on the northwestern side of the Sejila mountain, is affected by the orographic precipitation enhancement, leading to greater rainfall, while Linzhi, located on the leeward side of the Sejila mountain, experiences a decrease in the water vapor content, resulting in less rainfall than that at Lulang (Figure 1b). These geographic characteristics cause a different performance of IMERGs for the two stations. Pailong and Luolong are located in the same grid of GPCC03; the GPCC03 grid values closely match Luolong's but differ significantly from Pailong's. The GPM-IMERG grid value is closer to the CMA-RG sites (Linzhi and Luolong) than our INVC-RG sites (Lulang and Pailong); this demonstrates that the CMA-RG site has benefited from GPCC and GPM-IMERG precipitation estimation. YGC canyons are deep and broad, and thus, the precipitation formation mechanism is complex. The establishment of stations such as INVC-RGs is of significant importance as they enrich the spatial patterns of the precipitation under complex terrains and enhance the quality of the GPCC and IMERG in the YGC region.

A high correlation between GPCC and IMERG in Table 6 indicates that the 1° adjustment causes the seasonal variations of IMERG within the same GPCC grid to homogenize. Corrections should be considered starting from the data sources, such as the sensors. Moreover, when different platform estimates of the GPM are calibrated with each other and when Morph compensates for the absence of PMW, new errors are introduced. If these errors can be controlled, the effectiveness of the calibration using the GPCC can be enhanced.

Moving forward, our endeavors will pivot around the following key directions: First, Enhancing Representativeness of rain gauge data in the YGC Region. Obtaining ground-based data in the YGC region has always posed challenges. Our current dataset, though valuable, is relatively short in terms of temporal coverage, potentially limiting its climatic representativeness. We aim to gather data spanning longer periods in our subsequent endeavors, with the hope of achieving a more representative evaluation of IMERG. Second, we are exploring a range of methods to refine the calibration of IMERG data. For instance, we're currently testing a bias removal method, cumulative distribution functions (CDF) matching, to enhance the IMERG data's accuracy. At last, we are geared towards observing precipitation microphysics characteristics to fine-tune parameters in the GPM-IMERG algorithm. By focusing on these aspects, we aim to further the accuracy of GPM-IMERG data in the YGC region.

5. Conclusions

Conventional precipitation observations face challenges in accurately estimating precipitation in complex terrain and at high temporal resolutions. Satellite-based precipitation retrievals, such as IMERG, provide an alternative data source for filling these gaps. In this study, IMERG improved accuracy and detection capabilities compared to its predecessor, TRMM. However, IMERG still exhibited limitations in certain regions and for some precipitation event types. In this study, the performance of the IMERG half-hourly and daily precipitation products in the YGC region was evaluated. Furthermore, ground-based precipitation data from observation stations, including the STEP-INVC program and the China Meteorological Administration, were used for validation. The evaluation results were as follows.

(1) The assessment of precipitation events in the YGC region revealed the limitations of IMERG in accurately detecting and estimating rainfall. The evaluation showed that IMERG exhibited errors in rainfall detection, including misses and false alarms, with misses being more prevalent than false alarms. This under-detection issue was found to persist across all sites in the YGC region. The comprehensive skill score of IMERG for daily rainfall events was less than ideal. Elevation affected IMERG's ability to detect precipitation events, with higher elevations leading to a decreased POD and increased underestimation. Valley areas exhibited lower POD, higher FAR, and lower HSS values, indicating less reliable precipitation event capture by IMERG in valleys. The station's position in a valley or on a slope influenced IMERG's performance, primarily impacting the detection of precipitation events rather than the estimated rainfall volume.
(2) We focused on rainfall events correctly hit by IMERG and assessed the rainfall amounts at the half-hour and daily scales. IMERG tended to underestimate rainfall amounts, particularly at lower altitudes. The negative bias decreased with increasing altitude and was influenced by the position of the station within the moisture transport pathways. Gauges located in tributaries had larger random errors and higher false alarm rates, contributing to the overall estimation errors of IMERG. Stations with dramatic changes in altitude gradient had high precipitation levels, leading to a lower correlation with IMERG. When rainfall accumulated on a daily scale, the negative bias was amplified, and the random errors increased. IMERG also tended to overestimate the occurrence of light and moderate rainfall events and underestimate events with higher intensities. This pattern was also reflected in the accumulated rainfall estimates.
(3) IMERG could capture the peak precipitation in the early morning diurnal variation in the YGC region but underestimated the amplitude of the diurnal variation. The datasets from various data sources (HQ, IR, UN, and Cal) exhibited negative biases, indicating a systemic underestimation of the rainfall at the sensor level. Combining HQ and IR data (Uncal) improved the negative bias and correlation and enhanced the depiction of the diurnal phase. The calibrated data (Cal) primarily adjusted the volume of the rainfall rather than its occurrence. IMERG was more proficient in capturing the DVP volume at stations in valleys than those on slopes or peaks.
(4) IMERG and GPCC could describe the seasonal characteristics at gauge sites while underestimating the seasonal variation at the sites. There was considerable variability in the NME and CC between the GPCC and various RGs, indicating representational issues with the GPCC's 1° resolution at the station level. Therefore, a higher-resolution GPCC product is needed for calibrating IMERG within the YGC region. The GPCC exhibited a significant negative bias in the region to the south of Galongla Mountain due to a lack of precipitation observation data in this region before the establishment of the INVC-RGs. This deficiency contributed to the larger bias in the GPCC data. In contrast, the presence of CMA-RGs in the region to the north of Galongla Mountain improved the quality of the GPCC and IMERG. The INVC program will continue choosing appropriate precipitation observation stations and providing grid-level observation data for the YGC region to enrich observations and improve satellite rainfall estimates.

The results of this study filled gaps in IMERG validation and provided insights into its strengths and weaknesses in complex terrains. The findings could contribute to enhancing satellite-based precipitation estimates and understanding the dynamics of the TP's water cycle.

Author Contributions: All authors contributed significantly to this work. Conceptualization, L.L., X.C. and Y.M.; methodology, L.L. and X.C.; software, L.L., X.C. and W.Z.; validation, L.L. and X.C.; formal analysis, L.L., X.C., D.C. and X.X.; investigation, L.L. and X.C.; resources, L.L. and X.C.; data curation, L.L., Y.L., D.C. and X.X.; writing—original draft preparation, L.L.; writing—review and editing, L.L., X.C. and Y.M.; supervision, X.C., H.Z. and Y.M.; project administration, X.C., H.Z. and Y.M; funding acquisition, X.C., H.Z. and Y.M. All authors have read and agreed to the published version of the manuscript.

Funding: This research was funded by the Second Tibetan Plateau Scientific Expedition and Research (STEP) Program (grant Nos. 2019QZKK0105, 2019QZKK0103), the National Natural Science Foundation of China (42230610, 41975009, 42275069).

Data Availability Statement: The data from INVC-RGs can be downloaded from the website: http://data.tpdc.ac.cn/zh-hans/disallow/e68f1de1-3a13-4ae1-90e0-9e3a3f57f912/, accessed on 26 February 2023. The data from CMA-RGs can be downloaded from http://www.cma.gov.cn, accessed on 27 January 2023. The IMERG product can be downloaded from https://disc.gsfc.nasa.gov/ accessed on 13 December 2022, and the GPCC product can be downloaded from https://opendata.dwd.de/climate_environment/GPCC/html/download_gate.html, accessed on 8 November 2022.

Acknowledgments: We thank the Foundation for its support and the developers of the operational satellite precipitation products and precipitation gauge measurements. We thank the reviewers and editors for their constructive comments.

Conflicts of Interest: The authors declare no conflict of interest.

Appendix A

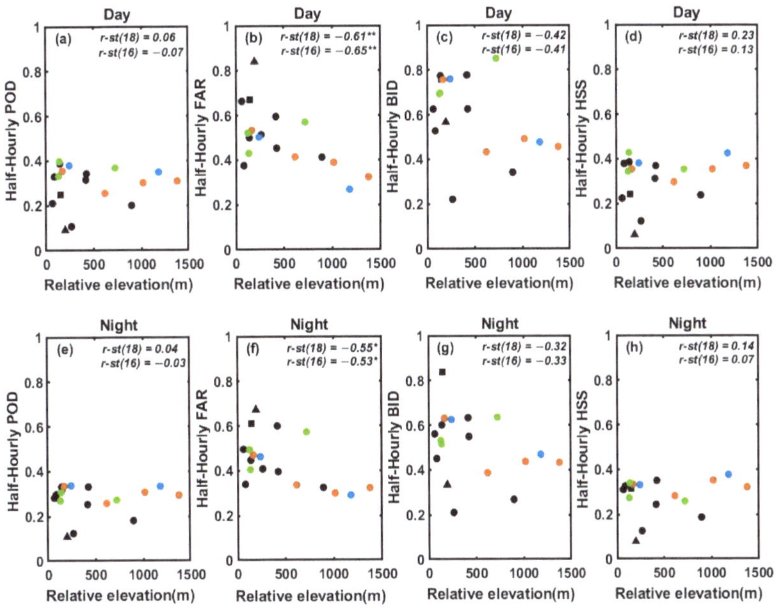

Figure A1. Scatter plots of the relative elevations (unit: m) of the sites versus the four scores (POD, FAR, BID, and HSS) at (**a–d**) day and (**e–h**) night on the half-hourly scales. The circles represent the sites in the GPCC01 grid, the squares represent sites in the GPCC02 grid, and the triangles represent the sites in the GPCC03 grid. The blue circles indicate the IMERG01 grid; the red circles indicate the IMERG02 grid; and the green circles indicate the IMERG03 grid. The correlations (r) between the four scores and the DEMs are noted in the upper right corner of each panel: r-all represents the correlations derived from all 22 sites at the daily scale, r-st18 represents the correlations derived from the 18 non-CMA sites, and r-st16 represents the correlations of the 16 non-CMA sites in the GPCC01 grid. * indicates a p-value of 0.05, and ** indicates a p-value of 0.01.

References

1. Qiu, J. China: The third pole. *Nature* **2008**, *454*, 393–396. [CrossRef] [PubMed]
2. Chen, X.; Cao, D.; Liu, Y.; Xu, X.; Ma, Y. An observational view of rainfall characteristics and evaluation of ERA5 diurnal cycle in the Yarlung Tsangbo Grand Canyon, China. *Q. J. R. Meteorol. Soc.* **2023**. [CrossRef]
3. Hou, A.Y.; Kakar, R.K.; Neeck, S.; Azarbarzin, A.A.; Kummerow, C.D.; Kojima, M.; Oki, R.; Nakamura, K.; Iguchi, T. The Global Precipitation Measurement Mission. *Bull. Am. Meteorol. Soc.* **2014**, *95*, 701–722. [CrossRef]
4. Kolbe, C.; Thies, B.; Egli, S.; Lehnert, L.; Schulz, H.; Bendix, J. Precipitation Retrieval over the Tibetan Plateau from the Geostationary Orbit—Part 1: Precipitation Area Delineation with Elektro-L2 and Insat-3D. *Remote Sens.* **2019**, *11*, 2302. [CrossRef]
5. Tan, J.; Petersen, W.A.; Tokay, A. A Novel Approach to Identify Sources of Errors in IMERG for GPM Ground Validation. *J. Hydrometeorol.* **2016**, *17*, 2477–2491. [CrossRef]
6. Hsu, K.-l.; Gao, X.; Sorooshian, S.; Gupta, H.V. Precipitation Estimation from Remotely Sensed Information Using Artificial Neural Networks. *J. Appl. Meteorol.* **1997**, *36*, 1176–1190. [CrossRef]
7. Joyce, R.J.; Janowiak, J.E.; Arkin, P.A.; Xie, P. CMORPH: A Method that Produces Global Precipitation Estimates from Passive Microwave and Infrared Data at High Spatial and Temporal Resolution. *J. Hydrometeorol.* **2004**, *5*, 487–503. [CrossRef]
8. Kubota, T.; Aonashi, K.; Ushio, T.; Shige, S.; Takayabu, Y.N.; Kachi, M.; Arai, Y.; Tashima, T.; Masaki, T.; Kawamoto, N. Global Satellite Mapping of Precipitation (GSMaP) products in the GPM era. In *Satellite Precipitation Measurement*; Springer: Berlin/Heidelberg, Germany, 2020; Volume 1, pp. 355–373.
9. Funk, C.; Peterson, P.; Landsfeld, M.; Pedreros, D.; Verdin, J.; Shukla, S.; Husak, G.; Rowland, J.; Harrison, L.; Hoell, A.; et al. The climate hazards infrared precipitation with stations—A new environmental record for monitoring extremes. *Sci. Data* **2015**, *2*, 150066. [CrossRef]
10. Beck, H.E.; Wood, E.F.; Pan, M.; Fisher, C.K.; Miralles, D.G.; van Dijk, A.I.J.M.; McVicar, T.R.; Adler, R.F. MSWEP V2 Global 3-Hourly 0.1° Precipitation: Methodology and Quantitative Assessment. *Bull. Am. Meteorol. Soc.* **2019**, *100*, 473–500. [CrossRef]
11. Skofronick-Jackson, G.; Petersen, W.A.; Berg, W.; Kidd, C.; Stocker, E.F.; Kirschbaum, D.B.; Kakar, R.; Braun, S.A.; Huffman, G.J.; Iguchi, T. The Global Precipitation Measurement (GPM) mission for science and society. *Bull. Am. Meteorol. Soc.* **2017**, *98*, 1679–1695. [CrossRef]
12. Zhu, Z.; Piao, S.; Myneni, R.B.; Huang, M.; Zeng, Z.; Canadell, J.G.; Ciais, P.; Sitch, S.; Friedlingstein, P.; Arneth, A.; et al. Greening of the Earth and its drivers. *Nat. Clim. Chang.* **2016**, *6*, 791–795. [CrossRef]
13. Zhang, S.; Wang, D.; Qin, Z.; Zheng, Y.; Guo, J. Assessment of the GPM and TRMM Precipitation Products Using the Rain Gauge Network over the Tibetan Plateau. *J. Meteorol. Res.* **2018**, *32*, 324–336. [CrossRef]
14. Huffman, G.J.; Bolvin, D.T.; Braithwaite, D.; Hsu, K.; Joyce, R.; Xie, P.; Yoo, S.-H. NASA Global Precipitation Measurement (GPM) Integrated Multi-Satellite Retrievals for GPM (IMERG). Algorithm Theoretical Basis Document (ATBD) Version 06. Available online: https://gpm.nasa.gov/sites/default/files/2020-05/IMERG_ATBD_V06.3.pdf (accessed on 13 December 2022).
15. Yu, S.; Lu, F.; Zhou, Y.; Wang, X.; Wang, K.; Song, X.; Zhang, M. Evaluation of Three High-Resolution Remote Sensing Precipitation Products on the Tibetan Plateau. *Water* **2022**, *14*, 2169. [CrossRef]
16. Zhao, K.; Peng, D.; Gu, Y.; Pang, B.; Zhu, Z. Daily precipitation dataset at 0.1 degrees for the Yarlung Zangbo River basin from 2001 to 2015. *Sci. Data* **2022**, *9*, 349. [CrossRef]
17. Lei, H.; Li, H.; Zhao, H.; Ao, T.; Li, X. Comprehensive evaluation of satellite and reanalysis precipitation products over the eastern Tibetan plateau characterized by a high diversity of topographies. *Atmos. Res.* **2021**, *259*, 105661. [CrossRef]
18. Ji, H.; Peng, D.; Gu, Y.; Liang, Y.; Luo, X. Evaluation of multiple satellite precipitation products and their potential utilities in the Yarlung Zangbo River Basin. *Sci. Rep.* **2022**, *12*, 13334. [CrossRef]
19. Chen, X. An observational view of rainfall characteristics and evaluation of rainfall products in the Yarlung Tsangbo Grand Canyon, China. In Proceedings of the EGU General Assembly Conference Abstracts, Vienna, Austria, 23–28 April 2023; p. EGU23-2196.
20. Chen, C.; Chen, Q.; Duan, Z.; Zhang, J.; Mo, K.; Li, Z.; Tang, G. Multiscale Comparative Evaluation of the GPM IMERG v5 and TRMM 3B42 v7 Precipitation Products from 2015 to 2017 over a Climate Transition Area of China. *Remote Sens.* **2018**, *10*, 944. [CrossRef]
21. Xu, R.; Tian, F.; Yang, L.; Hu, H.; Lu, H.; Hou, A. Ground validation of GPM IMERG and TRMM 3B42V7 rainfall products over southern Tibetan Plateau based on a high-density rain gauge network. *J. Geophys. Res. Atmos.* **2017**, *122*, 910–924. [CrossRef]
22. Zhang, Y.; Ju, Q.; Zhang, L.; Xu, C.-Y.; Lai, X. Evaluation and Hydrological Application of Four Gridded Precipitation Datasets over a Large Southeastern Tibetan Plateau Basin. *Remote Sens.* **2022**, *14*, 2936. [CrossRef]
23. Chen, X.; Xu, X. *Scientific Expedition and Research Report for the Yarlung Tsangpo Canyon—Implication for Water Vapor Transmission*; Science Press: Beijing, China, 2022; pp. 1–189.
24. Ma, L.; Zhao, L.; Tian, L.-M.; Yuan, L.-M.; Xiao, Y.; Zhang, L.-L.; Zou, D.-F.; Qiao, Y.-P. Evaluation of the integrated multi-satellite retrievals for global precipitation measurement over the Tibetan Plateau. *J. Mt. Sci.* **2019**, *16*, 1500–1514. [CrossRef]
25. Yang, Y.; Gao, D. A great geographical discovery in the end of 20th century—Demonstration on the Yarlung Zangbo river valley as the grandest canyon in the World. *Geogr. Res.* **1996**, *4*, 1–9.
26. Lai, H.-W.; Chen, H.W.; Kukulies, J.; Ou, T.; Chen, D. Regionalization of Seasonal Precipitation over the Tibetan Plateau and Associated Large-Scale Atmospheric Systems. *J. Clim.* **2021**, *34*, 2635–2651. [CrossRef]

27. Chen, X.; Xu, X.; Wang, G.; Chen, D.; Ma, Y.; Liu, L.; Hu, X.; Liu, Y.; Li, L.; Li, M. INVC-Investigation of the water vapor channel within the Yarlung Zangbo Grand Canyon, China. In Proceedings of the EGU General Assembly Conference Abstracts, Vienna, Austria, 3–8 April 2022; p. EGU22-2088.
28. Ma, Y.; Tang, G.; Long, D.; Yong, B.; Zhong, L.; Wan, W.; Hong, Y. Similarity and Error Intercomparison of the GPM and Its Predecessor-TRMM Multisatellite Precipitation Analysis Using the Best Available Hourly Gauge Network over the Tibetan Plateau. *Remote Sens.* **2016**, *8*, 569. [CrossRef]
29. Maranan, M.; Fink, A.H.; Knippertz, P.; Amekudzi, L.K.; Atiah, W.A.; Stengel, M. A Process-Based Validation of GPM IMERG and Its Sources Using a Mesoscale Rain Gauge Network in the West African Forest Zone. *J. Hydrometeorol.* **2020**, *21*, 729–749. [CrossRef]
30. Bookhagen, B.; Burbank, D.W. Topography, relief, and TRMM-derived rainfall variations along the Himalaya. *Geophys. Res. Lett.* **2006**, *33*, L13402. [CrossRef]
31. Fu, Y.; Pan, X.; Xian, T.; Liu, G.; Zhong, L.; Liu, Q.; Li, R.; Wang, Y.; Ma, M. Precipitation characteristics over the steep slope of the Himalayas in rainy season observed by TRMM PR and VIRS. *Clim. Dyn.* **2017**, *51*, 1971–1989. [CrossRef]
32. Chen, H.; Yuan, W.; Li, J.; Yu, R. A possible cause for different diurnal variations of warm season rainfall as shown in station observations and TRMM 3B42 data over the southeastern Tibetan plateau. *Adv. Atmos. Sci.* **2011**, *29*, 193–200. [CrossRef]
33. Thiemig, V.; Rojas, R.; Zambrano-Bigiarini, M.; Levizzani, V.; De Roo, A. Validation of Satellite-Based Precipitation Products over Sparsely Gauged African River Basins. *J. Hydrometeorol.* **2012**, *13*, 1760–1783. [CrossRef]
34. Tang, L.; Tian, Y.; Yan, F.; Habib, E. An improved procedure for the validation of satellite-based precipitation estimates. *Atmos. Res.* **2015**, *163*, 61–73. [CrossRef]
35. Sharifi, E.; Steinacker, R.; Saghafian, B. Multi time-scale evaluation of high-resolution satellite-based precipitation products over northeast of Austria. *Atmos. Res.* **2018**, *206*, 46–63. [CrossRef]
36. Yu, L.; Zhang, Y.; Yang, Y. Using High-Density Rain Gauges to Validate the Accuracy of Satellite Precipitation Products over Complex Terrains. *Atmosphere* **2020**, *11*, 633. [CrossRef]
37. Arulraj, M.; Barros, A.P. Improving quantitative precipitation estimates in mountainous regions by modelling low-level seeder-feeder interactions constrained by Global Precipitation Measurement Dual-frequency Precipitation Radar measurements. *Remote Sens. Environ.* **2019**, *231*, 111213. [CrossRef]
38. Zhou, Y.; Kim, H.; Guan, B. Life Cycle of Atmospheric Rivers: Identification and Climatological Characteristics. *J. Geophys. Res. Atmos.* **2018**, *123*, 12715–12725. [CrossRef]
39. Zhou, R.; Wang, G.; Zhaxi, S. Cloud vertical structure measurements from a ground-based cloud radar over the southeastern Tibetan Plateau. *Atmos. Res.* **2021**, *258*, 105629. [CrossRef]
40. Bergeron, T. Operation and results of "Project Pluvius". *Phys. Precip. Geophys. Monogr.* **1960**, *5*, 152–157.
41. Bergeron, T. *Problems and Methods of Rainfall Investigation: Address of the Honorary Chairman of the Conference*; American Geophysical Union: Washington, DC, USA, 1960.
42. Wang, G.; Zhou, R.; Zhaxi, S.; Liu, S. Raindrop size distribution measurements on the Southeast Tibetan Plateau during the STEP project. *Atmos. Res.* **2021**, *249*, 105311. [CrossRef]

remote sensing

Article

Distinct Impacts of Two Types of Developing El Niño–Southern Oscillations on Tibetan Plateau Summer Precipitation

Minghong Liu, Hong-Li Ren *, Run Wang, Jieru Ma and Xin Mao

State Key Laboratory of Severe Weather, and Institute of Tibetan Plateau Meteorology, Chinese Academy of Meteorological Sciences, Beijing 100081, China
* Correspondence: renhl@cma.gov.cn

Abstract: El Niño–Southern Oscillation (ENSO) has remarkable impacts on Tibetan Plateau (TP) summer precipitation. However, recently identified ENSO spatial diversity brings complexity to these impacts. This study investigates the distinct impacts of the Eastern Pacific (EP) and Central Pacific (CP) ENSOs on TP summer precipitation based on numerous precipitation data and satellite-observed and reanalyzed circulation data. The results show that the EP El Niño and the CP La Niña have opposite effects on summer precipitation in the southwestern TP, with significant decreases and increases, respectively, indicating a trans-type inversion. In contrast, the CP El Niño causes significant decreases in summer precipitation in the central-eastern TP. No significant anomaly occurs during the EP La Niña. Moisture budget and circulation analyses suggest that these distinct precipitation characteristics can be attributed to different atmospheric teleconnections, which provide varying vertical motion and moisture conditions. The EP El Niño triggers an atmospheric response similar to the Indian Summer Monsoon–East Asian Summer Monsoon teleconnection, and the CP La Niña generates a teleconnection in the opposite phase. In contrast, the CP El Niño mainly causes a teleconnection resembling the East Asian–Pacific pattern. This study may deepen our understanding of ENSO impacts on TP summer precipitation and have implications for regional climate predictions.

Keywords: Tibetan Plateau precipitation; El Niño–Southern Oscillation; ENSO diversity; ENSO teleconnection

Citation: Liu, M.; Ren, H.-L.; Wang, R.; Ma, J.; Mao, X. Distinct Impacts of Two Types of Developing El Niño–Southern Oscillations on Tibetan Plateau Summer Precipitation. *Remote Sens.* **2023**, *15*, 4030. https://doi.org/10.3390/rs15164030

Academic Editor: Gang Zheng

Received: 29 June 2023
Revised: 30 July 2023
Accepted: 11 August 2023
Published: 14 August 2023

1. Introduction

The Tibetan Plateau (TP), known as the "third pole", is the largest and highest plateau in the world, with an average elevation of approximately 4000 m and the largest non-polar ice reservoir [1,2]. Given its unique elevation and geographic position, TP plays a critical role in shaping and interacting with the global climate via dynamical blocking and thermal effects on large-scale atmospheric circulations [3–6]. The TP is also well-known as the "Asian water tower" because of its substantial role in the hydroclimate, which supplies water resources to the streamflow and runoff of several major rivers in Asia, including the Yangtze River, the Yellow River, and the Mekong River. Thus, the TP has a notable impact on the balance of downstream ecosystems and anthropogenic activities and livelihoods [7,8].

Precipitation is one of the most important factors modulating the TP hydroclimate, especially during its rainy season [9,10]. The variation in TP precipitation significantly impacts large-scale atmospheric circulations and climate variability in downstream East Asia and even globally through enormous latent heat releases and associated heating feedback [4,5,11]. Therefore, it is of great importance to investigate the interannual variability of TP precipitation and understand the underlying mechanisms.

The rainy season of the TP generally starts in May at the onset of the boreal (omitted in the following) summer monsoon and lasts until September, accounting for around 80 per-cent of the total annual precipitation [9,12]. Despite the limited data available on TP precipitation, many efforts have been devoted to studying the external factors that contribute to the interannual variation of TP summer precipitation, such as the summer

North Atlantic Oscillation [13,14] and the Indian summer monsoon (ISM) [15–17]. Sea surface temperature (SST) anomalies in the Atlantic Ocean and Indian Ocean also modify TP summer precipitation [9,18–20]. The El Niño–Southern Oscillation (ENSO), the dominant mode of interannual variability in the tropical Pacific Ocean, has also been revealed to impact TP summer precipitation remotely [21–23]. However, the impact of ENSO on TP summer precipitation is typically indirect and mediated by various factors. Previous studies mainly focused on TP precipitation anomalies during the decaying phase of ENSO. El Niño-associated basin-wide warming of the Indian Ocean SST may significantly delay the outbreak of the South Asian summer monsoon, thus reducing precipitation over the southeastern TP in the early rainy season [9,18,24,25]. The Indian Ocean basin-wide SST anomaly can also impact TP precipitation by adjusting the meridional position of the South Asian High (SAH) [26,27]. Moreover, the SST anomaly over the tropical northern Indian Ocean and western Pacific Ocean can persist during summer through the local air–sea interaction. This persistent SST anomaly has been revealed to play an important role in maintaining and enhancing the anomalous anticyclonic circulation over the western North Pacific during the decaying summer of El Niño, known as the Indo-western Pacific Ocean capacitor mechanism [28,29]. This enhanced anticyclonic circulation leads to increased water vapor transported to the TP, thus increasing summer TP precipitation [30,31].

Furthermore, some studies link the TP summer precipitation anomaly to the Pacific SST variation during the developing summer of ENSO [21,22]. ENSO-induced anomalous convective heating directly modifies the upper-tropospheric thermal status and influences the subtropical westerly winds covering the TP, thereby impacting its precipitation [32,33]. The developing ENSO also influences TP precipitation through its teleconnection impacts in the Indian Ocean. Many studies have suggested that ENSO can influence the intensity of ISM precipitation by altering the Walker circulation, and the suppressed ISM precipitation stimulates an upper-tropospheric cyclonic circulation anomaly to the west of the TP as part of the ISM-East Asian Summer Monsoon (EASM) teleconnection [34–36], which significantly reduces precipitation in the southwestern TP (SWTP) [15,37]. Meanwhile, the altered Walker circulation can also strengthen the India–Burma monsoon trough through regional air–sea interaction over the Indo-western Pacific Ocean [38,39], modifying the moisture transported to the TP and thereby affecting TP precipitation. In addition, ENSO-associated variations in TP local variables, such as the snow cover and water reserve, also modulate TP precipitation through coupled feedback [40,41].

Besides the various direct and indirect mechanisms of climate impacts, ENSO itself has characteristics of spatial diversity, which adds complexity to the topic of ENSO impacts [42–45]. In recent decades, a new type of El Niño, namely, the Central–Pacific (CP) El Niño, has frequently occurred. Compared with the canonical Eastern–Pacific (EP) El Niño, the SST anomaly center of the CP El Niño is westward shifted. Many studies have emphasized the substantial influence of ENSO diversity on its climate impacts [46–50]. For example, evident differences have been found between the impacts of the two types of El Niños on precipitation and air temperature in China [51–53]. Similarly, La Niña can also be classified into two types that exhibit distinct local and teleconnection responses based on their SST anomaly patterns [54], although whether these two types of La Niña are intrinsically distinguished remains controversial [55].

Some intermediate factors and processes linking ENSO and TP precipitation, such as the Indian Ocean air–sea interaction, exhibit distinct responses to different types of ENSOs [56,57]. However, the resulting variation of the TP precipitation response to ENSO has received less attention. A quasi-three-year cycle of interannual variability in precipitation (less–more–more) during the Tibetan rainy season was once proposed, which corresponds to an evolution consisting of the development of an EP El Niño, a shift from a CP El Niño to an EP La Niña, and the decaying of a CP La Niña [22]. However, realistic ENSO events do not always follow such an evolution. The precipitation response to different types of ENSOs remains unclear. This study aims to investigate TP summer precipitation variation due to

ENSO spatial diversity and the underlying mechanisms from the perspective of historical ENSO events. For practice, this study focuses on the developing summer of ENSO.

The remainder of the paper is organized as follows. Section 2 introduces the data and method utilized in this study. Section 3.1 presents the characteristics of the TP summer precipitation anomaly during different types of ENSOs, and the causes for these distinct characteristics are explored in Section 3.2 by examining large-scale circulation and moisture anomalies. Section 4 provides a brief discussion. The main conclusions are finally listed in Section 5.

2. Data and Methods

Four high-resolution gridded precipitation datasets are employed in this study, including Climate Research Unit Time-Series (CRU) version 4.05 [58], Global Precipitation Climatology Centre (GPCC) full data reanalysis version 2020 [59], CN05.1 precipitation [60], and Asian Precipitation-Highly–Resolved Observational Data Integration Towards Evaluation (APHRODITE) gridded precipitation versions 1101 and 1101EX-R1 [61]. These datasets are mainly based on dense gauge data networks, and some have undergone satellite-based quality control. We also perform an ensemble mean of these precipitation datasets using equal weights to remove inter-dataset bias. Observational precipitation data from 2400 stations of the China Meteorological Data Sharing Service System are used as a reference for the grid datasets.

The data for the outgoing longwave radiation (OLR), which is commonly used as a remote sensing proxy for convective activity, are obtained from the Interpolated Outgoing Longwave Radiation dataset provided by the National Oceanic and Atmospheric Administration (NOAA) Physical Sciences Laboratory [62]. Other atmospheric variables are from the European Centre for Medium-Range Weather Forecasts reanalysis generation 5 (ERA5) [63], and the SST data is obtained from the HadISST supplied by the Met Office Hadley Center [64]. All data utilized in this study are re-interpolated into a horizontal resolution of $1^\circ \times 1^\circ$ in a longitude/latitude grid. The analyzing period ranges from January 1961 to December 2020 (to December 2015 for the APHRODITE precipitation data), ensuring sufficient coverage of historical ENSO events and relatively reliable data. Note that the OLR data from NOAA starts in 1975, and data from ERA5 complements the earlier period. A linear trend was removed before analysis. Anomalies are obtained by removing a climatological mean derived from the whole analyzing period and conducting a 3-month running average to filter the noise. A composite analysis is utilized to investigate the common features of specific events. The statistical significance of our results is determined using the two-tailed Student's t-test. The "summer" in this study is defined as June–July–August–September (JJAS), the significant rainy season of the TP. The qualitative conclusions remain the same if using June–July–August or May–September.

ENSO events are identified when the 3-month running mean of the Niño3.4 index (SST anomalies averaged in the region of 5°S–5°N, 170°–120°W) exceeds a threshold of 0.5 °C for a minimum of five consecutive months [53]. Then we classify the identified ENSO events based on the spatial distribution of SST anomalies during JJAS preceding the ENSO peak, from which both spatial types of El Niño or La Niña can be effectively distinguished [54]. Specifically, events with the largest SST anomaly center located east (or west) of 150°W are classified into the EP (or CP)-type developing ENSO. In the analyzing period, a total of ten EP-type developing El Niños (1963, 1965, 1969, 1972, 1976, 1982, 1987, 1997, 2009, and 2015) and seven CP-type developing El Niños (1968, 1977, 1991, 1994, 2002, 2004, and 2006) are selected. Meanwhile, there are ten EP-type developing La Niñas (1964, 1970, 1973, 1984, 1988, 1995, 2007, 2010, 2017, and 2020) and five CP-type developing La Niñas (1975, 1999, 2000, 2008, 2011). The El Niño event in 2018 did not develop until autumn, and the La Niña event in 1983 developed accompanying a strong warming SST anomaly in the far-eastern tropical Pacific Ocean; these two events are excluded for clearly displaying the TP summer precipitation response to the developing ENSO. Moreover, to quantitatively measure the spatial characteristics of ENSO, the cold-tongue index (CTI) and warm-pool index (WPI) are calculated [45]. CTI is defined as $N_3 - \alpha \times N_4$, and WPI is $N_4 - \alpha \times N_3$, where N_3 is the mean SST anomalies over the Niño3 region (5°S–5°N, 150°–90°W) and N_4 is over the Niño4 region (5°S–5°N, 160°E–150°W). Parameter α is 0.4 if $N_3 \times N_4 > 0$; otherwise, α is set to 0.

An atmospheric moisture budget analysis [37] is conducted to understand the processes responsible for regional precipitation anomalies based on the following equation:

$$P' = E' - < u\frac{\partial q}{\partial x} >' - < v\frac{\partial q}{\partial y} >' - < \omega\frac{\partial q}{\partial p} >' + \xi \quad (1)$$

where the prime denotes the monthly anomaly, and the bracket denotes the vertical integration from the surface to the top of the atmosphere. P, E and ξ are the precipitation, evaporation and residual terms, and the others are advection terms. u, v, ω, and q, separately represent the zonal and meridional winds, the vertical pressure velocity, and the specific humidity. The term $- < \omega\frac{\partial q}{\partial p} >'$ can be further decomposed into two linear advection terms ($- < \omega'\frac{\partial \bar{q}}{\partial p} >$ and $- < \bar{\omega}\frac{\partial q'}{\partial p} >$) corresponding to the anomalous vertical motion and local moisture variation separately, and a nonlinear term consisting of nonlinear advection and a sub-monthly transient ($- < \omega'\frac{\partial q'}{\partial p} >$ + sub). The other two advection terms can be decomposed similarly.

3. Results

3.1. TP Summer Precipitation Anomalies during Different Types of ENSOs

We first show composites of TP precipitation anomalies during the developing summer of different types of ENSOs (Figure 1). The figures presented in the main text are all based on the ensemble mean of four gridded precipitation datasets, and the main conclusions of this study remain consistent across all datasets (Figure S1). If not distinguishing spatial types of ENSOs, TP summer precipitation generally decreases relative to the climatological mean during El Niño events, especially in the southwest and east (Figure 1a). This result is consistent with previous studies suggesting that SWTP summer precipitation is negatively correlated with simultaneous SST warming anomalies in the tropical eastern Pacific, e.g., [22]. Figure 1b,c further show the precipitation anomaly pattern corresponding to the two spatial types of El Niños, and it is clear that the two anomaly centers in Figure 1a are attributed to different types of El Niño. Specifically, the negative summer precipitation anomaly in the SWTP mainly occurs during the EP El Niños (Figure 1b), while the extensive precipitation decrease in the central-eastern TP (CETP) primarily exhibits during the CP El Niño (Figure 1c). The latter is rarely noted in previous studies.

The situation differs for TP precipitation anomalies during the developing summer of La Niña. Significant positive anomalies only occur in the SWTP in the composite for all La Niña events, while nothing is observed for the CETP (Figure 1d). After further classifying their spatial types, no significant anomalies can be observed during the EP La Niña, while CP events accompany statistically enhanced SWTP summer precipitation. Combining the results during El Niños, we can conclude that both EP and CP types of ENSO exhibit asymmetric impacts on TP summer precipitation between El Niño and La Niña phases. Moreover, the TP summer precipitation anomaly pattern during the CP La Niña appears to be opposite to that observed during the EP El Niño rather than the CP El Niño, suggesting a trans-type inversion. The anomaly in the CETP is unique to the CP El Niño.

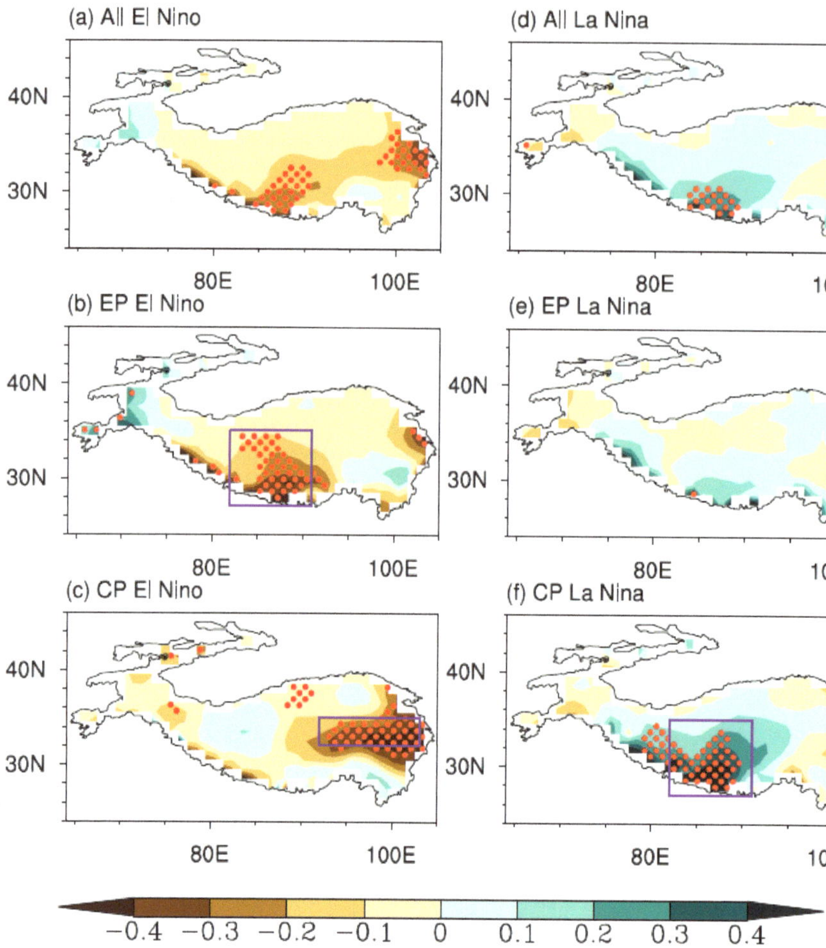

Figure 1. Composites of JJAS precipitation (unit: mm·day^{-1}) over the TP for different types of ENSO events. The red dotted areas are statistically significant at the 90% confidence level.

The asymmetric impacts due to ENSO diversity can be directly reflected in the spatial distribution of correlation between TP summer precipitation and the simultaneous Niño3.4 index. The correlation is high in the SWTP, while it is very low in the CETP (Figure 2a). We further define two regional precipitation indices to examine the relationship between TP summer precipitation and different types of ENSOs event by event. One is an SWTP precipitation index calculated as the area mean of the summer precipitation anomaly over the region located at 27°–35°N, 82°–91°E, and another is a CETP precipitation index similarly defined as the mean anomaly over the region located at 32°–35°N, 92°–103°E. Purple boxes mark these two regions in Figure 1. Even though SWTP summer precipitation strongly depends on ENSO intensity with a linear coefficient of around −0.66 within all ENSO events (Figure 2b), the role of ENSO diversity can still be observed. It is mainly EP ENSO events that contribute to the linear correlation, while the CP ENSO's impacts are nonlinear. Specifically, SWTP summer precipitation decreases during all EP El Niño events except that in 1963, but no regularity presents during CP El Niño events. For La Niñas, SWTP summer precipitation increases during only some strong EP events, while it increases during all CP events even with relatively weak ENSO intensity.

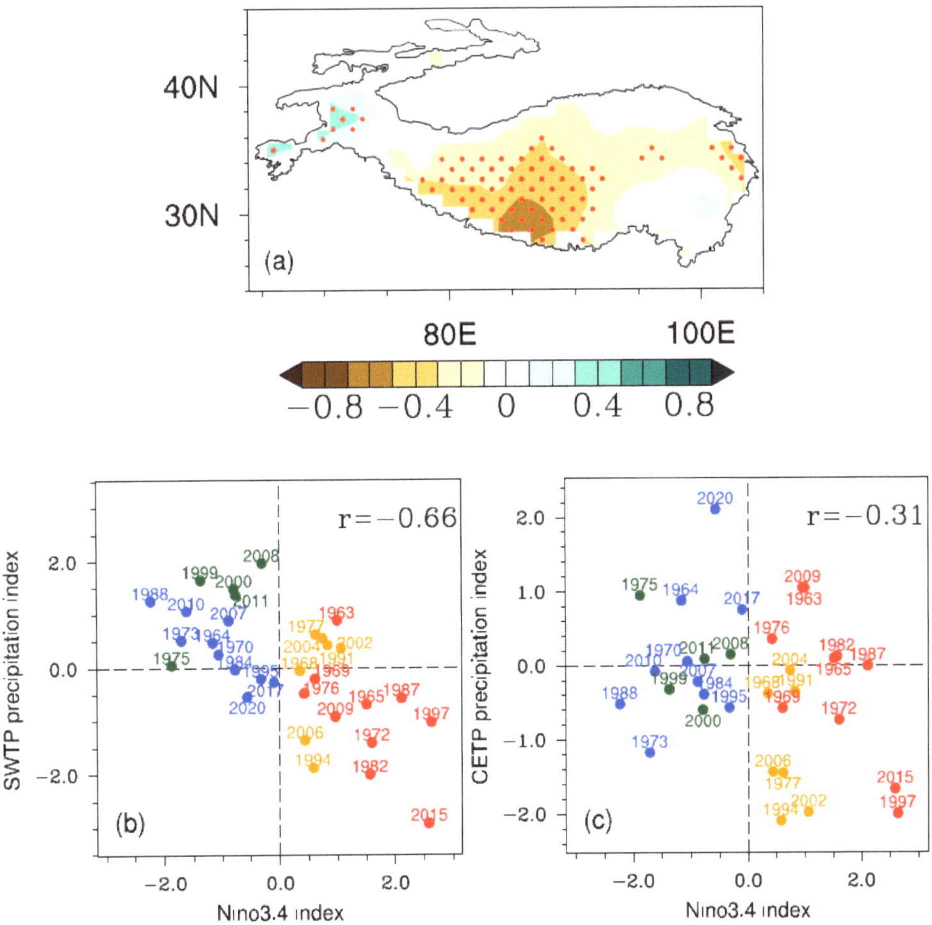

Figure 2. (**a**) Correlation distribution between TP summer precipitation anomalies and the Niño3.4 index (unit: 1). Red dots indicate correlation coefficients statistically significant at the 90% confidence level. The scatterplots show the normalized (**b**) SWTP precipitation index (unit: 1) and (**c**) CETP precipitation index (unit: 1) as a function of the normalized Niño3.4 index (unit: 1) in JJAS for all ENSO events. The red, orange, blue, and green solid circles separately indicate EP El Niño, CP El Niño, EP La Niña, and CP La Niña events. The colored text notes the developing year of ENSO events. The text in the bottom right presents the linear correlation coefficient between the TP precipitation index and the Niño3.4 index for all ENSO events.

In contrast, CETP summer precipitation is less connected with ENSO intensity, with a linear correlation coefficient of just −0.31 (Figure 2c). It only regularly decreases during CP El Niño events but randomly varies during other types of developing ENSOs. The above analyses highlight the role of ENSO spatial diversity in impacting TP summer precipitation, which makes these ENSO impacts asymmetric and trans-type inverse.

3.2. Circulation and Moisture Anomalies Associated with Different Types of ENSOs

How do these distinct TP summer precipitation anomalies form during different types of developing ENSOs? On an interannual timescale, vertical motion and large-scale moisture transport are crucial factors controlling TP summer precipitation variation [37].

Figure 3 illustrates anomalous summer upper-tropospheric circulation during different types of ENSOs. During the EP El Niño, a pair of cyclonic circulation anomalies characterize the subtropical upper troposphere. These anomalies are located to the west of the TP and over northern China, respectively, and are associated with evident negative geopotential anomalies (Figure 3a). These cyclonic circulation anomalies are equivalent-barotropic, with cyclonic circulation anomalies occurring at the low level (Figure 4a). These two equivalent-barotropic cyclonic circulation anomalies relate the ISM to the EASM, referred to as the ISM-EASM teleconnection [34,35]. Meanwhile, positive geopotential anomalies exist over the northern Indian Ocean and western North Pacific regions. These low-latitude anomalies are the Kelvin and Rossby wave responses directly induced by the enhanced convection in the tropical Pacific associated with El Niño [32]. In this case, significant westerly wind anomalies occur to the south of the two cyclonic circulation anomalies, crossing the SWTP (Figure 3a). This anomaly pattern suggests a southward-shifted SAH and a subtropical westerly jet that cause descending motion anomalies over the SWTP, thus decreasing rainfall [24–27].

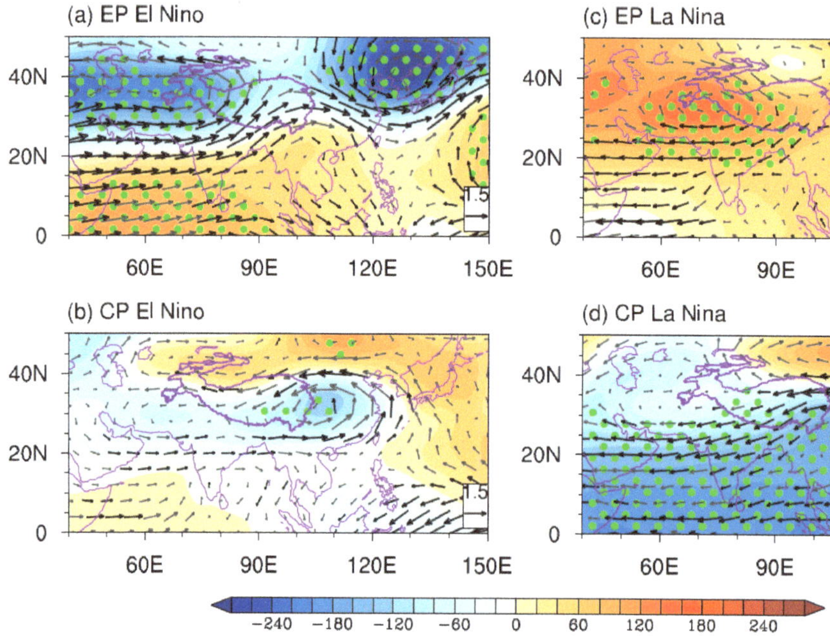

Figure 3. Composites of JJAS-mean 200-hPa geopotential (shading, unit: $m^2 \cdot s^{-2}$) and wind anomalies (vectors, unit: $m \cdot s^{-1}$) for different types of ENSO events. Green dots indicate geopotential anomalies statistically significant at the 90% confidence level. Dark-colored vectors denote the wind anomalies whose zonal or meridional components are significant at the 90% confidence level.

However, the situation changes during the developing summer of CP El Niño. Compared with those of EP El Niño, upper-tropospheric positive geopotential anomalies over the northern Indian Ocean and anomalous cyclonic circulation to the west of the TP disappear. Instead, the CP-El Niño-associated circulation pattern manifests as a significant meridional equivalent-barotropic dipole over mid-latitude East Asia, which features a significant cyclonic circulation anomaly over eastern China and an anticyclonic circulation anomaly north of it (Figures 3b and 4b). This circulation pattern resembles the so-called East Asia–Pacific (EAP) pattern (also known as the Pacific–Japan pattern) [65–67]. This result is reasonable, as previous studies have implied that the EAP pattern is more clearly shown in the developing summer of CP ENSO rather than EP events [65,68]. This EAP-like pattern's upper-tropospheric cyclonic circulation anomaly covers the CETP with significant easterly wind anomalies, suggesting an abnormal descending motion. Therefore, the significant region of the precipitation anomaly switches to the CETP during the CP El Niño (Figure 1c).

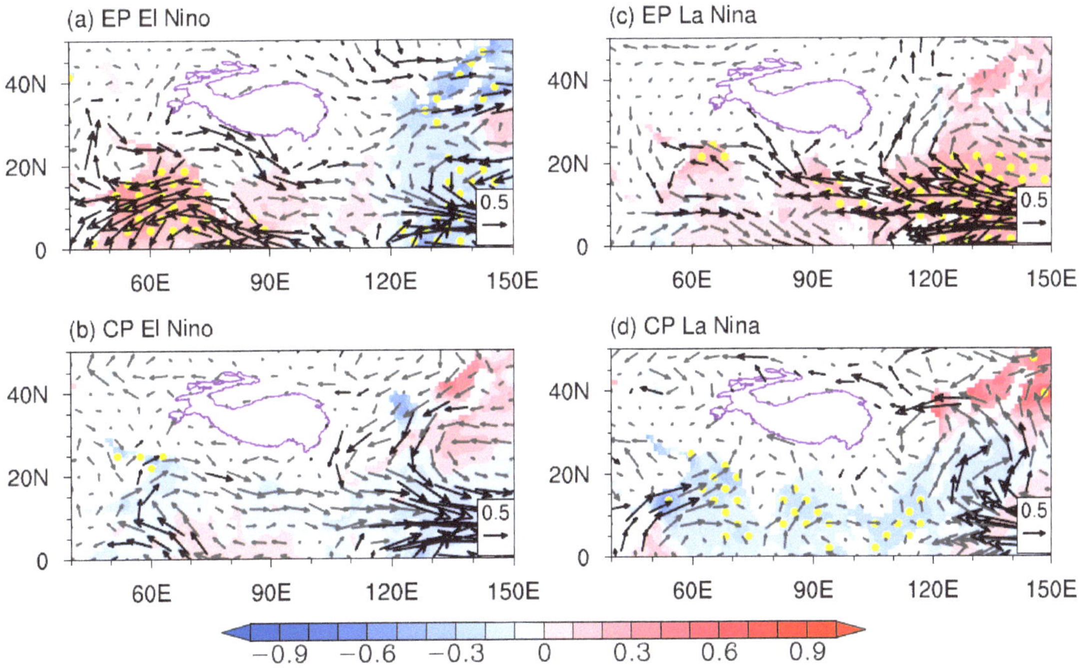

Figure 4. Composites of JJAS-mean SST (shading, unit: °C) and 850-hPa wind (vectors, unit: 10 $m \cdot s^{-1}$) anomalies for different types of ENSO events. Yellow dots indicate SST anomalies statistically significant at the 90% confidence level. Dark-colored vectors denote the wind anomalies whose zonal or meridional components are significant at the 90% confidence level.

During the developing summer of EP La Niña events, a pair of significant equivalent-barotropic anticyclonic circulation anomalies locate west of the TP and over northern China, respectively, and the western one covers the SWTP (Figures 3c and 4c). Therefore, the SWTP might be controlled by anomalous ascending motion. This upper-tropospheric circulation anomaly can also be seen as an ISM-EASM teleconnection pattern featuring anticyclonic circulation anomalies. However, it should be noted that there are few negative geopotential anomalies in the upper troposphere over the subtropical Indo-western Pacific Ocean region, suggesting a relatively weak response of the tropical upper troposphere to the EP La Niña. In this case, less easterly wind anomalies occur over subtropical Eurasia, mainly northerly wind anomalies dominating the SWTP. Several previous studies have suggested that the variation in the subtropical westerly jet related to the zonal wind anomaly is a key factor that affects anomalous vertical motion and precipitation over the SWTP [25,26,37]. Thus, the circulation anomaly associated with EP La Niña events has little influence on SWTP precipitation.

During the CP La Niña, the upper-tropospheric circulation anomaly pattern features intense negative geopotential anomalies over the low-latitude Indo-western Pacific Ocean region and a pair of equivalent-barotropic anticyclonic circulation anomalies located to the west of the TP and over northern China, although the one west of the TP is less significant (Figures 3d and 4d). This pattern is essentially an ISM-EASM teleconnection pattern of the opposite phase to its counterpart during the EP El Niño. In this case, strong easterly wind anomalies dominate subtropical Eurasia, including the SWTP. This circulation anomaly pattern may suggest a northward-shifted SAH [26] with an abnormal ascending motion over the SWTP.

We also examine anomalous vertically-integrated vapor flux during the developing summer of different types of ENSOs. As shown in Figure 5a, an intense anomalous southeastward vapor flux occurs south of the TP during EP El Niño events. This anomalous flux suppresses the ISM precipitation and hinders the import of moisture to the TP from its southwestern boundary. Meanwhile, anomalous eastward vapor flux dominates moisture transport over the TP under strong upper-tropospheric westerly wind anomalies, also contributing to a significant moisture deficiency over the SWTP. Therefore, besides the effect of anomalous descending motion, the anomalous moisture transport during EP El Niño events also benefits the decrease in SWTP precipitation. In addition, the suppressed ISM results in an extensive negative convection anomaly over the Indian subcontinent (Figure 6a), which has been revealed as a key factor in triggering the ISM-EASM teleconnection [15,35].

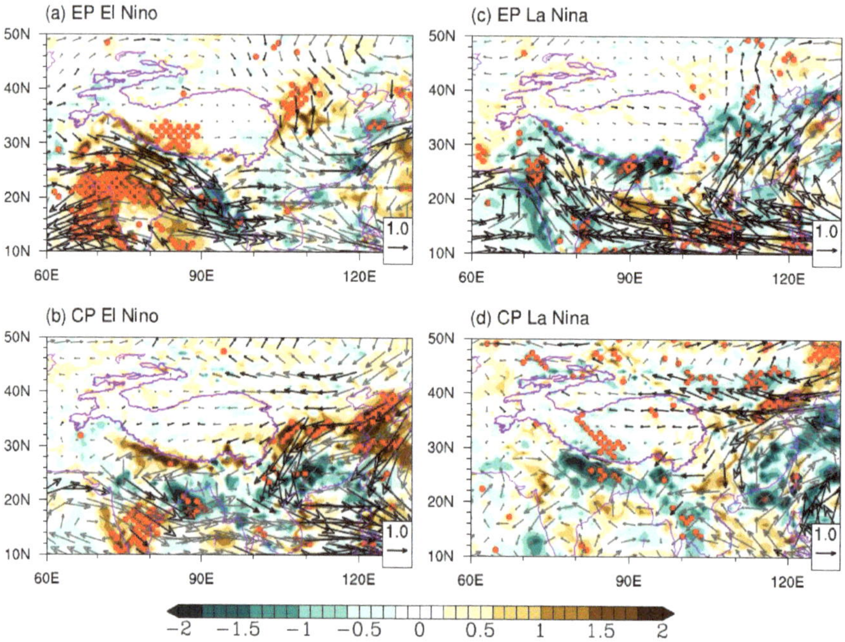

Figure 5. Composites of JJAS-mean vertically integrated vapor flux (vectors, unit: 10 kg·m^{-1}·s^{-1}) and its divergence (shading, unit: 10^{-5} kg·m^{-2}·s^{-1}) anomalies for different types of ENSO events. Red dots indicate divergence anomalies statistically significant at the 90% confidence level. Dark-colored vectors denote the flux anomalies whose zonal or meridional components are significant at the 90% confidence level.

During the CP El Niño, in contrast, the anomalous moisture transport from the Indian Ocean and the suppressed convection on the Indian subcontinent are not significant (Figures 5b and 6b). The vapor flux anomaly imported to the TP mainly comes from eastern China and the North Pacific, consistent with the upper-tropospheric easterly wind anomalies over the TP (Figure 3b). However, the moisture convergence anomaly over the CETP is negligible, partly because the regional descending motion anomaly may suppress the moisture convergence there. This result suggests that unlike during the EP El Niño, anomalous moisture transport may play a minor role in causing negative CETP precipitation anomalies during the CP El Niño.

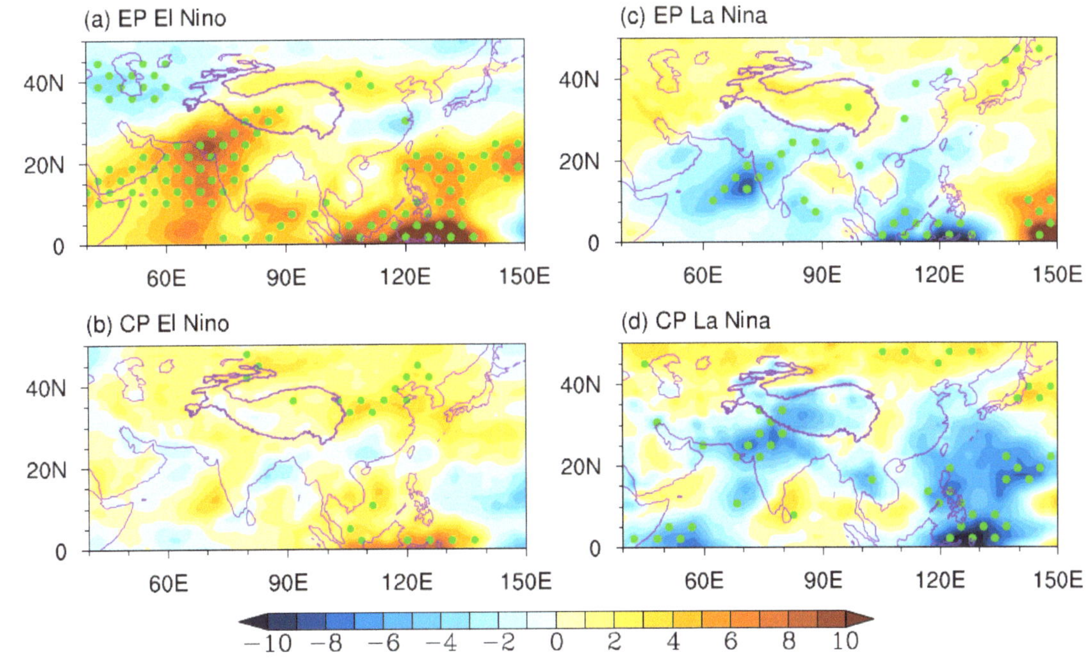

Figure 6. Composites of JJAS-mean OLR (shading, unit: W·m^{-2}) anomalies for different types of ENSO events. Green dots indicate anomalies statistically significant at the 90% confidence level.

The moisture conditions during the two types of La Niña are not simply symmetric to those observed during El Niños. As shown in Figure 5c, there is no significant anomalous vapor flux over the TP. South of the TP, intense northward vapor flux anomalies occur, but they mostly converge over the Himalayan foothills. Only a tiny amount of anomalous moisture can cross the Himalayas and reach the SWTP. Therefore, the lack of a significant precipitation increase in the SWTP can be attributed to unfavorable moisture conditions, despite the potential presence of abnormal ascending motion. In contrast, during the CP La Niña, subtropical easterly winds associated with equivalent-barotropic cyclonic circulation anomalies lead to strong westward vapor flux over the TP. This flux transports abundant moisture anomalies from eastern China and the western North Pacific to the SWTP (Figure 5d). In addition to the regional anomalous ascending motion, the resulting moisture convergence anomaly also contributes to the enhanced summer precipitation in the SWTP during the CP La Niña.

Based on the above analysis, we have concluded that various vertical motion and moisture conditions are responsible for the distinct TP precipitation anomaly during different types of ENSOs. We further apply an atmospheric moisture budget analysis within the domain region of the precipitation anomaly to examine the specific role of these vertical motion and moisture conditions. As shown in Figure 7a, SWTP summer precipitation reduction associated with the EP El Niño can be primarily attributed to anomalous vertical moisture advection: $-<\omega\frac{\partial q}{\partial p}>'$. Moreover, the climatological moisture advected by descending anomalies, $-<\omega'\frac{\partial \bar{q}}{\partial p}>$, and the anomalous moisture advected by climatological vertical motion, $-<\bar{\omega}\frac{\partial q'}{\partial p}>$, are comparable, suggesting that EP El Niño-related descending motion and moisture deficiency contribute to the precipitation decline in the SWTP to a similar extent. For the positive CETP precipitation anomaly associated with CP El Niño, anomalous vertical moisture advection, $-<\omega\frac{\partial q}{\partial p}>'$, remains the dominant contributor, but the climatological moisture advected by descending anomalies, $-<\omega'\frac{\partial\bar{q}}{\partial p}>$, singly contributes to the CETP precipitation decrease (Figure 7b). This diagnostic result agrees with the above analysis based on the composites; that is, the descending anomalies induced by the CP El Niño play a dominant role in precipitation reduction, but the effect of moisture transport is negligible (Figures 5b and 6b). Evaporation also makes a positive contribution.

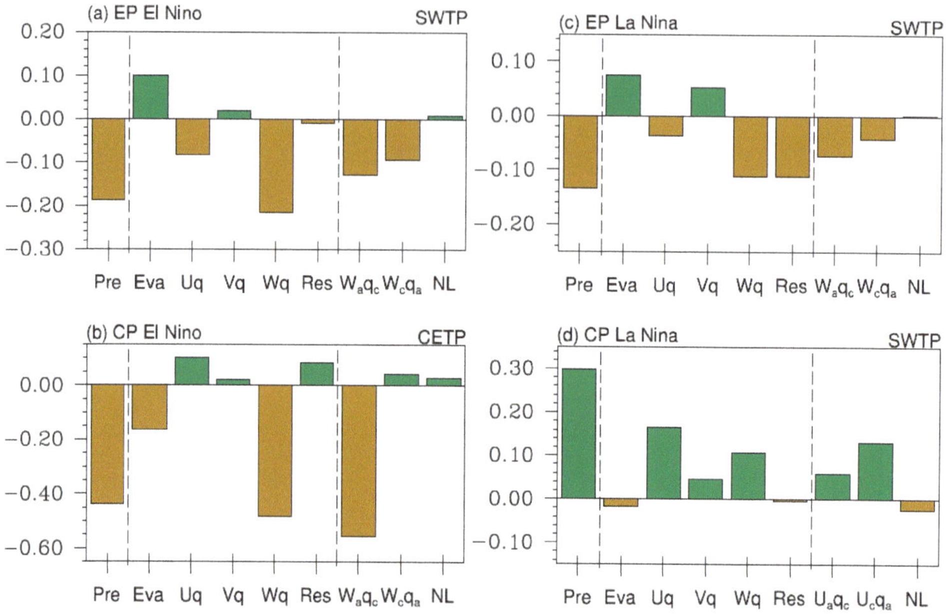

Figure 7. Moisture budget analysis of JJAS-mean atmospheric moisture (**a**) for the SWTP region associated with the EP El Niño, (**b**) the CETP region associated with the CP El Niño, (**c**) the SWTP region associated with the EP La Niña, and (**d**) the SWTP region associated with the CP La Niña. The left six terms in turn represent the precipitation anomaly, the evaporation anomaly, the anomalous zonal moisture advection, the anomalous meridional moisture advection, the anomalous vertical moisture advection, and the residual. The three rightmost terms in (**a**–**c**) are the decomposition of the anomalous vertical moisture advection, namely, the linear term related to the anomalous vertical motion, the linear term related to the moisture variation, and the nonlinear term. Similarly, in (**d**), they represent the linear and nonlinear decomposition terms of anomalous zonal moisture advection. The green and brown colors denote positive and negative anomalies, respectively. Terms associated with the EP (CP) El Niño are obtained by regressing them against the JJAS-mean CTI (WPI) index in all El Niño years, and those for La Niña are similarly obtained and multiplied by a factor of −1. Units are mm·day^{-1}·°C^{-1}.

The EP La Niña-associated residual term in the moisture budget within the SWTP is almost equivalent to the precipitation variability (Figure 7c), suggesting a less robust relationship between SWTP precipitation and the EP La Niña. This result is reasonable as the precipitation anomaly in the composite is insignificant (Figure 1e). The primary cause of the increased summer precipitation in the SWTP associated with the CP La Niña is the anomalous zonal moisture advection $-<u\frac{\partial q}{\partial x}>'$, and the most significant positive contributor to this advection is the anomalous moisture advected by climatological zonal wind $-<\bar{u}\frac{\partial q'}{\partial x}>$ (Figure 7d). This result confirms the above analysis that emphasizes the role of large-scale moisture transport in enhancing SWTP precipitation during the CP La Niña (Figure 5d). In addition, the anomalous vertical moisture advection, $-<\omega\frac{\partial q}{\partial p}>'$, also has a positive effect on the precipitation, which mainly results from the anomalous ascending motion.

To summarize, different types of ENSOs trigger various teleconnections during their developing summers, which alter the vertical motion and moisture conditions that affect TP precipitation. Generally, the EP El Niño can induce a circulation pattern resembling the ISM-EASM teleconnection over subtropical Eurasia, which results in significant descending motion as well as moisture deficiency over the SWTP region, together reducing the regional precipitation. Similarly, the CP La Niña can induce an ISM-EASM teleconnection in the opposite phase, which transports a large amount of anomalous moisture to the SWTP region and causes ascending motion, increasing the regional precipitation. In contrast, the CP El Niño mainly stimulates an EAP-like teleconnection pattern over East Asia, which tends to reduce the summer precipitation in the CETP region mainly by leading to a strong regional descending motion anomaly. Finally, the relationship between the EP La Niña and the vertical motion and moisture anomalies over the TP is less robust, although there may also be an ISM-EASM-like teleconnection.

4. Discussion

The causes of these ENSO teleconnection variations during different types of ENSOs are complex and multifaceted. As mentioned above, the ISM convection anomaly is critical to triggering the ISM-EASM teleconnection. It is well known that the EP El Niño can induce a Walker circulation anomaly, leading to strong downward motion over the Indo-western Pacific regions (Figure 6a), thus weakening the ISM convection [36]. Meanwhile, the EP El Niño tends to be accompanied by a positive Indian Ocean Dipole (IOD)-like pattern, which contributes to the suppressed ISM as its warming SST anomaly reduces the meridional land-sea thermal contrast (Figure 4a). However, the CP El Niño has less impact on the Indian Ocean and subcontinent (Figures 4b and 6b) because of its weak intensity and westward-shifted warming centre [56,57], which may account for the absence of a clear ISM-EASM teleconnection pattern. Meanwhile, compared with the EP El Niño, the CP El Niño induces a very westward extended western Pacific subtropical high (Figure 4a,b) [47,69], which may lead to anomalous convective activity over the western North Pacific and trigger an EAP-like teleconnection [65].

The EP La Niña also accompanies air-sea anomalies in the Indian Ocean and subcontinent (Figure 4c) because of an enhanced ascending branch of Walker circulation over the Indo-Pacific warm-pool region (Figure 6c). As a result, the enhanced convection associated with the ISM over the Indian subcontinent will excite an ISM-EASM teleconnection featuring equivalent barotropic anticyclonic anomalies in the subtropics. However, as shown in Figure 3c, there is no significant upper-tropospheric negative geopotential anomaly over the northern Indian Ocean during the EP La Niña. This may be because the Pacific cooling anomaly of the EP La Niña is located east, and some of the selected samples are relatively weak. Therefore, it cannot trigger a strong Kelvin wave response in the upper troposphere over the tropical Indian Ocean.

In contrast, the CP La Niña-associated Pacific SST anomaly is located further west and has a much broader meridional scale [47]. Despite having a similar or weaker intensity compared with the EP La Niña, the CP La Niña can induce extensive negative geopotential anomalies over the Indo-Pacific warm-pool region. Meanwhile, although the ascending branch of the CP La Niña-associated anomalous Walker circulation is confined to the Pacific sector (Figure 6d), there are still negative SST anomalies in the northern Indian Ocean, which serve to enhance the ISM and induce an ISM-EASM teleconnection featuring anticyclonic circulation anomalies. We note that all CP La Niña events selected in this study follow another La Niña in the preceding winter. Thus, those negative SST anomalies in the northern Indian Ocean may also be attributed to the preceding La Niñas through the Indo-western Pacific Ocean capacitor mechanism [28,29]. The above discussion provides some possible explanations for the varying teleconnection patterns during different types of ENSOs based on composites of limited historical events. However, these processes are not mutually exclusive, and sometimes, these teleconnection patterns may coexist. Therefore, further studies are needed to identify the critical factor that determines whether the ISM-EASM teleconnection or the EAP teleconnection dominates during different types of ENSOs.

Previous studies have revealed that the ENSO regime and its connection with the East Asian climate have experienced a decadal shift [70–72]. Whether the teleconnection impacts on TP summer precipitation also exhibit similar variations and what the underlying mechanisms are need to be investigated in the future. This study highlights the importance and necessity of considering spatial diversity in investigating ENSO impacts on the TP climate. The newly found significant impact of the CP El Niño on CETP precipitation provides more sources of predictability for TP summer precipitation, which may have implications for regional climate prediction, especially since the CP El Niño tends to occur frequently under global warming [44]. In fact, this predictability has been suggested by some numerical simulations and climate prediction products [73,74]. However, whether these distinct ENSO teleconnection impacts and underlying mechanisms can be reasonably reproduced in climate models or numerical climate prediction systems remains unclear and deserves further investigation.

5. Conclusions

Based on multiple precipitation data and satellite-observed and reanalyzed circulation data, this study investigated the distinct impacts of the two spatial types of ENSOs on TP summer precipitation and analyzed the associated atmospheric teleconnections, reaching the following conclusions.

(1) During their developing summers, both the EP and CP types of ENSOs exhibit asymmetric impacts on TP summer precipitation among El Niños and La Niñas, and the impacts of the EP El Niño and the CP La Niña are trans-type inverse.

(2) SWTP summer precipitation significantly decreases during the EP El Niño but increases during the CP La Niña, while CETP summer precipitation decreases intensely during the CP El Niño. No significant anomaly occurs during the EP La Niña.

(3) The distinct variation in TP summer precipitation during different types of ENSOs can be attributed to various teleconnections that provide altered vertical motion and moisture conditions. The EP El Niño and the CP La Niña both trigger an atmospheric response resembling the ISM-EASM teleconnection but in opposite phases, while the CP El Niño induces an EAP-like teleconnection pattern.

Supplementary Materials: The following supporting information can be downloaded at: https://www.mdpi.com/article/10.3390/rs15164030/s1, Figure S1: Tibetan Plateau summer precipitation anomaly during different types of ENSOs across multiple datasets.

Author Contributions: Conceptualization, H.-L.R.; methodology, H.-L.R. and M.L.; software, M.L.; validation, M.L.; formal analysis, H.-L.R. and M.L.; investigation, H.-L.R., M.L. and R.W.; writing—original draft preparation, M.L.; writing—review and editing, M.L., H.-L.R., J.M. and X.M.; supervision, H.-L.R.; funding acquisition, H.-L.R., M.L. and R.W. All authors have read and agreed to the published version of the manuscript.

Funding: This research was jointly funded by China National Natural Science Foundation (grant numbers U2242206, 42205047, 42105067, and 41975094) and the Special operating expenses of scientific research institutions for "Key Technology Development of Numerical Forecasting" of Chinese Academy of Meteorological Sciences.

Data Availability Statement: CRU data are openly available at https://www.uea.ac.uk/web/groups-and-centres/climatic-research-unit/data (accessed on 18 May 2023), and APHRODITE precipitation data are at http://aphrodite.st.hirosaki-u.ac.jp/index.html (accessed on 18 May 2023). GPCC precipitation and NOAA OLR data can be found at https://psl.noaa.gov/data/gridded/index.html (accessed on 21 May 2023). HadISST data can be found at https://www.metoffice.gov.uk/hadobs/index.html (accessed on 20 February 2023). ERA5 data are at https://cds.climate.copernicus.eu/ (accessed on 18 March 2023). CN05.1 precipitation and the station-observed precipitation were obtained by the authors through an agreement with the China Meteorological Administration for the current study and are not publicly available.

Acknowledgments: We thank the National Meteorological Information Center of China Meteorological Administration for providing precipitation data.

Conflicts of Interest: The authors declare no conflict of interest.

References

1. Qiu, J. China: The third pole. *Nature* **2008**, *454*, 393–396. [CrossRef] [PubMed]
2. Yao, T.; Xue, Y.; Chen, D.; Chen, F.; Thompson, L.; Cui, P.; Koike, T.; Lau, W.K.-M.; Lettenmaier, D.; Mosbrugger, V.; et al. Recent Third Pole's Rapid Warming Accompanies Cryospheric Melt and Water Cycle Intensification and Interactions between Monsoon and Environment: Multidisciplinary Approach with Observations, Modeling, and Analysis. *Bull. Am. Meteor. Soc.* **2019**, *100*, 423–444. [CrossRef]
3. Son, J.H.; Seo, K.H.; Wang, B. How does the Tibetan Plateau dynamically affect downstream monsoon precipitation? *Geophys. Res. Lett.* **2020**, *47*, e2020GL090543. [CrossRef]
4. Hu, J.; Duan, A. Relative contributions of the Tibetan Plateau thermal forcing and the Indian Ocean sea surface temperature basin mode to the interannual variability of the East Asian summer monsoon. *Clim. Dym.* **2015**, *45*, 2697–2711. [CrossRef]
5. Wu, G.; Liu, Y.; He, B.; Bao, Q.; Duan, A.; Jin, F.-F. Thermal Controls on the Asian Summer Monsoon. *Sci. Rep.* **2012**, *2*, 404. [CrossRef]
6. Wu, G.; Liu, Y.; Zhang, Q.; Duan, A.; Wang, T.; Wan, R.; Liu, X.; Li, W.; Wang, Z.; Liang, X. The Influence of Mechanical and Thermal Forcing by the Tibetan Plateau on Asian Climate. *J. Hydrometeorol.* **2007**, *8*, 770–789. [CrossRef]
7. Xu, X. World water tower: An atmospheric perspective. *Geophys. Res. Lett.* **2008**, *35*, L20815. [CrossRef]
8. Immerzeel, W.W.; van Beek, L.P.H.; Bierkens, M.F.P. Climate change will affect the Asian water towers. *Science* **2010**, *328*, 1382–1385. [CrossRef]
9. Chen, X.; You, Q. Effect of Indian Ocean SST on Tibetan Plateau precipitation in the early rainy season. *J. Clim.* **2017**, *30*, 8973–8985. [CrossRef]
10. Zhang, L.; Su, F.; Yang, D.; Hao, Z.; Tong, K. Discharge regime and simulation for the upstream of major rivers over Tibetan Plateau. *J. Geophys. Res. Atmos.* **2013**, *118*, 8500–8518. [CrossRef]
11. Yanai, M.; Li, C.; Song, Z. Seasonal heating of the Tibetan Plateau and its effects on the evolution of the Asian Summer Monsoon. *J. Meteorol. Soc. Japan. Ser. II* **1992**, *70*, 319–351. [CrossRef]
12. Feng, L.; Zhou, T. Water vapor transport for summer precipitation over the Tibetan Plateau: Multidata set analysis. *J. Geophys. Res.* **2012**, *117*, D20114. [CrossRef]
13. Wang, Z.Q.; Yang, S.; Lau, N.-C.; Duan, A. Teleconnection between summer NAO and East China rainfall variations: A bridge effect of the Tibetan Plateau. *J. Clim.* **2018**, *31*, 6433–6444. [CrossRef]
14. Liu, X.; Yin, Z.-Y. Spatial and temporal variation of summer precipitation over the eastern Tibetan Plateau and the North Atlantic Oscillation. *J. Clim.* **2001**, *14*, 2896–2909. [CrossRef]
15. Yang, J.; Chen, H. Influences of spring land surface thermal anomalies over West Asia on Indian early Summer Monsoon activity and its pathway. *J. Clim.* **2022**, *35*, 6051–6074. [CrossRef]
16. Jiang, X.; Ting, M. A dipole pattern of summertime rainfall across the Indian subcontinent and the Tibetan Plateau. *J. Clim.* **2017**, *30*, 9607–9620. [CrossRef]
17. Dong, W.; Lin, Y.; Wright, J.S.; Ming, Y.; Xie, Y.; Wang, B.; Luo, Y.; Huang, W.; Huang, J.; Wang, L.; et al. Summer rainfall over the southwestern Tibetan Plateau controlled by deep convection over the Indian subcontinent. *Nat. Commun.* **2016**, *7*, 10925. [CrossRef]
18. Zhao, Y.; Duan, A.; Wu, G. Interannual variability of late-spring circulation and diabatic heating over the Tibetan Plateau associated with Indian Ocean forcing. *Adv. Atmos. Sci.* **2018**, *35*, 927–941. [CrossRef]
19. Gao, Y.; Wang, H.; Li, S. Influences of the Atlantic Ocean on the summer precipitation of the southeastern Tibetan Plateau. *J. Geophys. Res. Atmos.* **2013**, *118*, 3534–3544. [CrossRef]
20. Li, F.; Wang, B.; He, Y.; Huang, W.; Xu, S.; Liu, L.; Liu, J.; Li, L. Important role of North Atlantic air–sea coupling in the interannual predictability of summer precipitation over the eastern Tibetan Plateau. *Clim. Dym.* **2021**, *56*, 1433–1448. [CrossRef]
21. Hu, S.; Wu, B.; Zhou, T.; Yu, Y. Dominant anomalous circulation patterns of Tibetan Plateau summer climate generated by ENSO-forced and ENSO-independent teleconnections. *J. Clim.* **2022**, *35*, 1679–1694. [CrossRef]
22. Wang, C.; Ma, Z. Quasi-3-yr cycle of rainy season precipitation in Tibet related to different types of ENSO during 1981–2015. *J. Meteorol. Res.-Prc.* **2018**, *32*, 181–190. [CrossRef]
23. Yang, X.; Yao, T.; Deji; Zhao, H.; Xu, B. Possible ENSO Influences on the Northwestern Tibetan Plateau Revealed by Annually Resolved Ice Core Records. *J. Geophys. Res. Atmos.* **2018**, *123*, 3857–3870. [CrossRef]
24. Choudhury, D.; Nath, D.; Wen, C. The role of surface air temperature over the east Asia on the early and late Indian Summer Monsoon Onset over Kerala. *Sci. Rep.* **2019**, *9*, 11756. [CrossRef]
25. Choudhury, D.; Nath, D.; Chen, W. The modulation of Indian summer monsoon onset processes during ENSO through equatorward migration of the subtropical jet stream. *Clim. Dym.* **2021**, *57*, 141–152. [CrossRef]

26. Xue, X.; Chen, W. Distinguishing interannual variations and possible impacted factors for the northern and southern mode of South Asia high. *Clim. Dym.* **2019**, *53*, 4937–4959. [CrossRef]
27. Wei, W.; Zhang, R.; Wen, M.; Rong, X.; Li, T. Impact of Indian summer monsoon on the South Asian High and its influence on summer rainfall over China. *Clim. Dym.* **2014**, *43*, 1257–1269. [CrossRef]
28. Xie, S.-P.; Kosaka, Y.; Du, Y.; Hu, K.; Chowdary, J.S.; Huang, G. Indo-western Pacific ocean capacitor and coherent climate anomalies in post-ENSO summer: A review. *Adv. Atmos. Sci.* **2016**, *33*, 411–432. [CrossRef]
29. Xie, S.-P.; Zhou, Z.-Q. Seasonal modulations of El Niño–related atmospheric variability: Indo–western Pacific Ocean feedback. *J. Clim.* **2017**, *30*, 3461–3472. [CrossRef]
30. Ren, Q.; Zhou, C.; He, J.; Cen, S.; Deng, M. Impact of preceding Indian Ocean sea surface temperature anomaly on water vapor content over the Tibetan Plateau moist pool in summer and its possible reason. *Chin. J. Atmos. Sci.* **2017**, *41*, 648–658. (In Chinese) [CrossRef]
31. Dong, N.; Xu, X.; Cai, W.; Zhao, T.; Sun, C. Comprehensive effects of interdecadal change of sea surface temperature increase in the Indo-Pacific Ocean on the warming-wetting of the Qinghai–Tibet Plateau. *Sci. Rep.* **2022**, *12*, 22306. [CrossRef] [PubMed]
32. Chiang, J.C.H.; Sobel, A.H. Tropical tropospheric temperature variations caused by ENSO and their influence on the remote tropical climate. *J. Clim.* **2002**, *15*, 2616–2631. [CrossRef]
33. Peng, L.; Sun, Z.; Ni, D.; Chen, H.; Tan, G. Interannual variation of summer South Asia high and its association with ENSO. *Chin. J. Atmos. Sci.* **2009**, *33*, 783–795. (In Chinese) [CrossRef]
34. Wang, B.; Wu, R.; Lau, K.M. Interannual variability of the Asian summer monsoon: Contrasts between the Indian and the western North Pacific–East Asian monsoons. *J. Clim.* **2001**, *14*, 4073–4090. [CrossRef]
35. Yang, J.; Chen, H.; Song, Y.; Zhu, S.; Zhou, B.; Zhang, J. Atmospheric circumglobal teleconnection triggered by spring land thermal anomalies over West Asia and its possible impacts on early summer climate over Northern China. *J. Clim.* **2021**, *34*, 5999–6021. [CrossRef]
36. Cherchi, A.; Navarra, A. Influence of ENSO and of the Indian Ocean Dipole on the Indian summer monsoon variability. *Clim. Dym.* **2013**, *41*, 81–103. [CrossRef]
37. Hu, S.; Zhou, T.; Wu, B. Impact of developing ENSO on Tibetan Plateau summer rainfall. *J. Clim.* **2021**, *34*, 3385–3400. [CrossRef]
38. Wu, R.; Hu, Z.Z.; Kirtman, B.P. Evolution of ENSO-related rainfall anomalies in East Asia. *J. Clim.* **2003**, *16*, 3742–3758. [CrossRef]
39. Jiang, X.; Li, Y.; Yang, S.; Yang, K.; Chen, J. Interannual variation of summer atmospheric heat source over the Tibetan Plateau and the role of convection around the western Maritime Continent. *J. Clim.* **2016**, *29*, 121–138. [CrossRef]
40. Jin, R.; Wu, Z.; Zhang, P. Tibetan Plateau capacitor effect during the summer preceding ENSO: From the Yellow River climate perspective. *Clim. Dym.* **2017**, *51*, 57–71. [CrossRef]
41. Lei, Y.; Zhu, Y.; Wang, B.; Yao, T.; Yang, K.; Zhang, X.; Zhai, J.; Ma, N. Extreme Lake Level Changes on the Tibetan Plateau Associated With the 2015/2016 El Niño. *Geophys. Res. Lett.* **2019**, *46*, 5889–5898. [CrossRef]
42. Ashok, K.; Behera, S.K.; Rao, S.A.; Weng, H.Y.; Yamagata, T. El Niño Modoki and its possible teleconnection. *J. Geophys. Res.-Oceans* **2007**, *112*, C11007. [CrossRef]
43. Kao, H.-Y.; Yu, J.-Y. Contrasting eastern-Pacific and central-Pacific types of ENSO. *J. Clim.* **2009**, *22*, 615–632. [CrossRef]
44. Capotondi, A.; Wittenberg, A.T.; Newman, M.; Di Lorenzo, E.; Yu, J.-Y.; Braconnot, P.; Cole, J.; Dewitte, B.; Giese, B.; Guilyardi, E.; et al. Understanding ENSO diversity. *Bull. Am. Meteor. Soc.* **2015**, *96*, 921–938. [CrossRef]
45. Ren, H.-L.; Jin, F.-F. Niño indices for two types of ENSO. *Geophys. Res. Lett.* **2011**, *38*, L04704. [CrossRef]
46. Feng, J.; Li, J. Contrasting impacts of two types of ENSO on the boreal spring Hadley Circulation. *J. Clim.* **2013**, *26*, 4773–4789. [CrossRef]
47. Karori, M.A.; Li, J.; Jin, F.-F. The asymmetric influence of the two types of El Niño and La Niña on summer rainfall over Southeast China. *J. Clim.* **2013**, *26*, 4567–4582. [CrossRef]
48. Hu, C.D.; Yang, S.; Wu, Q.G.; Li, Z.N.; Chen, J.W.; Deng, K.Q.; Zhang, T.T.; Zhang, C.Y. Shifting El Niño inhibits summer Arctic warming and Arctic sea-ice melting over the Canada Basin. *Nat. Commun.* **2016**, *7*, 11721. [CrossRef]
49. Beyene, M.T.; Jain, S. North American wintertime temperature anomalies: The role of El Niño diversity and differential teleconnections. *Clim. Dym.* **2017**, *50*, 4365–4377. [CrossRef]
50. Wu, R.; Chen, J.; Chen, W. Different types of ENSO influences on the Indian summer monsoon variability. *J. Clim.* **2012**, *25*, 903–920. [CrossRef]
51. Zhang, W.; Jin, F.-F.; Turner, A. Increasing autumn drought over southern China associated with ENSO regime shift. *Geophys. Res. Lett.* **2014**, *41*, 4020–4026. [CrossRef]
52. Feng, J.; Li, J.; Zheng, F.; Xie, F.; Sun, C. Contrasting impacts of developing phases of two types of El Niño on southern China rainfall. *J. Meteorol. Soc. Japan. Ser. II* **2016**, *94*, 359–370. [CrossRef]
53. Ren, H.-L.; Lu, B.; Wan, J.; Tian, B.; Zhang, P. Identification standard for ENSO events and its application to climate monitoring and prediction in China. *J. Meteorol. Res.-Prc.* **2018**, *32*, 923–936. [CrossRef]
54. Zhang, W.; Wang, L.; Xiang, B.; Qi, L.; He, J. Impacts of two types of La Niña on the NAO during boreal winter. *Clim. Dym.* **2015**, *44*, 1351–1366. [CrossRef]
55. Chen, D.K.; Lian, T.; Fu, C.B.; Cane, M.A.; Tang, Y.M.; Murtugudde, R.; Song, X.S.; Wu, Q.Y.; Zhou, L. Strong influence of westerly wind bursts on El Niño diversity. *Nat. Geosci.* **2015**, *8*, 339–345. [CrossRef]

56. Zhang, W.J.; Wang, Y.L.; Jin, F.F.; Stuecker, M.F.; Turner, A.G. Impact of different El Niño types on the El Niño/IOD relationship. *Geophys. Res. Lett.* **2015**, *42*, 8570–8576. [CrossRef]
57. Wang, X.; Wang, C.Z. Different impacts of various El Niño events on the Indian Ocean Dipole. *Clim. Dym.* **2014**, *42*, 991–1005. [CrossRef]
58. Harris, I.; Osborn, T.J.; Jones, P.; Lister, D. Version 4 of the CRU TS monthly high-resolution gridded multivariate climate dataset. *Sci. Data* **2020**, *7*, 109. [CrossRef]
59. Schneider, U.; Becker, A.; Finger, P.; Meyer-Christoffer, A.; Rudolf, B.; Ziese, M. *GPCC Full Data Reanalysis Version 6.0 at 1.0°: Monthly Land-SURFACE Precipitation from Rain-Gauges Built on GTS-Based and Historic Data*; Global Precipitation Climatology Centre: Offenbach am Main, Germany, 2011. [CrossRef]
60. Wu, J.; Gao, X.-J. A gridded daily observation dataset over China region and comparison with the other datasets. *Chin. J. Geophys.* **2013**, *56*, 1102–1111.
61. Yatagai, A.; Kamiguchi, K.; Arakawa, O.; Hamada, A.; Yasutomi, N.; Kitoh, A. APHRODITE: Constructing a long-term daily gridded precipitation dataset for Asia based on a dense network of rain gauges. *Bull. Am. Meteor. Soc.* **2012**, *93*, 1401–1415. [CrossRef]
62. Liebmann, B.; Smith, C.A. Description of a complete (interpolated) outgoing longwave radiation dataset. *Bull. Am. Meteor. Soc.* **1996**, *77*, 1275–1277.
63. Hersbach, H.; Bell, B.; Berrisford, P.; Hirahara, S.; Horányi, A.; Muñoz-Sabater, J.; Nicolas, J.; Peubey, C.; Radu, R.; Schepers, D.; et al. The ERA5 global reanalysis. *Quart. J. R. Meteor. Soc.* **2020**, *146*, 1999–2049. [CrossRef]
64. Rayner, N.A.; Parker, D.E.; Horton, E.B.; Folland, C.K.; Alexander, L.V.; Rowell, D.P.; Kent, E.C.; Kaplan, A. Global analyses of sea surface temperature, sea ice, and night marine air temperature since the late nineteenth century. *J. Geophys. Res. Atmos.* **2003**, *108*, 4407. [CrossRef]
65. Xu, P.; Wang, L.; Chen, W.; Feng, J.; Liu, Y. Structural changes in the Pacific–Japan Pattern in the late 1990s. *J. Clim.* **2019**, *32*, 607–621. [CrossRef]
66. Nitta, T. Convective activities in the tropical western Pacific and their impact on the Northern Hemisphere summer circulation. *J. Meteor. Soc. Jpn.* **1987**, *65*, 373–390. [CrossRef]
67. Huang, R.H.; Li, W.J. Influence and physical mechanism of heat source anomaly over the tropical western Pacific on the subtropical high over East Asia. *Chin. J. Atmos. Sci.* **1988**, *12*, 107–116. (In Chinese) [CrossRef]
68. Yim, S.-Y.; Jhun, J.-G.; Yeh, S.-W. Decadal change in the relationship between east Asian–western North Pacific summer monsoons and ENSO in the mid-1990s. *Geophys. Res. Lett.* **2008**, *35*, L20711. [CrossRef]
69. Wang, B.; Wu, R.G.; Fu, X.H. Pacific-East Asian teleconnection: How does ENSO affect East Asian climate? *J. Clim.* **2000**, *13*, 1517–1536. [CrossRef]
70. Lin, R.; Zheng, F.; Dong, X. ENSO Frequency Asymmetry and the Pacific Decadal Oscillation in Observations and 19 CMIP5 Models. *Adv. Atmos. Sci.* **2018**, *35*, 495–506. [CrossRef]
71. Dong, X.; Xue, F. Phase transition of the Pacific decadal oscillation and decadal variation of the East Asian summer monsoon in the 20th century. *Adv. Atmos. Sci.* **2016**, *33*, 330–338. [CrossRef]
72. Zhou, T.; Yu, R.; Zhang, J.; Drange, H.; Cassou, C.; Deser, C.; Hodson, D.L.R.; Sanchez-Gomez, E.; Li, J.; Keenlyside, N.; et al. Why the western Pacific subtropical high has extended westward since the late 1970s. *J. Clim.* **2009**, *22*, 2199–2215. [CrossRef]
73. Wang, L.; Ren, H.-L.; Xu, X.; Huang, B.; Wu, J.; Liu, J. Seasonal-interannual predictions of summer precipitation over the Tibetan Plateau in North American Multimodel Ensemble. *Geophys. Res. Lett.* **2022**, *49*, e2022GL100294. [CrossRef]
74. Jiang, X.; Li, Y.; Yang, S.; Shu, J.; He, G. Interannual variation of mid-summer heavy rainfall in the eastern edge of the Tibetan Plateau. *Clim. Dym.* **2015**, *45*, 3091–3102. [CrossRef]

remote sensing

MDPI

Article

Regional Climate Effects of Irrigation under Central Asia Warming by 2.0 °C

Liyang Wu [1,2] and Hui Zheng [1,2,*]

[1] College of Geography and Environmental Science, Henan University, Kaifeng 475004, China; wuly@henu.edu.cn

[2] Key Laboratory of Geospatial Technology for the Middle and Lower Yellow River Regions, Henan University, Ministry of Education, Kaifeng 475004, China

* Correspondence: zhenghui@vip.henu.edu.cn; Tel.: +86-183-3784-9530

Abstract: There has been a severe shortage of water resources in Central Asia and agriculture has been highly dependent on irrigation because of the scarce precipitation in the croplands. Central Asia is also experiencing climate warming in the context of global warming; however, few studies have focused on changes in the amount of irrigation in Central Asia under future climate warming and their regional climate effects. In this study, we adopted the Weather Research and Forecasting (WRF) model to design three types of experiments: historical experiments (Exp01); warming experiments using future driving fields (Exp02); and warming experiments that involved increasing the surface energy (Exp03). In each type of experiment, two experiments (considering and not considering irrigation) were carried out. We analyzed the regional climate effects of irrigation under the warming of Central Asia by 2.0 °C through determining the differences between the two types of warming experiments and the historical experiments. For surface variables (irrigation amount; sensible heat flux; latent heat flux; and surface air temperature), the changes (relative to Exp01) in Exp03 were thought to be reasonable. For precipitation, the changes (relative to Exp01) in Exp02 were thought to be reasonable. The main conclusions were as follows: in Central Asia, after warming by 2.0 °C, the irrigation amount increased by 10–20%; in the irrigated croplands of Central Asia, the irrigation-caused increases (decreases) in latent heat flux (sensible heat flux) further expanded; and then the irrigation-caused decreases in surface air temperature also became enhanced; during the irrigation period, the irrigation-caused increases in precipitation in the mid-latitude mountainous areas were reduced. This study also showed that, in the WRF model, the warming experiments caused by driving fields were not suitable to simulate the changes in irrigation amount affected by climate warming.

Keywords: Central Asia; irrigation; climate warming; land-atmosphere interaction; regional climate; numerical simulation

Citation: Wu, L.; Zheng, H. Regional Climate Effects of Irrigation under Central Asia Warming by 2.0 °C. *Remote Sens.* **2023**, *15*, 3672. https://doi.org/10.3390/rs15143672

Academic Editors: Massimo Menenti, Yaoming Ma, Li Jia and Lei Zhong

Received: 15 May 2023
Revised: 22 June 2023
Accepted: 21 July 2023
Published: 23 July 2023

1. Introduction

As an important source of energy and moisture in the atmosphere, land surface is an important component of the climate system [1,2]. In the past hundreds of years, 30–50% of the Earth's land surface has been altered by human activities [3–5], and studies have shown that these changes in land surface exerted important impacts on climate through interactions between the land and atmosphere [6–11]. Therefore, the changes in land surface have become the important driving factors for local, regional, and global climate changes. Among the anthropogenic activities that alter the underlying surface characteristics, irrigation is an important land management practice. Through adding water to the soil, irrigation increases soil moisture and maintains the normal growth of crops. Irrigation accounts for 90% of the consumptive water usage and 70% of global freshwater withdrawals [12]. Irrigated croplands account for over 18% of global croplands, corresponding to approximately 2% of the total land surface [13]. Since irrigation can increase water vapor in the atmosphere and affect the partitioning of surface energy, irrigation has been recognized as an important anthropogenic factor that affects the local and regional climate [11,14–19].

Thus far, there have been many researchers who studied the climate effects of irrigation, in which most of the studies adopted numerical simulation means. For numerical simulations, the crucial part in the model is the irrigation scheme. In the past, default climate models did not consider irrigation processes, leading to irrigation schemes that were diverse in different studies. At the core of irrigation schemes is determining how to increase soil moisture in the model. Wu et al. [11] summarized three commonly used types of irrigation schemes: (a) the soil moisture of croplands being set to fixed values (such as saturation, field capacity) during an irrigation period [20–24]; (b) the estimated irrigation amount based on observations was applied into croplands during the irrigation period [25–28]; (c) the dynamic irrigation scheme, which basically regarded the soil moisture of the root zone as the irrigation trigger, has been adopted by increasing the number of studies [11,14–16,19,29–34]. In this dynamic irrigation scheme, the amount of each irrigation is calculated on the basis of the soil's current moisture. The third scheme mentioned above considered the impacts of climatic conditions on the irrigation amount, but the first and second scheme did not. Obviously, the dynamic irrigation scheme was more in line with reality. The WRF (Weather Research and Forecasting) model has been widely adopted to study the regional climate effects of irrigation. In the official WRF model version 4.2 and above, Valmassoi et al. [35] developed and incorporated the irrigation schemes. However, these schemes did not consider well the impacts of climatic conditions on the irrigation amount. Global warming has become an indisputable fact, and it will continue; however, few studies have focused on the regional climate effects of irrigation under future climate warming. Since the dynamic irrigation scheme considered the effects of climatic conditions, this scheme was very suitable for simulating the changes in irrigation amount caused by future climate warming, including their regional climate effects.

Central Asia is in the hinterland of Eurasia, including Kazakhstan, Uzbekistan, Turkmenistan, Tajikistan, Kyrgyzstan, and Xinjiang province of China [36]. As the central area of the Silk Road Economic Belt, Central Asia has become a hot study area. The topography of Central Asia generally shows high in the southeast and low in the northwest, which is characterized with large contrast in elevation, from 154 m below the sea level to 8611 m above the sea level. The precipitation is less affected by the ocean due to Central Asia being far away from it. Therefore, precipitation was scarce in Central Asia (not including its mountainous region). However, affected by the orographic lifting in mountains, the high-value areas of precipitation in Central Asia were in the high-altitude mountainous regions [11]. For example, the annual precipitation is below 50 mm in the Tarim Basin, but the annual precipitation can be more than 800 mm in the Tianshan Mountains. Unfortunately, the croplands of Central Asia are widely distributed in low-altitude areas, where precipitation is scarce. Therefore, the agricultural production here was highly dependent on irrigation, and a large amount of water resources was inevitably consumed by irrigation. Due to water sources for irrigation, the major irrigation areas in Central Asia are mainly located near the rivers (such as Syr Darya, Amu Darya, and Yarkand River). Under global warming, Central Asia is also experiencing an area-averaged surface air temperature that had significantly increased by 0.18 °C per decade in the past century (1901–2003), an amount that was greater than the global averaged value [37]. From 1979 to 2011, the temperature in Central Asia increased by up to 0.36–0.42 °C per decade [38]. Climate warming in Central Asia was projected to continue [39]. Central Asia has been facing severe water stress, and irrigation consumes most of its water resources [40]. Therefore, in Central Asia, for the sustainable development of irrigated agriculture, it was very meaningful to study the influences of future climate warming on irrigation amount and their regional climate effects.

Wu et al. [11] have adopted the numerical simulation method to investigate the impacts of irrigation over Central Asia on regional climate during the historical period (1995–2014), and they drew some important conclusions. Compared to the official irrigation schemes in the WRF model, the advantage of the irrigation scheme constructed by Wu et al. [11] was obvious, which provided a more suitable irrigation scheme for this study. As mentioned above, although Central Asia has been facing a severe water crisis, irrigation consumes a large amount of water resources. However, in Central Asia, under climate warming, few studies have focused on the impacts of future climate warming on water resources use (such as irrigation). Therefore, the impacts of future climate warming on irrigation amount were unknown, and the impacts of future climate warming on irrigation-induced changes in the related physical variables were also unknown. The purpose of this study was to address these scientific questions. In addition, were the warming experiments caused by driving fields suitable to simulate the influences of climate warming on irrigation amount? This was also unclear. Therefore, based on the partial settings of Wu et al. [11], in this study, we designed three types of experiments (historical experiments, warming experiments using future driving fields, and warming experiments by increasing the surface energy), in order to analyze the impacts of irrigation on the regional climate under Central Asia warming by 2.0 °C, including changes caused by climate warming to irrigation amount.

2. Methods

2.1. WRF Setup and Irrigation Scheme

Wu et al. [11] have simulated the impacts of irrigation over Central Asia on regional climate via the WRF model. We also adopted the WRF model, with version 4.0.3 as the regional climate model in this study. The adopted physical parameterization schemes were the same as those selected by Wu et al. [11]. The schemes were as follows: the Thompson graupel microphysics scheme, the Tiedtke cumulus scheme, the RRTMG shortwave radiation and longwave radiation scheme, the Noah-mosaic land surface model, and the Yonsei University planetary boundary layer scheme.

Figure 1 showed the simulated region. The land use type used was MODIS 20-category (Figure 1a). The model domain and the settings of the 3D grid points were consistent with a previous study [11]. The center of model domain was (44.5°N, 67°E), and the horizontal grid points were 196 (meridional) × 146 (zonal) with a horizontal resolution of 20 km. In the vertical coordinates, 33 eta levels from the surface to 50 mb were used. The lateral boundary adopted an exponential relaxation scheme and the buffer set 10 grid points. The driving fields of the WRF model were the bias-corrected CMIP6 global dataset, which were constructed by Xu et al. [41]. The reference data required for bias correction were the ERA5 reanalysis data. The model data were the outputs of 18 models from CMIP6. This dataset contained the historical period (1979–2014) and future scenarios (2015–2100) with 1.25° × 1.25° grid spacing and 6-hourly intervals, and the future emission scenarios of this dataset were SSP2-4.5 and SSP5-8.5.

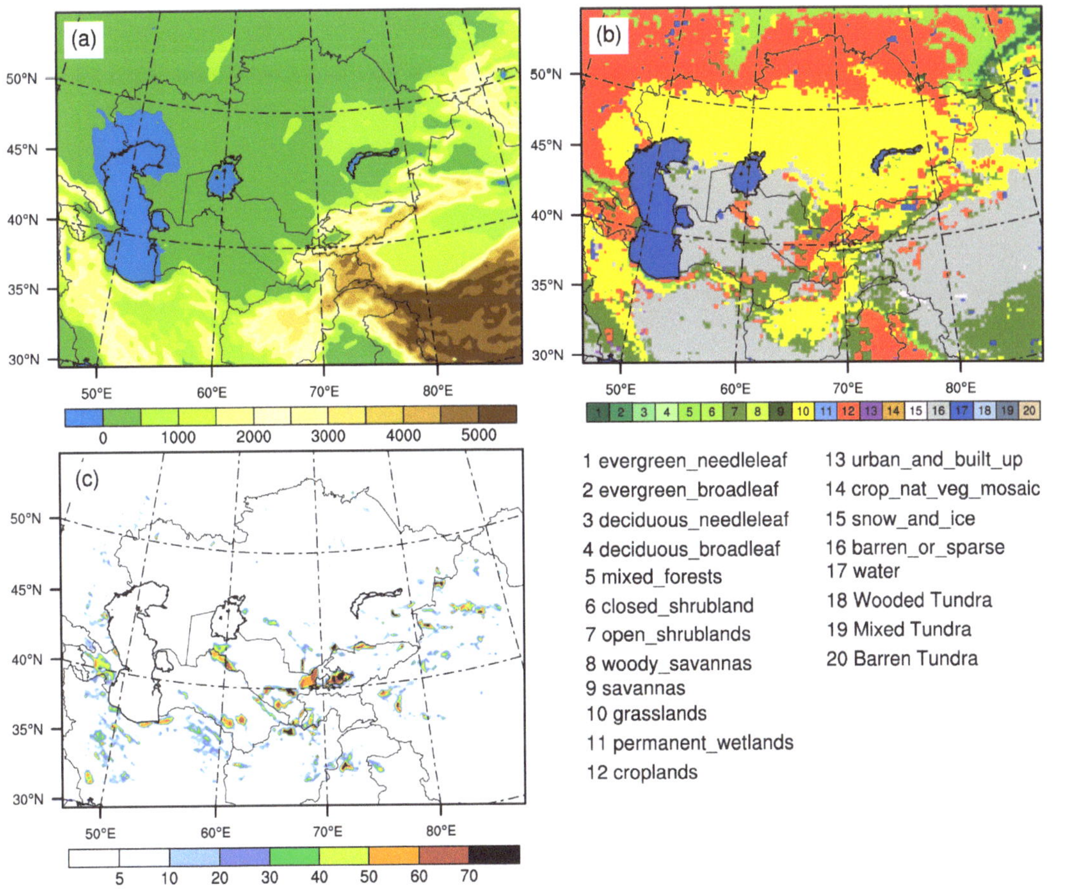

Figure 1. (**a**) The altitude (unit: m) of the simulated region. (**b**) Land use type in the simulated domain. (**c**) Irrigation fraction (unit: %) in the simulated domain.

In the dynamic irrigation scheme, the irrigation trigger is the root-zone soil moisture, which was more realistic [11]. We also adopted the dynamic irrigation scheme in this study, and this scheme considered irrigation only in subgrid-scale. According to the irrigation fraction (Figure 1b), the irrigated areas consisted of the irrigated portions within each grid cell. Referred to the irrigation period of Wu et al. [11] and Zhang et al. [33], the irrigation period was also set from April 10 to September 10, while the other time of the year was the non-irrigation period. To avoid that the start day of irrigation every year was the same in the whole simulation region, the start day of irrigation in different croplands was determined according to the land surface temperature at 06:00 UTC on 10 April every year [11]. Since the core of the dynamic irrigation scheme is the irrigation trigger, we again exhibited the irrigation trigger (Equation (4)) and the calculation of irrigation amount each irrigation (Equation (5)) in this study. If the soil water content of the upper 0.6 m layers satisfies Equation (4) at 08:00 UTC on each day, croplands start irrigation, and irrigation continues for 4 h and ends at 12:00 UTC, in which the irrigation rate (I_{RATE}, unit: mm s^{-1}) is calculated by Equation (5).

$$A_{\mathrm{SM}} = \sum_{i=1}^{3} SM(i) \times D(i) \times 1000 \tag{1}$$

$$A_{SMCWLT} = \sum_{i=1}^{3} S_{WLT} \times D(i) \times 1000 \quad (2)$$

$$A_{SMCREF} = \sum_{i=1}^{3} S_{REF} \times D(i) \times 1000 \quad (3)$$

$$A_{SM} < A_{SMCWLT} + 0.5 \times (A_{SMCREF} - A_{SMCWLT}) \quad (4)$$

$$I_{RATE} = \left(\sum_{i=1}^{3} (S_{MAX} - SM(i)) \times D(i) \times 1000\right) / (4 \times 3600) \quad (5)$$

where $D(i)$ is the thicknesses of soil layers for layer i: $D(1)$ = 0.1 m, $D(2)$ = 0.3 m, and $D(3)$ = 0.2 m (only taken from 0.4 to 0.6 m). $SM(i)$ refers to the soil moisture for layer i; S_{WLT}, S_{REF}, and S_{MAX} are the soil wilting point, field capacity, and saturated value, respectively, and their units are $m^3\ m^{-3}$. A_{SM} is the total soil water content of the three layers soil thickness (0.6 m); T_{SMCWLT} and T_{SMCREF} are the total soil water content of 0.6 m soil thickness when soil moisture is at the soil wilting point and field capacity, respectively, and their units are mm.

2.2. Experimental Design

We designed three types of experiments (Table 1). (a) The simulated period started at 00:00 UTC on 1 January 1994 and ended at 18:00 UTC on 31 December 2014, which was driven by the historical period (1994–2014) CMIP6 dataset. (b) The adopted driving fields were the CMIP6 SSP5-8.5 data of the future scenarios; the simulated experiments started from 00:00 UTC on 1 January 2035 and ended at 18:00 UTC on 31 December 2055. (c) The simulated period and driving fields were the same as those in (a), with the only difference being that the surface upward longwave radiation in the Noah land surface module was set to decrease by 8% in (c). The purpose of setting an 8% reduction rate was to increase the simulated surface air temperature by about 2.0 °C (mentioned in Section 3.2 later). The determination of the decrease rate had undergone a series of experiments with different reduction rates. The first type of experiments were historical experiments, which were referred to as Exp01; the second type of experiments were the warming experiments using future scenario (SSP5-8.5) data and were referred to as Exp02; the third type of experiments were the warming experiments with increasing surface energy, which were referred to as Exp03. For each type of experiment, two experiments were designed: the control experiment (CTL) that did not consider irrigation, and the sensitivity experiment (SEN) that considered irrigation. The adopted scheme was the dynamic irrigation scheme mentioned previously. The simulated results were outputted every 6 hrs. The first year (1994 or 2035) of each simulation was considered as spin-up time and was not used for analysis.

Table 1. The experimental design.

Type of Experiment	Driving Fields	Simulated Period
Historical experiments (Exp01)	The historical period (1994–2014) CMIP6 dataset	1 January 1994–31 December 2014
Warming experiments using driving fields (Exp02)	The CMIP6 SSP5-8.5 (2035–2055) data of the future scenarios	1 January 2035–31 December 2055
Warming experiments by increasing surface energy (Exp03)	The historical period (1994–2014) CMIP6 dataset	1 January 1994–31 December 2014. The surface upward longwave radiation during simulation was set to decrease by 8%

Note: there were two experiments (considering and not considering irrigation) in each type of experiment.

3. Results

3.1. The Warming Experiments Using Driving Fields

The simulated results of the historical period were evaluated by adopting the validation data [11], which were not repeated in this study. For future simulations in Exp02, we adopted the high emission scenario (SSP5-8.5) data as the driving fields of the WRF model, and thus the simulated surface air temperature during the future period should be greater than that during the historical period. As shown in Figure 2a, the simulations indicated that the future multi-year (2036–2055) annual mean surface air temperature obviously increased, relative to the historical period (1995–2014), in which the regional averaged value of the increased temperature was about 2.0 °C. Therefore, in this study, based on the high emission scenario (SSP5-8.5), we treated 2035–2055 as the time period for future warming of Central Asia by 2.0 °C, and then, we simulated the impacts of irrigation over Central Asia on regional climate against the background of future warming. It should be noted that, in our study, the changes (increases or decreases) in the related variables in Exp02 and Exp03 were referred to as the results of Exp02 minus Exp01 and Exp03 minus Exp01, respectively.

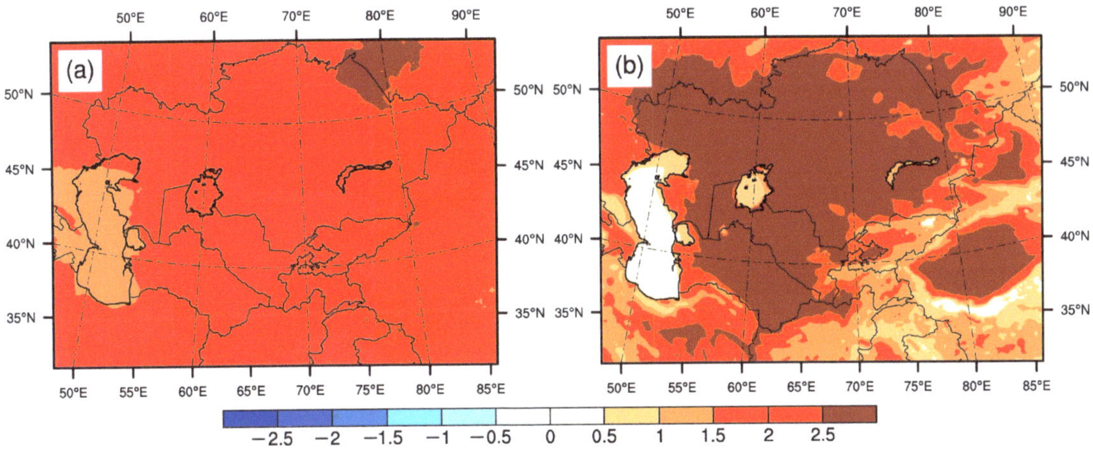

Figure 2. Spatial differences ((**a**) CTL in Exp02 minus CTL in Exp01; (**b**) CTL in Exp03 minus CTL in Exp01) in multi-year (20 years) annual mean surface air temperature (unit: °C). The 20 years denoted were from 1995 to 2014 in Exp 01 and Exp03, and from 2036 to 2055 in Exp02 (the same below).

In Central Asia, the irrigation period was the warm season, and the non-irrigation period was the cold season, meaning that the temperatures of these two periods were totally different. However, the magnitude of precipitation in the non-irrigation period was roughly consistent with that in the irrigation period, with some areas where the former was greater than the latter (Figure 3a,b). Therefore, after warming by 2.0 °C in Central Asia, it was necessary to analyze the changes in precipitation during the non-irrigation period and the irrigation period, respectively. There was a very interesting phenomenon where the direction of warming-induced changes in precipitation in mid-latitude mountainous areas was opposite during the non-irrigation period and the irrigation period (Figure 3c,d). In mid-latitude mountainous areas, the precipitation affected by warming increased during the non-irrigation period (Figure 3c) and decreased during the irrigation period (Figure 3d).

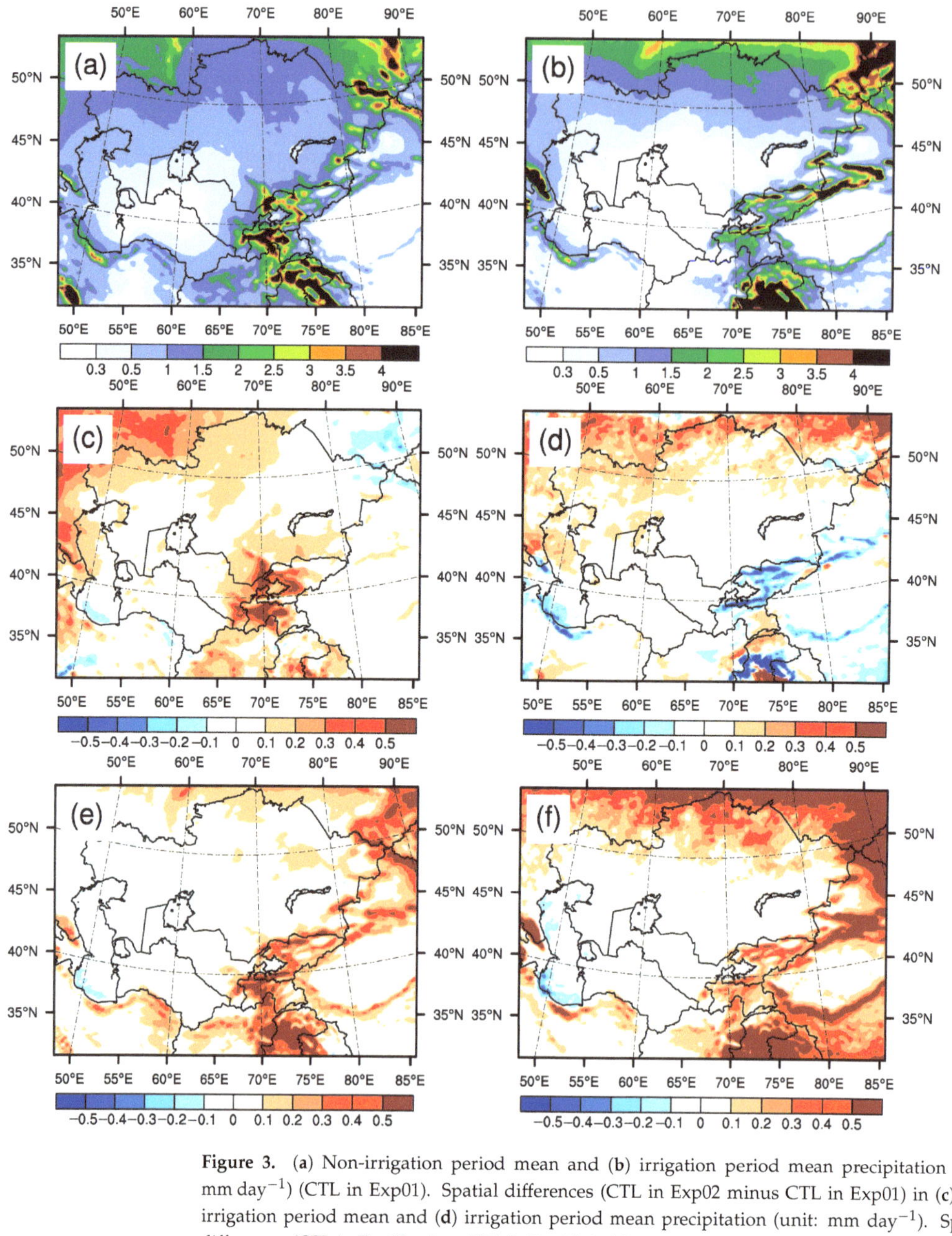

Figure 3. (**a**) Non-irrigation period mean and (**b**) irrigation period mean precipitation (unit: mm day^{-1}) (CTL in Exp01). Spatial differences (CTL in Exp02 minus CTL in Exp01) in (**c**) non-irrigation period mean and (**d**) irrigation period mean precipitation (unit: mm day^{-1}). Spatial differences (CTL in Exp03 minus CTL in Exp01) in (**e**) non-irrigation period mean and (**f**) irrigation period mean precipitation (unit: mm day^{-1}). The results shown were all the multi-year (20 years) mean values.

Figure 4a,b showed the spatial characteristics of the annual mean latent heat flux and sensible heat flux, respectively. It could be seen that, in addition to mountainous and water areas, and high-latitude regions, the surface energy was mainly allocated to sensible heat flux (Figure 4a,b). Although the surface air temperature of Central Asia was raised by 2.0 °C under the future high emission scenario, our simulations showed that the changes in latent heat flux and sensible heat flux were small (Figure 4c,d). Obviously, in the simulations of Exp02, the increased surface air temperature was only induced by the driving fields and had nothing to do with the changes in surface energy. In Exp02, for the non-irrigation period mean and the irrigation period mean values, the changes in latent heat flux and sensible heat flux caused by warming were also small, which was the reason that we only analyzed the annual mean values here.

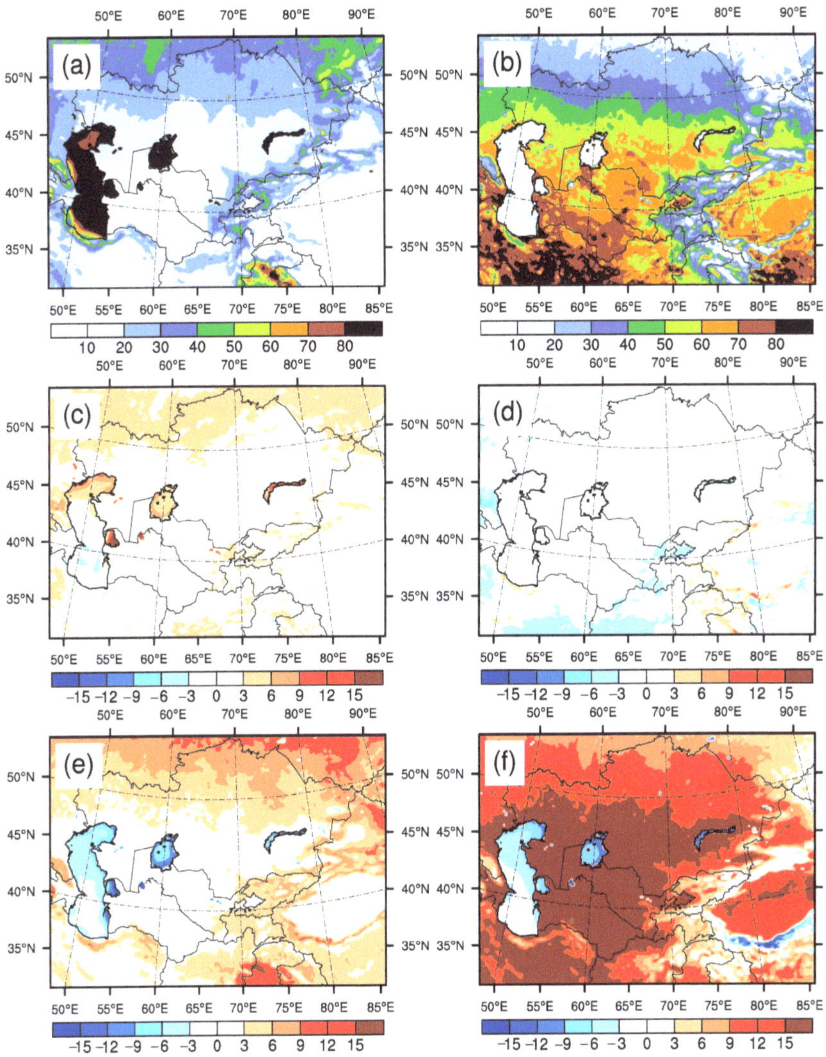

Figure 4. Annual mean (**a**) latent heat flux and (**b**) sensible heat flux (unit: W m^{-2}) (CTL in Exp01). Spatial differences (CTL in Exp02 minus CTL in Exp01) in annual mean (**c**) latent heat flux and (**d**) sensible heat flux (unit: W m^{-2}). Spatial differences (CTL in Exp03 minus CTL in Exp01) in annual mean (**e**) latent heat flux and (**f**) sensible heat flux (unit: W m^{-2}). The results shown were all the multi-year (20 years) mean values.

According to general knowledge, a rise in temperature should lead to increases in the amount of irrigation. However, the simulations showed that the irrigation amount (in Exp02) affected by warming did not increase, and even decreased slightly in the main irrigation areas except for Xinjiang province of China (Figure 5b), which was obviously unreasonable. From the physical point of view, the problem of global warming is actually an imbalance in the Earth's energy budget. Since the greenhouse gases can absorb a large amount of longwave radiation, the increased greenhouse gases in the atmosphere reduce the energy emitted from the top of the atmosphere, which increases the energy on Earth, and eventually leads to increases in the surface air temperature. Unfortunately, in Exp02, the increased surface air temperature (relative to Exp01) was only caused by driving fields and had nothing to do with changes in surface energy (as previously mentioned, the changes in latent heat flux and sensible heat flux were small). Small changes in the surface energy were not conducive to increases in surface evapotranspiration, which was also the reason that the irrigation amount (in Exp02) affected by warming did not increase. Therefore, in the WRF model, it was not feasible to study the impacts of future climate warming on the irrigation amount through the warming experiments with future driving fields. The impacts of future climate warming on irrigation amount would be again analyzed in the warming experiments with increasing surface energy (Exp03).

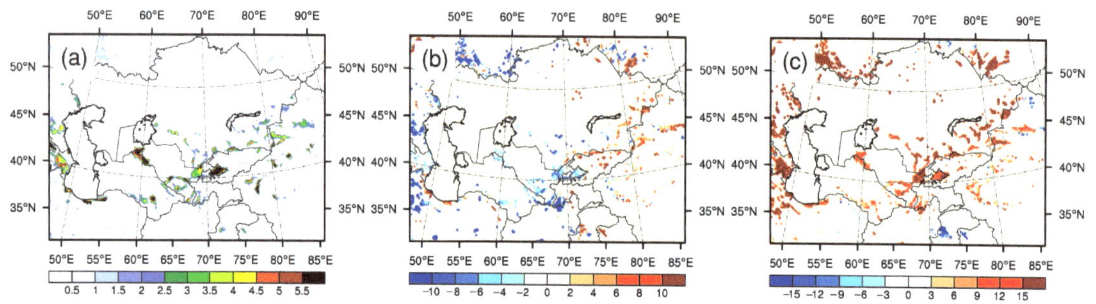

Figure 5. Spatial distributions of multi-year (20 years) irrigation period mean irrigation rate in (**a**) Exp01 (unit: mm day^{-1}), (**b**) (Exp02 minus Exp01)/Exp01 (unit: %), and (**c**) (Exp03 minus Exp01)/Exp01 (unit: %).

For Exp01, in croplands after irrigation, the latent heat flux increased (Figure 6a), and the sensible heat flux and the surface air temperature decreased (Figure 6b,c), which was completely consistent with the results of study [11]. Although the differences (Exp02 minus Exp01) in the irrigation-caused changes in the latent heat flux, sensible heat flux, and surface air temperature were not obvious (Figure 6d–f), the latent heat flux affected by warming did not increase (even decreased) in the main irrigation areas (such as the Fergana basin) (Figure 6d), which was also unreasonable. Here, the non-increases in the latent heat flux affected by warming corresponded to the non-increases in the irrigation amount (Figures 5b and 6d).

For Exp01, precipitation affected by irrigation obviously increased in the mountainous regions of the dotted box (Xinjiang province of China and Kyrgyzstan) (Figure 7a), which was completely consistent with a previous study [11]. As mentioned above, the precipitation affected by warming in the mid-latitude mountainous areas (dotted box) decreased during the irrigation period (Figure 3d), which led to that, in Exp02, the irrigation-caused increases in precipitation in the mountainous areas of the dotted box also decreased by 0.05–0.3 mm day^{-1}, relative to Exp01 (Figure 7b).

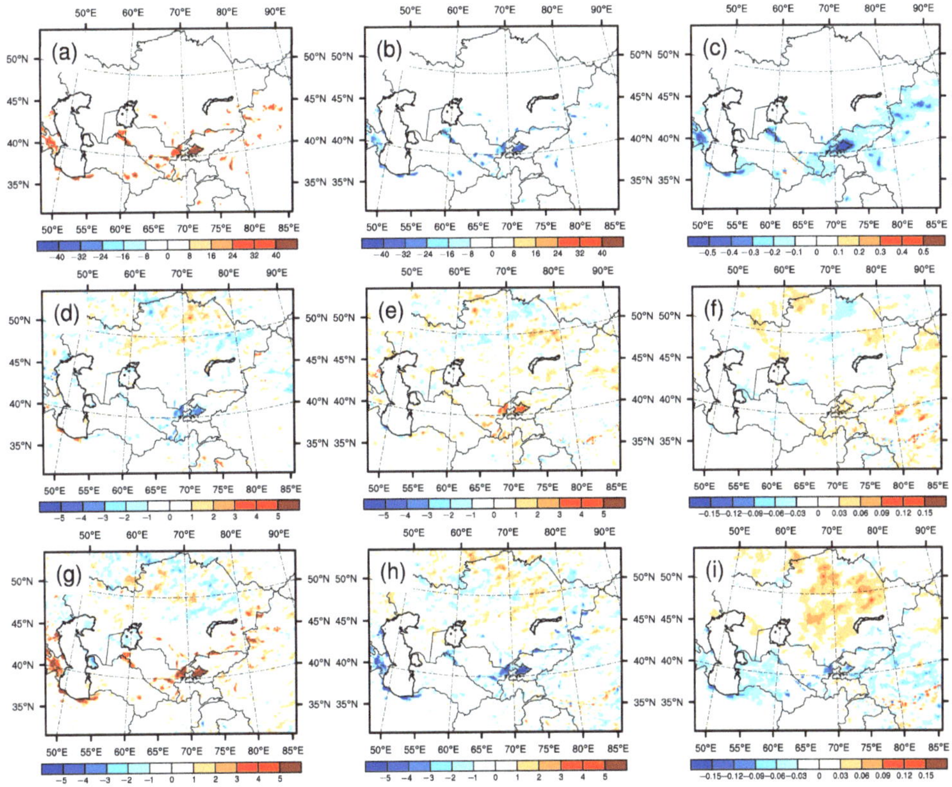

Figure 6. Spatial differences (SEN in Exp01 minus CTL in Exp01) in (**a**) latent heat flux (unit: W m^{-2}), (**b**) sensible heat flux (unit: W m^{-2}), and (**c**) surface air temperature (unit: °C). Spatial differences ((SEN in Exp02 minus CTL in Exp02) minus (SEN in Exp01 minus CTL in Exp01)) in (**d**) latent heat flux (unit: W m^{-2}), (**e**) sensible heat flux (unit: W m^{-2}), and (**f**) surface air temperature (unit: °C). Spatial differences ((SEN in Exp03 minus CTL in Exp03) minus (SEN in Exp01 minus CTL in Exp01)) in (**g**) latent heat flux (unit: W m^{-2}), (**h**) sensible heat flux (unit: W m^{-2}), and (**i**) surface air temperature (unit: °C). The results shown were the multi-year (20 years) irrigation period mean values.

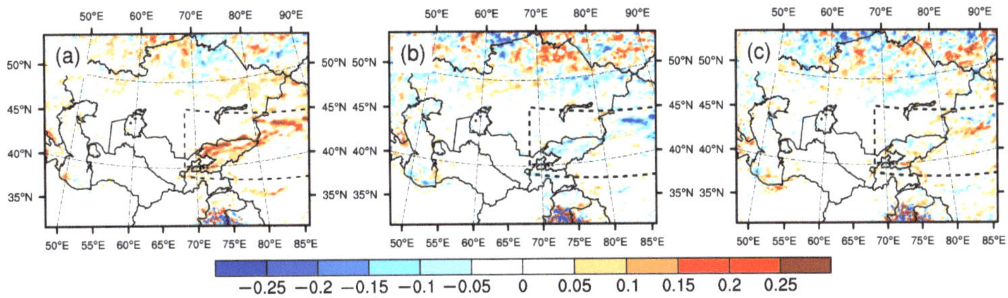

Figure 7. (**a**) Spatial differences (SEN in Exp01 minus CTL in Exp01) in multi-year (20 years) irrigation period mean precipitation (unit: mm day^{-1}). (**b**) Spatial differences ((SEN in Exp02 minus CTL in Exp02) minus (SEN in Exp01 minus CTL in Exp01)) in multi-year (20 years) irrigation period mean precipitation (unit: mm day^{-1}). (**c**) Spatial differences ((SEN in Exp03 minus CTL in Exp03) minus (SEN in Exp01 minus CTL in Exp01)) in multi-year (20 years) irrigation period mean precipitation (unit: mm day^{-1}).

3.2. The Warming Experiments by Increasing Surface Energy

As mentioned in Section 3.1, although the surface air temperature in Exp02 was increased by 2.0 °C (relative to Exp01), the changes in surface energy (latent heat flux, sensible heat flux) were small (Figure 4c,d), and the irrigation amount (in Exp02) affected by warming also did not increase (they even decreased in some irrigation areas) (Figure 5b). The main reason for climate warming is that the increased greenhouse gases can absorb more longwave radiation [42,43]. Therefore, the third type of experiments (referred to as Exp03) was designed by decreasing surface upward longwave radiation by 8% in the land surface model, with the purpose of more reasonably simulating the impacts of climate warming on the irrigation amount. As shown in Figure 2b, the multi-year (1995–2014) annual mean surface air temperature in Exp03 was obviously greater than that in Exp01, and the regional averaged value of their differences was also about 2.0 °C, which was the reason that the surface upward longwave radiation was set to decrease by 8% in Exp03. It should be mentioned that, in this study, the warming up of 2.0 °C in Exp03 referred to the mean surface air temperature of the whole simulation domain. The calculation of surface upward longwave radiation in the WRF model is proportional to the fourth power of land surface temperature. The surface upward longwave radiation in different areas was different because of the different land surface temperature. The same reduction rate (8%) led to the size of the decreased surface upward longwave radiation to be different in different areas. In the low-altitude areas, the difference in the mean surface air temperature between Exp03 and Exp01 was greater than 2.0 °C (Figure 2b) because of larger decreases in surface upward longwave radiation.

In contrast to the warming-induced changes in precipitation in Exp02, in Exp03, the precipitation affected by warming both increased in the irrigation period and the non-irrigation period (Figure 3e,f). In Exp03, the surface upward longwave radiation decreased by 8%, meaning that the surface net radiation was increased. The surface net radiation was mainly allocated to sensible heat flux and latent heat flux. Therefore, the differences showed that the latent heat flux and sensible heat flux in Exp03 were greater than those in Exp01 (Figure 4e,f). Due to the characteristics of drought in Central Asia, the increased surface net radiation was more allocated to sensible heat flux. Therefore, in Exp03, the warming-induced increases in sensible heat flux were obviously greater than those in latent heat flux (Figure 4e,f). Compared to the small changes in latent heat flux and sensible heat flux after warming by 2.0 °C in Exp02 (Figure 4c,d), the differences in latent heat flux and sensible heat flux between Exp03 and Exp01 were more obvious (Figure 4e,f), especially sensible heat flux (Figure 4f), whose reason was the increased surface net radiation in Exp03.

The increased surface energy was beneficial to increases in surface evapotranspiration, which affected soil moisture. If the irrigation amount remained unchanged with the increased surface energy, the increased surface evapotranspiration would reduce the soil moisture of croplands, which was not conducive to the growth of crops. To ensure the normal growth of crops, the amount of irrigation should be increased. As shown in Figure 5c, the irrigation amount in Exp03 increased by 10–20%, relative to Exp01. Therefore, the third type of experiments (Exp03) reasonably simulated the impacts of future climate warming on irrigation amount.

In correspondence to the increases in irrigation amount in Exp03, the irrigation-induced increases in latent heat flux were increased by 2–8 W m^{-2} (Figure 6g). According to the surface energy balance, the decreases in sensible heat flux caused by irrigation also increased (1–7 W m^{-2}) (Figure 6h). Unlike Exp02, in Exp03, the changes in surface energy were thought to be reasonable, and the reason was that climate warming was caused by the increased energy of the climate system [42]. In other words, irrigation had a greater influence on the partitioning of surface energy after warming. The above changes in surface energy led to intensifying irrigation-induced decreases (0.03–0.12 °C) in surface air temperature in the irrigated croplands (Figure 6i), which indicated that the cooling effect of irrigation would enhance after warming. As mentioned above, in Exp03, the precipitation affected by warming increased during the irrigation period (Figure 3f), and the irrigation amount also increased (Figure 5c), both of which helped to increase the irrigation-caused increases in precipitation (Figure 7c). In Exp03, the surface net radiation increased because of the decreased surface upward longwave radiation, and the increased surface net radiation led to increases in surface air temperature. Therefore, the warming in Exp03 was only driven by decreasing surface upward longwave radiation, and then the temperature only increased in the lower atmosphere (discussed later). Based on the high emission scenario (SSP5-8.5), the future temperature increased in the whole atmosphere [44], which was consistent with our conclusion in Exp02 (Figure 8a). In this study, the whole atmosphere referred to 0–12km and mainly located in troposphere. The formation of precipitation is also mainly in the troposphere; thus, the temperature of the whole atmosphere should affect the precipitation. Therefore, the increased irrigation-induced increases in precipitation in Exp03 might not be reasonable.

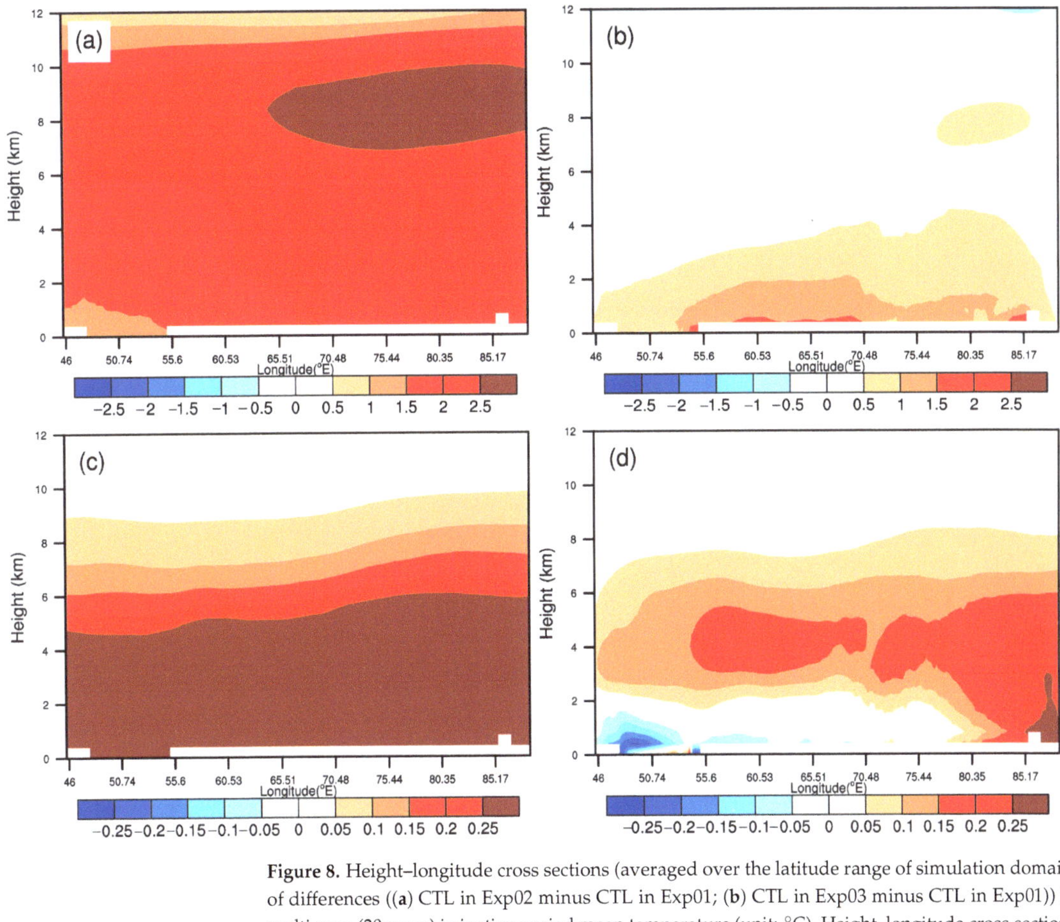

Figure 8. Height–longitude cross sections (averaged over the latitude range of simulation domain) of differences ((**a**) CTL in Exp02 minus CTL in Exp01; (**b**) CTL in Exp03 minus CTL in Exp01)) in multi-year (20 years) irrigation period mean temperature (unit: °C). Height–longitude cross sections (averaged over the latitude range of simulation domain) of differences ((**c**) CTL in Exp02 minus CTL in Exp01; (**d**) CTL in Exp03 minus CTL in Exp01)) in multi-year (20 years) irrigation period mean water vapor mixing ratio (unit: g kg^{-1}).

4. Discussion

In Exp02, the surface air temperature obviously increased by about 2.0 °C because of future driving fields, but the changes in the simulated surface energy (latent heat flux and sensible heat flux) were very small. The small changes in surface energy would lead to changes in surface evapotranspiration that were also small. Therefore, in Exp02, the irrigation amount affected by warming did not increase. Obviously, in Exp02, the influences of warming on surface variables (latent heat flux, sensible heat flux, and irrigation amount) were unreasonable. To solve the shortage of Exp02, we designed other warming experiments (Exp03) by decreasing the surface upward longwave radiation. The surface net radiation increased because of the decreased surface upward longwave radiation, that is, the surface energy was increased by the experimental setup in Exp03. In Exp03, not only did the surface air temperature obviously increase by about 2.0 °C, but the simulated surface energy also obviously increased. The increased surface energy was favorable to increases in surface evapotranspiration; thus, in Exp03, the irrigation amount affected by warming (2.0 °C) increased by 10–20%, which was thought to be reasonable. In Exp03, the changes in latent heat flux and sensible heat flux were thought to be reasonable, indicating that the impacts of warming on irrigation-caused changes in surface variables were also reasonable. In the irrigated croplands of Central Asia, after warming by 2.0 °C, the irrigation-induced increases in latent heat flux and the irrigation-induced decreases in sensible heat flux both expanded further, and then the irrigation-induced decreases in surface air temperature became further enhanced.

Although the regional averaged values of the increased surface air temperature were both about 2.0 °C in Exp02 and Exp03 (Figure 2), the warming mechanisms of Exp02 and Exp03 were completely different. In Exp02, the warming was induced by adopting the future high emission scenario data (SSP5-8.5) to drive the WRF model; in Exp03, the warming was only driven by decreasing the surface upward longwave radiation. The different warming mechanisms leading to the changes in temperature of the whole atmosphere were also different. In Exp02, the temperature of the whole atmosphere increased (Figure 8a); however, in Exp03, the temperature only increased in the lower atmosphere (Figure 8b). Banerjee and Singh [44] also drew the cross sections of the differences (SSP5-8.5 minus Historical) in temperature, and the results indicated that the temperature increased in the whole atmosphere, which was consistent with our conclusion. Therefore, we thought that the warming-induced increases (Exp02 minus Exp01) in the temperature of the whole atmosphere were reasonable. The temperature affected the atmospheric water vapor content, and their changes were basically consistent [45–47]. Therefore, the increased temperature found in Exp02 and Exp03 should lead to increases in water vapor. As shown in Figure 8c, the water vapor in Exp02 obviously increased by more than 0.1 g kg^{-1} from the surface to a height of 7 km; however, in Exp03, the height range of obvious increases in water vapor was from approximately 2 km to 6 km (Figure 8d). These different changes in water vapor were caused by the aforementioned different warming mechanisms.

The warming-induced changes in temperature and water vapor of the whole atmosphere in Exp02 were thought to be reasonable from the above analysis. The changes in temperature and water vapor of the whole atmosphere affected the precipitation. During the irrigation period, although the water vapor affected by warming obviously increased from the surface to 7 km high (Figure 8c), the low cloud cover decreased (Figure 9a), which corresponded to decreases in precipitation after warming in the mid-latitude mountainous areas (Figure 3d). Obviously, in Exp02, the increased water vapor did not lead to increases in precipitation, and the reason was that the temperature increased by about 2.0 °C in the whole atmosphere, which was not conducive to the water vapor condensation and the formation of precipitation. The increased temperature of the whole atmosphere was also not favorable for the irrigation-caused increases in precipitation. Therefore, as mentioned in Section 3.1, the irrigation-induced increases in precipitation in the mid-latitude mountainous areas decreased (Figure 7b), which was thought to be reasonable. In Exp03, the warming of only the lower atmosphere favored the water vapor condensation; that is, the low cloud cover increased (Figure 9b), which could explain the changes in precipitation after warming during the irrigation period (Figures 3f and 7c). However, in Exp03, the warming-induced changes in precipitation might not be reasonable.

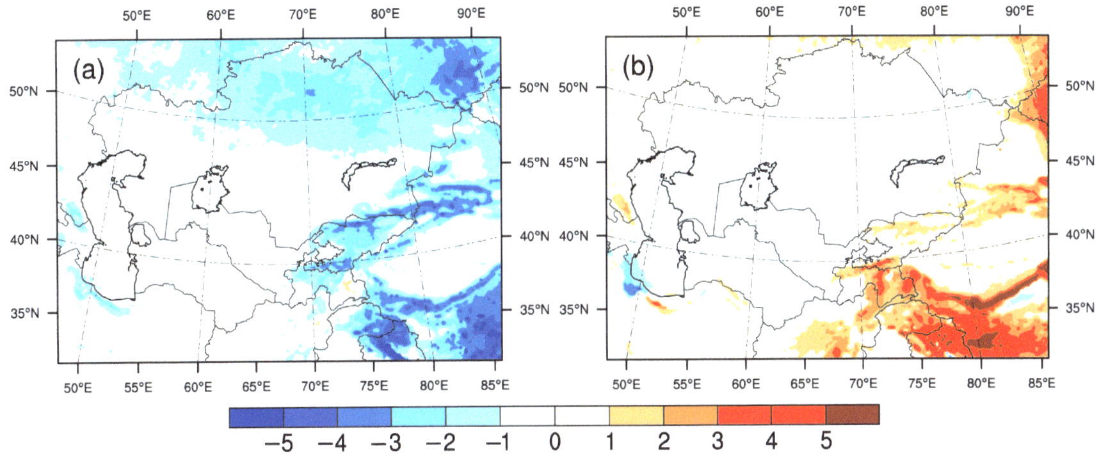

Figure 9. Spatial differences ((**a**) CTL in Exp02 minus CTL in Exp01; (**b**) CTL in Exp03 minus CTL in Exp01) in multi-year (20 years) irrigation period mean low cloud cover (unit: %). Low cloud cover is for 0.8 < sigma < 1.0 (sigma = pressure/surface pressure).

In summary, in Exp03, the impacts of warming on irrigation amount and irrigation-induced changes in surface variables (latent heat flux, sensible heat flux, and surface air temperature) were reasonable; in Exp02, the influences of warming on precipitation and irrigation-induced changes in precipitation were reasonable. The advantages of these two types of experiments (Exp02 and Exp03) were complementary, which achieved the objective of running reasonable simulations.

5. Conclusions

In this study, we designed three types of experiments: historical experiments (Exp01), warming experiments using future driving fields (Exp02), and warming experiments that increased surface energy (Exp03). We analyzed the impacts of irrigation on the regional climate under Central Asia warming by 2.0 °C by drawing on the differences between two types of warming experiments and the historical experiments, respectively. For surface variables (irrigation amount, latent heat flux, sensible heat flux, and surface air temperature), the changes (relative to Exp01) were thought to be reasonable in Exp03. For precipitation, the changes (relative to Exp01) were thought to be reasonable in Exp02. The main conclusions were as follows: (1) in Central Asia, after warming by 2.0 °C, the irrigation amount of the croplands increased by 10–20%. (2) In the irrigated croplands of Central Asia, after warming by 2.0 °C, the irrigation-induced increases in latent heat flux and the irrigation-induced decreases in sensible heat flux both expanded further, and then the irrigation-induced decreases in surface air temperature became further enhanced. (3) After warming up by 2.0 °C, the irrigation-induced increases in precipitation during the irrigation period in the mid-latitude mountainous areas were reduced.

Before conducting this study, we intended to simulate the impacts of irrigation on regional climate under Central Asia warming by 2.0 °C by adopting the future high emission scenario data (SSP5-8.5) as the driving fields. However, things did not happen as we wished, and the simulated results of some surface variables were not reasonable. To achieve more reasonable simulations, we designed another warming experiment that involved decreasing the surface upward longwave radiation. Although the advantages of these two types of warming experiments were complementary, the most ideal warming experiments were those that increased the temperature by increasing greenhouse gases in the atmosphere. However, the increased surface air temperature was less than 0.9 °C in the experiment where the CO_2 concentration was set to 1000 ppmv, relative to the experiment where the CO_2 concentration was set to 400 ppmv. This simulation showed that the size of the increased surface air temperature was too low in the high CO_2 concentration scenario. The reason for this might be that the WRF is a regional model, and the simulated results were affected by the lateral boundary conditions.

Author Contributions: Conceptualization, L.W. and H.Z.; methodology, L.W.; investigation, L.W. and H.Z.; visualization, L.W.; writing—original draft preparation, L.W.; data curation, L.W. and H.Z.; writing—review and editing, L.W. and H.Z.; supervision, L.W. and H.Z.; funding acquisition, L.W. and H.Z. All authors have read and agreed to the published version of the manuscript.

Funding: This research was funded by the National Natural Science Foundation of China (Grant No. 42005031), the Strategic Priority Research Program of Chinese Academy of Sciences (Grant No. XDA20020201), and the Science and Technology Development Project of Henan Province China (Grant No. 222102110419).

Data Availability Statement: The irrigation fraction data are obtained from http://www.fao.org/nr/water/aquastat/irrigationmap/index10.stm (accessed on 11 June 2020). The Bias-corrected CMIP6 global dataset (historical and SSP585) are available at http://www.doi.org/10.11922/sciencedb.00487 (accessed on 20 May 2021). The WRF version 4.0.3 (Software) source codes are obtained from https://github.com/wrf-model/ (accessed on 30 October 2020).

Conflicts of Interest: The authors declare no conflict of interest.

References

1. Schneider, N.; Eugster, W. Historical land use changes and mesoscale summer climate on the Swiss Plateau. *J. Geophys. Res. Atmos.* **2005**, *110*, D19102. [CrossRef]
2. Feng, D.; Wang, G.; Wei, X.; Amankwah, S.O.Y.; Hu, Y.; Luo, Z.; Hagan, D.F.T.; Ullah, W. Merging and Downscaling Soil Moisture Data From CMIP6 Projections Using Deep Learning Method. *Front. Environ. Sci.* **2022**, *10*, 847475. [CrossRef]
3. Crutzen, P. Geology of mankind. *Nature* **2002**, *415*, 23. [CrossRef] [PubMed]
4. Foley, J.A.; DeFries, R.; Asner, G.P.; Barford, C.; Bonan, G.; Carpenter, S.R.; Chapin, F.S.; Coe, M.T.; Daily, G.C.; Gibbs, H.K.; et al. Global consequences of land use. *Science* **2005**, *309*, 570–574. [CrossRef]
5. Bonan, G.B.; Doney, S.C. Climate, ecosystems, and planetary futures: The challenge to predict life in Earth system models. *Science* **2018**, *359*, eaam8328. [CrossRef] [PubMed]
6. Pielke, R.A., Sr.; Marland, G.; Betts, R.A.; Chase, T.N.; Eastman, J.L.; Niles, J.O. The influence of land-use change and landscape dynamics on the climate system: Relevance to climate-change policy beyond the radiative effect of greenhouse gases. *Philos. Trans. R. Soc. Lond. Ser. A Math. Phys. Eng. Sci.* **2002**, *360*, 1705–1719. [CrossRef] [PubMed]
7. Feddema, J.J.; Oleson, K.W.; Bonan, G.B.; Mearns, L.O.; Buja, L.E.; Meehl, G.A.; Washington, W.M. The importance of land-cover change in simulating future climates. *Science* **2005**, *310*, 1674–1678. [CrossRef] [PubMed]
8. Feng, J.M.; Wang, Y.L.; Ma, Z.G.; Liu, Y.H. Simulating the regional impacts of urbanization and anthropogenic heat release on climate across China. *J. Clim.* **2012**, *25*, 7187–7203. [CrossRef]
9. Cao, Q.; Yu, D.; Georgescu, M.; Han, Z.; Wu, J. Impacts of land use and land cover change on regional climate: A case study in the agro-pastoral transitional zone of China. *Environ. Res. Lett.* **2015**, *10*, 124025. [CrossRef]
10. Li, D.; Tian, P.; Luo, H.; Hu, T.; Dong, B.; Cui, Y.; Khan, S.; Luo, Y. Impacts of land use and land cover changes on regional climate in the Lhasa River basin, Tibetan Plateau. *Sci. Total Environ.* **2020**, *742*, 140570. [CrossRef]
11. Wu, L.; Feng, J.; Qin, F.; Qiu, Y. Regional climate effects of irrigation over Central Asia using Weather Research and Forecasting model. *J. Geophys. Res. Atmos.* **2022**, *127*, e2021JD036210. [CrossRef]
12. Siebert, S.; Burke, J.; Faures, J.M.; Frenken, K.; Hoogeveen, J.; Döll, P. Groundwater use for irrigation-a global inventory. *Hydrol. Earth Syst. Sci.* **2010**, *14*, 1863–1880. [CrossRef]
13. Siebert, S.; Döll, P.; Hoogeveen, J.; Faures, J.M.; Frenken, K.; Feick, S. Development and validation of the global map of irrigation areas. *Hydrol. Earth Syst. Sci.* **2005**, *9*, 535–547. [CrossRef]
14. Yang, B.; Zhang, Y.; Qian, Y.; Tang, J.; Liu, D. Climatic effects of irrigation over the Huang-Huai-Hai Plain in China simulated by the weather research and forecasting model. *J. Geophys. Res. Atmos.* **2016**, *121*, 2246–2264. [CrossRef]
15. Yang, Z.; Dominguez, F.; Zeng, X.; Hu, H.; Gupta, H.; Yang, B. Impact of irrigation over the California Central Valley on regional climate. *J. Hydrometeorol.* **2017**, *18*, 1341–1357. [CrossRef]
16. Wu, L.; Feng, J.; Miao, W. Simulating the impacts of irrigation and dynamic vegetation over the North China Plain on regional climate. *J. Geophys. Res. Atmos.* **2018**, *123*, 8017–8034. [CrossRef]
17. Kang, S.; Eltahir, E.A.B. Impact of irrigation on regional climate over Eastern China. *Geophys. Res. Lett.* **2019**, *46*, 5499–5505. [CrossRef]
18. Yang, Q.; Huang, X.; Tang, Q. Irrigation cooling effect on land surface temperature across China based on satellite observations. *Sci. Total Environ.* **2020**, *705*, 135984. [CrossRef]

19. Liu, G.; Wang, W.; Shao, Q.; Wei, J.; Zheng, J.; Liu, B.; Chen, Z. Simulating the Climatic Effects of Irrigation over China by Using the WRF-Noah Model System with Mosaic Approach. *J. Geophys. Res. Atmos.* **2021**, *126*, e2020JD034428. [CrossRef]
20. Adegoke, J.O.; Pielke Sr, R.A.; Eastman, J.; Mahmood, R.; Hubbard, K.G. Impact of irrigation on midsummer surface fluxes and temperature under dry synoptic conditions: A regional atmospheric model study of the US High Plains. *Mon. Weather Rev.* **2003**, *131*, 556–564. [CrossRef]
21. Lobell, D.B.; Bonfils, C.J.; Kueppers, L.M.; Snyder, M.A. Irrigation cooling effect on temperature and heat index extremes. *Geophys. Res. Lett.* **2008**, *35*, L09705. [CrossRef]
22. Kanamaru, H.; Kanamitsu, M. Model diagnosis of nighttime minimum temperature warming during summer due to irrigation in the California Central Valley. *J. Hydrometeorol.* **2008**, *9*, 1061–1072. [CrossRef]
23. Kueppers, L.M.; Snyder, M.A. Influence of irrigated agriculture on diurnal surface energy and water fluxes, surface climate, and atmospheric circulation in California. *Clim. Dyn.* **2012**, *38*, 1017–1029. [CrossRef]
24. Huber, D.B.; Mechem, D.B.; Brunsell, N.A. The effects of Great Plains irrigation on the surface energy balance, regional circulation, and precipitation. *Climate* **2014**, *2*, 103–128. [CrossRef]
25. Sacks, W.J.; Cook, B.I.; Buenning, N.; Levis, S.; Helkowski, J.H. Effects of global irrigation on the near-surface climate. *Clim. Dyn.* **2009**, *33*, 159–175. [CrossRef]
26. Lo, M.H.; Famiglietti, J.S. Irrigation in California's Central Valley strengthens the southwestern US water cycle. *Geophys. Res. Lett.* **2013**, *40*, 301–306. [CrossRef]
27. Zou, J.; Xie, Z.; Yu, Y.; Zhan, C.; Sun, Q. Climatic responses to anthropogenic groundwater exploitation: A case study of the Haihe River Basin, Northern China. *Clim. Dyn.* **2014**, *42*, 2125–2145. [CrossRef]
28. Zeng, Y.; Xie, Z.; Zou, J. Hydrologic and climatic responses to global anthropogenic groundwater extraction. *J. Clim.* **2017**, *30*, 71–90. [CrossRef]
29. Qian, Y.; Huang, M.; Yang, B.; Berg, L.K. A modeling study of irrigation effects on surface fluxes and land–air–cloud interactions in the Southern Great Plains. *J. Hydrometeorol.* **2013**, *14*, 700–721. [CrossRef]
30. Sorooshian, S.; AghaKouchak, A.; Li, J. Influence of irrigation on land hydrological processes over California. *J. Geophys. Res. Atmos.* **2014**, *119*, 13137–13152. [CrossRef]
31. Pei, L.; Moore, N.; Zhong, S.; Kendall, A.D.; Gao, Z.; Hyndman, D.W. Effects of irrigation on summer precipitation over the United States. *J. Clim.* **2016**, *29*, 3541–3558. [CrossRef]
32. Zhang, X.; Xiong, Z.; Tang, Q. Modeled effects of irrigation on surface climate in the Heihe River Basin, Northwest China. *J. Geophys. Res. Atmos.* **2017**, *122*, 7881–7895. [CrossRef]
33. Zhang, M.; Luo, G.; Cao, X.; Hamdi, R.; Li, T.; Cai, P. Numerical simulation of the irrigation effects on surface fluxes and local climate in typical mountain-oasis-desert systems in the Central Asia arid area. *J. Geophys. Res. Atmos.* **2019**, *124*, 12485–12506. [CrossRef]
34. Wang, W.; Liu, G.; Wei, J.; Chen, Z.; Ding, Y.; Zheng, J. The climatic effects of irrigation over the middle and lower reaches of the Yangtze River, China. *Agric. For. Meteorol.* **2021**, *308*, 108550. [CrossRef]
35. Valmassoi, A.; Dudhia, J.; Di Sabatino, S.; Pilla, F. Evaluation of three new surface irrigation parameterizations in the WRF-ARW v3. 8.1 model: The Po Valley (Italy) case study. *Geosci. Model Dev.* **2020**, *13*, 3179–3201. [CrossRef]
36. Li, C.; Zhang, C.; Luo, G.; Chen, X.; Maisupova, B.; Madaminov, A.A.; Han, Q.; Djenbaev, B.M. Carbon stock and its responses to climate change in Central Asia. *Glob. Chang. Biol.* **2015**, *21*, 1951–1967. [CrossRef] [PubMed]
37. Chen, F.; Wang, J.; Jin, L.; Zhang, Q.; Li, J.; Chen, J. Rapid warming in mid-latitude central Asia for the past 100 years. *Front. Earth Sci. China* **2009**, *3*, 42–50. [CrossRef]
38. Hu, Z.; Zhang, C.; Hu, Q.; Tian, H. Temperature changes in Central Asia from 1979 to 2011 based on multiple datasets. *J. Clim.* **2014**, *27*, 1143–1167. [CrossRef]
39. Ma, X.; Zhu, J.; Yan, W.; Zhao, C. Projections of desertification trends in Central Asia under global warming scenarios. *Sci. Total Environ.* **2021**, *781*, 146777. [CrossRef]
40. Tian, J.; Zhang, Y. Detecting changes in irrigation water requirement in Central Asia under CO_2 fertilization and land use changes. *J. Hydrol.* **2020**, *583*, 124315. [CrossRef]
41. Xu, Z.; Han, Y.; Tam, C.Y.; Yang, Z.L.; Fu, C. Bias-corrected CMIP6 global dataset for dynamical downscaling of the historical and future climate (1979–2100). *Sci. Data* **2021**, *8*, 293. [CrossRef]
42. Donohoe, A.; Armour, K.C.; Pendergrass, A.G.; Battisti, D.S. Shortwave and longwave radiative contributions to global warming under increasing CO_2. *Proc. Natl. Acad. Sci. USA* **2014**, *111*, 16700–16705. [CrossRef] [PubMed]
43. Kweku, D.W.; Bismark, O.; Maxwell, A.; Desmond, K.A.; Danso, K.B.; Oti-Mensah, E.A.; Quachie, A.T.; Adormaa, B.B. Greenhouse effect: Greenhouse gases and their impact on global warming. *J. Sci. Res. Rep.* **2018**, *17*, 1–9. [CrossRef]
44. Banerjee, D.; Singh, C. An appraisal of seasonal precipitation dynamics over the North-West Himalayan region under future warming scenarios. *Int. J. Climatol.* **2022**, *42*, 2328–2350. [CrossRef]
45. Zhao, T.; Dai, A.; Wang, J. Trends in tropospheric humidity from 1970 to 2008 over China from a homogenized radiosonde dataset. *J. Clim.* **2012**, *25*, 4549–4567. [CrossRef]

46. Zhang, J.; Zhao, T. Historical and future changes of atmospheric precipitable water over China simulated by CMIP5 models. *Clim. Dyn.* **2019**, *52*, 6969–6988. [CrossRef]
47. He, W.; Meng, H.; Han, J.; Zhou, G.; Zheng, H.; Zhang, S. Spatiotemporal PM2.5 estimations in China from 2015 to 2020 using an improved gradient boosting decision tree. *Chemosphere* **2022**, *296*, 134003. [CrossRef]

remote sensing

MDPI

Communication

Comparison of Spring Wind Gusts in the Eastern Part of the Tibetan Plateau and along the Coast: The Role of Turbulence

Xingxu Zhou [1,2], Chao Zhang [1,2], Yunying Li [1,2,*] and Zhiwei Zhang [3]

1 College of Meteorology and Oceanography, National University of Defense Technology, Changsha 410073, China; zhouxingxu@nudt.edu.cn (X.Z.); zchao@nudt.edu.cn (C.Z.)
2 High Impact Weather Key Laboratory of CMA, Changsha 410073, China
3 Unit No. 92192 of Chinese People's Liberation Army, Ningbo 315000, China; zhangzhiwei@stu.ouc.edu.cn
* Correspondence: ghlyy@mail.iap.ac.cn

Abstract: Wind gusts are sudden, brief increases in wind speed that have important implications for wind power generation, building design, aviation and marine safety. However, wind gusts in the Plateau and coastal plain are very different. In this paper, the gust characteristics are explored and compared at two sites in the same latitude—Xining, a city in the eastern Tibetan Plateau, and Qingdao, a city in the coast in China—using Doppler lidar data. The results indicate that the wind gusts in Xining are more intense and occur at a higher height than those in Qingdao. Though mean winds and turbulence significantly influence gusts, the turbulence intensity is responsible for the differences in gust, and high turbulence in the eastern part of the Tibetan Plateau is inferred. These results provide observational evidence for wind gusts over the complex terrain of the Tibetan Plateau and are useful for studying their impact on important aspects, such as flight safety.

Keywords: wind gusts; Tibetan Plateau; turbulence; Doppler lidar

Citation: Zhou, X.; Zhang, C.; Li, Y.; Zhang, Z. Comparison of Spring Wind Gusts in the Eastern Part of the Tibetan Plateau and along the Coast: The Role of Turbulence. *Remote Sens.* **2023**, *15*, 3655. https://doi.org/10.3390/rs15143655

Academic Editor: Yuanjian Yang

Received: 8 June 2023
Revised: 8 July 2023
Accepted: 20 July 2023
Published: 21 July 2023

1. Introduction

Wind gusts are sudden, brief increases in wind speed that play a critical role in wind power generation, building design, aviation, and marine safety [1,2]. Wind gusts and extreme winds can cause extensive damage to coastal areas and urban communities, and this damage is even more powerful on the Tibetan Plateau [3].

Wind gusts are closely related to atmospheric variables such as mean wind speed, boundary layer turbulence, bottom surface and atmospheric stability, as many studies have found in recent years with the increasing number of ground-based observations [4–6]. Many studies [7,8] have indicated that turbulence is one of the most important factors affecting gust formation. For example, Hu et al. [5] showed that near-surface wind gusts are closely related to mean wind speed and turbulence intensity. Luchetti et al. [9] further assessed the changes in meteorological elements during gust passage and found that vertical airflow and turbulence kinetic energy were enhanced in most cases. He et al. [10] used meteorological tower data to analyze the structure of wind and turbulence and showed a tendency for turbulence to intensify prior to the passage of wind gusts. These results highlight the important role of turbulence in the development of wind gusts.

Turbulence may arise from complex topography and may contribute to the strong local characteristics of the wind gusts [11–13]. Harper et al. [14] reported that the various bottom surface properties of the ocean and land lead to distinct turbulence and wind gusts under diverse substrates (e.g., inland, off-land, off-sea and offshore), indicating that wind gusts have significant sea–land differences. Similarly, Wang et al. [13] classified the Bohai Sea and its coastal areas into four regions based on topography and analyzed every region's peak gust climatology through observations. They found that a rough bottom surface stimulated stronger turbulence and increased the peak gust and gust factor. Moreover, wind speed can be affected by the unevenness of the terrain, which creates complex flow patterns [15]. For example, wind can be channeled or accelerated by topographic features such as hills or valleys [16–18]. Complex terrain also leads to more frequent and stronger wind gusts [6,19,20]. Wind gusts are amplified by the effects of topography on the formation and structure of large-scale weather systems and mountain waves [21,22]. However, the lack of adequate observational data in complex terrain limits the understanding and prediction of wind gusts, and more research is needed to explore the features and causes of wind gusts in different environments [23–26].

The Qinghai–Tibet Plateau, the highest plateau on Earth, has complex topography and is one of the key drivers of regional atmospheric circulation and climate [27]. Thus far, the characterization of wind gusts on the Tibetan Plateau remains inadequate, leading to difficulties in forecasting blowing snow and low-visibility events [28–31]. The Tibetan Plateau has complex topography that may cause turbulence and wind gusts with strong local features. However, it is unknow how these wind gusts differ from those in a typical coastal plain because of the lack of wind data in the Tibetan Plateau [3]. To date, the estimation of local wind gusts in weather and climate prediction algorithms remains challenging due to the lack of understanding of the spatial and temporal characteristics of wind gusts in plateau areas with complex terrain [25]. An analysis of the characteristics of wind gusts influenced by the complex terrain in the Plateau based on measured data could help to improve the forecasting of wind gusts, assessment of potential hazards and aircraft safety.

Long-term wind gust observations in plateau areas with complex terrain are challenging to conduct. In addition, meteorological towers are costly and limited in height, while conventional radiosondes cannot provide continuous wind data. Doppler lidar can overcome the limitations of weather towers and radiosondes with its large vertical detection range and high temporal resolution. That makes it more suitable for observing complex turbulence and wind gusts [32,33]. Indeed, Doppler lidar has been used to estimate wind gusts and turbulence in many pioneering works [16,34], demonstrating its potential ability to support gust and turbulence studies in complex environments.

This study analyzed wind gusts using data from the same type of Doppler lidar in Xining, on the eastern part of the Tibetan Plateau, and Qingdao, a coastal city at a similar latitude. The aim was to compare the wind gust characteristics of these two typical locations and to explore the influence of complex topography on wind gusts. We also discuss how turbulence affects the development of wind gusts over the eastern part of the Tibetan Plateau.

2. Materials and Methods

2.1. Scope

This paper chose two sites with similar latitudes and surface wind speed (Figure 1d) but very different topography to avoid the influence of latitude. Xining (at approximately 102.0°E, 36.5°N, 2594 m above mean sea level; Figure 1b) is located in the eastern part of the Tibetan Plateau and is surrounded by two mountains about 4000 m in height. In contrast, Qingdao (at approximately 120.4°E, 36.1°N, 11 m above mean sea level; Figure 1c) is near the northern coast of the Yellow Sea with no tall steep peaks nearby. Moreover, the high concentration of human activities and transportation in Xining and Qingdao makes the wind gusts in these two sites influential and noteworthy.

The study used a dataset derived from Doppler lidar observations at both sites between 24 March and 10 June 2022, the period examined by our experiment. During this period, although the western Pacific subtropical high was located between 15 and 30°N and the rainband covered South Central China, they did not directly influence the weather in Qingdao. Moreover, neither typhoons nor strong fronts directly affected Qingdao during the study period; only local convection occurred occasionally. In April, the Xining site experienced several cold air activities and strong convection due to the high pressure from the eastern part of Lake Baikal. However, these factors did not significantly affect the statistical results. Most of the time, the winds at these two sites were free from strong weather system disturbances, which facilitated the study of wind gusts and turbulence in the lower troposphere.

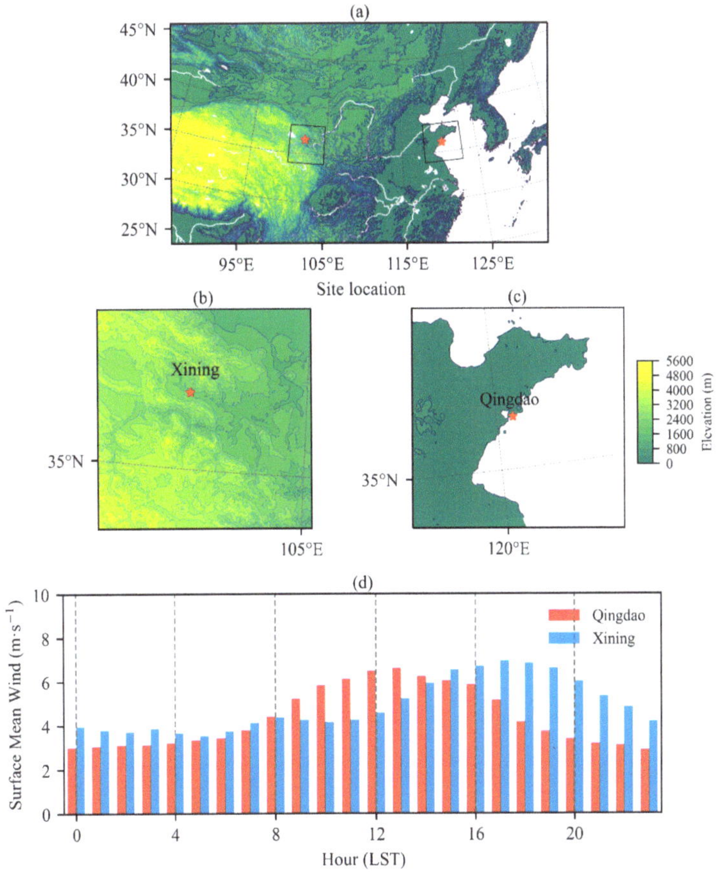

Figure 1. (**a–c**) Locations of the Xining site in the eastern Tibetan Plateau and the coastal site in Qingdao (background color represents elevation), where Doppler lidar was deployed on open ground. (**d**) Diurnal variation in the mean surface wind speed at the two sites during the study period, with the horizontal axis showing the local standard time (LST).

2.2. Doppler Lidar Data Quality Control

Doppler lidar can detect low-level tropospheric winds continuously, unlike conventional weather towers and radio sounders. Therefore, it is increasingly used in aviation weather services and local weather studies [35–37]. To compare the data from the two sites and ensure their consistency, we used the same model of Doppler lidar (FC-II) in the experiment. The device uses a 1.55 μm wavelength Doppler beam swing (DBS) technique. The lidar measures the wind with four beams that are oriented at an elevation angle of 70°. By varying the zenith angle of each beam at four different azimuths, the lidar scans the radial wind field and estimates the horizontal and vertical components of the wind speed and direction at various heights. The device has a 50 m vertical resolution and a 5 s temporal resolution, with a wind speed detection error of less than 0.5 $m \cdot s^{-1}$.

The efficacy and quantity of obtainable data for lidar are influenced by the rapid attenuation of aerosol particles, which are the indicators for lidar, with increasing height [38,39]. As depicted in Figure 2a, the data availability diminishes considerably with increasing height, but it is satisfactory below 2000 m. Erroneous data from Doppler lidar can contaminate gust studies and need to be eliminated. We applied an unsupervised machine learning technique based on the isolated forest algorithm to ensure the quality of the raw data.

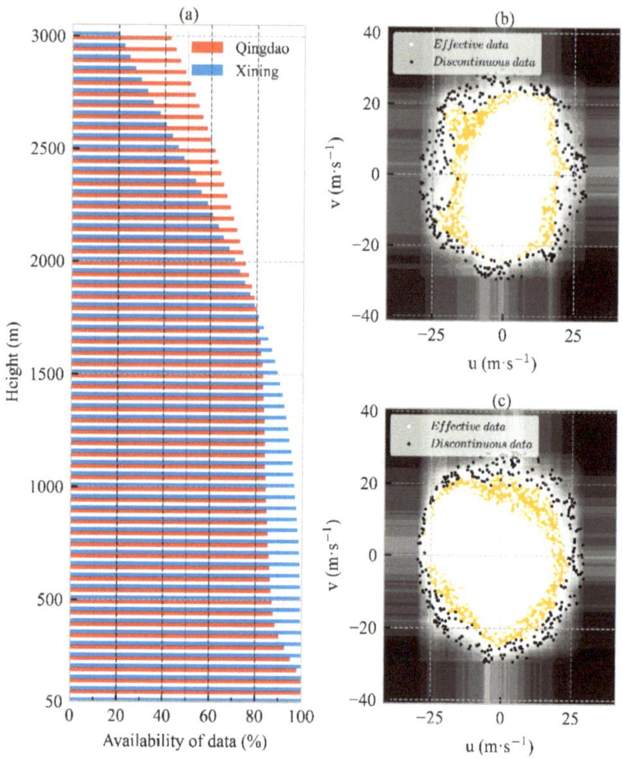

Figure 2. (**a**) Data availability at different heights during the study period, with red and blue colors representing the Qingdao and Xining sites, respectively. Partial data quality control results (at 250 m) for (**b**) Qingdao and (**c**) Xining based on the isolated forest algorithm, with white dots indicating valid data, black dots indicating rejected data and yellow ranges indicating the distribution of the valid area (horizontal axis shows the u-component of wind and vertical axis shows the v-component of wind).

The isolated forest algorithm works by splitting the data space with a random hyperplane into two subspaces and repeating this step until each subspace has only one data node. Each leaf node contains a single data node, forming an isolated tree [40].

We split the wind data into longitudinal and latitudinal components and normalized their sum. The outlier points had sparse data density and remained in a subspace after the hyperspace was repeatedly divided by a multidimensional plane. We set the threshold to the path length from the longest leaf node to the root node in the first 0.15% of each height layer and considered any data longer than this length as singular. The wind data at a certain height should be continuous with high resolution. The isolated forest algorithm tends to eliminate isolated points in the multidimensional space that have poor continuity. The black points in Figure 2b,c represent some of the data that were eliminated. This method eliminates data that may be affected by other noise disturbances while preserving gust turbulence disturbances. A possible limitation of our method is that removing isolated points of poor continuity may also filter out strong turbulence signals. However, as a previous experiment [41] has discussed, this does not significantly affect the results.

We note that the conventional carrier-to-noise ratio (CNR) thresholding method can also produce reliable results after adjusting the thresholds [42,43]. However, the CNR of lidar can vary considerably under different seasonal, weather, or terrain conditions, and different thresholds are needed for data from both sites. Machine learning methods for quality control based on data continuity have better portability and can offer a convenient basis for experiments under different environmental conditions.

2.3. Parameters and Definitions

Wind gusts can be measured using several parameters, and different sampling intervals can have an impact on the gust factor values [44]. While the WMO recommendations do not specify a strict limit, most studies on near-surface wind gusts adopted 10 min and 5 s sampling intervals to measure mean and instantaneous winds.

The mean wind speed (V_T) is the average of all 5 s wind speeds in the same interval, while the instantaneous wind speed ($V_{t,T}$) is any 5 s running mean wind speed observed. And the peak gust ($V_{t,T}^{\max}$) is the highest 5 s running mean wind speed in a 10 min interval.

The gust amplitude ($V_{t,T}^{a}$) is described as the difference between the peak gust ($V_{t,T}^{\max}$) and the 10 min mean wind speed:

$$V_{t,T}^{a} = V_{t,T}^{\max} - V_T \quad (1)$$

The gust factor (G_T) is the ratio of the peak gust to the 10 min mean wind speed:

$$G_T = \frac{V_{t,T}^{\max}}{V_T} \quad (2)$$

The turbulence intensity (I_T) is defined as

$$I_T = \frac{\sigma_T}{V_T} \quad (3)$$

where σ_T is the standard deviation of the wind component and is expressed as

$$\sigma_T = \sqrt{\sigma_u^2 + \sigma_v^2 + \sigma_w^2} \quad (4)$$

3. Comparison of Wind Gusts Parameters at Two Sites: Phenomenon

To compare the wind gusts in the eastern part of the Qinghai–Tibet Plateau and the wind gusts along the coast at the same latitude, we took Xining and Qingdao as examples, respectively. These two locations show different characteristics of wind gusts, as illustrated in Figure 3.

Figure 3 shows the different characteristics of wind gusts at the two sites with increasing height. Above 200 m, the average wind speed in Qingdao is significantly higher than that in Xining (Figure 3a). This may be because the complex canyon topography on both sides of Xining hinders the wind, while the sea–land wind in Qingdao strengthens it [45,46]. However, the difference in peak gust ($V_{t,T}^{\max}$) over the two sites is small (Figure 3b). This means that Xining may experience transient wind gusts of the same magnitude as Qingdao, even though the background wind field is weaker. Moreover, Xining has a larger gust amplitude ($V_{t,T}^{a}$) and gust factor (G_T) (Figure 3c,d). These are the difference and ratio between instantaneous and mean winds when wind gusts occur. These results suggest that the gusts in the eastern part of the Tibetan Plateau are more complex and unpredictable than those at the coast. They also imply that wind gusts in Xining may cause more powerful damage when they occur.

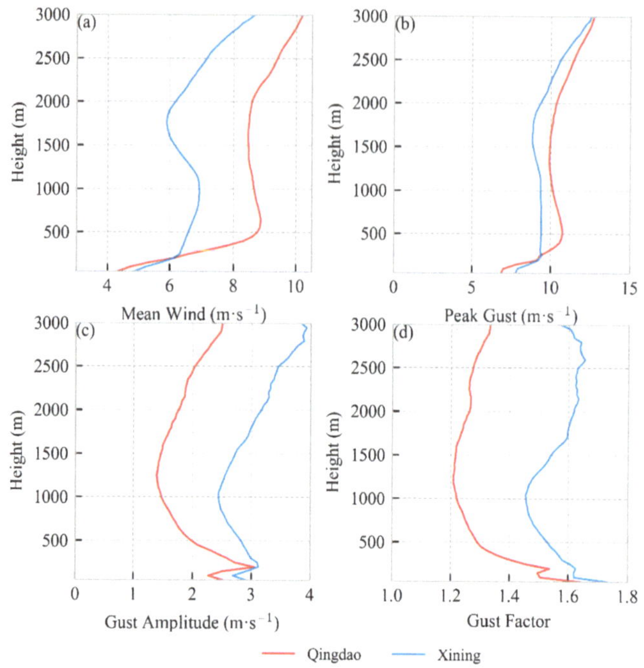

Figure 3. Comparison of wind and gust parameters at the Qingdao and Xining sites during the study period. The figure shows the variation in (**a**) mean wind speed V_T, (**b**) peak gust $V_{t,T}^{max}$, (**c**) gust amplitude $V_{t,T}^{a}$ and (**d**) gust factor G_T with increasing height. The red line represents Qingdao, and the blue line represents Xining.

Previous studies [43] have shown that wind gusts vary significantly with diurnal conditions. We studied the diurnal variation in gust parameters at different heights for both sites (Figure 4).

Figure 4 shows the diurnal variation in gust parameters at two typical height layers: 100 m and 1000 m. These layers represent the near-surface mass and low height, respectively. For Qingdao, the gust parameters vary more at the bottom layer than at the upper layer diurnally. The most notable difference is that the peak gust ($V_{t,T}^{max}$) at 100 m reaches its maximum at noon and is weakest at night. However, this feature disappears at 1000 m. Moreover, the gust amplitude ($V_{t,T}^{a}$), gust factor (G_T) and turbulence intensity (I_T) are also weaker at 1000 m than at 100 m. This result confirms Suomi et al.'s [43] report that the gust phenomenon becomes less prevalent with height.

In Xining, the gust parameters and turbulence intensity still vary significantly with time of day at 1000 m. In fact, the peak of the gust amplitude ($V_{t,T}^{a}$) and peak gust ($V_{t,T}^{max}$) at 1000 m are higher than those at 100 m (Figure 4a,b). This indicates that strong wind gusts occur above the near-surface mass in Xining. The gust factor (G_T) and turbulence intensity (I_T) at 1000 m in Xining are lower than those at 100 m, but still much higher than those in Qingdao (Figure 4c,d). This indicates that the gust phenomenon in Xining can extend to higher heights and is not limited to the near-surface mass, which poses a threat to aviation activities.

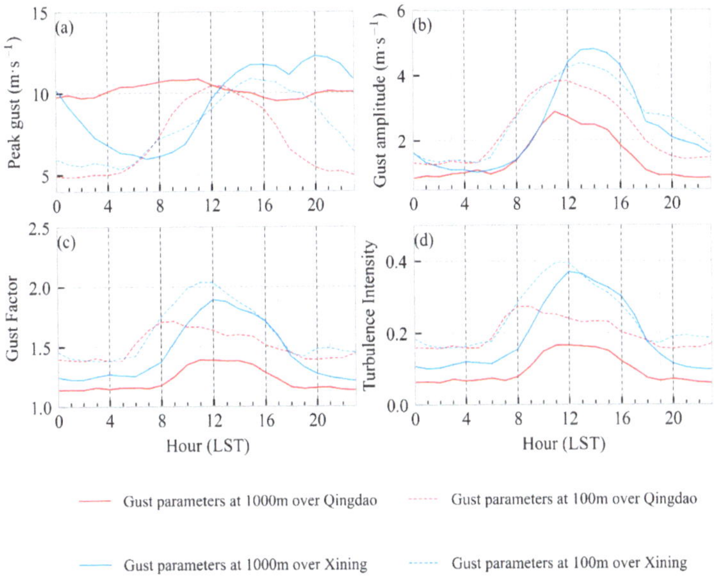

Figure 4. Diurnal variation in gust parameters and turbulence intensity at Qingdao and Xining sites at two heights. The figure shows the hourly averages of (**a**) peak gust $V_{t,T}^{max}$, (**b**) gust amplitude $V_{t,T}^{a}$, (**c**) gust factor G_T and (**d**) turbulence intensity I_T at 100 m and 1000 ms. The red line represents Qingdao, and the blue line represents Xining. The solid line indicates 1000 m and the dashed line indicates 100 m.

The possible reasons for the significant differences in the gust characteristics between Xining and Qingdao are worth exploring. We note that the gust factor (G_T) and turbulence intensity (I_T) vary similarly with time of day for both Xining and Qingdao, at both 1000 m and 100 m (Figure 4c,d). This may imply that turbulence influences the gust development. We will discuss this in the next section.

4. The Role of Turbulence: A Possible Explanation

As discussed in Section 3, the gust characteristics in Xining and Qingdao are very different, and turbulence may play some role in gust development. In studying this, Yus-Díez et al. [47] indicated that wind gusts are influenced not only by the mean wind speed but also by turbulence intensity. To investigate further, we analyzed the correlation of gust parameters with mean wind speed and turbulence intensity for both sites, respectively (Figure 5).

As shown in Figure 5, the mean wind speed (V_T) had a good correlation with the peak gust ($V_{t,T}^{max}$) in Qingdao and Xining, but a weak correlation with the gust amplitude ($V_{t,T}^{a}$). Likewise, the gust factor (G_T) had a weakly negative correlation with the mean wind speed (V_T). On the other hand, turbulence intensity (I_T) has a significant positive correlation with both gust amplitude ($V_{t,T}^{a}$) and gust factor (G_T), while its correlation with peak gust ($V_{t,T}^{max}$) is weak.

Based on the approach of Zeng et al. [48], we consider wind gusts a superposition of mean winds and turbulence. The peak gust ($V_{t,T}^{max}$) correlates well with the mean wind speed (V_T), but not with the turbulence intensity (I_T), suggesting that the peak gust may be influenced by background wind. Moreover, turbulence intensity acts on the magnitude of instantaneous wind, and the stronger the turbulence intensity (I_T), the higher the gust factor (G_T) and gust amplitude ($V_{t,T}^{a}$). These results indicate that the mean wind and turbulence intensity may be responsible for the differences in wind gusts.

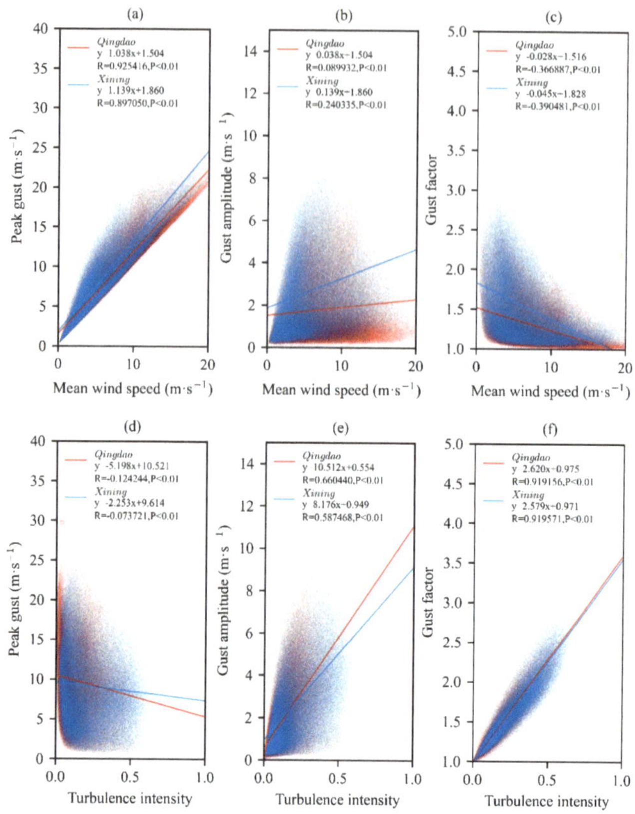

Figure 5. Correlations between gust parameters (peak gust $V_{t,T}^{max}$, gust amplitude $V_{t,T}^{a}$ and gust factor G_T) and (**a–c**) mean wind speed V_T or (**d–f**) turbulence intensity I_T for Qingdao (blue) and Xining (red). R and P are the correlation coefficient and the extreme probability value, respectively.

To examine how turbulence affects gust formation, we classified the 50 m winds in Xining into two groups based on their turbulence intensity (I_T): strong-turbulence winds ($0.365 < I_T$) and weak-turbulence winds ($I_T < 0.161$). The lower and upper quartiles of I_T at this height are 0.161 and 0.365, respectively. We then analyzed the wind gusts for each group separately and compared them in Figure 6.

Figure 6 shows that the peak gust ($V_{t,T}^{max}$) increases rapidly with the mean wind speed (V_T) under strong turbulence conditions. However, this does not mean that the peak gust ($V_{t,T}^{max}$) are always higher under strong turbulence conditions. Under strong turbulence conditions, the mean wind speed (V_T) is mostly below 5 m·s^{-1} and the peak gust ($V_{t,T}^{max}$) is below 10 m·s^{-1}. Under weak turbulence conditions, however, the mean wind speed (V_T) can reach up to 10 m·s^{-1} and the peak gust ($V_{t,T}^{max}$) can reach up to 15 m·s^{-1}. A possible explanation is that a strong mean wind reduces the turbulence intensity.

This indicates that both the mean wind and the turbulence affect the wind gusts. However, strong turbulence conditions can produce higher wind gusts than weak turbulence conditions at the same mean wind speed, which supports our previous experiments [41].

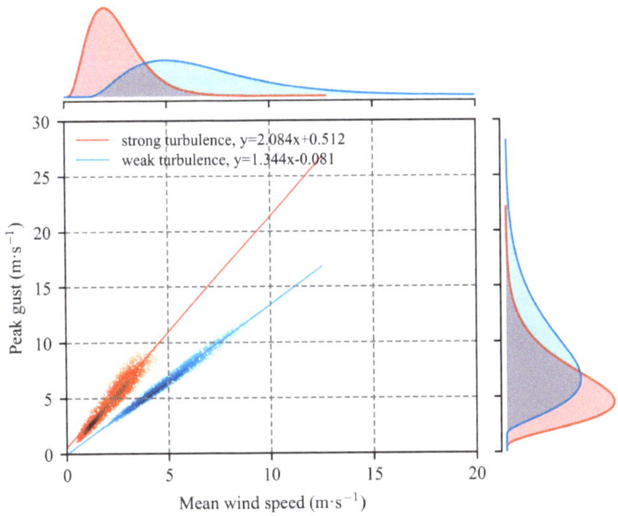

Figure 6. Relationship between mean wind speed (V_T) and peak gust ($V_{t,T}^{max}$) under different turbulence intensity (I_T) conditions. The vertical axis shows the peak gust and the horizontal axis shows the mean wind speed. The red points represent strong turbulence conditions and the blue points represent weak turbulence conditions.

As discussed above, turbulence and mean winds jointly influence the development of wind gusts, which may explain the different gust characteristics between Xining and Qingdao. These two sites are in the eastern part of the Tibetan Plateau and the coastal area, respectively, and their background wind and turbulence activities need to be analyzed to further explain the reasons for the differences in wind gusts between Xining and Qingdao.

Figures 7 and 8 show the average wind speed and wind direction at different height levels in Qingdao and Xining, respectively. In Qingdao, the lower layer has a smaller wind speed and a southwesterly flow, being influenced by the secondary circulation of the sea and land. The upper layer has a strong westerly wind, consistent with the geostrophic wind theory. The wind speed increases and the wind direction shifts westerly with the increase in height.

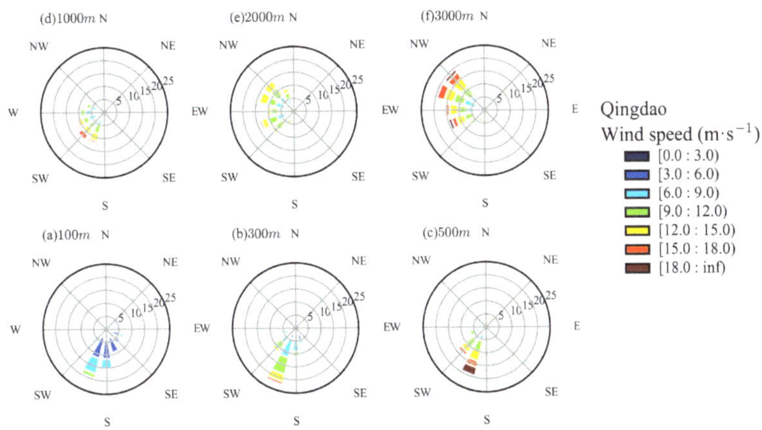

Figure 7. *Cont.*

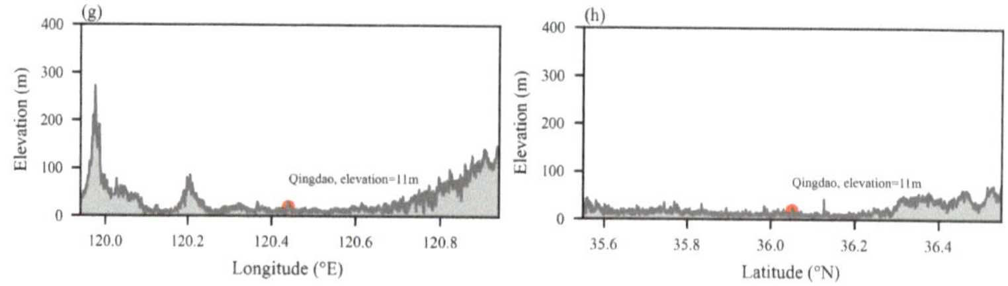

Figure 7. Average wind at different heights in Qingdao. (**a–f**) Wind speeds at 100 m, 300 m, 500 m, 1000 m, 2000 m and 3000 m, respectively. The circles show the 5% frequency increment wind speed of lidar, with wind speed divided into seven grades. Topographic elevations along (**g**) 120.43°E and (**h**) 36.11°N, respectively.

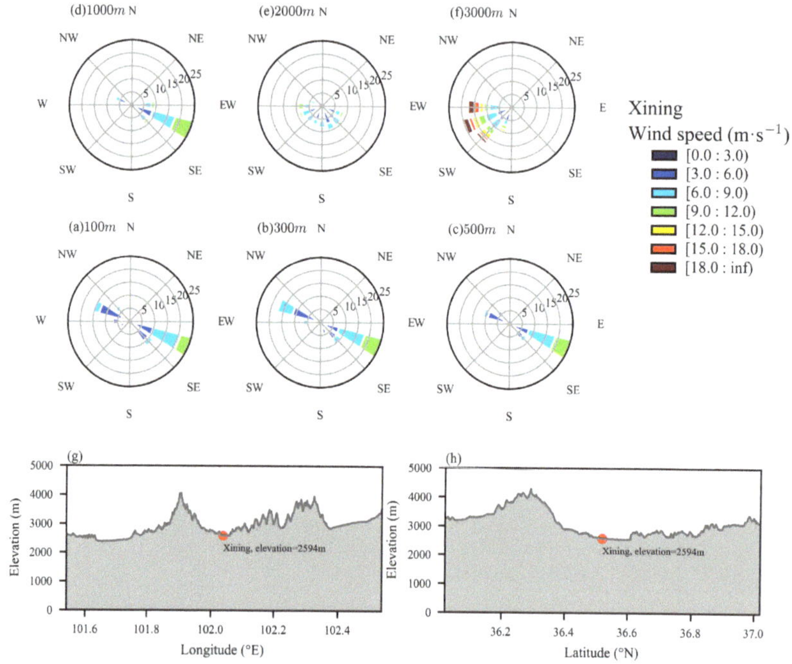

Figure 8. Average wind at different heights in Xining. (**a–f**) Wind speeds at 100 m, 300 m, 500 m, 1000 m, 2000 m and 3000 m, respectively. The circles show the 5% frequency increment wind speed of lidar, with wind speed divided into seven grades. Topographic elevations along (**g**) 102.01°E and (**h**) 36.50°N, respectively.

The wind field in Xining is strongly influenced by the complex topography of the plateau. The mean winds are weak and constant in direction (either southeast or northwest) below 1000 m, but they change abruptly to strong westerlies above 2000 m. This suggests that the high mountains on the north and south sides of Xining increase friction and obstruct air flow.

Figure 9 shows the diurnal variation in turbulence intensity at different height layers in Qingdao and Xining. The turbulence intensity (I_T) of the near-surface mass increases sharply after sunrise in Qingdao and starts to decrease in the late afternoon, at up to 500 m in height (Figure 9a). In contrast, very strong turbulence is observed in Xining during the daytime, and its intensity is much greater than that in Qingdao. Furthermore, the turbulence development in Xining is very vigorous and can reach very high heights of up to 3000 m (Figure 9b).

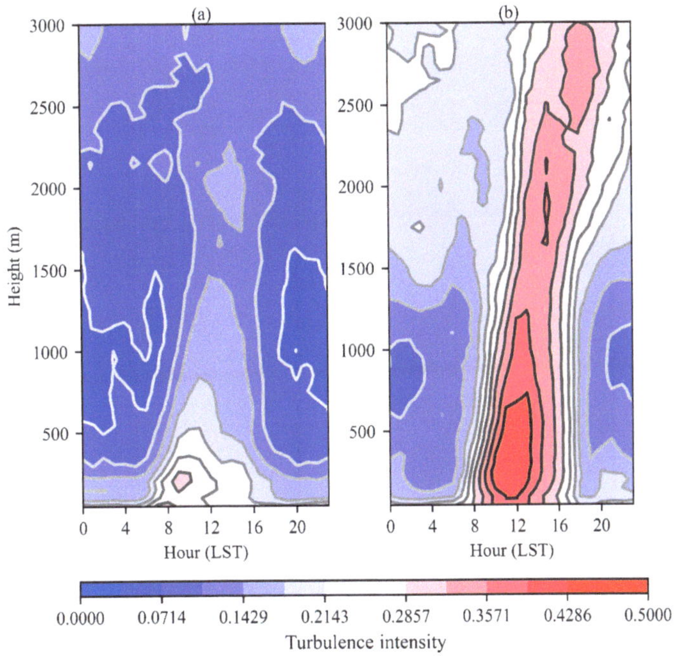

Figure 9. Turbulence intensity I_T at different heights and times for (**a**) Qingdao and (**b**) Xining. The horizontal axis is the local standard time (LST), and the vertical axis is the height. The background color shows the turbulence intensity.

One possible cause of the special gust characteristics of Xining is the strong turbulence intensity at this site. As shown in Figure 5, turbulence intensity (I_T) has a strong positive correlation with gust factor (G_T) and gust amplitude ($V^a_{t,T}$). The strong turbulence in Xining may result from various factors, such as strong radiation and topographic friction in the eastern part of the Tibetan Plateau. Further work is needed to analyze the exact causes.

We propose that the bulk shear, which is the ratio of the vertical airflow to the height [49,50], is the main driver of the strong turbulence near the surface of Xining. To examine this hypothesis, we compared the bulk shear characteristics at different heights between the two sites, Xining and Qingdao (Figure 10).

According to Figure 10, the bulk shear in both Xining and Qingdao is strong near the surface and decreases rapidly with increasing height. However, the bulk shear in Xining is always stronger than that in Qingdao. Moreover, the bulk shear has a significant positive correlation with turbulence intensity at 100 m, but not at 1000 m or 2000 m. This result supports the work of Yus-Díez et al. [47], which demonstrates that low-level turbulence is mainly driven by shear. Therefore, we infer that due to the complex topography in the eastern part of the Qinghai–Tibet Plateau, there can be stronger friction and shear, thus driving stronger near-surface turbulence. To measure terrain complexity, the standard deviation of the elevation based on the ASTER Global Digital Elevation Model was calculated [51]. The results show that the standard deviation of topography within 1° centered on Xining (624.59) is much higher than that of Qingdao (66.02), which supports our inference.

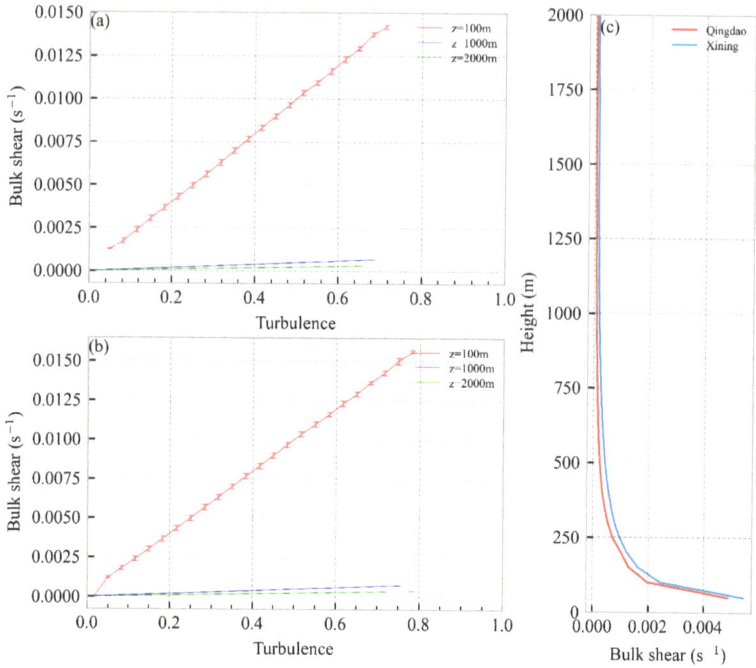

Figure 10. Bulk shear under different turbulence intensities in (**a**) Qingdao and (**b**) Xining for three heights: 100 m (red), 1000 m (blue) and 2000 m (green). (**c**) Bulk shear as a function of height for Qingdao (red) and Xining (blue).

However, this result still does not explain why turbulence can develop at very high heights in Xining. This could be related to mountain waves, gravity waves and density flow [52–54], but more evidence is still needed.

5. Conclusions and Discussion

Wind gusts and their hazards have been extensively studied in recent years. However, the characteristics of wind gusts under the influence of complex topography on the Tibetan Plateau are still poorly understood [3,43]. This paper compares the wind gust characteristics of the eastern Qinghai–Tibet Plateau and the coastal areas at the same latitude, using Doppler lidar data from Xining and Qingdao stations. We also analyze the causes of these differences.

Although Doppler lidar is highly reliable, some erroneous data still exist. In this paper, data quality control is performed by machine learning methods to remove significant erroneous data while preserving turbulence perturbations. The gust parameters and turbulence were analyzed using the data after data quality control, and the conclusions are as follows.

Previous studies have shown that wind gusts have different properties in the vertical direction and vary greatly with diurnal conditions [6,36,41,55]. In this study, we analyzed the gust parameters and their diurnal variation at different heights at the two sites (Figures 3 and 4).

The results show that the vertical wind speed over Qingdao is significantly higher than that over Xining, but the difference in peak gust between the two sites is small (Figure 3b). Moreover, the gust amplitude and gust factor are larger for Xining than for Qingdao. For Qingdao, the gust phenomenon decreases with height, confirming the report of Suomi et al. [43], which showed that wind gusts near the surface are more influenced by surface properties and subject to stronger frictional effects. However, for Xining, the gust parameters and turbulence intensity are still strong at 1000 m above ground.

The complex canyon topography on both sides of Xining impedes the winds, while the sea–land winds from Qingdao strengthen the winds [45,46]. However, the results indicate that the gust phenomenon in Xining is stronger and can extend to higher altitudes, not being limited to near the surface. This poses a threat to aviation activities. Therefore, wind gusts on the eastern Tibetan Plateau may be more complex and unpredictable than wind gusts on the coast, and wind gusts in Xining may cause more powerful damage when they occur.

We analyzed the relationship between mean wind and turbulence intensity and its effect on gust parameters (Figure 5), referring to the work of Yus-Díez et al. [47]. The aim was to identify the factors that influence the differences in wind gusts between the two sites. The results show that the peak gust is significantly correlated with mean wind speed, while gust factors and gust amplitudes are strongly correlated with turbulence intensity. These results suggest that the mean wind and turbulence intensity may be responsible for the differences in gust. Moreover, strong turbulence conditions can generate higher wind gusts than weak turbulence conditions at the same mean wind speed. However, a high mean wind speed may also reduce the turbulence (Figure 6).

Finally, we analyzed the background wind and turbulence at both sites. We found that the wind field in Qingdao is consistent with the ground-rotation wind theory, where the wind speed increases with increasing height and the wind direction shifts to the west. In contrast, the wind in Xining is strongly influenced by the complex topography of the plateau, and the high mountains to the north and south of Xining increase friction and impede the air flow. We also observed that very strong turbulence is present in Xining during the daytime, and its intensity is much greater than that in Qingdao. In addition, turbulence development in Xining is very strong and can reach very high altitudes of up to 3000 m (Figure 9b). Turbulence and mean wind jointly influence the development of wind gusts, and the strong differences in turbulence may explain the different gust characteristics between Xining and Qingdao.

However, the cause of such strong turbulence in the eastern part of the Tibetan Plateau is not well understood. We compared the bulk shear at different heights between two sites, Xining and Qingdao, and found that Xining had stronger bulk shear than Qingdao. This result is consistent with that of Yus-Díez et al. [47], who suggested that shear is the main driver of low-level turbulence. We hypothesize that the complex topography of the eastern Tibetan Plateau creates more friction and shear, which enhances the near-surface turbulence.

In conclusion, this study compares the different characteristics of wind gusts at two sites, Xining and Qingdao, and analyzes the reasons behind them. We found that complex topography and turbulence may contribute greatly to the development of wind gusts over the eastern part of the Tibetan Plateau. These results provide observational evidence for wind gusts over the complex terrain of the Tibetan Plateau. That will contribute to the understanding of the mechanism of wind gust development, regional differences in wind gusts and the study of the impact of wind gusts on flight safety in the eastern part of the Tibetan Plateau. However, the generalizability of these findings needs to be tested on a larger scale, and the causes of strong turbulence at high altitude over the eastern Tibetan Plateau need to be further analyzed. In addition, topography and bulk shear do not adequately account for the complex wind gusts and turbulence on the Tibetan Plateau. It also depends on thermal and radiative factors that need more comprehensive data analysis.

Author Contributions: The study was completed with cooperation among all authors. C.Z. and Y.L. designed this study; X.Z. and Z.Z. conducted the experiment and wrote the manuscript with input from all coauthors; C.Z. and Y.L. checked the experimental results. All authors have read and agreed to the published version of the manuscript.

Funding: This research was supported by the National Natural Science Foundation of China (grant U2242201, 42075077), the Hunan Provincial Natural Science Foundation of China (grant 2021JC0009, 2023JJ30629) and the Fengyun Application Pioneering Project (FY-APP-2022.0605).

Acknowledgments: We gratefully acknowledge Jie Zhang from the College of Atmospheric Science, Chengdu University of Information Technology, Plateau Atmosphere and Environment Key Laboratory of Sichuan Province for providing the Doppler lidar data on Xining with stable performance.

Conflicts of Interest: The authors declare no conflict of interest.

References

1. Black, A.W.; Ashley, W.S. Fatalities Associated with Nonconvective High-Wind Events in the United States. *J. Appl. Meteorol. Climatol.* **2008**, *47*, 717–725. [CrossRef]
2. Pryor, S.C.; Conrick, R.; Miller, C.; Tytell, J.; Barthelmie, R.J. Intense and Extreme Wind Speeds Observed by Anemometer and Seismic Networks: An Eastern U.S. Case Study. *J. Appl. Meteorol. Climatol.* **2014**, *53*, 2417–2429. [CrossRef]
3. Yao, Z.; Li, X.; Xiao, J. Characteristics of daily extreme wind gusts on the Qinghai-Tibet Plateau, China. *J. Arid Land* **2018**, *10*, 673–685. [CrossRef]
4. Cheng, X.; Wu, L.; Hu, F.; Zeng, Q.-C. Parameterizations of some important characteristics of turbulent fluctuations and gusty wind disturbances in the atmospheric boundary layer. *J. Geophys. Res. Atmos.* **2012**, *117*. [CrossRef]
5. Hu, W.; Letson, F.; Barthelmie, R.J.; Pryor, S.C. Wind Gust Characterization at Wind Turbine Relevant Heights in Moderately Complex Terrain. *J. Appl. Meteorol. Climatol.* **2018**, *57*, 1459–1476. [CrossRef]
6. Letson, F.; Barthelmie, R.J.; Hu, W.; Pryor, S.C. Characterizing wind gusts in complex terrain. *Atmos. Chem. Phys.* **2019**, *19*, 3797–3819. [CrossRef]
7. Monahan, A.H.; Rees, T.; He, Y.; McFarlane, N. Multiple Regimes of Wind, Stratification, and Turbulence in the Stable Boundary Layer. *J. Atmos. Sci.* **2015**, *72*, 3178–3198. [CrossRef]
8. Xie, J.; Lan, C.; Yang, H.; Gao, R.; Lu, C.; Wang, B.; Chan, P.W.; Fan, S.; Li, L. Tower-observed structural evolution of the low-level boundary layer before, during, and after gust front passage in a coastal area at low latitude. *Weather Clim. Extrem.* **2022**, *36*, 100429. [CrossRef]
9. Luchetti, N.T.; Friedrich, K.; Rodell, C.E.; Lundquist, J.K. Characterizing Thunderstorm Gust Fronts near Complex Terrain. *Mon. Weather Rev.* **2020**, *148*, 3267–3286. [CrossRef]
10. He, J.Y.; Chan, P.W.; Li, Q.S.; Li, L.; Zhang, L.; Yang, H.L. Observations of wind and turbulence structures of Super Typhoons Hato and Mangkhut over land from a 356 m high meteorological tower. *Atmos. Res.* **2022**, *265*, 105910. [CrossRef]
11. Lee, Y.-H.; Lee, G.; Joo, S.; Ahn, K.-D. Observational study of surface wind along a sloping surface over mountainous terrain during winter. *Adv. Atmos. Sci.* **2018**, *35*, 276–284. [CrossRef]
12. Xia, J.; Li, H.; Kang, Y.; Yu, C.; Ji, L.; Wu, L.; Lou, X.; Zhu, G.; Wang, Z.; Yan, Z.; et al. Machine Learning-based Weather Support for the 2022 Winter Olympics. *Adv. Atmos. Sci.* **2020**, *37*, 927–932. [CrossRef]
13. Wang, K.; Lyu, X.; Huang, J.; Luo, M.; Xu, F. Influence of Topography and the Underlying Surface of the Bohai Sea on Wind and Gust Forecasts. *Earth Space Sci.* **2022**, *10*, e2022EA002705. [CrossRef]
14. Harper, B.A.; Kepert, J.D.; Ginger, J.D. Guidelines for Converting between Various Wind Averaging Periods in Tropical Cyclone Conditions. World Meteorological Organization Technical Document WMO/TD-1555. Available online: https://library.wmo.int/index.php?lvl=notice_display&id=135 (accessed on 1 January 2010).
15. Wood, N. Wind Flow Over Complex Terrain: A Historical Perspective and the Prospect for Large-Eddy Modelling. *Bound. Layer Meteorol.* **2000**, *96*, 11–32. [CrossRef]
16. Barthelmie, R.J.; Wang, H.; Doubrawa, P.; Giroux, G.; Pryor, S.C. Effects of an escarpment on flow parameters of relevance to wind turbines. *Wind Energy* **2016**, *19*, 2271–2286. [CrossRef]
17. Wagenbrenner, N.S.; Forthofer, J.M.; Lamb, B.K.; Shannon, K.S.; Butler, B.W. Downscaling surface wind predictions from numerical weather prediction models in complex terrain with WindNinja. *Atmos. Chem. Phys.* **2016**, *16*, 5229–5241. [CrossRef]
18. Jubayer, C.M.; Hangan, H. A hybrid approach for evaluating wind flow over a complex terrain. *J. Wind Eng. Ind. Aerodyn.* **2018**, *175*, 65–76. [CrossRef]
19. Hasager, C.B.; Nielsen, N.W.; Jensen, N.O.; Boegh, E.; Christensen, J.H.; Dellwik, E.; Soegaard, H. Effective Roughness Calculated from Satellite-Derived Land Cover Maps and Hedge-Information used in a Weather Forecasting Model. *Bound. Layer Meteorol.* **2003**, *109*, 227–254. [CrossRef]
20. Earl, N.; Dorling, S.; Starks, M.; Finch, R. Subsynoptic-scale features associated with extreme surface gusts in UK extratropical cyclone events. *Geophys. Res. Lett.* **2017**, *44*, 3932–3940. [CrossRef]
21. Tieleman, H.W. Wind characteristics in the surface layer over heterogeneous terrain. *J. Wind Eng. Ind. Aerodyn.* **1992**, *41*, 329–340. [CrossRef]
22. Markowski, P.M.; Dotzek, N. A numerical study of the effects of orography on supercells. *Atmos. Res.* **2011**, *100*, 457–478. [CrossRef]
23. Bechmann, A.; Sørensen, N.N.; Berg, J.; Mann, J.; Réthoré, P.E. The Bolund Experiment, Part II: Blind Comparison of Microscale Flow Models. *Bound. Layer Meteorol.* **2011**, *141*, 245–271. [CrossRef]

24. Berg, J.; Mann, J.; Bechmann, A.; Courtney, M.S.; Jørgensen, H.E. The Bolund Experiment, Part I: Flow Over a Steep, Three-Dimensional Hill. *Bound. Layer Meteorol.* **2011**, *141*, 219–243. [CrossRef]
25. Butler, B.W.; Wagenbrenner, N.S.; Forthofer, J.M.; Lamb, B.K.; Shannon, K.S.; Finn, D.; Eckman, R.M.; Clawson, K.; Bradshaw, L.; Sopko, P.; et al. High-resolution observations of the near-surface wind field over an isolated mountain and in a steep river canyon. *Atmos. Chem. Phys.* **2015**, *15*, 3785–3801. [CrossRef]
26. Suomi, I.; Vihma, T. Wind Gust Measurement Techniques—From Traditional Anemometry to New Possibilities. *Sensors* **2018**, *18*, 1300. [CrossRef]
27. Guo, J.; Zhai, P.; Wu, L.; Cribb, M.; Li, Z.; Ma, Z.; Wang, F.; Chu, D.; Wang, P.; Zhang, J. Diurnal variation and the influential factors of precipitation from surface and satellite measurements in Tibet. *Int. J. Climatol.* **2014**, *34*, 2940–2956. [CrossRef]
28. Fang, X.; Han, Y.; Ma, J.; Song, L.; Yang, S.; Zhang, X. Dust storms and loess accumulation on the Tibetan Plateau: A case study of dust event on 4 March 2003 in Lhasa. *Chin. Sci. Bull.* **2004**, *49*, 953–960. [CrossRef]
29. Han, Y.; Fang, X.; Kang, S.; Wang, H.; Kang, F. Shifts of dust source regions over central Asia and the Tibetan Plateau: Connections with the Arctic oscillation and the westerly jet. *Atmos. Environ.* **2008**, *42*, 2358–2368. [CrossRef]
30. Li, J.; Liu, H.; Su, Z.; Fan, X. Changes in wind activity from 1957 to 2011 and their possible influence on aeolian desertification in northern China. *J. Arid Land* **2015**, *7*, 755–764. [CrossRef]
31. Ding, J.; Cuo, L.; Zhang, Y.; Zhang, C. Varied spatiotemporal changes in wind speed over the Tibetan Plateau and its surroundings in the past decades. *Int. J. Climatol.* **2021**, *41*, 5956–5976. [CrossRef]
32. Yang, Y.; Fan, S.; Wang, L.; Gao, Z.; Zhang, Y.; Zou, H.; Miao, S.; Li, Y.; Huang, M.; Yim, S.H.L.; et al. Diurnal Evolution of the Wintertime Boundary Layer in Urban Beijing, China: Insights from Doppler Lidar and a 325-m Meteorological Tower. *Remote Sens.* **2020**, *12*, 3935. [CrossRef]
33. De Arruda Moreira, G.; Guerrero-Rascado, J.L.; Benavent-Oltra, J.A.; Ortiz-Amezcua, P.; Román, R.; Bedoya-Velásquez, A.E.; Bravo-Aranda, J.A.; Olmo Reyes, F.J.; Landulfo, E.; Alados-Arboledas, L. Analyzing the turbulent planetary boundary layer by remote sensing systems: The Doppler wind lidar, aerosol elastic lidar and microwave radiometer. *Atmos. Chem. Phys.* **2019**, *19*, 1263–1280. [CrossRef]
34. Steinheuer, J.; Detring, C.; Beyrich, F.; Löhnert, U.; Friederichs, P.; Fiedler, S. A new scanning scheme and flexible retrieval for mean winds and gusts from Doppler lidar measurements. *Atmos. Meas. Tech.* **2022**, *15*, 3243–3260. [CrossRef]
35. Liu, J.; Song, X.; Long, W.; Fu, Y.; Yun, L.; Zhang, M. Structure Analysis of the Sea Breeze Based on Doppler Lidar and Its Impact on Pollutants. *Remote Sens.* **2022**, *14*, 324. [CrossRef]
36. Suomi, I.; Gryning, S.E.; O'Connor, E.J.; Vihma, T. Methodology for obtaining wind gusts using Doppler lidar. *Q. J. R. Meteorol. Soc.* **2017**, *143*, 2061–2072. [CrossRef]
37. Clifton, A.; Clive, P.; Gottschall, J.; Schlipf, D.; Simley, E.; Simmons, L.; Stein, D.; Trabucchi, D.; Vasiljevic, N.; Würth, I. IEA Wind Task 32: Wind Lidar Identifying and Mitigating Barriers to the Adoption of Wind Lidar. *Remote Sens.* **2018**, *10*, 406. [CrossRef]
38. Hill, C. Coherent Focused Lidars for Doppler Sensing of Aerosols and Wind. *Remote Sens.* **2018**, *10*, 466. [CrossRef]
39. Yuan, J.; Wu, K.; Wei, T.; Wang, L.; Shu, Z.; Yang, Y.; Xia, H. Cloud Seeding Evidenced by Coherent Doppler Wind Lidar. *Remote Sens.* **2021**, *13*, 3815. [CrossRef]
40. Liu, F.T.; Ting, K.M.; Zhou, Z.-H. Isolation Forest. In Proceedings of the 2008 Eighth IEEE International Conference on Data Mining, Pisa, Italy, 15–19 December 2008; pp. 413–422. [CrossRef]
41. Zhou, X.; Zhang, C.; Li, Y.; Chen, Z.; Zhang, J.; Ding, X. Wind Gust Parameters in the Lower Troposphere Based on Doppler Lidar Data. *J. Geophys. Res. Atmos.* **2023**, *128*, e2022JD038156. [CrossRef]
42. O'Connor, E.J.; Illingworth, A.J.; Brooks, I.M.; Westbrook, C.D.; Hogan, R.J.; Davies, F.; Brooks, B.J. A Method for Estimating the Turbulent Kinetic Energy Dissipation Rate from a Vertically Pointing Doppler Lidar, and Independent Evaluation from Balloon-Borne In Situ Measurements. *J. Atmos. Ocean. Technol.* **2010**, *27*, 1652–1664. [CrossRef]
43. Suomi, I.; Gryning, S.E.; Floors, R.; Vihma, T.; Fortelius, C. On the vertical structure of wind gusts. *Q. J. R. Meteorol. Soc.* **2014**, *141*, 1658–1670. [CrossRef]
44. Harris, A.R.; Kahl, J.D.W. Gust Factors: Meteorologically Stratified Climatology, Data Artifacts, and Utility in Forecasting Peak Gusts. *J. Appl. Meteorol. Climatol.* **2017**, *56*, 3151–3166. [CrossRef]
45. Miller, S.T.K.; Keim, B.D.; Talbot, R.W.; Mao, H. Sea breeze: Structure, forecasting, and impacts. *Rev. Geophys.* **2003**, *41*. [CrossRef]
46. Zhang, L.; Xin, J.; Yin, Y.; Wang, Z.; Wang, D.; Ma, Y.; Jia, D.; Jiang, Y.; Wu, L.; Pan, X. Adaptability evaluation of boundary layer schemes for simulation of sea and land breeze circulation in the west coast of the Yellow Sea. *Atmos. Res.* **2022**, *278*, 106354. [CrossRef]
47. Yus-Díez, J.; Udina, M.; Soler, M.R.; Lothon, M.; Nilsson, E.; Bech, J.; Sun, J. Nocturnal boundary layer turbulence regimes analysis during the BLLAST campaign. *Atmos. Chem. Phys.* **2019**, *19*, 9495–9514. [CrossRef]
48. Zeng, Q.; Cheng, X.; Hu, F.; Peng, Z. Gustiness and coherent structure of strong winds and their role in dust emission and entrainment. *Adv. Atmos. Sci.* **2009**, *27*, 1–13. [CrossRef]
49. Sun, J.; Mahrt, L.; Banta, R.M.; Pichugina, Y.L. Turbulence Regimes and Turbulence Intermittency in the Stable Boundary Layer during CASES-99. *J. Atmos. Sci.* **2012**, *69*, 338–351. [CrossRef]
50. Sun, J.; Lenschow, D.H.; LeMone, M.A.; Mahrt, L. The Role of Large-Coherent-Eddy Transport in the Atmospheric Surface Layer Based on CASES-99 Observations. *Bound. Layer Meteorol.* **2016**, *160*, 83–111. [CrossRef]

51. Lv, Y.; Guo, J.; Li, J.; Cao, L.; Chen, T.; Wang, D.; Chen, D.; Han, Y.; Guo, X.; Xu, H.; et al. Spatiotemporal characteristics of atmospheric turbulence over China estimated using operational high-resolution soundings. *Environ. Res. Lett.* **2021**, *16*, 054050. [CrossRef]
52. Viana, S.; Terradellas, E.; Yagüe, C. Analysis of Gravity Waves Generated at the Top of a Drainage Flow. *J. Atmos. Sci.* **2010**, *67*, 3949–3966. [CrossRef]
53. Udina, M.; Soler, M.R.; Viana, S.; Yagüe, C. Model simulation of gravity waves triggered by a density current. *Q. J. R. Meteorol. Soc.* **2013**, *139*, 701–714. [CrossRef]
54. Soler, M.R.; Udina, M.; Ferreres, E. Observational and Numerical Simulation Study of a Sequence of Eight Atmospheric Density Currents in Northern Spain. *Bound. Layer Meteorol.* **2014**, *153*, 195–216. [CrossRef]
55. Davis, F.K.; Newstein, H. The Variation of Gust Factors with Mean Wind Speed and with Height. *J. Appl. Meteorol.* **1968**, *7*, 372–378. [CrossRef]

remote sensing

Technical Note

Assessing the Sensitivity of Snow Depth Simulations to Land Surface Parameterizations within Noah-MP in Northern Xinjiang, China

Yuanhong You [1], Chunlin Huang [2,*] and Yuhao Zhang [1]

1 School of Geography and Tourism, Anhui Normal University, Wuhu 241002, China; youyuanhong@lzb.ac.cn (Y.Y.)
2 Key Laboratory of Remote Sensing of Gansu Province, Northwest Institute of Eco-Environment and Resources, Chinese Academy of Sciences, Lanzhou 730000, China
* Correspondence: huangcl@lzb.ac.cn

Abstract: Snow cover plays a crucial role in the surface energy balance and hydrology and serves as a key indicator of climate change. In this study, we conducted an ensemble simulation comprising 48 members generated by randomly combining the parameterizations of five physical processes within the Noah-MP model. Utilizing the variance-based Sobol total sensitivity index, we quantified the sensitivity of regional-scale snow depth simulations to parameterization schemes. Additionally, we analyzed the spatial patterns of the parameterization sensitivities and assessed the uncertainty of the multi-parameterization scheme ensemble simulation. The results demonstrated that the differences in snow depth simulation results among the 48 scheme combinations were more pronounced in mountain regions, with melting mechanisms being the primary factor contributing to uncertainty in ensemble simulation. Contrasting mountain regions, the sensitivity index for the physical process of partitioning precipitation into rainfall and snowfall was notably higher in basin areas. Unexpectedly, the sensitivity index of the lower boundary condition of the physical process of soil temperature was negligible across the entire region. Surface layer drag coefficient and snow surface albedo parameterization schemes demonstrated meaningful sensitivity in localized areas, while the sensitivity index of the first snow layer or soil temperature time scheme exhibited a high level of sensitivity throughout the entire region. The uncertainty of snow depth ensemble simulation in mountainous areas is predominantly concentrated between 0.2 and 0.3 m, which is significantly higher than that in basin areas. This study aims to provide valuable insights into the judicious selection of parameterization schemes for modeling snow processes.

Keywords: multi-parameterization ensemble simulation; Sobol total sensitivity; sensitivity analysis; uncertainty analysis

Citation: You, Y.; Huang, C.; Zhang, Y. Assessing the Sensitivity of Snow Depth Simulations to Land Surface Parameterizations within Noah-MP in Northern Xinjiang, China. *Remote Sens.* **2024**, *16*, 594. https://doi.org/10.3390/rs16030594

Academic Editor: Alexander Marshak

Received: 22 December 2023
Revised: 23 January 2024
Accepted: 29 January 2024
Published: 5 February 2024

1. Introduction

As a fundamental element of the cryosphere, snow cover exerts a substantial influence on various aspects of the earth's systems; it plays a significant role in modulating the surface energy balance, hydrological processes, and atmospheric circulation [1–3]. Moreover, snow cover stands as a critical bellwether of climate change, exhibiting a rapid response to its evolving patterns. In many regions, it plays a crucial role as a vital freshwater resource essential for both human sustenance and socioeconomic development, thereby holding a central position in the pursuit of global sustainability [4–7]. Furthermore, the inherent characteristics of this specialized land surface exert a significant influence on the energy and water cycles within the terrestrial environment [8–10]. For instance, the high albedo of snow-covered surfaces adjusts the amount of solar radiation reaching the Earth's surface, consequently influencing the exchange of energy between the land and the atmosphere [11,12]. The low thermal conductivity of snow has a profound effect on soil temperatures by insulating the soil surface from direct atmospheric contact [13]. Additionally, the melting patterns of snow are critical factors in the occurrence of natural disasters, such as flooding and avalanches [14,15]. Thus, a comprehensive understanding of the temporal and spatial variations in snow cover is crucial for estimating snowmelt runoff, issuing snow-related disaster warnings, and advancing our knowledge of atmospheric circulation.

Land surface models (LSMs) serve as invaluable tools for examining the temporal and spatial dynamics of snow cover. The incorporation of a snow process module into many LSMs and hydrology models has proven effective in achieving satisfactory snow simulation results [16,17]. However, it is noteworthy that LSMs exhibit varying levels of simulation accuracy, with disparities often observed even within the same model across different regions. This variability can be attributed to the inherent characteristics and applicability of individual LSMs. The use of parameterizations is a common strategy to represent diverse physical processes, including snow surface albedo, runoff, and dynamic vegetation. Notably, selecting different parameterizations for the same physical process can lead to discernible differences in simulation outputs [10,18,19]. Essentially, parameterization comprises a set of mathematical equations designed to capture a special physical process, with the parameters within these equations serving as adjustable numerical values [6,20]. Understanding and accounting for these inherent model characteristics and parameterizations are critical considerations in ensuring the reliability and applicability of LSMs in diverse geographical contexts [13]. Divergent simulation outputs resulting from various mathematical equations for a given physical process pose a substantial obstacle to a comprehensive understanding of the cryosphere's response to climate change. This limitation impedes the elucidation of intricate interactions between the land surface and the atmosphere. As a result, the selection of appropriate parameterizations for LSMs becomes a pressing concern.

The Noah-MP model, an advanced iteration building upon the Noah model, addresses this challenge by incorporating two to four innovative parameterization schemes for identical physical processes, enhancing the model's applicability [10,12,21]. The inclusion of multiple parameterization schemes in the Noah-MP model offers a convenient avenue for exploring the sensitivity of model outputs to the option of parameterization schemes [18,20]. This has garnered attention in the hydrology and meteorological communities, with researchers keen on unraveling the impact of alternative parameterization schemes on critical physical processes. A study comprising 288 experiments utilized multi-parameterization schemes for six physical processes across ten hydrologic regions in China, revealing varying sensitivities and optimal parameterization schemes for different variables and regions [22]. On a site scale, the benchmarking method was employed to identify the optimal combination of parameterization schemes within the Noah-MP model based on sensitivity experiments of heat simulations [23]. Sensitivity analysis methods, such as those investigating heat flux simulations, have revealed that sensitive parameterization schemes introduce greater uncertainties, enabling researchers to identify schemes that could potentially impact model performance [18]. Various sensitivity analysis methods, including qualitative assessments and the Sobol total sensitivity index, have been employed to evaluate the impact of parameterizations on sensible heat flux, latent heat flux, and net ecosystem exchange [18,20]. Results demonstrated inconsistent sensitivities across sites, emphasizing the need for region-specific evaluations [24]. Among numerous sensitivity analysis methods, the Sobol variance analysis method [25] stands out as a particularly effective choice for exploring parameterization sensitivity, both in small watersheds and large-scale regions. This method not only demonstrates exceptional robustness but also produces reliable sensitivity calculation results [20,26]. Another notable advantage of this method is its independence from observational data in assessing the sensitivity of parameterization schemes; sensitivity calculations rely solely on ensemble simulation results. In the context of snow processes, the Noah-MP model was employed to simulate snow, and a physics ensemble simulation experiment was conducted to analyze the sensitivity of snow simulations to parameterization schemes [10,27]. Although sensitive parameterization schemes were identified and optimal combinations determined, addressing the regional-scale sensitivity of parameterization schemes remained a pressing concern. To address this gap, we utilized the Sobol total sensitivity index in 48 Noah-MP physics-ensemble simulations driven by the China Meteorological Forcing Dataset (CMFD) in Northern Xinjiang, China. Conducting sensitivity analysis, we determined uncertainty intervals and investigated their origins, with the goal of offering a valuable reference for the simulation of snow processes.

The paper is structurally organized as follows: Section 2 provides an overview of the study area, the regional atmospheric forcing data, the design of numerical experiments, and the employed sensitivity analysis method. In Section 3, we showcase the simulation results of snow depth using the default parameterization scheme combination, delve into the sensitivity of parameterization schemes, and establish the uncertainty interval within ensemble simulations. Finally, Section 4 succinctly summarizes the findings derived from this study.

2. Materials and Methods

2.1. Study Area and Data

The research is focused on the northern Xinjiang region, spanning from 79.0° to 92.0°E and 42.0° to 50.0°N, situated along the northwest border area of China. This region boasts diverse topography, with the Altay Mountains in the north and the Tianshan Mountains in the south, reaching a maximum elevation of 7440 m. The central portion comprises the Junggar basin, maintaining an average elevation of 400 m. Terrain heights across Northern Xinjiang are depicted in Figure 1. The climate in this area is characterized as a temperate continental climate, owing to its unique geographical location in the Eurasian continent's hinterland [28]. Influenced by Siberian circulation, the region experiences a cold and snowy winter, with snow cover persisting for approximately 120 days from November to March. In certain areas, this snow cover can endure until May [28,29]. This climatic profile renders the study area ideal for investigating the regional-scale sensitivity of snow simulations to parameterization schemes, given its abundant snow resources.

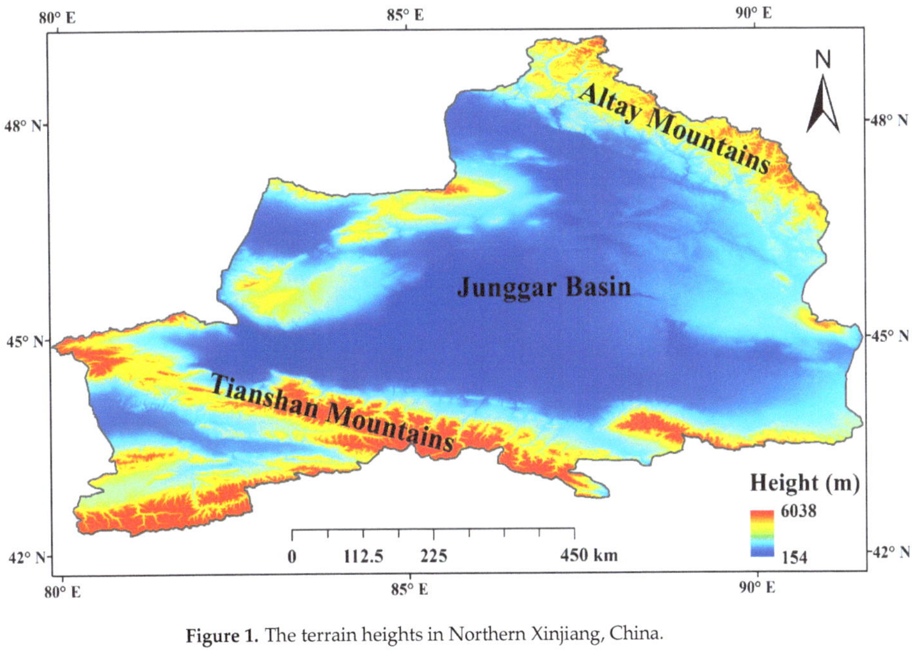

Figure 1. The terrain heights in Northern Xinjiang, China.

In this study, the Noah-MP model was driven by the China Meteorological Forcing Dataset (CMFD) [30], covering the period from 1 October 2012 to 30 September 2013. Developed by the hydrometeorological research group at the Institute of Tibetan Plateau Research, Chinese Academy of Sciences, the CMFD is a comprehensive dataset that includes seven critical meteorological variables: pressure, precipitation, wind speed, specific humidity, near-surface air temperature, and downward shortwave and longwave radiation. Renowned for its precision, this dataset integrates observations from 740 operational stations of the China Meteorological Administration with data from Princeton meteorological forcing, the Global Land Data Assimilation Systems (GLDAS) reanalysis, the precipitation products of the Tropical Rainfall Measuring Mission (TRMM), and the radiation data obtained from the Global Energy and Water Cycle Experiment—Surface Radiation Budget (GEWEXSRB) [31]. The CMFD utilized in this study provides comprehensive coverage across the entire geographical expanse of China, boasting a spatial resolution of 0.1 degree and a temporal resolution of 3 h. To enhance experimental precision, the MicroMet approach [32] was employed to downscale the atmospheric forcing data. This approach relies on the correlations between meteorological variables and topography. Through this methodology, we successfully derived a refined set of forcing data with an improved spatial resolution of 0.05 degrees and a heightened temporal resolution of 1 h.

2.2. Simulation Setups

The Noah-MP model was enhanced by incorporating recent advancements in physics from the land surface models (LSMs) of Noah [33,34]. Various parameterization schemes were integrated into distinct physical processes, with details provided in Table 1, making the model have stronger applicability [10,21]. Our prior research revealed that simulated snow depth and snow water equivalent exhibit sensitivity to parameterization schemes, particularly those related to surface layer drag coefficient (SFC), snow surface albedo (ALB), partitioning of precipitation into rainfall and snowfall (PCP), lower boundary condition of soil temperature (TBOT), and the first snow layer or soil temperature time scheme (TEMP) [10,27]. Although we have determined that the physical processes SFC and TEMP demonstrated high sensitivity across various snow climates and that the choice of different parameterization schemes for these processes led to substantial variations in simulation results, the distribution of parameterization scheme sensitivity on a regional scale is still unknown. Therefore, we generated a total of 48 scheme combinations by randomly combining parameterization schemes of these five physical processes, allowing us to quantify the sensitivity of simulated snow depth to parameterization schemes on a regional scale.

Table 1. The main physical options within the Noah-MP model.

Parameterization Description	Scheme Options
Dynamic vegetation (DVEG)	1. off [*Default*], 2. on
Canopy stomatal resistance (CRS)	1. Ball-berry [*Default*], 2. Jarvis
Soil moisture factor controlling stomatal resistance (BTR)	1. Noah scheme [*Default*], 2. CLM scheme, 3. SSiB scheme
Runoff and groundwater (RUN)	1. SIMGM [*Default*], 2. SIMTOP, 3. Schaake96, 4. BATS
Surface layer drag coefficient (SFC)	1. M-O [*Default*], 2. Original Noah (Chen97)
Frozen soil permeability (INF)	1. NY06 [*Default*], 2. Koren99
Super-cooled liquid water in frozen soil (FRZ)	1. NY06 [*Default*], 2. Koren99
Radiation transfer (RAD)	1. gap = F(3D, cosz) [*Default*], 2. gap = 0, 3. gap = 1 − FVEG
Snow surface albedo (ALB)	1. BATS, 2. CLASS [*Default*]
Partitioning of precipitation into rainfall and snowfall (PCP)	1. Jordan91 [*Default*], 2. BATS, 3. Noah
Lower boundary condition of soil temperature (TBOT)	1. Zero-flux scheme, 2. Noah [*Default*]
First snow layer or soil temperature time scheme (TEMP)	1. Semi-implicit [*Default*], 2. Fully implicit

The meteorological forcing data used to drive the Noah-MP model were from the CMFD dataset. Standard soil and vegetation parameters, including the monthly leaf area index, were derived from the soil and vegetation type data using the Noah-MP default lookup tables [35]. Additional hard-coded parameters of Noah-MP, as specified by reference [35], adopted default values manually calibrated by parameterization developers [36,37]. These datasets have been extensively used in continental hydrological simulations and have demonstrated reasonable accuracy [37–39]. Building upon our previous research revealing the sensitivity of snow simulations to various parameterization schemes for physical processes SFC, ALB, PCP, TBOT, and TEMP across diverse climates [10,27], this study randomly combined parameterization schemes for these five physical processes while maintaining default settings for the other seven physical processes, resulting in a total of 48 scheme combinations. To ensure soil-state equilibrium—a prerequisite for effective initialization—we employed forcing data from 1 October 2011 to 30 September 2012. The equilibrium criterion, stipulating that the time when the difference in annual means between consecutive single-year simulations is less than 0.1% of the mean, was rigorously applied [37,40]. All simulations spanned one snow period, with snow depth output recorded at an hourly interval. The hourly outputs were aggregated into monthly values, and then the monthly means were analyzed using Sobol's total sensitivity index.

2.3. Analysis and Evaluation Methods

In this study, the sensitivity analysis utilized the variance-based Sobol's total sensitivity index as proposed by Saltelli et al. [41] to quantify the impact of parameterization schemes for the five aforementioned physical processes on snow depth simulations. The calculation of sensitivity indices involved a structured series of steps.

Initially, we employed the various parameterization scheme combinations to derive 48 individual snow season averaged snow depth (SD) values, denoted as Y_i, $i = 1, \ldots, 48$. Simultaneously, the mean and variance of the 48 values were promptly calculated.

$$E(Y) = \sum_{i=1}^{48} Y_i / 48. \quad (1)$$

$$Var(Y) = \sum_{i=1}^{48} (Y_i - E(Y))^2 / 48. \quad (2)$$

Secondly, the Sobol total sensitivity of the parameterization schemes for a given physical process, denoted as A, was computed using the following equation:

$$S_A = E_A(Var_A(Y| \sim A)) / Var(Y). \quad (3)$$

Here S_A represents the Sobol total sensitivity for the parameterization of a given process, A. Additionally, $\sim A$ denotes all processes except A, Y signifies the snow depth simulations of the 48 scheme combinations, $Var(Y)$ is the variance of the 48 snow depth simulations, $Var_A(Y| \sim A)$ represents the variance caused by different parameterization schemes of A, and $E \sim_A$ represents the arithmetic average across all scheme combinations of the parameterizations other than A. To illustrate, consider the physical process of PCP, which comprises three parameterization schemes. The 48 values were categorized into 16 groups, where the parameterization schemes for SFC, ALB, TBOT, and TEMP processes remained constant, differing only in the PCP parameterization. Each group yielded three values corresponding to the three PCP schemes (Jordan91, BATS, and Noah). For instance, if the parameterization schemes for SFC, ALB, TBOT, and TEMP are M-O, BATS, Jordan 91, Noah, and Semi-implicit, respectively, in one group, the variance of the three PCP schemes ($Var_A(Y| \sim A)$) was calculated, with a denominator value of 3 for each of the 16 groups, $(Y|\sim A)_j$, $j = 1, \ldots, 16$. The mean of these 16 values was calculated and denoted as $E_A(Var_A(Y| \sim A))$. Further details about this method can be found in reference [20]. The Sobol total sensitivity index's greatest strength lies in accounting for the variance of model outputs resulting from different parameterization schemes for a given physical process. The normalized value, ranging from 0 to 1, indicates the level of sensitivity of simulations to parameterization schemes. A larger value implies high sensitivity, while approaching zero suggests that the parameterization has minimal influence on the simulations.

3. Results and Discussion

3.1. Results of Snow Depth Ensemble Simulations

The CMFD dataset served as the driving force for the 48 scheme combinations, resulting in snow depth simulations across Northern Xinjiang. As illustrated in Figure 2, it displays the annual average snow depth from the 48 ensemble simulations and the variance in the snow depth ensemble simulations. The annual average snow depth reveals a predominant distribution in the mountain regions, particularly evident in the Tianshan and Altay Mountains, where the maximum annual average snow depth approaches 2 m, primarily concentrated in the southern part of Tianshan Mountain. Conversely, the Junggar Basin exhibits significantly lower snow depths, attributed to the arid and semi-arid climate characteristics in Northern Xinjiang. The possible reasons for this phenomenon could be that the colder temperatures in the mountainous regions facilitate snow preservation, while the warmer and drier conditions in the basin contribute to snow sublimation and melting. The simulation results employing the 48 scheme combinations align with the spatial distribution of average snow depth, corresponding to the snow distribution characteristics in Northern Xinjiang. This alignment underscores the Noah-MP model's ability to accurately simulate snow processes in the region. Importantly, we refrain from comparing the averaged snow depth results of multiple scheme combinations with the corresponding snow depth observational data in this study. Model performance evaluation has been addressed in previous research, specifically assessing the snow simulation capabilities of the Noah-MP model in Northern Xinjiang, China [27]. Our focus here is solely on discerning differences in simulation results among different combination schemes within the Noah-MP model.

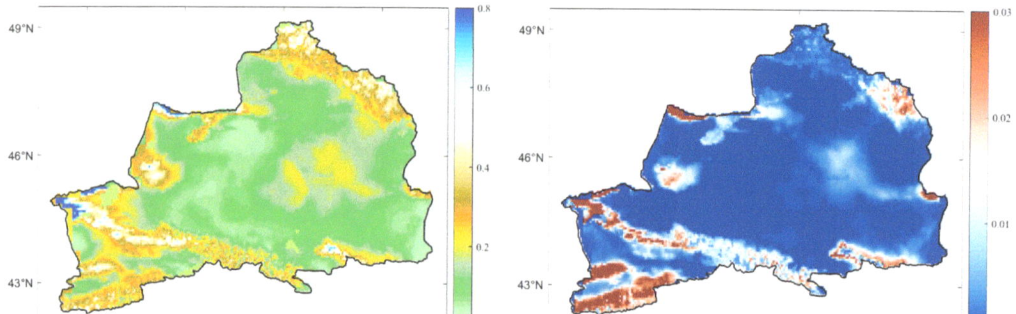

Figure 2. The snow depth simulation results from multi-parameterization scheme combinations. (**Left**) is the annual average snow depth of 48 ensemble simulations, and the (**right**) figure is the variance of 48 ensemble simulations.

The right subfigure of Figure 2 displays the variance of the 48 ensemble simulations. Examining the figure reveals a maximum variance of 0.10 m (the values of individual pixels are not shown in the figure) and a minimum of 0 m. Most areas exhibit a variance in snow depth simulation results below 0.02 m, particularly notable in the Junggar basin region. In contrast, the variance in the Tianshan Mountains region is relatively pronounced, with many areas having variance values around 0.03 m. Additionally, significant variances are observed in the southern part of the Tianshan Mountains and the southeastern part of the Altay Mountains. These findings indicate substantial differences in snow depth simulation results within the Northern Xinjiang mountainous area among different combinations of the Noah-MP model. This phenomenon can be attributed to two factors. Firstly, thicker snow in mountainous areas corresponds to larger errors. For instance, with a 10% error, the resulting errors for snow depths of 10 cm and 50 cm are 1 cm and 5 cm, respectively, illustrating significant differences. Secondly, distinct scheme combinations governing snow accumulation and melting mechanisms contribute to notable differences in snow depth simulation results. Despite these variations, the right subfigure highlights a relatively small variance in the northern region of the Altay Mountains. While located in a high-altitude mountainous region, the snow depth variance in this area is notably smaller than in the southeastern region of the Altay Mountains. This phenomenon may be attributed to relatively minor differences in simulated snow depth results among the 48 scheme combinations in this region, resulting in a reduced variance. Consequently, it underscores the importance of further investigating the sensitivity of different parameterization schemes at the regional scale.

3.2. Spatial Patterns of the Parameterization Sensitivities

In our previous study, a qualitative analysis was undertaken, establishing that the modeled snow depth exhibited sensitivity to the parameterizations of the physical processes SFC, ALB, PCP, TBOT, and TEMP. In this current study, we extend our analysis to quantify the sensitivity of modeled snow depth to these parameterizations in Northern Xinjiang. The 48 scheme combinations were randomly generated by incorporating options for the five sensitive physical processes. Figure 3 illustrates the Sobol total sensitivity of the modeled snow depth to these five physical processes, showcasing the parameterization sensitivity patterns across Northern Xinjiang. Higher values in the figure indicate greater sensitivities. As depicted in Figure 3, the sensitivity patterns of these parameterizations display significant differences. Two parameterization schemes are assigned for the physical process of SFC, and the choice between these two schemes notably influences the simulated surface layer drag coefficient. The first scheme incorporates the zero-displacement height, while the second considers the difference between roughness length for heat and momentum in calculating the surface layer drag coefficient, leading to substantial variations in t water and energy computations in the model. In Figure 3, Sobol sensitivity indices in most areas of Northern Xinjiang are below 0.1, with many regions showing indices at zero, indicating weak sensitivity of surface physical processes in the majority of the region. However, higher Sobol sensitivity indices are observed in the eastern part of the Tian Shan region, reaching a maximum of 0.5. This indicates a stronger sensitivity of surface physical processes in that area, highlighting significant variations in simulated snow depth results when utilizing different parameterization schemes for the surface. Additionally, localized areas near the western part of the Altay Mountains exhibit slightly higher Sobol indices than other regions, averaging around 0.3. This suggests stronger sensitivity of the physical process SFC, specifically in the eastern part of the Tianshan Mountain region, while its sensitivity is weaker in other areas of Northern Xinjiang.

The control of solar radiation entering the snowpack by snow surface albedo significantly influences the energy balance, potentially leading to snowmelt. Snow surface albedo schemes within Noah-MP, specifically the BATS and CLASS schemes, play a distinct role in regulating snow depth, with the CLASS scheme set as the default due to its presumed smaller snow surface albedo resulting from a stronger aging effect. Figure 3 shows notable spatial variations in the Sobol sensitivity index for the ALB process, with a maximum index value of 0.4, primarily in the Tianshan Mountains. Sensitivity indices for most areas in Northern Xinjiang are below 0.1, while the northeast of the Junggar Basin and the east of the Tianshan Mountains show relatively high indices, ranging from 0.3 to 0.4. Overall, ALB exhibits strong sensitivity in mountainous areas at higher altitudes, particularly concentrated in the eastern and southern parts of Northern Xinjiang. In our previous study, the physical process of PCP demonstrated sensitivity at the site scale across various snowfall climates [10]. Altering the parameterization scheme for this particular process can significantly impact the simulation outcomes of snow depth. PCP encompasses three distinct parameterization schemes that categorize precipitation into rain and snow, contingent on diverse temperature thresholds. The first scheme of PCP is the Jordan scheme, which incorporates relatively complex functions. The second scheme divides all precipitation into snowfall when $T_{air} < T_{frz} + 2.2K$ (the T_{air} and T_{frz} are the air temperature and freezing point, respectively), while in other cases, it is divided into rainfall in the second scheme. The third scheme of PCP divides precipitation into snowfall when $T_{air} < T_{frz}$, while all other situations are classified as rainfall. While these three parameterization schemes share a minimal distinction in essence, their respective calculation methods reveal noteworthy disparities in the model's snow simulation outcomes, especially in situations where near-surface air temperatures frequently oscillate around the freezing point. As illustrated in Figure 3, regions with relatively high Sobol sensitivity indices are concentrated in the Junggar Basin, north of the Tianshan Mountains, and in the northern part of Xinjiang, with the maximum index value reaching approximately 0.4. Notably, Sobol sensitivity indices in most areas of northern Xinjiang are low, approaching zero in many regions, particularly in the mountainous areas of the Tianshan Mountains. This observation indicates that unlike the physical process ALB, the sensitivity of the physical process PCP does not escalate with altitude. The primary reason is that the three parameterization schemes of the physical process PCP classify precipitation into varying proportions of rain and snow based on temperature range. In the mountainous regions of northern Xinjiang during the snow cover period, where temperatures are significantly lower than the freezing point, all three schemes classify precipitation as snow. Consequently, the physical process of PCP does not exhibit significant sensitivity in higher-altitude areas. However, in the Junggar Basin, located in a lower-altitude, dry area, temperatures during the snowmelt period fluctuate around the freezing point, leading to notable differences in snow simulation results when different parameterization schemes are chosen. In summary, selecting the appropriate parameterization scheme for the physical process PCP is crucial for accurately simulating snow depth in Northern Xinjiang.

While Tukey's test sensitivity method was previously utilized to assess TBOT sensitivity under diverse snow climate conditions at specific sites, a tentative sense emerged that simulation results for snow depth might not be notably sensitive to the physical process of TBOT. Our earlier statistical investigations, despite detecting sensitivity at all sites using Tukey's test method, suggested that the differences in snow depth simulation results arising from two distinct TBOT parameterization schemes were not statistically significant. Figure 3 presents quantitative sensitivity outcomes in Northern Xinjiang, yielding genuinely surprising results. The maximum value of the Sobol sensitivity index across the entire region is 0.05, with only a few pixels achieving the maximum sensitivity index. Remarkably, the sensitivity index for the physical process TBOT throughout the entire Northern Xinjiang region is essentially zero. This finding appears to contradict the results of the study by You et al. [10], where Tukey's test method examined TBOT sensitivity in eight sites under different snow climates. Further analysis reveals that, on one hand, Tukey's test scrutinizes the sensitivity of parameterization schemes at the point scale from a microscopic perspective, overlooking spatial heterogeneity. On the other hand, a meticulous comparison of the sample ensembles used for Tukey's test in the study by You et al. [10] shows that the mean, median, and kernel density distribution of the sample ensemble at most sites are highly similar, with minimal differences between sets. Consequently, it can be concluded that the choice of different parameterization schemes for the physical process TBOT is unlikely to significantly impact the simulation results of snow depth at the regional scale, at least in Northern Xinjiang.

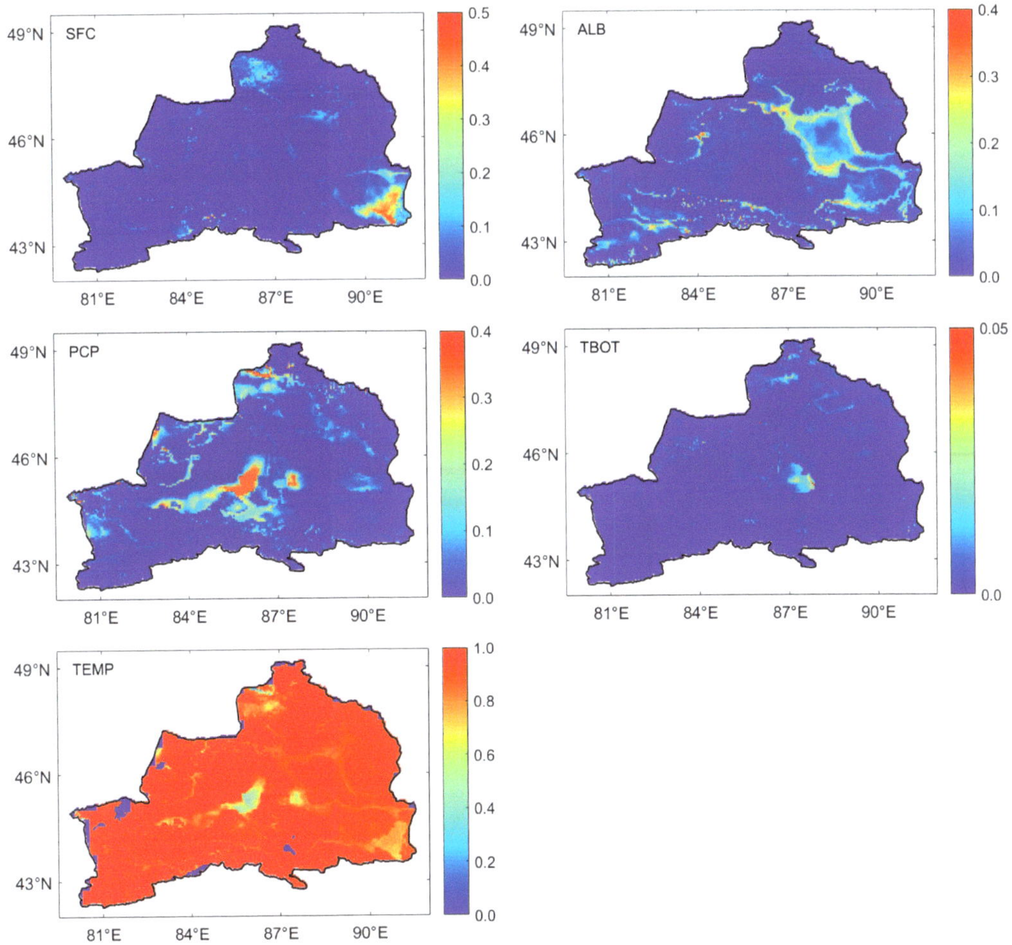

Figure 3. The Sobol total sensitivity of the simulated snow depth to the five parameterizations; higher values indicate higher sensitivities.

Figure 3 highlights the significant sensitivity of the modeled snow depth to the TEMP parameterization, evident across nearly the entire area. Sobol sensitivity indices in Figure 3 consistently approach 1 for the physical process TEMP throughout the region, with indices reaching approximately 0.6 in specific areas of the Junggar Basin. Notably, areas where these indices approach zero coincide with significant water bodies, falling outside the domain of land surface models. TEMP has long been recognized as a crucial factor influencing snow depth simulation outcomes, contributing significantly to the uncertainty in multi-parameterization scheme ensemble simulations. The parameterization schemes in question revolve around solving the heat conduction equation for soil/snow temperature in the first layer, with two options employing different boundary conditions based on temperature and fluxes. Our prior investigations reveal that opting for TEMP (1) results in rapid snow ablation, while opting for TEMP (2) results in slower snow ablation.

3.3. Uncertainty Analysis of Physical Parameterization Schemes

Expanding on the sensitivity analysis of quantified parameterization schemes, we proceeded to quantify uncertainty in the simulated outcomes of a multi-parameterized scheme ensemble for snow depth in Northern Xinjiang. Figure 4 illustrates the uncertainty in snow depth simulation results for 48 combination schemes. Notably, most areas in Northern Xinjiang exhibit small uncertainties, gradually increasing with altitude. For instance, the Tianshan Mountains region displays significantly higher uncertainty compared to the Junggar Basin, ranging mostly from 0.2 to 0.4 m. In local mountainous areas, the uncertainty in snow depth can exceed 0.5 m. This discrepancy is attributed to both the greater snow depth in mountainous regions, leading to relatively larger errors, and the selection of different parameterization schemes for specific physical processes, causing significant variations in snowmelt mechanisms. The 48 combinations, derived through the permutation and combination of parameterization schemes for five sensitive physical processes, underscore the impact of model structure on simulation results. Figure 4 emphasizes that regions with higher uncertainties are concentrated in high-altitude areas such as the Tianshan Mountains and Altay Mountains, suggesting that distinctions between parameterization schemes are primarily manifested during the snowmelt stage. While different combinations of schemes result in diverse snowmelt mechanisms, areas with lower altitudes and less snow accumulation exhibit comparatively smaller uncertainties in the overall simulation results for the multi-parametric scheme ensemble. Figure 5 delves into the number of pixels within different uncertainty intervals in Northern Xinjiang. The majority of pixels have an uncertainty range for snow depth below 0.1 m, with the highest number falling within the 0.02–0.04 m range, indicating generally low uncertainty in simulated snow depth. This may stem from the small amount of accumulated snow in these pixels, resulting in minimal errors or negligible differences among the 48 scheme combinations. Additionally, the figure reveals a considerable number of pixels with a snow depth simulation uncertainty range exceeding 0.1 m, signifying substantial variations in simulated snow depth results in certain areas due to the sensitivity of parameterization schemes. These differences may arise from the varied snow accumulation and melting mechanisms associated with the diverse combination schemes.

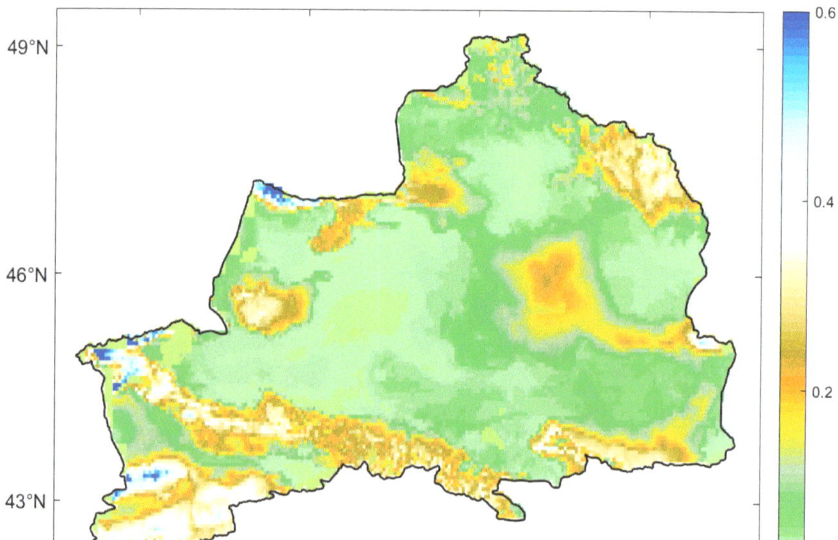

Figure 4. The uncertainty of the simulated snow depth using parameterization scheme combinations, in total 48.

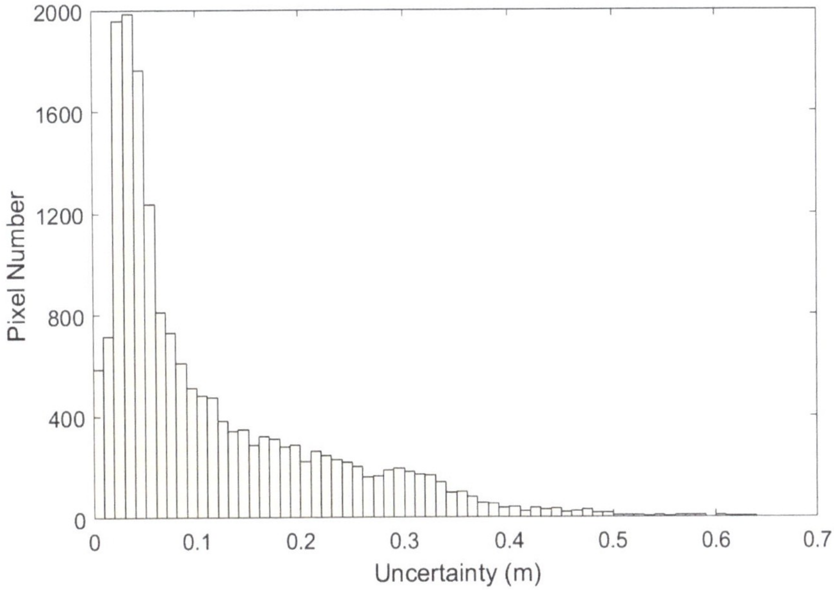

Figure 5. Statistics of pixel counts in different uncertainty range intervals.

4. Conclusions

In this study, our focus was on quantifying the spatial patterns of parameterization sensitivities across Northern Xinjiang, coupled with an examination of the uncertainty in simulated snow depth arising from different parameterization scheme combinations. We accomplished this by creating a simulation ensemble that combined various parameterization schemes and used the Sobol total sensitivity index. The subsequent section outlines the primary findings derived from this study.

Notably, in contrast to the basin region, the variations in snow depth simulation outcomes among the 48 combination schemes exhibit more pronounced distinctions in the mountainous region. Additionally, the ensemble's snow depth simulation results demonstrate a greater variance in the mountainous region compared to the basin area. While the sensitivity of parameterization schemes is well established in single-point experiments, the quantified results in northern Xinjiang reveal noteworthy distinctions in key parameterization schemes influencing snow simulation. The sensitivity index of the PCP parameterization scheme is consistently high across northern Xinjiang, with sensitivity indices approaching 1 in all areas, with notably higher sensitivity in the basin region compared to the mountains, especially in the central area of the Junggar Basin. In contrast, the sensitivity index of the TBOT parameterization scheme reaches a maximum value of 0.05 across the entire region, indicating low sensitivity. Additionally, the parameterization schemes for SFC and ALB exhibit sensitivity in localized areas. In contrast to the basin region, the ensemble simulation results of snow depth in mountainous regions exhibit greater uncertainty, concentrated mainly between 0.2 and 0.3 m. In the basin, the uncertainty is primarily below 0.1 m, with the majority of pixels falling within the 0.03–0.04 range. These disparities underscore substantial variations in snow accumulation and melting mechanisms among different combination schemes.

This study explores the sensitivity of regional parameterization schemes using the Noah-MP model's ensemble simulation results and the Sobol sensitivity index calculation method. It is necessary to further investigate whether the sensitivity index calculation of parameterization schemes at the regional level is influenced by the temporal and spatial resolution of the driving data. Future research will employ data assimilation techniques to reduce the uncertainty of ensemble snow simulation results and clarify the impact of different parameterization schemes on snow process simulation. The findings of this study can serve as a reference for future ensemble simulation experiments.

Author Contributions: Methodology, Y.Y.; Software, Y.Y.; Investigation, Y.Y.; Data curation, Y.Z.; Supervision, C.H. All authors have read and agreed to the published version of the manuscript.

Funding: This research was funded by the National Natural Science Foundation of China (grant number 42130113 and 42101361) and the Key Natural Science Research Project of Anhui Provincial Colleges (grant number 2023AH050143). And The APC was funded by 42130113.

Data Availability Statement: The China Meteorological Forcing Dataset used in this study can be downloaded online in the website: https://data.tpdc.ac.cn.

Conflicts of Interest: The authors declare no conflict of interest.

References

1. Barnett, T.P.; Adam, J.C.; Lettenmaier, D.P. Potential impacts of a warming climate on water availability in snow-dominated regions. *Nature* **2005**, *438*, 303–309. [CrossRef]
2. Immerzeel, W.W.; Van Beek, L.P.; Bierkens, M.F. Climate change will affect the Asian water towers. *Science* **2010**, *328*, 1382–1385. [CrossRef]
3. Xu, L.; Dirmeyer, P. Snow-atmosphere coupling strength in a global atmospheric model. *Geophys. Res. Lett.* **2011**, *38*, L13401. [CrossRef]
4. Kumar, S.V.; Peters-Lidard, C.D.; Arsenault, K.R.; Getirana, A.; Mocko, D.; Liu, Y.Q. Quantifying the added value of snow cover area observations in passive microwave snow depth data assimilation. *J. Hydrometeorol.* **2015**, *16*, 1736–1741. [CrossRef]
5. Verbunt, M.; Gurtz, J.; Jasper, K.; Lang, H.; Warmerdam, P.; Zappa, M. The hydrological role of snow and glaciers in alpine river basins and their distributed modeling. *J. Hydrol.* **2003**, *282*, 36–55. [CrossRef]
6. Clark, M.P.; Kavetski, D.; Fenicia, F. Pursuing the method of multiple working hypotheses for hydrological modeling. *Water Resour. Res.* **2011**, *47*, W09301. [CrossRef]
7. Clark, M.P.; Hendrikx, J.; Slater, A.G.; Kavetski, D.; Anderson, B.; Cullen, N.J.; Kerr, T.; Hreinsson, E.Ö.; Woods, R.A. Representing spatial variability of snow water equivalent in hydrologic and land-surface models: A review. *Water Resour. Res.* **2011**, *47*, W07539. [CrossRef]
8. Che, T.; Li, X.; Jin, R.; Huang, C.L. Assimilating passive microwave remote sensing data into a land surface model to improve the estimation of snow depth. *Remote Sens. Environ.* **2014**, *143*, 54–63. [CrossRef]
9. Kwon, Y.; Yang, Z.L.; Hoar, T.J.; Toure, A.M. Improving the radiance assimilation performance in estimating snow water storage across snow and land-cover types in North America. *J. Hydrometeorol.* **2017**, *18*, 651–668. [CrossRef]
10. Yang, Z.L.; Niu, G.Y.; Mitchell, K.E.; Chen, F.; Ek, M.B.; Barlage, M.; Longuevergne, L.; Manning, K.; Niyogi, D.; Tewari, M.; et al. The community Noah land surface model with multiparameterization options (Noah-MP): 2. Evaluation over global river basins. *J. Geophys. Res.* **2011**, *116*, D12110. [CrossRef]
11. Huang, C.L.; Margulis, S.A.; Durand, M.T.; Musselman, K.N. Assessment of snow grain-size model and stratigraphy representation impacts on snow radiance assimilation: Forward Modeling Evaluation. *IEEE Trans. Geosci. Remote Sens.* **2012**, *50*, 4551–4564. [CrossRef]
12. You, Y.H.; Huang, C.L.; Hou, J.L.; Zhang, Y.; Wang, Z.; Zhu, G.F. Improving the estimation of snow depth in the Noah-MP model by combining particle filter and Bayesian model averaging. *J. Hydrol.* **2023**, *617*, 128877. [CrossRef]
13. You, Y.H.; Huang, C.L.; Wang, Z.; Hou, J.L.; Zhang, Y.; Xu, P.P. A genetic particle filter scheme for univariate snow cover assimilation into Noah-MP model across snow climates. *Hydrol. Earth Syst. Sci.* **2023**, *27*, 2919–2933. [CrossRef]
14. Gao, J.L.; Huang, X.D.; Ma, X.F.; Feng, Q.S.; Liang, T.G.; Xie, H.J. Snow disaster early warning in pastoral areas of Qinghai province, China. *Remote Sens.* **2017**, *9*, 475. [CrossRef]
15. Zhu, L.L.; Zhang, Y.H.; Wang, J.G.; Tian, W.; Liu, Q.; Ma, G.Y.; Kan, X.; Chu, Y. Downscaling snow depth mapping by fusion of microwave and optical remote-sensing data based on deep learning. *Remote Sens.* **2021**, *13*, 584. [CrossRef]
16. Saha, S.K.; Sujith, K.; Chaudhari, H.S.; Hazra, A. Effects of multilayer snow scheme on the simulation of snow: Offline Noah and coupled with NCEP CFSv2. *J. Adv. Model. Earth Syst.* **2017**, *9*, 271–290. [CrossRef]
17. Moiz, A.; Wei, Z.W.; Naseer, A.; Kawasaki, A.; Acierto, R.A.; Koike, T. Improving snow-process modeling by evaluating reanalysis vertical temperature profiles using a distributed hydrological model. *J. Geophys. Res. Atmos.* **2022**, *127*, e2021JD036174. [CrossRef]
18. Zhang, G.; Chen, F.; Gan, Y.J. Assessing uncertainties in the Noah-MP ensemble simulations of a cropland site during the Tibet Joint International Cooperation program field campaign. *J. Geophys. Res. Atmos.* **2016**, *121*, 9576–9596. [CrossRef]
19. You, Y.H.; Huang, C.L.; Yang, Z.L.; Zhang, Y.; Bai, Y.L.; Gu, J. Assessing Noah-MP parameterization sensitivity and uncertainty interval across snow climates. *J. Geophys. Res. Atmos.* **2020**, *125*, e2019JD030417. [CrossRef]
20. Zheng, H.; Yang, Z.L.; Lin, P.R.; Wei, J.F.; Wu, W.Y.; Li, L.C.; Zhao, L.; Wang, S. On the sensitivity of the precipitation partitioning into evapotranspiration and runoff in land surface parameterizations. *Water Resour. Res.* **2019**, *55*, 95–111. [CrossRef]

21. Niu, G.Y.; Yang, Z.L.; Mitchell, K.E.; Chen, F.; Ek, M.B.; Barlage, M.; Kumar, A.; Manning, K.; Niyogi, D.; Rosero, E.; et al. The community Noah land surface model with multi-parameterization options (Noah-MP): 1. Model description and evaluation with local scale measurements. *J. Geophys. Res.* **2011**, *116*, D12109. [CrossRef]
22. Gan, Y.J.; Liang, X.Z.; Duan, Q.Y.; Chen, F.; Li, J.D.; Zhang, Y. Assessment and reduction of the physical parameterization uncertainty for Noah-MP land surface model. *Water Resour. Res.* **2019**, *55*, 5518–5538. [CrossRef]
23. Chang, M.; Liao, W.H.; Wang, X.M.; Zhang, Q.; Chen, W.H.; Wu, Z.Y.; Hu, Z.C. An optimal ensemble of the Noah-MP land surface model for simulating surface heat fluxes over a typical subtropical forest in South China. *Agric. For. Meteorol.* **2020**, *281*, 107815. [CrossRef]
24. Yang, Q.D.; Dan, L.; Lv, M.Z.; Wu, J.; Li, W.D.; Dong, W.J. Quantitative assessment of the parameterization sensitivity of the Noah-MP land surface model with dynamic vegetation using China FLUX data. *Agric. For. Meteorol.* **2021**, *307*, 108542. [CrossRef]
25. Sobolprime, I.M. Sensitivity analysis for non-linear mathematical models. *Math. Model. Comput. Exp.* **1993**, *1*, 407–414.
26. Dai, H.; Ye, M.; Walker, A.P.; Chen, X.Y. A new process sensitivity index to identify important system processes under process model and parametric uncertainty. *Water Resour. Res.* **2017**, *53*, 3476–3490. [CrossRef]
27. You, Y.H.; Huang, C.L.; Gu, J.; Li, H.Y.; Hao, X.H.; Hou, J.L. Assessing snow simulation performance of typical combination schemes within Noah-MP in northern Xinjiang, China. *J. Hydrol.* **2020**, *581*, 124380. [CrossRef]
28. Xu, J.H.; Shu, H.; Dong, L. DEnKF-Variational Hybrid Snow Cover Fraction Data Assimilation for Improving Snow Simulations with the Common Land Model. *Remote Sens.* **2014**, *6*, 10612–10635. [CrossRef]
29. Huang, X.D.; Liang, T.G.; Zhang, X.T.; Guo, Z.G. Validation of MODIS snow cover products using Landsat and ground measurements during the 2001-2005 snow seasons over northern Xinjiang, China. *Int. J. Remote Sens.* **2011**, *32*, 133–152. [CrossRef]
30. He, J.; Yang, K.; Tang, W.J.; Lu, H.; Qin, J.; Chen, Y.Y.; Li, X. The first high-resolution meteorological forcing dataset for land process studies over China. *Sci. Data* **2020**, *7*, 25. [CrossRef]
31. Chen, Y.; Yang, K.; He, J.; Qin, J.; Shi, J.C.; Du, J.Y.; He, Q. Improving land surface temperature modeling for dry land of China. *J. Geophys. Res.* **2011**, *116*, D20104. [CrossRef]
32. Liston, G.E.; Elder, K. A meteorological distribution system for high-resolution terrestrial modeling (MicroMet). *J. Hydrometeorol.* **2006**, *7*, 217–234. [CrossRef]
33. Ek, M.B.; Mitchell, K.E.; Lin, Y.; Rogers, E.; Grunmann, P.; Koren, V.; Gayno, G.; Tarpley, J.D. Implementation of Noah land surface model advances in the National Centers for Environmental Prediction operational mesoscale Eta model. *J. Geophys. Res. Atmos.* **2003**, *108*, 8851. [CrossRef]
34. Chen, F.; Manning, K.W.; LeMone, M.A.; Trier, S.B.; Alfieri, J.G.; Roberts, R.; Tewari, M.; Niyogi, D.; Horst, T.W.; Oncley, S.P. Description and evaluation of the characteristics of the NCAR high-resolution land data assimilation system. *J. Appl. Meteorol. Climatol.* **2007**, *46*, 694–713. [CrossRef]
35. Cuntz, M.; Mai, J.; Samaniego, L.; Clark, M.; Wulfmeyer, V.; Branch, O.; Attinger, S.; Thober, S. The impact of standard and hard-coded parameters on the hydrologic fluxes in the Noah-MP land surface model. *J. Geophys. Res. Atmos.* **2016**, *121*, 10676–10700. [CrossRef]
36. Chen, F.; Dudhia, J. Coupling an advanced land surface-hydrology model with the Penn State-NCAR MM5 modeling system. Part I: Model implementation and sensitivity. *Mon. Weather Rev.* **2001**, *129*, 569–585. [CrossRef]
37. Cai, X.T.; Yang, Z.L.; Xia, Y.L.; Huang, M.Y.; Wei, H.L.; Leung, L.R.; Ek, M.B. Assessment of simulated water balance from Noah, Noah-MP, CLM, and VIC over CONUS using the NLDAS test bed. *J. Geophys. Res. Atmos.* **2014**, *119*, 13751–13770. [CrossRef]
38. Xia, Y.; Cosgrove, B.A.; Mitchell, K.E.; Peters-Lidard, C.D.; Ek, M.B.; Kumar, S.; Mocko, D.; Wei, H.L. Basin-scale assessment of the land surface energy budget in the National Centers for Environmental Prediction operational and research NLDAS-2 systems. *J. Geophys. Res. Atmos.* **2016**, *121*, 196–220. [CrossRef]
39. Xia, Y.; Ek, M.B.; Wu, Y.; Ford, T.; Quiring, S.M. Comparison of NLDAS-2 simulated and NASMD observed daily soil moisture. Part I: Comparison and analysis. *J. Hydrometeorol.* **2015**, *16*, 1962–1980. [CrossRef]
40. Gao, Y.H.; Li, K.; Chen, F.; Jiang, Y.S.; Lu, C.G. Assessing and improving Noah-MP land model simulations for the central Tibetan Plateau. *J. Geophys. Res. Atmos.* **2015**, *120*, 9258–9278. [CrossRef]
41. Saltelli, A.; Annoni, P.; Azzini, I.; Campolongo, F.; Ratto, M.; Tarantola, S. Variance based sensitivity analysis of model output. Design and estimator for the total sensitivity index. *Comput. Phys. Commun.* **2010**, *181*, 259–270. [CrossRef]